T0122456

Modeling and Simulation in Science, Engineering and Technology

More information about this series at http://www.springer.com/series/4960

Ananth Grama • Ahmed H. Sameh
Editors

Parallel Algorithms in Computational Science and Engineering

Editors
Ananth Grama
Purdue University
West Lafayette, IN, USA

Ahmed H. Sameh
Purdue University
West Lafayette, IN, USA

ISSN 2164-3679   ISSN 2164-3725   (electronic)
Modeling and Simulation in Science, Engineering and Technology
ISBN 978-3-030-43738-1   ISBN 978-3-030-43736-7   (eBook)
https://doi.org/10.1007/978-3-030-43736-7

Mathematics Subject Classification: 15-02, 62Pxx, 65F50, 68M15, 68Nxx, 68W10, 97Mxx

This book is published under the imprint Birkhäuser, www.birkhauser-science.com by the registered company Springer Nature Switzerland AG.
The registered company address is: Gewerbestrasse 11, 6330 Cham, Switzerland

Preface

High-performance computing (HPC) has been a critical enabling technology for a variety of scientific and engineering applications over the past five decades. As a result of advances in parallel computing, one is now able to handle large-scale and large-scope applications in diverse domains such as fluid dynamics, structural mechanics, fluid–structure interaction, weather forecasting, materials modeling and design, electromagnetics, and computational biology. As HPC methods have continued to mature for traditional applications, emerging application domains pose new challenges for modeling paradigms, numerical techniques, and their parallel implementations.

The past decade has also witnessed significant advances in our ability to collect massive datasets from physical, engineered, social, and computational processes. There has been a concomitant realization of the immense value that can be unlocked by these datasets through the use of data analytics and machine learning techniques. The scale of the underlying datasets and computational costs of common analytics techniques strongly motivate the use of HPC platforms for these applications. Where a number of core computational techniques such as linear and nonlinear system solvers, eigenvalue problem solvers, and optimization techniques were traditionally motivated by scientific and engineering applications, these kernels are now ubiquitous in data science. However, the structure and scale of data science applications pose new challenges for parallel formulations of even traditional compute kernels. Statistical and randomized approaches play important roles in these applications as well. These approaches present additional opportunities for parallel execution, in the form of ensemble methods, sampling techniques, and asynchronous (or loosely synchronous) computations. Finally, the role of data is central to these approaches, since the overheads associated with accessing massive datasets (typically from disks or over the network) are significant. The size of these datasets, which typically do not fit in the main memory, also influences the nature of the solution methods. For this reason, solution techniques are often designed to make several passes over the data so as to process a subset that can fit in the main memory in each pass. All of these characteristics motivate new algorithms even for traditional compute kernels.

The past two decades have also seen significant changes in computing platforms. Processors have gone from single-core scalar execution units to multicore superscalar issue units. It is common to have single processors with 16 processing cores in conventional configurations. These chip multiprocessors typically have private L1-caches and shared L2- or L3-caches, connected to large DRAMs. This deep memory hierarchy puts additional pressure on optimizing for locality of data access. Current data centers have in the range of 10^5 such processors and beyond, connected through high-speed interconnects, typically meshes or fat-tree topologies. At these scales, the energy footprint of computations is an important consideration. To this end, accelerators such as GPUs and FPGAs have emerged as alternatives for enabling high-speed computations with lower energy consumption (on a per-FLOP basis). For this reason, typical supercomputing installations and many data centers have nodes with one or more accelerators. The I/O subsystems of these platforms have also evolved significantly over time. The addition of nonvolatile RAMs (NVRAMs) provides significant opportunities for staging data that may not fit in memory. The distinct read–write overhead characteristics of NVRAMs pose additional challenges. Finally, emerging network concepts allow for limited in-network computations, which allow for much faster aggregate communication operations.

In addition to science and engineering applications and hardware platforms, the system software stack has also evolved significantly in recent times. While traditional APIs such as MPI, pthreads, and OpenMP continue to be used for parallel software, emerging distributed frameworks such as MapReduce and Spark find increasing utilization. Higher-level programming models such as parallel Matlab and Julia hold the promise for significantly higher programmer productivity in parallel software. In terms of the runtime, cloud systems that support schedulers, resource managers, and higher-level I/O primitives are increasingly used. Virtualization and containerization technologies have enabled a level of portability and ease of execution that has transformed many engineering applications.

The evolution of classical applications, emergence of new applications, and changes to underlying hardware platforms motivate significant innovations in algorithms and software for scalable high-performance computations. At the single node level, use of accelerators, managing memory hierarchies for high node performance, and extracting concurrency to leverage on-chip parallelism are major challenges. At the data center level, extracting all available parallelism, minimizing communication and idling, leveraging algorithmic asynchrony, and fundamentally redesigning algorithms for extreme scalability are major challenges. Finally, at the level of applications, integrating disparate compute kernels into a single scalable application, integrating I/O, and analyzing large amounts of data from high-resolution simulations pose significant challenges. These themes are explored in various application contexts in this book.

This book covers two major areas in high-performance computing: core algorithms and compute kernels, and high-performance science and engineering applications. These kernels and applications span the space of conventional scientific and engineering domains, in addition to emerging data analytics problems.

Part 1: High-Performance Algorithms

The chapter "State of the Art of Sparse Direct Solvers" by Bollhofer et al. provides an overview of how parallel sparse direct linear system solvers should be developed. Specifically, for PARDISO 6.2, the preprocessing stage consists of combinatorial algorithms (maximum weight matching and multilevel nested dissection) that enhance diagonal dominance, reduce fill-in, and improve concurrency. The effectiveness of PARDISO 6.2 is demonstrated for solving challenging linear systems that arise in integrated circuit simulation. PARDISO 6.2 proved to be far superior to the sequential direct solver KLU and the parallel direct sparse solver in Intel's Math Kernel Library (MKL). Further, the authors point out two special features of PARDSO 6.2 that proved to be extremely effective in several situations: (1) computing only those elements of the solution vector that correspond to the few nonzero elements of a sparse right-hand side and (2) computing only the diagonal elements of the inverse of the sparse coefficient matrix. Using these two features, PARDSO 6.2 realizes remarkable savings compared to the obvious brute-force procedures. For example, the authors show that for a sparse coefficient matrix A (that arises in circuit simulation) of order almost 5.6 million, PARDISO 6.2 computes the diagonal elements of the inverse of A in 2.1 s, while computing all the elements of the inverse of A on the same parallel computing platform consumes 371 h.

Reordering of general sparse matrices plays a crucial role in both direct and preconditioned iterative sparse linear system solvers. This topic is addressed in the chapter "The Effect of Various Sparsity Structures on Parallelism and Algorithms to Reveal those Structures" by Selvitopi et al. This chapter deals with graph and hypergraph algorithms for reordering general sparse matrices into structured forms. The focus is on four structures: single-bordered block diagonal, double-bordered block diagonal, nonempty off-diagonal block minimization, and overlapped diagonal blocks. The authors, who produced the well-known sparse matrix reordering software package "PArtitioning TOol for Hypergraphs" (PATOH), also demonstrate the advantage of such forms in sparse matrix–vector multiplication as well as the transpose of a sparse matrix–vector multiplication.

In the chapter "Structure-Exploiting Interior Point Methods" by Kardso et al., the authors address interior point (IP) methods, which proved to be quite effective for large-scale nonlinear optimization problems in computational science and engineering applications. In many of these applications, however, the resulting optimization problems possess certain structures. This chapter reviews parallel variants of IP methods that take advantage of such structure. In particular, the chapter provides efficient parallel algorithms for solving the resulting saddle-point problems (KKT systems) that consume most of the computing time in handling these optimization problems. Specifically, it highlights the vital role of the sparse direct solver PARDISO in solving KKT systems, associated arrowhead systems, and determining the inertia of symmetric matrices via the LDL^T factorization, in case correction is needed to guarantee that the Hessian matrix projected on the null space of the constraint Jacobian is positive definite. Exploiting the structure of large-scale optimization problems that arise in the management of modern power grids, those parallel variants of IP methods realize significant savings in both memory requirements and computing time compared to those achieved in IPOPT (the well-known software library for "Interior Point OPTimizer").

In the chapter "Parallel Hybrid Sparse Linear System Solvers" by Manguoglu et al., the authors describe a highly versatile class of parallel algorithms for banded linear systems and their generalization for solving general sparse linear systems. A crucial preprocessing stage consists of reordering the coefficient matrix so as to strengthen the main diagonal and encapsulate as many of the heaviest off-diagonal elements in a central band. This central band, which may be dense or sparse, can be used as an effective preconditioner. Based on the bandwidth of the central band, variations of the SPIKE algorithm are referred to as members of the PSPIKE solver. The "P" in "PSPIKE" denotes the use of the direct sparse solver PARDISO as an essential kernel. This chapter describes the banded solver SPIKE and its generalization, PSPIKE, in detail, outlining various algorithmic choices, their parallel implementation on shared and distributed memory platforms, and comparison of their parallel performance with other state-of-the-art solvers. The chapter also provides a historical context for these hybrid solvers.

Part 2: High-Performance Computational Science and Engineering Applications

The chapter "Computational Materials Science and Engineering" by Polizzi and Saad gives a summary of the state of the art in computational materials science and engineering. It focuses on density functional theory and the novel algorithms developed for the solution of the underlying symmetric eigenvalue problem, which are suitable for parallel computing. More specifically, the authors outline algorithmic advances introduced in the software package NESSIE and the eigensolvers EVSL and FEAST. This chapter is not only of interest to researchers in nonoelectronics but also of interest to researchers in parallel numerical linear algebra.

The following three chapters deal with critical applications that involve fluid–structure Interaction: "Computational Cardiovascular Analysis with the Variational Multiscale Methods and Isogeometric Discretization," by Hughes et al., "ALE and Space-Time Variational Multiscale Isogeometric Analysis of Wind Turbines and Turbomachinery," by Bazilevs et al., and "Variational Multiscale Flow Analysis in Aerospace, Energy, and Transportation Technologies," by Takizawa et al.

These chapters consider challenging problems in the above applications that are characterized by: (a) fluid–structure interaction; (b) complex geometries; (c) moving boundaries and interfaces; (d) contact between moving solid surfaces; and (e) turbulent and rotational flows. These challenges are addressed by techniques developed mainly by the authors of these chapters. Such techniques include space-time variational multiscale methods, the arbitrary Lagrangian–Eulerian method, and isogeometric discretization. The simulations presented in these chapters demonstrate agreements with experiments, thus proving the power and scope of the methods developed for computational analysis of a variety of applications: (1) cardiovascular flows, surgical planning, and virtual stent placement; (2) wind turbines (including two back-to-back wind turbines); and (3) aerodynamics of ram-air parachutes.

The chapter "Multiscale Crowd Dynamics Modeling and Safety Problems: Towards Parallel Computing" by Bellomo and Aylaj deals with multiscale crowd dynamics modeling and safety problems, along with issues concerning implementation on parallel computing platforms. It specifically deals with the modeling and simulation of human crowds. This is motivated not only by scientific curiosity but also by safety of individuals in emergency situations. It proposes three components of a research strategy: (1) multiscale vision of crowd dynamics varying from motion of an individual to clusters of individuals; (2) development of a systems approach to movements across different areas with specific geometries; and (3) development of modeling and simulation to support crisis managers in evacuation dynamics. In each of the above three components, the authors describe the use of artificial intelligence and parallel computing to ensure the effectiveness of rapid evacuation strategies in crisis situations.

The chapter "HPC for Weather Forecasting" by Michalakes discusses the use of high-performance computing platforms for weather forecasting. It starts with a history of weather forecasting based on numerical models that go back from its roots in the year 1922 to current day petascale models and concludes with an extrapolation beyond the year 2025. The chapter describes various issues associated with modeling, complexity of grids laid atop spherical surfaces, and issues of parallel implementation. The need for spectral models and dynamics is motivated in this context. The underlying operations are mapped to common library calls (in LAPACK) and the associated complexities are discussed. The use of semi-implicit semi-Lagrangian transport for advection is motivated in this context as well. An important issue in modeling is the development of suitable meshing techniques. These are discussed in detail, for Cartesian, non-Cartesian, structured, and unstructured grids. In terms of parallel processing, issues of domain decomposition and load balancing are considered. Use of suitable finite element formulations that incorporate appropriate physics models and constraints is then discussed. The chapter concludes with an outline of the emerging challenges in weather models on the next generation of HPC platforms. These challenges include single-node and scale-out considerations, along with issues in algorithm and software design at extreme scales.

In the chapter "A Simple Study of Pleasing Parallelism on Multicore Computers" by Gleich et al., the authors present a detailed study of parallel graph computations with PageRank and personalized PageRank as the canonical representative computation. This kernel appears in different forms in other graph computations as well, notably all-pairs type computations. The chapter sets up the problems of PageRank and personalized PageRank and describes variants that are used in practice (e.g., those that do not generate the entire $O(n^2)$ data associated with personalized PageRank). A number of commonly used algorithms are described, along with considerations of computational cost and memory. Parallel processing issues of these methods are discussed, both in the context of shared address space and message passing platforms. The need for graph reordering and partitioning is discussed, along with scheduling, load balancing, and communication optimizations. Finally, the chapter presents a comprehensive set of experimental results evaluating the algorithms discussed. The code (in Julia) is made available to the readers in order to conduct their own experiments.

The chapter "Parallel Fast Time-Domain Integral-Equation Methods for Transient Electromagnetic Analysis" by Liu and Michielssen presents a comprehensive overview of time-domain integral equation methods for analyzing electromagnetic systems. The chapter starts with an introduction to Marching-On-in-Time (MOT) based integral equation formulation for modeling radiation and scattering. The formulation is contrasted with time-domain differential equation models (lack of need for absorbing boundary conditions, fewer degrees of freedom), along with challenges presented by time-domain integral equation (TDIE) formulations. It also presents a concise survey of recent efforts aimed at addressing these challenges and the need for use of HPC platforms with accelerators. Mathematical formulations for TDIE solvers in the presence of different scatterers and media are then discussed, including PEC scatterers, as well as homogeneous and inhomogeneous dielectrics. This is followed by a discussion of solvers for the underlying mathematics models. These solvers focus on higher-order methods, time-stability, and conditioning. Issues of discretization including higher-order spatial basis functions and meshing are also discussed. The chapter then considers techniques for dealing with high-frequency, DC, and resonant instabilities. A number of fast solvers are also considered, along with a discussion of parallel implementation techniques for TDIE solvers. Issues of partitioning workload (rays) and associated computations are presented and their computational costs are characterized. Optimizations for load balancing and communication are described, and experimental results are presented for several parallel solvers.

The chapter "Parallel Optimization Techniques for Machine Learning" by Grama et al. deals with the development of parallel optimization techniques for machine learning problems involving massively large datasets. Such datasets arise in applications like autonomous vehicles, artificial intelligence, image classification, and cybersecurity. These applications are modeled as either convex or non-convex optimization problems. This chapter focuses on parallel algorithms for finite sum minimization problems in the context of convex and non-convex formulations.

This book is meant to provide a state-of-the-art reference for researchers and practitioners. The thirteen chapters present a survey of various topics while providing a comprehensive methodological coverage of algorithms and applications. In many instances, these chapters reference, or are accompanied by parallel software, which the readers can download and use to conduct their own experiments.

West Lafayette, IN, USA   Ananth Grama
West Lafayette, IN, USA   Ahmed H. Sameh

Contents

Part I
High Performance Algorithms

State-of-the-Art Sparse Direct Solvers

**Matthias Bollhöfer, Olaf Schenk, Radim Janalik, Steve Hamm,
and Kiran Gullapalli**

1 Introduction

Solving large sparse linear systems is at the heart of many application problems
arising from computational science and engineering applications. Advances in
combinatorial methods in combination with modern computer architectures have
massively influenced the design of the state-of-the-art direct solvers that are
feasible for solving larger systems efficiently in a computational environment with
rapidly increasing memory resources and cores. Among these advances are novel
combinatorial algorithms for improving diagonal dominance which pave the way to
a static pivoting approach, thus improving the efficiency of the factorization phase
dramatically. Besides, partitioning and reordering the system such that a high level
of concurrency is achieved, the objective is to simultaneously achieve the reduction
of fill-in and the parallel concurrency. While these achievements already significantly improve the factorization phase, modern computer architectures require one
to compute as many operations as possible in the cache of the CPU. This in turn
can be achieved when dense subblocks that show up during the factorization can be

M. Bollhöfer
Institute for Numerical Analysis, TU Braunschweig, Braunschweig, Germany
e-mail: m.bollhoefer@tu-bs.de

O. Schenk (✉) · R. Janalik
Institute of Computational Science, Faculty of Informatics, Università della Svizzera Italiana,
Lugano, Switzerland
e-mail: olaf.schenk@usi.ch; radim.janalik@usi.ch

S. Hamm · K. Gullapalli
NXP, Austin, TX, USA
e-mail: steve.hamm@nxp.com; kiran.gullapalli@nxp.com

© Springer Nature Switzerland AG 2020
A. Grama, A. H. Sameh (eds.), *Parallel Algorithms in Computational Science and
Engineering*, Modeling and Simulation in Science, Engineering and Technology,
https://doi.org/10.1007/978-3-030-43736-7_1

grouped together into dense submatrices which are handled by multithreaded and cache-optimized dense matrix kernels using level-3 BLAS and LAPACK [3].

This chapter will review some of the basic technologies together with the latest developments for sparse direct solution methods that have led to the state-of-the-art LU decomposition methods. The paper is organized as follows. In Sect. 2 we will start with maximum weighted matchings which is one of the key tools in combinatorial optimization to dramatically improve the diagonal dominance of the underlying system. Next, Sect. 3 will review multilevel nested dissection as a combinatorial method to reorder a system symmetrically such that fill-in and parallelization can be improved simultaneously, once pivoting can be more or less ignored. After that, we will review established graph-theoretical approaches in Sect. 4, in particular the elimination tree, from which most of the properties of the LU factorization can be concluded. Among these properties is the prediction of dense submatrices in the factorization. In this way several subsequent columns of the factors L and U^T are collected in a single dense block. This is the basis for the use of dense matrix kernels using optimized level-3 BLAS as well to exploit fast computation using the cache hierarchy which is discussed in Sect. 5. Finally, we show in Sect. 6 how the ongoing developments in parallel sparse direct solution methods have advanced integrated circuit simulations. We assume that the reader is familiar with some elementary knowledge from graph theory, see, e.g., [15, 21] and some simple computational algorithms based on graphs [1].

2   Maximum Weight Matching

In modern sparse elimination methods the key to success is ability to work with efficient data structures and their underlying numerical templates. If we can increase the size of the diagonal entries as much as possible in advance, pivoting during Gaussian elimination can often be bypassed and we may work with static data structures and the numerical method will be significantly accelerated. A popular method to achieve this goal is the maximum weight matching method [16, 37] which permutes, e.g., the rows of a given nonsingular matrix $A \in \mathbb{R}^{n,n}$ by a permutation matrix $\Pi \in \mathbb{R}^{n,n}$ such that $\Pi^T A$ has a *nonzero diagonal*. Moreover, it maximizes the product of the absolute diagonal values and yields diagonal scaling matrices $D_r, D_c \in \mathbb{R}^{n,n}$ such that $\tilde{A} = \Pi^T D_r A D_c$ satisfies $|\tilde{a}_{ij}| \leqslant 1$ and $|\tilde{a}_{ii}| = 1$ for all $i, j = 1, \ldots, n$. The original idea on which these nonsymmetric permutations and scalings are based is to find a *maximum weighted matching* of a *bipartite graphs*. Finding a maximum weighted matching is a well known assignment problem in operation research and combinatorial analysis.

Definition 1 A graph $G = (V, E)$ with vertices V and edges $E \subset V^2$ is called *bipartite* if V can be partitioned into two sets V_r and V_c, such that no edge $e = (v_1, v_2) \in E$ has both ends v_1, v_2 in V_r or both ends v_1, v_2 in V_c. In this case we denote G by $G_b = (V_r, V_c, E)$.

Definition 2 Given a matrix A, then we can associate with it a canonical bipartite graph $G_b(A) = (V_r, V_c, E)$ by assigning the labels of $V_r = \{r_1, \ldots, r_n\}$ with the row indices of A and $V_c = \{c_1, \ldots, c_n\}$ being labeled by the column indices. In this case E is defined via $E = \{(r_i, c_j) | a_{ij} \neq 0\}$.

For the bipartite graph $G_b(A)$ we see immediately that if $a_{ij} \neq 0$, then we have that $r_i \in V_r$ from the row set is connected by an edge $(r_i, c_j) \in E$ to the column $c_j \in V_c$, but neither rows are connected with each other nor do the columns have interconnections.

Definition 3 A *matching* $\mathcal{M}$ of a given graph $G = (V, E)$ is a subset of edges $e \in E$ such that no two of which share the same vertex.

If $\mathcal{M}$ is a matching of a bipartite graph $G_b(A)$, then each edge $e = (r_i, c_j) \in \mathcal{M}$ corresponds to a row i and a column j and there exists no other edge $\hat{e} = (r_k, c_l) \in \mathcal{M}$ that has the same vertices, neither $r_k = r_i$ nor $c_l = c_j$.

Definition 4 A matching $\mathcal{M}$ of $G = (V, E)$ is called *maximal*, if no other edge from E can be added to $\mathcal{M}$.

If for an $n \times n$ matrix A a *matching* $\mathcal{M}$ of $G_b(A)$ with maximum cardinality n is found, then by definition the edges must be $(i_1, 1), \ldots, (i_n, n)$ with $i_1, \ldots, i_n$ being the numbers $1, \ldots, n$ in a suitable order and therefore we obtain $a_{i_1,1} \neq 0$, $\ldots a_{i_n,n} \neq 0$. In this case we have established that the matrix A is at least structurally nonsingular and we can use a row permutation matrix Π^T associated with row ordering $i_1, \ldots, i_n$ to place a nonzero entry on each diagonal location of $\Pi^T A$.

Definition 5 A *perfect matching* is a maximal matching with cardinality n.

It can be shown that for a structurally nonsingular matrix A there always exists a perfect matching $\mathcal{M}$.

Perfect Matching
In Fig. 1, the set of edges $\mathcal{M} = \{(1, 2), (2, 4), (3, 5), (4, 1), (5, 3), (6, 6)\}$ represents a perfect maximum matching of the bipartite graph $G_b(A)$.

The most efficient combinatorial methods for finding maximum matchings in bipartite graphs make use of an *augmenting path*. We will introduce some graph terminology for the construction of perfect matchings.

Definition 6 If an edge $e = (u, v)$ in a graph $G = (V, E)$ joins a vertices $u, v \in V$, then we denote it as uv. A path then consists of edges $u_1u_2, u_2u_3, u_3u_4 \ldots, u_{k-1}u_k$, where each $(u_i, u_{i+1}) \in E, i = 1, \ldots, k - 1$.

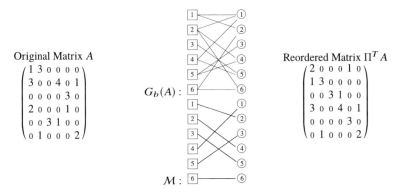

Fig. 1 Perfect matching. Left side: original matrix A. Middle: bipartite representation $G_b(A) = (V_r, V_c, E)$ of the matrix A and perfect matching M. Right side: permuted matrix $\Pi^T A$

If $G_b = (V_r, V_c, E)$ is a bipartite graph, then by definition of a path, any path is alternating between the vertices of V_r and V_c, e.g., paths in G_b could be such as $r_1c_2, c_2r_3, r_3c_4, \ldots$.

Definition 7 Given a graph $G = (V, E)$, a vertex is called *free* if it is not incident to any other edge in a matching M of G. An *alternating path* relative to a matching M is a path $P = u_1u_2, u_2u_3, \ldots, u_{s-1}u_s$ where its edges are alternating between $E \setminus M$ and M. An *augmenting path* relative to a matching M is an alternating path of odd length and both of it vertex endpoints are free.

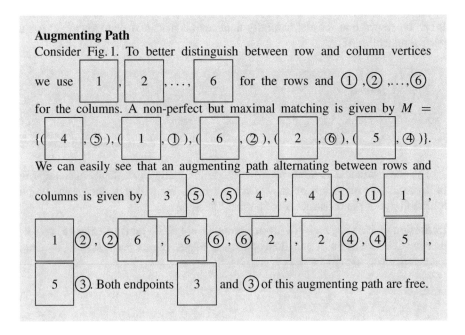

In a bipartite graph $G_b = (V_r, V_c, E)$ one vertex endpoint of any augmenting path must be in V_r, whereas the other one must be in V_c. The symmetric difference, $A \oplus B$ of two edge sets A, B, is defined to be $(A \setminus B) \cup (B \setminus A)$.

Using these definitions and notations, the following theorem [5] gives a constructive algorithm for finding perfect matchings in bipartite graphs.

Theorem 1 *If M is non-maximum matching of a bipartite graph $G_b = (V_r, V_c, E)$, then there exists an augmenting path P relative to M such that $P = \tilde{M} \oplus M$ and $\tilde{M}$ is a matching with cardinality $|M| + 1$.*

According to this theorem, a combinatorial method of finding perfect matching in a bipartite graph is to seek augmenting paths.

The perfect matching as discussed so far only takes the nonzero structure of the matrix into account. For their use as static pivoting methods prior to the LU decomposition one requires in addition to maximize the absolute value of the product of the diagonal entries. This is referred to as *maximum weighted matching*. In this case a permutation π has to be found, which maximizes

$$\prod_{i=1}^{n} |a_{\pi(i)i}|. \tag{1}$$

The maximization of this product is transferred into a minimization of a sum as follows. We define a matrix $C = (c_{ij})$ via

$$c_{ij} = \begin{cases} \log a_i - \log |a_{ij}| & a_{ij} \neq 0 \\ \infty & \text{otherwise}, \end{cases}$$

where $a_i = \max_j |a_{ij}|$ is the maximum element in row i of matrix A. A permutation π which minimizes the sum

$$\sum_{i=1}^{n} c_{\pi(i)i}$$

also maximizes the product (1). The minimization problem is known as linear-sum assignment problem or bipartite weighted matching problem in combinatorial optimization. The problem is solved by a sparse variant of the Hungarian method. The complexity is $O(n\tau \log n)$ for sparse matrices with τ entries. For matrices, whose associated graph fulfill special requirements, this bound can be reduced further to $O(n^{\alpha}(\tau + n \log n))$ with $\alpha < 1$. All graphs arising from finite-difference or finite element discretizations meet the conditions [24]. As before, we finally get a perfect matching which in turn defines a nonsymmetric permutation.

When solving the assignment problem, two dual vectors $u = (u_i)$ and $v = (v_i)$ are computed which satisfy

$$u_i + v_j = c_{ij} \qquad (i, j) \in \mathcal{M}, \tag{2}$$

$$u_i + v_j \leq c_{ij} \qquad \text{otherwise.} \tag{3}$$

Using the exponential function these vectors can be used to scale the initial matrix. To do so define two diagonal matrices D_r and D_c through

$$D_r = \text{diag}(d_1^r, d_2^r, \ldots, d_n^r), \qquad d_i^r = \exp(u_i), \tag{4}$$

$$D_c = \text{diag}(d_1^c, d_2^c, \ldots, d_n^c), \qquad d_j^c = \exp(v_j)/a_j. \tag{5}$$

Using Eqs. (2) and (3) and the definition of C, it immediately follows that $\tilde{A} = \Pi^T D_r A D_c$ satisfies

$$|\tilde{a}_{ii}| = 1, \tag{6}$$

$$|\tilde{a}_{ij}| \leq 1. \tag{7}$$

The permuted and scaled system $\tilde{A}$ has been observed to have significantly better numerical properties when being used for direct methods or for preconditioned iterative methods, cf., e.g., [4, 16]. Olschowka and Neumaier [37] introduced these scalings and permutation for reducing pivoting in Gaussian elimination of full matrices. The first implementation for sparse matrix problems was introduced by Duff and Koster [16]. For symmetric matrices $|A|$, these nonsymmetric matchings can be converted to a symmetric permutation P and a symmetric scaling $D_s = (D_r D_c)^{1/2}$ such that $P^T D_s A D_s P$ consists mostly of diagonal blocks of size 1×1 and 2×2 satisfying a similar condition as (6) and (7), where in practice it rarely happens that 1×1 blocks are identical to 0 [17]. Recently, successful parallel approaches to compute maximum weighted matchings have been proposed [28, 29].

Example 1: Maximum Weight Matching

To conclude this section we demonstrate the effectiveness of maximum weight matchings using a simple sample matrix "west0479" from the SuiteSparse Matrix Collection. The matrix can also directly be loaded in MATLAB using `load west0479`. In Fig. 2 we display the matrix before and after applying maximum weighted matchings. To illustrate the improved diagonal dominance we further compute $r_i = |a_{ii}|/\sum_{j=1}^{n} |a_{ij}|$ for each row of A and $\tilde{A} = \Pi^T D_r A D_s, i = 1, \ldots, n$. r_i can be read as relative diagonal dominance of row i and yields a number between 0 and 1. Moreover, whenever $r_i > \frac{1}{2}$, the row is strictly diagonal dominant, i.e., $|a_{ii}| > \sum_{j:j \neq i} |a_{ij}|$. In Fig. 3 we display for both matrices r_i by sorting its values in increasing order and taking $\frac{1}{2}$ as reference line. We can see the dramatic impact of maximum weighted

(continued)

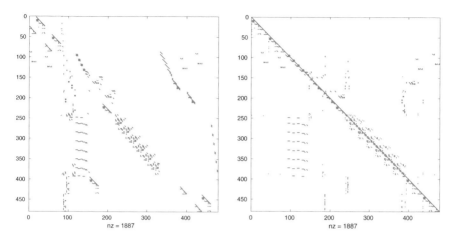

Fig. 2 Maximum weight matching. Left side: original matrix A. Right side: permuted and rescaled matrix $\tilde{A} = \Pi^T D_r A D_c$

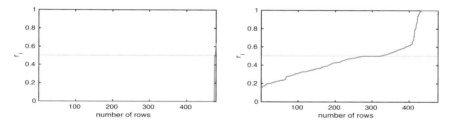

Fig. 3 Diagonal dominance. Left side: r_i for A. Right side: r_i $\tilde{A} = \Pi^T D_r A D_c$

matchings in improving the diagonal dominance of the given matrix and thus paving the way to a static pivoting approach in incomplete or complete LU decomposition methods.

3 Symbolic Symmetric Reordering Techniques

When dealing with large sparse matrices a crucial factor that determines the computation time is the amount of fill that is produced during the factorization of the underlying matrix. To reduce the complexity there exist many mainly symmetric reordering techniques that attempt to reduce the fill-in heuristically. Here we will demonstrate only one of these methods, the so-called nested dissection method.

The main reason for selecting this method is that it can be easily used for parallel computations.

3.1 Multilevel Nested Dissection

Recursive multilevel nested dissection methods for direct decomposition methods were first introduced in the context of multiprocessing. If parallel direct methods are used to solve a sparse system of equations, then a graph partitioning algorithm can be used to compute a fill-reducing ordering that leads to a high degree of concurrency in the factorization phase.

Definition 8 For a matrix $A \in \mathbb{R}^{n,n}$ we define the associated (directed) graph $G_d(A) = (V, E)$, where $V = \{1, \ldots, n\}$ and the set of edges $E = \{(i, j)| a_{ij} \neq 0\}$. The (undirected) graph is given by $G_d(|A| + |A|^T)$ and is denoted simply by $G(A)$.

In graph terminology for a sparse matrix A we simply have a directed edge (i, j) for any nonzero entry a_{ij} in $G_d(A)$, whereas the orientation of the edge is ignored in $G(A)$.

The research on graph partitioning methods in the mid-nineties has resulted in high-quality software packages, e.g., METIS [25]. These methods often compute orderings that on the one hand lead to small fill-in for (incomplete) factorization methods, while on the other hand they provide a high level of concurrency. We will briefly review the main idea of multilevel nested dissection in terms of graph partitioning.

Definition 9 Let $A \in \mathbb{R}^{n,n}$ and consider its graph $G(A) = (V, E)$. A *k-way graph partitioning* consists of partitioning V into k disjoint subsets $V_1, V_2, \ldots, V_k$ such that $V_i \cap V_j = \emptyset$ for $i \neq j \cup_i V_i = V$. The subset $E_s = E \cap \bigcup_{i \neq j}(V_i \times V_j)$ is called *edge separator*.

Typically we want a k-way partitioning to be balanced, i.e., each V_i should satisfy $|V_i| \approx n/k$. The edge separator E_s refers to the edges that have to be taken away from the graph in order to have k separate subgraphs associated with $V_1, \ldots, V_k$ and the number of elements of E_s is usually referred to as edge-cut.

Definition 10 Given $A \in \mathbb{R}^{n,n}$, a *vertex separator* V_s of $G(A) = (V, E)$ is a set of vertices such that there exists a k-way partitioning $V_1, V_2, \ldots, V_k$ of $V \setminus V_s$ having no edge $e \in V_i \times V_j$ for $i \neq j$.

A useful vertex separator V_s should not only separate $G(A)$ into k independent subgraphs associated with $V_1, \ldots, V_k$, it is intended that the number of edges $\cup_{i=1}^{k}|\{e_{is} \in V_i, s \in V_s\}|$ is also small.

Nested dissection recursively splits a graph $G(A) = (V, E)$ into almost equal parts by constructing a vertex separator V_s until the desired number k of partitionings are obtained. If k is a power of 2, then a natural way of obtaining a vertex separator is to first obtain a 2-way partitioning of the graph, a so-called *graph*

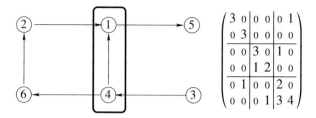

Fig. 4 A 2-way partition with vertex separator $V_s = \{1, 4\}$ and the associated reordered matrix placing the two rows and columns associated with V_s to the end

bisection with its associated edge separator E_s. After that a vertex separator V_s is computed from E_s, which gives a 2-way partitioning V_1, V_2 of $V \setminus V_s$. This process is then repeated separately for the subgraphs associated with V_1, V_2 until eventually a $k = 2^l$-way partitioning is obtained. For the reordering of the underlying matrix A, the vertices associated with V_1 are taken first followed by V_2 and V_s. This reordering is repeated similarly during repeated bisection of each V_i. In general, vertex separators of small size result in low fill-in.

Example 2: Vertex Separators

To illustrate vertex separators, we consider the reordered matrix $\Pi^T A$ from Fig. 1 after a matching is applied. In Fig. 4 we display its graph $G(\Pi^T A)$ ignoring the orientation of the edges. A 2-way partitioning is obtained with $V_1 = \{3, 5\}$, $V_2 = \{2, 6\}$, and a vertex separator $V_s = \{1, 4\}$. The associated reordering refers to taking the rows and the columns of $\Pi^T A$ in the order $3, 5, 2, 6, 1, 4$.

Since a naive approach to compute a recursive graph bisection is typically computationally expensive, combinatorial *multilevel graph bisection* has been used to accelerate the process. The basic structure is simple. The multilevel approach consists of three phases: at first there is a *coarsening phase* which compresses the given graph successively level by level by about half of its size. When the coarsest graph with about a few hundred vertices is reached, the second phase, namely the so-called *bisection*, is applied. This is a high-quality partitioning algorithm. After that, during the *uncoarsening phase*, the given bisection is successively refined as it is prolongated towards the original graph.

3.1.1 Coarsening Phase

The initial graph $G_0 = (V_0, E_0) = G(A)$ of $A \in \mathbb{R}^{n,n}$ is transformed during the coarsening phase into a sequence of graphs $G_1, G_2, \ldots, G_m$ of decreasing size such that $|V_0| \gg |V_1| \gg |V_2| \gg \cdots \gg |V_m|$. Given the graph $G_i = (V_i, E_i)$, the next coarser graph G_{i+1} is obtained from G_i by collapsing adjacent vertices. This can be done, e.g., by using a maximal matching $\mathcal{M}_i$ of G_i (cf. Definitions 3 and 4). Using $\mathcal{M}_i$, the next coarser graph G_{i+1} is constructed from G_i collapsing the vertices being matched into multinodes, i.e., the elements of $\mathcal{M}_i$ together with the unmatched vertices of G_i become the new vertices V_{i+1} of G_{i+1}. The new edges E_{i+1} are the remaining edges from E_i connected with the collapsed vertices. There are various differences in the construction of maximal matchings [9, 25]. One of the most popular and efficient methods is heavy edge matching [25].

3.1.2 Partitioning Phase

At the coarsest level m, a 2-way partitioning $V_{m,1} \dot{\cup} V_{m,2} = V_m$ of $G_m = (V_m, E_m)$ is computed, each of them containing about half of the vertices of G_m. This specific partitioning of G_m can be obtained by using various algorithms such as spectral bisection [19] or combinatorial methods based on Kernighan–Lin variants [18, 27]. It is demonstrated in [25] that for the coarsest graph, combinatorial methods typically compute smaller edge-cut separators compared with spectral bisection methods. However, since the size of the coarsest graph G_m is small (typically $|V_m| < 100$), this step is negligible with respect to the total amount of computation time.

3.1.3 Uncoarsening Phase

Suppose that at the coarsest level m, an edge separator $E_{m,s}$ of G_m associated with the 2-way partitioning has been computed that has led to a sufficient edge-cut of G_m with $V_{m,1}, V_{m,2}$ of almost equal size. Then $E_{m,s}$ is prolongated to G_{m-1} by reversing the process of collapsing matched vertices. This leads to an initial edge separator $E_{m-1,s}$ for G_{m-1}. But since G_{m-1} is finer, $E_{m-1,s}$ is sub-optimal and one usually decreases the edge-cut of the partitioning by local refinement heuristics such as the Kernighan–Lin partitioning algorithm [27] or the Fiduccia–Mattheyses method [18]. Repeating this refinement procedure level-by-level we obtain a sequence of edge separators $E_{m,s}, E_{m-1,s}, \ldots, E_{0,s}$ and eventually and edge separator $E_s = E_{0,s}$ of the initial graph $G(A)$ is obtained. If one is seeking for a vertex separator V_s of $G(A)$, then one usually computes V_s from E_s at the end.

There have been a number of methods that are used for graph partitioning, e.g., METIS [25], a parallel MPI version PARMETIS [26], or a recent multithreaded approach MT-METIS [30]. Another example for a parallel partitioning algorithm is SCOTCH [9].

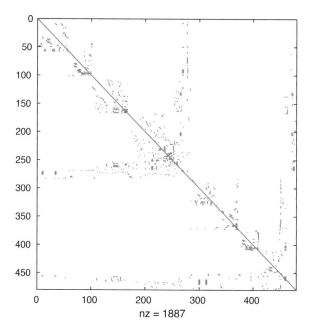

Fig. 5 Application of multilevel nested dissection after the matrix is already rescaled and permuted using maximum weight matching

Multilevel Nested Dissection
We will continue Example 1 using the matrix $\tilde{A} = \Pi^T D_r A D_s$ that has
been rescaled and permuted using maximum weight matching. We illustrate
in Fig. 5 how multilevel nested dissection changes the pattern $\hat{A} = P^T \tilde{A} P$,
where P refers to the permutation matrix associated with the partitioning of
$G(\tilde{A})$.

3.2 Other Reordering Methods

One of the first methods to reorder the system was the reverse Cuthill–McKee
(RCM) methods [10, 34] which attempts to reduce the bandwidth of a given matrix.
Though this algorithm is still attractive for sequential methods and incomplete
factorization methods, its use for direct solvers is considered as obsolete. An attractive alternative to nested dissection as reordering method for direct factorization
methods is the minimum degree algorithm (MMD) [20, 40] and its recent variants,

in particular the approximate minimum degree algorithm (AMD) [2, 12] with or without constraints. The main objective of the minimum degree algorithm is to simulate the Gaussian elimination process symbolically by investigating the update process $a_{ij} \rightarrow a_{ij} - a_{ik}a_{kk}^{-1}a_{kj}$ by means of graph theory, at least in the case of the undirected graph. The name-giving degree refers to the number of edges connected to a vertex and how the graph and therefore the degrees of its vertices change during the factorization process. Over the years this has led to an evolution of the underlying minimum degree algorithm using the so-called *external degree* for selecting vertices as pivots and further techniques like *incomplete degree update*, *element absorption*, and *multiple elimination* as well as data structures based on cliques. For an overview see [20]. One of the most costly parts in the minimum degree algorithm is to update of the degrees. Instead of computing the exact external degree, in the approximate minimum degree algorithm [2], an approximate external degree is computed that significantly saves time while producing comparable fill-in for the LU decomposition.

We like to conclude this section by mentioning that if nested dissection is computed to produce a vertex separator V_s and a related k-way partitioning $V_1, \ldots, V - k$ for the remaining vertices of $V \setminus V_s$ of $G(A) = (V, E)$ which allow for parallel computations, then the entries of each V_i, $i, \ldots, k$ could be taken in any order. Certainly, inside V_i one could use nested dissection as well, which is the default choice in multilevel nested dissection methods. However, as soon as the coarsest graph G_m is small enough (typically about 100 vertices), not only the separator is computed, but in addition the remaining entries of G_m are reordered to lead to a fill-reducing ordering. In both cases, for G_m as well as $V_1, \ldots, V_k$ one could alternatively use different reordering methods such as variants of the minimum degree algorithm. Indeed, for G_m this is what the METIS software is doing. Furthermore, a reordering method such as the constrained approximate minimum degree algorithm is also suitable as local reordering for $V_1, \ldots, V_k$ as alternative to nested dissection, taking into account the edges connected with V_s (also referred to as HALO structure), see, e.g., [38].

4 Sparse *LU* Decomposition

In this section we will assume that the given matrix $A \in \mathbb{R}^{n,n}$ is nonsingular and that it can be factorized as $A = LU$, where L is a lower triangular matrix with unit diagonal and U is an upper triangular matrix. It is well-known [21], if $A = LU$, where L and $U^{\top}$ are lower triangular matrices, then in the generic case we will have $G_d(L + U) \supset G_d(A)$, i.e., we will only get additional edges unless some entries cancel by "accident" during the elimination. In the sequel we will ignore cancellations. Throughout this section we will always assume that the diagonal entries of A are nonzero as well. We also assume that $G_d(A)$ is connected.

In the preceding sections we have argued that maximum weight matching often leads to a rescaled and reordered matrix such that static pivoting is likely

Fig. 6 Fill-in with respect to
$L + U$ is denoted by $\times$

$$\begin{pmatrix} 3 & 0 & 0 & 0 & 0 & 1 \\ 0 & 3 & 0 & 0 & 0 & 0 \\ 0 & 0 & 3 & 0 & 1 & 0 \\ 0 & 0 & 1 & 2 & \times & 0 \\ 0 & 1 & 0 & 0 & 2 & 0 \\ 0 & 0 & 0 & 1 & 3 & 4 \end{pmatrix}$$

to be enough, i.e., pivoting is restricted to some dense blocks inside the LU factorization. Furthermore, reordering strategies such as multilevel nested dissection have further symmetrically permuted the system such that the fill-in that occurs during Gaussian elimination is acceptable and even parallel approaches could be drawn from this reordering. Thus assuming that A does not need further reordering and a factorization $A = LU$ exists is a realistic scenario in what follows.

4.1 The Elimination Tree

The basis of determining the fill-in in the triangular factors L and U as by-product of the Gaussian elimination can be characterized as follows (see [23] and the references therein).

Theorem 2 *Given $A = LU$ with the aforementioned assumptions, there exists an edge (i, j) in $G_d(L + U)$ if and only if there exists a path*

$$i x_1, x_2 x_3, \ldots, x_k j$$

in $G_d(A)$ such that $x_1, \ldots, x_k < \min(i, j)$.

In other words, during Gaussian elimination we obtain a fill edge (i, j) for every path from i to j through vertices less than $\min(i, j)$.

Fill-in
We will use the matrix $\Pi^T A$ from Example 2 and sketch the fill-in obtained during Gaussian elimination in Fig. 6.

The fastest known method for predicting the filled graph $G_d(L + U)$ is Gaussian elimination. The situation is simplified if the graph is undirected. In the sequel we ignore the orientation of the edges and simply consider the undirected graph $G(A)$ and $G(L + U)$, respectively.

Fig. 7 Entries of $G(A)$ are
denoted by filled circle, fill-in
is denoted by times

Definition 11 The undirected graph $G(L + U)$ that is derived from the undirected graph $G(A)$ by applying Theorem 2 is called the *filled graph* and it will be denoted by $G_f(A)$.

Fill-in with Respect to the Undirected Graph

When we consider the undirected graph $G(A)$ in Example 4.1, the pattern of $|\Pi^T A| + |\Pi^T A|^T$ and its filled graph $G_f(A)$ now equals $G(A)$ up to positions $(5, 4)$ and $(4, 5)$ (cf. Fig. 7).

The key tool to predict the fill-in easily for the undirected graph is the *elimination tree* [33].

Recall that an undirected and connected graph is called a *tree*, if it does not contain any cycle. Furthermore, one vertex is identified as *root*. As usual we call a vertex j *parent* of i, if there exists an edge (i, j) in the tree such that j is closer to the root. In this case i is called *child* of j. The subtree rooted at vertex j is denoted by $T(j)$ and the vertices of this subtree are called *descendants* of j, whereas j is called their *ancestor*. Initially we will define the elimination tree algorithmically using the depth-first-search algorithm [1]. Later we will state a much simplified algorithm.

Definition 12 Given the filled graph $G_f(A)$ the *elimination tree* $T(A)$ is defined by the following algorithm.

Perform a depth-first-search in $G_f(A)$ starting from vertex n.

When vertex m is visited, choose from its unvisited neighbors $i_1, \ldots, i_k$ the index j with the largest number $j = \max\{i_1, \ldots, i_k\}$ and continue the search with j.

A leaf of the tree is reached, when all neighbors have already been visited.

We like to point out that the application of the depth-first-search to $G_f(A)$ starting at vertex n behaves significantly different from other graphs. By Theorem 2 it follows that as soon as we visit a vertex m, all its neighbors $j > m$ must have been visited prior to vertex m. Thus the labels of the vertices are strictly decreasing until we reach a leaf node.

Depth-First-Search

We illustrate the depth-first-search using the (filled) graph in Fig. 8 and the pattern from Example 4.1. The extra fill edge is marked by a bold line.

The ongoing depth-first-search visits the vertices in the order $6 \to 5 \to 4 \to 3$. Since at vertex 3, all neighbors of 3 are visited (and indeed have a larger number), the algorithm backtracks to 4 and to 5 and continues the search in the order $5 \to 2$. Again all neighbors of vertex 2 are visited (and have larger number), thus the algorithm backtracks to 5 and to 6 and continues by $6 \to 1$. Then the algorithm terminates.

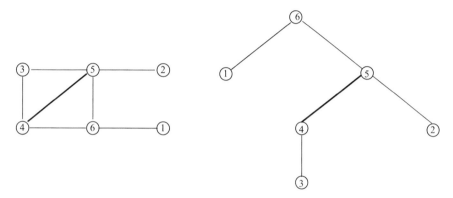

Fig. 8 Filled graph (left) and elimination tree (right)

Remark 1 It follows immediately from the construction of $T(A)$ and Theorem 2 that additional edges of $G_f(A)$ which are not covered by the elimination tree can only show up between a vertex and some of its ancestors (referred to as "back-edges"). In contrast to that, "cross-edges" between unrelated vertices do not exist.

Remark 2 One immediate consequence of Remark 1 is that triangular factors can be computed independently starting from the leaves until the vertices meet a common parent, i.e., column j of L and U^T only depend on those columns s of L and U^T such that s is a descendant of j in the elimination tree $T(A)$.

Elimination Tree

We use the matrix "west0479" from Example 3.1.3, after maximum weight matching and multilevel nested dissection have been applied. We use MATLAB's etreeplot to display its elimination tree (see Fig. 9). The

(continued)

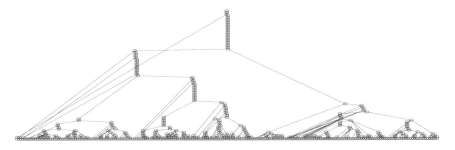

Fig. 9 Elimination tree of "west0479" after maximum weight matching and nested dissection are applied

elimination tree displays the high level of concurrency that is induced by nested dissection, since by Remark 2 the computations can be executed independently at each leaf node towards the root until a common parent vertex is reached.

Further conclusions can be easily derived from the elimination tree, in particular Remark 2 in conjunction with Theorem 2.

Remark 3 Consider some $k \in \{1, \ldots, n\}$. Then there exists a (fill) edge (j, k) with $j < k$ if and only if there exists a common descendant i of k, j in $T(A)$ such that $a_{ik} \neq 0$. This follows from the fact that once $a_{ik} \neq 0$, by Theorem 2 this induces (fill) edges (j, k) in the filled graph $G_f(A)$ for all nodes j between i and k in the elimination tree $T(A)$, i.e., for all ancestors of i that are also descendants of k. This way, i propagates fill-edges along the branch from i to k in $T(A)$ and the information $a_{ik} \neq 0$ can be used as path compression to advance from i towards k along the elimination tree.

Path Compression

Consider the graph and the elimination tree from Fig. 8. Since there exists the edge $(3, 5)$ in $G(A)$, therefore another (fill) edge $(4, 5)$ must exist. Similarly, the same conclusion can be drawn from the existence of the edge $(4, 6)$ (here not a fill edge, but a regular edge).

The elimination tree itself can be easily described by a vector p of length n such that for any $i < n$, p_i denotes the parent node, while $p_n = 0$ corresponds to the root. Consider some step k with $a_{ik} \neq 0$, for some $i < k$. By Remark 3, i must be a descendant of k and there could be further ancestors j of i which are also descendants of k. Possibly not all ancestors of i have been assigned a parent node so far. Thus we can replace i by $j = p_i$ until we end up with $p_j = 0$ or $p_j \geq k$. This way we traverse $T(A)$ from i towards to k until we have found the child node j of k. If the parent of j has not been assigned to j yet, then $p_j = 0$ and k must be the parent of j. If some $l < k$ were the parent of j, then we would have assigned l as parent of j in an earlier step $l < k$. In this case we set $p_j \leftarrow k$. Otherwise, if $p_j \geq k$, then we have already assigned j's parent in an earlier step $l < k$.

Computation of Parent Nodes

Consider the elimination tree $T(A)$ from Fig. 8. Unless $k = 4$, no parents have been assigned, i.e., $p_i = 0$ for all i.

Now for $k = 4$ we have $a_{34} \neq 0$ and using the fact that $p_3 = 0$ implies that we have to set $a_3 = p_3 \leftarrow 4$.

For $k = 5$, $a_{25} \neq 0$ and again $p_2 = 0$ requires to set $a_2 = p_2 \leftarrow 5$. Next, $a_{35} \neq 0$, path compression enables $a_3 \leftarrow 5$ and after another loop we obtain $a_4 = p_4 \leftarrow 5$.

Finally, if $k = 6$, we have $a_{16} \neq 0$ and immediately obtain $a_1 = p_1 \leftarrow 6$. Since $a_{46} \neq 0$, a path compression is applied which yields $a_4 \leftarrow 6$ and in the next step we set $a_5 = p_5 \leftarrow 6$. At last $a_{56} \neq 0$ does not cause further changes.

In total we have $p = [6, 5, 4, 5, 6, 0]$ which perfectly reveals the parent properties of the elimination trees in Fig. 8.

By Remark 3 (cf. [12, 43]), we can also make use of path compression. Since our goal is to traverse the branch of the elimination tree from i to k as fast as possible, any ancestor $j = a_i$ of i would be sufficient. With the same argument as before, an ancestor $a_j = 0$ would refer to a vertex that does not have a parent yet. In this case we can again set $p_j \leftarrow k$. Moreover, k is always an ancestor of a_i.

The algorithm including path compression can be summarized as follows (see also [12, 33]).

Computation of the Elimination Tree

Input: $A \in \mathbb{R}^{n,n}$ such that A has the same pattern as $|A| + |A|^T$.

Output: vector $p \in \mathbb{R}^n$ such that p_i is the parent of i, $i = 1, \ldots, n - 1$, except $p_n = 0$.

1: let $a \in \mathbb{R}^n$ be an auxiliary vector used for path compression.
2: $p \leftarrow 0, a \leftarrow 0$
3: **for** $k = 2, \ldots, n$ **do**
4:     **for all** $i < k$ such that $a_{ik} \neq 0$ **do**

(continued)

```
 5:       while i ≠ 0 and i < k do
 6:           j ← aᵢ
 7:           aᵢ ← k
 8:           if j = 0 then
 9:               pᵢ ← k
10:           end if
11:           i ← j
12:       end while
13:     end for
14: end for
```

4.2 The Supernodal Approach

We have already seen that the elimination tree reveals information about concurrency. It is further useful to determine the fill-in L and U^T. This information can be computed from the elimination tree $T(A)$ together with $G(A)$. The basis for determining the fill-in in each column is again Remark 3. Suppose we are interested in the nonzero entries of column j of L and U^T. Then for all descendants of j, i.e., the nodes of the subtree $T(j)$ rooted at vertex j, a nonzero entry $a_{ik} \neq 0$ also implies $l_{kj} \neq 0$. Thus, starting at any leaf i, we obtain its fill by all $a_{ik} \neq 0$ such that $k > i$ and when we move forward from i to its parent j, vertex j will inherit the fill from node i for all $k > j$ plus the nonzero entries given by $a_{jk} \neq 0$ such that $k > j$. When we reach a common parent node k with multiple children, the same argument applies using the union of fill-in greater than k from its children together with the nonzero entries $a_{kl} \neq 0$ such that $l > k$. We summarize this result in a very simple algorithm

Computation of Fill-in

Input: $A \in \mathbb{R}^{n,n}$ such that A has the same pattern as $|A| + |A|^T$.
Output: sparse strict lower triangular pattern $P \in \mathbb{R}^{n,n}$ with same pattern as L, U^T.

```
1: compute parent array p of the elimination tree T(A)
2: for j = 1, ..., n do
3:     supplement nonzeros of column j of P with all i > j such that aᵢⱼ ≠ 0
4:     k = pⱼ
5:     if k > 0 then
```

(continued)

> 6:   supplement nonzeros of column k of P with nonzeros of column j of P greater than k
> 7:   **end if**
> 8: **end for**

Algorithm 4.2 only deals with the fill pattern. One additional aspect that allows to raise efficiency and to speed up the numerical factorization significantly is to detect dense submatrices in the factorization. Block structures allow to collect parts of the matrix in dense blocks and to treat them commonly using dense matrix kernels such as level-3 BLAS and LAPACK [13, 14].

Dense blocks can be read off from the elimination tree employing Algorithm 4.2.

Definition 13 Denote by $\mathcal{P}_j$ the nonzero indices of column j of P as computed by Algorithm 4.2. A sequence $k, k+1, \ldots, k+s-1$ is called *supernode* of size s if the columns of $\mathcal{P}_j = \mathcal{P}_{j+1} \cup \{j+1\}$ for all $j = k, \ldots, k+s-2$.

In simple words, Definition 13 states that for a supernode s subsequent columns can be grouped together in one dense block with a triangular diagonal block and a dense subdiagonal block since they perfectly match the associated trapezoidal shape. We can thus easily supplement Algorithm 4.2 with a supernode detection.

Computation of Fill-in and Supernodes

Input: $A \in \mathbb{R}^{n,n}$ such that A has the same pattern as $|A| + |A|^T$.
Output: sparse strict lower triangular pattern $P \in \mathbb{R}^{n,n}$ with same pattern as L, U^T as well as column size $s \in \mathbb{R}^m$ of each supernode.
1: compute parent array p of the elimination tree $T(A)$
2: $m \leftarrow 0$
3: **for** $j = 1, \ldots, n$ **do**
4:   supplement nonzeros of column j of P with all $i > j$ such that $a_{ij} \neq 0$
5:   denote by r the number of entries in column j of P
6:   **if** $j > 1$ and $j = p_{j-1}$ and $s_m + r = l$ **then**
7:     $s_m \leftarrow s_m + 1$ ▷ continue current supernode
8:   **else**
9:     $m \leftarrow m + 1, s_m \leftarrow 1, l \leftarrow r$ ▷ start new supernode
10:   **end if**
11:   $k = p_j$

(continued)

Fig. 10 Supernodes in the
triangular factor

12:    **if** $k > 0$ **then**
13:        supplement nonzeros of column k of P with nonzeros of column j
       of P greater than k
14:    **end if**
15: **end for**

Supernode Computation
To illustrate the use of supernodes, we consider the matrix pattern from Fig. 7
and illustrate the underlying dense block structure in Fig. 10. Supernodes are
the columns 1, 2, 3 as scalar columns as well as columns 4–6 as one single
supernode.

Supernodes form the basis of several improvements, e.g., a supernode can be
stored as one or two dense matrices. Beside the storage scheme as dense matrices,
the nonzero row indices for these blocks need only be stored once. Next the use
of dense submatrices allows the usage of dense matrix kernels using level-3 BLAS
[13, 14].

Supernodes
We use the matrix "west0479" from Example 3.1.3, after maximum weight
matching and multilevel nested dissection have been applied. We use its undirected graph to compute the supernodal structure. Certainly, since the matrix
is nonsymmetric, the block structure is only sub-optimal. We display the
supernodal structure for the associated Cholesky factor, i.e., for the Cholesky
factor of a symmetric positive definite matrix with same undirected graph as
our matrix (see left part of Fig. 11). Furthermore, we display the supernodal
structure for the factors L and U computed from the nonsymmetric matrix
without pivoting (see right part of Fig. 11).

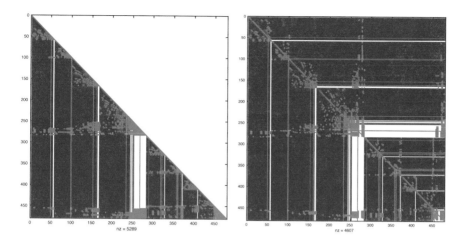

Fig. 11 Supernodal structure. Left: vertical lines display the blocking of the supernodes with respect to the associated Cholesky factor. Right: vertical and horizontal lines display the blocking of the supernodes applied to L and U

While the construction of supernodes is fairly easy in the symmetric case, its generalization for the nonsymmetric case is significantly harder, since one has to deal with pivoting in each step of Gaussian elimination. In this case one uses the column elimination tree [22].

5 Sparse Direct Solvers—Supernodal Data Structures

High-performance sparse solver libraries have been a very important part of scientific and engineering computing for years, and their importance continues to grow as microprocessor architectures become more complex and software libraries become better designed to integrate easily within applications. Despite the fact that there are various science and engineering applications, the underlying algorithms typically have remarkable similarities, especially those algorithms that are most challenging to implement well in parallel. It is not too strong a statement to say that these software libraries are essential to the broad success of scalable high-performance computing in computational sciences. In this section we demonstrate the benefit of supernodal data structures within the sparse solver package PARDISO [42]. We illustrate it by using the triangular solution process. The forward and backward substitution is performed column wise with respect to the columns of L, starting with the first column, as depicted in Fig. 12. The data dependencies here allow to store vectors y, z, b, and x in only one vector r. When column j is reached, r_j contains the solution for y_j. All other elements of L in this column, i.e. L_{ij} with $i = j + 1, \ldots, N$, are used to update the remaining entries in r by

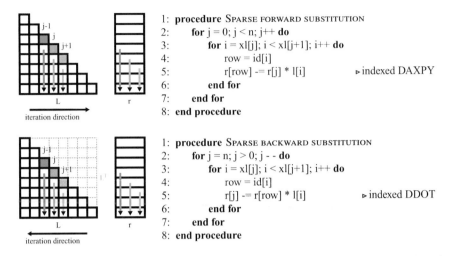

Fig. 12 Sparse triangular substitution in CSC format based on indexed DAXPY/DDOT kernel operations

$$r_i = r_i - r_j L_{ij}. \tag{8}$$

The backward substitution with L^T will take place row wise, since we use L and perform the substitution column wise with respect to L, as shown in the lower part of Fig. 12. In contrast to the forward substitution the iteration over columns starts at the last column N and proceeds to the first one. If column j is reached, then r_j, which contains the j-component of the solution vector x_j, is computed by subtracting the dot-product of the remaining elements in the column L_{ij} and the corresponding elements of r_i with $i = j + 1, \ldots, N$ from it:

$$r_j = r_j - r_i L_{ij}. \tag{9}$$

After all columns have been processed r contains the required solution x. It is important to note that line 5 represents in both substitutions an indexed DAXPY and indexed DDOT kernel operations that has to be computed during the streaming operations of the vector r and the column j of the numerical factor L. As we are dealing with sparse matrices it makes no sense to store the lower triangular matrix L as a dense matrix. Hence, PARDISO uses its own data structure to store L, as shown in Fig. 13.

Adjacent columns exhibiting the same row sparsity structure form a *panel*, also known as *supernode*. A panel's column count is called the *panel size* n_p. The columns of a panel are stored consecutively in memory excluding the zero entries. Note that columns of panels are padded in the front with zeros so they get the same length as the first column inside their panel. The padding is of utmost performance for the PARDISO solver to use Level-3 BLAS and LAPACK functionalities [41]. Furthermore panels are stored consecutively in the 1 array. Row and column

Fig. 13 Sparse matrix data structures in PARDISO. Adjacent columns of L exhibiting the same structure form panels also known as supernodes. Groups of panels which touch independent elements of the right-hand side r are parts. The last part in the lower triangular matrix L is called separator

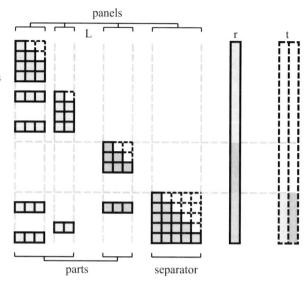

information is now stored in accompanying arrays. The `xsuper` array stores for each panel the index of its first column. Also note that here column indices are the running count of nonzero columns. Column indices are used as indices into `xl` array to lookup the start of the column in the `l` array which contains the numerical values of the factor L. To determine the row index of a column's element an additional array `id` is used, which holds for each panel the row indices. The start of a panel inside `id` is found via `xid` array. The first row index of panel p is `id[xid[p]]`. For serial execution this information is enough. However, during parallel forward/backward substitution concurrent updates to the same entry of r must be avoided. The *parts* structure contains the start (and end) indices of the panels which can be updated independently as they do not touch the same entries of r. Two parts, colored blue and orange, are shown in Fig. 13. The last part in the bottom right corner of L is special and is called the *separator* and is colored green. Parts which would touch entries of r in the range of the separator perform their updates into separate temporary arrays t. Before the separator is then serially updated, the results of the temporary arrays are gathered back into r. The backward substitution works the same, just reversed and only updates to different temporary arrays are not required. The complete forward substitution and backward substitution is listed in Algorithms 1 and 2.

6   Application—Circuit Simulation

In this section we demonstrate how these developments in sparse direct linear solvers have advanced integrated circuit simulations. Integrated circuits are composed of interconnected transistors. The interconnects are modeled primarily with

---

Algorithm 1 Forward substitution in PARDISO. Note that in case of serial execution separated updates to temporary arrays in Lines 10–13 are not necessary and can be handled via the loop in Lines 6–9

```
 1: procedure FORWARD
 2:     for part o in parts do                                              ▷ parallel execution
 3:         for panel p in part p do
 4:             for column j in panel do                                            ▷ unroll
 5:                 i = xid[p] + offset
 6:                 for k = xl[j] + offset; k < sep; ++k do
 7:                     row = id[i++]
 8:                     r[row] - = r[j] l[k]                                  ▷ indexed DAXPY
 9:                 end for
10:                 for k = sep + 1; k < xl[j+1]; ++k do
11:                     row = id[i++]
12:                     t[row,p] -= r[j] l[k]                                 ▷ indexed DAXPY
13:                 end for
14:             end for
15:         end for
16:     end for
17:     r[i] = r[i] - sum(t[i,:])                                    ▷ gather temporary arrays
18:     for panel p in separator do                                        ▷ serial execution
19:         for column j in panel do                                            ▷ unroll
20:             i = xid[p] + offset
21:             for k = xl[j] + offset; k < xl[j+1]; ++k do
22:                 row = id[i++]
23:                 r[row] -= r[j] l[k]                                   ▷ indexed DAXPY
24:             end for
25:         end for
26:     end for
27: end procedure
```

---

resistors, capacitors, and inductors. The interconnects route signals through the circuit, and also deliver power. Circuit equations arise out of Kirchhoff's current law, applied at each node, and are generally nonlinear differential-algebraic equations. In transient simulation of the circuit, the differential portion is handled by discretizing the time derivative of the node charge by an implicit integration formula. The associated set of nonlinear equations is handled through use of quasi-Newton methods or continuation methods, which change the nonlinear problem into a series of linear algebraic solutions. Each component in the circuit contributes only to a few equations. Hence, the resulting systems of linear algebraic equations are extremely sparse, and most reliably solved by using direct sparse matrix techniques. Circuit simulation matrices are peculiar in the universe of matrices, having the following characteristics [11]:

- they are nonsymmetric, although often nearly structurally symmetric;
- they have a few dense rows and columns (e.g., power and ground connections);
- they are *very* sparse and the straightforward usage of BLAS routines (as in SuperLU[32]) may be ineffective;

Algorithm 2 Backward substitution in PARDISO. Separator (sep.), parts, and panels are iterated over in reversed (rev.) order

---

```
 1: procedure BACKWARD
 2:     for panel p in sep. rev. do                                    ▷ serial execution
 3:         for col. j in panel p rev. do                                      ▷ unroll
 4:             i = xid[p] + offset
 5:             for k = xl[j] + offset; k < xl[j+1]; ++k do
 6:                 row = id[i++]
 7:                 r[j] -= r[row] l[k]                                  ▷ indexed DDOT
 8:             end for
 9:             offset = offset - 1
10:         end for
11:     end for
12:     for part in parts do                                        ▷ parallel execution
13:         for panel p in part rev. do
14:             for col. j in panel p rev. do                                  ▷ unroll
15:                 i = xid[p] + offset
16:                 for k = xl[j] + offset; k < xl[j+1]; ++k do
17:                     row = id[i++]
18:                     r[j] -= r[row] l[k]                              ▷ indexed DDOT
19:                 end for
20:                 offset = offset - 1
21:             end for
22:         end for
23:     end for
24: end procedure
```

- their LU factors remain sparse if well-ordered;
- they can have high fill-in if ordered with typical strategies;
- and being unstructured, the highly irregular memory access causes factorization to proceed only at a few percent of the peak flop-rate.

Circuit simulation matrices also vary from being positive definite to being *extremely* ill-conditioned, making pivoting for stability important also. As circuit size increases, and depending on how much of the interconnect is modeled, sparse matrix factorization is the dominant cost in the transient analysis.

To overcome the complexity of matrix factorization a new class of simulators arose in the 1990s, called fast-SPICE [39]. These simulators partition the circuit into subcircuits and use a variety of techniques, including model order reduction and multirate integration, to overcome the matrix bottleneck. However, the resulting simulation methods generally incur unacceptable errors for analog and tightly coupled circuits. As accuracy demands increase, these techniques become much slower than traditional SPICE methods. Even so, since much of the research effort was directed at fast-SPICE simulators, it brought some relief from impossibly slow simulations when some accuracy trade-off was acceptable. Because these simulators partitioned the circuit, and did not require the simultaneous solution of the entire system of linear equations at any given time, they did not push the state of the art in sparse matrix solvers.

Starting in the mid-2000s, increasing demands on accuracy, due to advancing semiconductor technology, brought attention back to traditional SPICE techniques. This was aided by the proliferation of multicore CPUs. Parallel circuit simulation, an area of much research focus in the 1980s and 1990s, but not particularly in practice, received renewed interest as a way to speed up simulation without sacrificing accuracy. Along with improved implementations to avoid cache misses, rearchitecture of code for parallel computing, and better techniques for exploitation of circuit latency, improved sparse matrix solvers, most notably the release of KLU [11], played a crucial role in expanding the utility of SPICE.

Along with the ability to simulate ever larger circuits with full SPICE accuracy came the opportunity to further improve sparse matrix techniques. A sparse matrix package for transient simulation needs to have the following features:

- must be parallel;
- fast matrix reordering;
- incremental update of the L and U factors when only a few nonzeros change;
- fast computation of the diagonal entries of the inverse matrix;
- fast computation of Schur-complements for a submatrix;
- allow for multiple LU factors of the same structure to be stored;
- use the best-in-class method across the spectrum of sparsity;
- use iterative solvers with fast construction of sparse preconditioners;
- run on various hardware platforms (e.g., GPU acceleration).

Some of these features must be available in a single package. Others, such as iterative solvers and construction of preconditioners, can be implemented with a combination of different packages. The PARDISO solver[1] stands out as a package that does most of these very well. Here we touch on a few of these features.

When applied in the simulation of very large circuits, the difference between a "good" and a "bad" matrix ordering can be the difference between seconds and days. PARDISO offers AMD and nested dissection methods for matrix ordering, as well as permitting user-defined ordering. Because the matrix reordering method which has been used most often in circuit simulation is due to Markowitz [35], and because modern sparse matrix packages do not include this ordering method, we briefly describe it here. The Markowitz method is quite well-adapted for circuit simulation. Some desirable aspects of the typical implementation of the Markowitz method, as opposed to the MD variants, are that it works for nonsymmetric matrices and combines pivot choice with numerical decomposition, such that a pivot choice is a numerically "good" pivot which generates in a local sense the least fill-in at that step of the decomposition. Choosing pivots based on the Markowitz score often produces very good results: near-minimal fill-in, unfortunately at the cost of an $O(n^3)$ algorithm (for dense blocks). Even though the Markowitz algorithm has some good properties when applied to circuit matrices, the complexity of the algorithm has become quite burdensome. When SPICE [36] was originally conceived, a hundred-

---

[1]The PARDISO solver is available from http://www.pardiso-project.org.

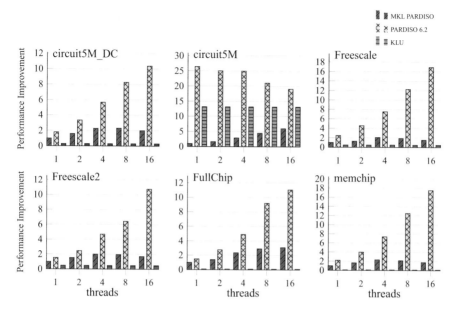

Fig. 14 Performance improvements of PARDISO 6.2 against Intel MKL PARDISO (one thread) for various circuit simulation matrices

node circuit was huge and the Markowitz algorithm was not a problem. Now we routinely see netlists with hundreds of thousands of nodes and post-layout netlists with millions of elements. As matrix order and element counts increase, Markowitz reordering time can become an obstruction. Even as improved implementations of the Markowitz method have extended its reach, AMD and nested dissection have become the mainstay of simulation of large denser-than-usual matrices.

Next we turn our attention to parallel performance. While KLU remains a benchmark for serial solvers, for parallel solvers, MKL-PARDISO is often cited as the benchmark [6, 8]. To give the reader a sense of the progress in parallel sparse matrix methods, in Fig. 14 we compare KLU, PARDISO (Version 6.2) to MKL-PARDISO on up to 16 cores on an Intel Xeon E7-4880 architecture with 2.5 GHz processors.

Some of the matrices here can be obtained from the SuiteSparse Matrix Collection, and arise in transistor level full-chip and memory array simulations. It is clear that implementation of sparse matrix solvers has improved significantly over the years.

Exploiting latency in all parts of the SPICE algorithm is very important in enabling accurate circuit simulation, especially as the circuit size increases. By latency, we mean that only a few entries in the matrix change from one Newton iteration to the next, and from one timepoint to the next. As the matrix depends on the time-step, some simulators hold the time-steps constant as much as feasible to allow increased reuse of matrix factorizations. The nonzero entries of a matrix

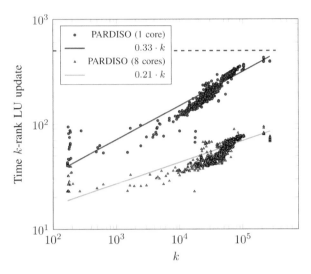

Fig. 15 Regression analysis on the rank-k update LU factorization in PARDISO

change only when the transistors and other nonlinear devices change their operation point. In most circuits, very few devices change state from one iteration to the next and from one time-step to the next. Nonzeros contributed by entirely linear components do not change value during the simulation. This makes incremental LU factorization a very useful feature of any matrix solver used in circuit simulation. As of April 2019 the version PARDISO 6.2 has a very efficient exploitation of incremental LU factorization, both serial and parallel. In Fig. 15 we show that PARDISO scales linearly with number of updated columns, and also scales well with number of cores. Here, the series of matrices were obtained from a full simulation of a post-layout circuit that includes all interconnects, power and ground networks. The factorization time is plotted against the number of columns that changed compared to the previous factorization. The scatter plot shows the number of rank-k update and the corresponding factorization time in milliseconds. The regression analysis clearly demonstrates a linear trend both for the single and the multiple core versions. The dashed line shows the time for the full factorization.

Another recent useful feature in PARDISO is parallel selective inverse matrix computation as demonstrated in Table 1. In circuit simulation, the diagonal of the inverse matrix is the driving point impedance. It is often required to flag nodes in the circuit with very high driving point impedance. Such nodes would indicate failed interfaces between different subcircuits, leading to undefined state and high current leakage and power dissipation. A naive approach to this is to solve for the driving point impedance, the diagonal of the inverse matrix, by N triangular solves. This is sometimes unacceptably expensive even with exploiting the sparsity of the right-hand side, and minimizing the number of entries needed in the diagonal of the inverse. To bypass this complexity, heuristics to compute the impedance of

Table 1 Details of the benchmark matrices

Matrix	N	nnz(A)	nnz($\frac{L+U}{A}$)	A^{-1}	Selected A^{-1}
circuit5M_DC	3,523,317	19,194,193	2.87	82.3 h	1.3 s
circuit5M	5,558,326	59,524,291	1.04	371.1 h	2.1 s
Freescale	3,428,755	18,920,347	2.94	89.8 h	1.0 s
Freescale2	2,999,349	23,042,677	2.92	8.5 h	1.2 s
FullChip	2,987,012	26,621,990	7.41	162.9 h	11.9 s
memchip	2,707,524	14,810,202	4.40	62.5 h	0.9 s

"N" is the number of matrix rows and "nnz" is the number of nonzeros
The table shows the fill-in factor related to the number of nonzeros in $\frac{L+U}{A}$, the time for computing all diagonal elements of the inverse A^{-1} using N multiple forward/backward substitution in hours, and using the selected inverse method in PARDISO for computing all diagonal elements of the inverse A^{-1} in seconds

connected components are used. But this is error prone with many false positives and also false negatives. In the circuit Freescale, PARDISO, e.g., finished the required impedance calculations in 11.9 s compared to the traditional computation that consumed 162.9 h.

The productivity gap in simulation continues to grow, and challenges remain. Signoff simulations demand $10\times$ speedup in sparse matrix factorization. Simply using more cores does not help unless the matrices are very large and complex. For a majority of simulations, scaling beyond eight cores is difficult. As a result, some of these simulations can take a few months to complete, making them essentially impossible. Some of the problems in parallelizing sparse matrix operations for circuit simulation are fundamental. Others may be related to implementation. Research on sparse matrix factorization for circuit simulation continues to draw attention, especially in the area of acceleration with Intel's many integrated core (MIC) architecture [6] and GPUs [7, 31]. Other techniques for acceleration include improved preconditioners for iterative solvers [44]. We are presently addressing the need for runtime selection of optimal strategies for factorization, and also GPU acceleration. Given that circuits present a wide spectrum of matrices, no matter how we categorize them, it is possible to obtain a solver that is $2–10\times$ better on a given problem. Improvements in parallel sparse matrix factorization targeted at circuit simulation is more necessary today than ever and will continue to drive applicability of traditional SPICE simulation methods. Availability of sparse matrix packages such as PARDISO that completely satisfy the needs of various circuit simulation methods is necessary for continued performance gains.

References

1. A. Aho, J. Hopcroft, and J. Ullman. *Data structures and algorithms*. Addison-Wesley, 1983.
2. P. Amestoy, T. A. Davis, and I. S. Duff. An approximate minimum degree ordering algorithm. *SIAM J. Matrix Anal. Appl.*, 17(4):886–905, 1996.

3. E. Anderson, Z. Bai, C. Bischof, J. Demmel, J. Dongarra, J. D. Croz, A. Greenbaum, S. Hammarling, A. McKenney, S. Ostrouchov, and D. Sorensen. *LAPACK Users' Guide, Second Edition*. SIAM Publications, 1995.
4. M. Benzi, J. Haws, and M. Tuma. Preconditioning highly indefinite and nonsymmetric matrices. *SIAM J. Sci. Comput.*, 22(4):1333–1353, 2000.
5. C. Berge. Two theorems in graph theory. In *Proceedings of National Academy of Science*, pages 842–844, USA, 1957.
6. J. D. Booth, N. D. Ellingwoodb, and S. R. Heidi K. Thornquist. Basker: Parallel sparse LU factorization utilizing hierarchical parallelism and data layouts. *Parallel Computing*, 68:17–31, 2017.
7. X. Chen, L. Ren, Y. Wang, and H. Yang. GPU-accelerated sparse LU factorization for circuit simulation with performance modeling. *IEEE Transactions on Parallel and Distributed Systems*, 26:786–795, 2015.
8. X. Chen, Y. Wang, and H. Yang. NICSLU: An adaptive sparse matrix solver for parallel circuit simulation. *IEEE Transactions on Computer-Aided Design of Integrated Circuits and Systems*, 32:261–274, 2013.
9. C. Chevalier and F. Pellegrini. PT-SCOTCH: a tool for efficient parallel graph ordering. *Parallel Comput.*, 34(6–8):318–331, 2008.
10. E. Cuthill and J. McKee. Reducing the bandwidth of sparse symmetric matrices. In *Proceedings of the 24th national conference of the ACM*. ACM, 1969.
11. T. Davis and K. Stanley. Sparse LU factorization of circuit simulation matrices. In *Numerical Aspects of Circuit and Device Modeling Workshop*, June 2004.
12. T. A. Davis. *Direct Methods for Sparse Linear Systems*. SIAM Publications, 2006.
13. D. Dodson and J. G. Lewis. Issues relating to extension of the basic linear algebra subprograms. *ACM SIGNUM Newslett.*, 20:2–18, 1985.
14. J. J. Dongarra, J. Du Croz, S. Hammarling, and R. J. Hanson. Issues relating to extension of the basic linear algebra subprograms. *ACM SIGNUM Newslett.*, 20:2–18, 1985.
15. I. S. Duff, A. Erisman, and J. Reid. *Direct Methods for Sparse Matrices*. Oxford University Press, 1986.
16. I. S. Duff and J. Koster. The design and use of algorithms for permuting large entries to the diagonal of sparse matrices. *SIAM J. Matrix Anal. Appl.*, 20:889–901, 1999.
17. I. S. Duff and S. Pralet. Strategies for scaling and pivoting for sparse symmetric indefinite problems. Technical Report TR/PA/04/59, CERFACS, Toulouse, France, 2004.
18. C. M. Fiduccia and R. M. Mattheyses. A linear-time heuristic for improving network partitions. In *Proceedings of the 19th Design Automation Conference*, pages 175–181. IEEE, 1997.
19. M. Fiedler. A property of eigenvectors of nonnegative symmetric matrices and its application to graph theory. *Czechoslovak Mathematical Journal*, 25(100):619–633, 1975.
20. A. George and J. W. H. Liu. The evolution of the minimum degree ordering algorithm. *SIAM Review*, 31:1–19, 1989.
21. J. A. George and J. W. Liu. *Computer Solution of Large Sparse Positive Definite Systems*. Prentice-Hall, Englewood Cliffs, NJ, USA, 1981.
22. J. A. George and E. Ng. An implementation of Gaussian elimination with partial pivoting for sparse systems. *SIAM J. Sci. Statist. Comput.*, 6(2):390–409, 1985.
23. J. R. Gilbert. Predicting structure in sparse matrix computations. *SIAM J. Matrix Anal. Appl.*, 15(1):162–79, 1994.
24. A. Gupta and L. Ying. On algorithms for finding maximum matchings in bipartite graphs. Technical Report RC 21576 (97320), IBM T. J. Watson Research Center, Yorktown Heights, NY, October 25, 1999.
25. G. Karypis and V. Kumar. A fast and high quality multilevel scheme for partitioning irregular graphs. *SIAM Journal on Scientific Computing*, 20(1):359–392, 1998.
26. G. Karypis, K. Schloegel, and V. Kumar. *ParMeTis: Parallel Graph Partitioning and Sparse Matrix Ordering Library, Version 2.0*. University of Minnesota, Dept. of Computer Science, September 1999.

27. B. Kernighan and S. Lin. An efficient heuristic procedure for partitioning graphs. *The Bell System Technical Journal*, 29(2):291–307, 1970.
28. J. Langguth, A. Azad, and F. Manne. On parallel push-relabel based algorithms for bipartite maximum matching. *Parallel Comput.*, 40(7):289–308, 2014.
29. J. Langguth, M. M. A. Patwary, and F. Manne. Parallel algorithms for bipartite matching problems on distributed memory computers. *Parallel Comput.*, 37(12):820–845, 2011.
30. D. LaSalle and G. Karypis. Multi-threaded graph partitioning. Technical report, Department of Computer Science & Engineering, University of Minnesota, Minneapolis, 2013.
31. W.-K. Lee, R. Achar, and M. S. Nakhla. Dynamic GPU parallel sparse LU factorization for fast circuit simulation. *IEEE Transactions on Very Large Scale Integration (VLSI) Systems*, pages 1–12, 2018.
32. X. S. Li. An overview of SuperLU: Algorithms, implementation, and user interface. *ACM Trans. Math. Softw.*, 31(3):302–325, 2005.
33. J. W. Liu. The role of elimination trees in sparse factorization. *SIAM J. Matrix Anal. Appl.*, 11(1):134–172, 1990.
34. J. W. Liu and A. Sherman. Comparative analysis of the Cuthill–McKee and the reverse Cuthill–McKee ordering algorithms for sparse matrices. *SIAM J. Numer. Anal.*, 13:198–213, 1976.
35. H. M. Markowitz. The elimination form of the inverse and its application to linear programming. *Management Science*, 3:255–269, April 1957.
36. L. W. Nagel. SPICE2: a computer program to simulate semiconductor circuits. Memorandum No. ERL-M520, University of California, Berkeley, California, May 1975.
37. M. Olschowka and A. Neumaier. A new pivoting strategy for Gaussian elimination. *Linear Algebra and its Applications*, 240:131–151, 1996.
38. F. Pellegrini, J. Roman, and P. Amestoy. Hybridizing nested dissection and halo approximate minimum degree for efficient sparse matrix ordering. *Concurrency: Practice and Experience*, 12:69–84, 2000.
39. M. Rewienski. A perspective on fast-spice simulation technology. Springer, Dordrecht, 2011.
40. D. J. Rose. A graph-theoretic study of the numerical solution of sparse positive definite systems of linear equations. In R. C. Read, editor, *Graph Theory and Computing*. Academic Press, 1972.
41. O. Schenk. *Scalable parallel sparse LU factorization methods on shared memory multiprocessors*. PhD thesis, ETH Zurich, 2000. Diss. Technische Wissenschaften ETH Zurich, Nr. 13515, 2000.
42. O. Schenk and K. Gärtner. Solving unsymmetric sparse systems of linear equations with PARDISO. *Journal of Future Generation Computer Systems*, 20(3):475–487, 2004.
43. R. E. Tarjan. Data structures and network algorithms. In *CBMS–NSF Regional Conference Series in Applied Mathematics*, volume 44, 1983.
44. X. Zhao, L. Han, and Z. Feng. A performance-guided graph sparsification approach to scalable and robust spice-accurate integrated circuit simulations. *IEEE Transactions on Computer-Aided Design of Integrated Circuits and Systems*, 34:1639–1651, 2015.

The Effect of Various Sparsity Structures on Parallelism and Algorithms to Reveal Those Structures

Oguz Selvitopi, Seher Acer, Murat Manguoğlu, and Cevdet Aykanat

1 Introduction

Various parallel numerical methods can greatly benefit from a structured matrix in order to enhance parallel performance and convergence. Examples of these methods include, but certainly not constrained to, QR factorization for solving linear least squares problems and linear systems, the direct methods such as LU or Cholesky factorizations for solving sparse linear systems of equations, or direct-iterative hybrid solvers. Our goal in this work is to present combinatorial algorithms to obtain certain sparse matrix forms that are beneficial for a wide variety of parallel numerical methods.

We do not follow a comprehensive approach for obtaining structured sparse matrices and limit ourselves to combinatorial graph and hypergraph models for that purpose. An effective way of attaining the desired form of a sparse matrix is to represent it with a graph/hypergraph, and then partition this graph/hypergraph with a relevant objective and constraint in mind that correlates to the metrics pertaining

O. Selvitopi
Lawrence Berkeley National Laboratory, Berkeley, CA, USA
e-mail: roselvitopi@lbl.gov

S. Acer
Center for Computing Research, Sandia National Laboratories, Albuquerque, NM, USA
e-mail: sacer@sandia.gov

M. Manguoğlu (✉)
Department of Computer Engineering, Middle East Technical University, Ankara, Turkey
e-mail: manguoglu@ceng.metu.edu.tr

C. Aykanat
Department of Computer Engineering, Bilkent University, Ankara, Turkey
e-mail: aykanat@cs.bilkent.edu.tr

© Springer Nature Switzerland AG 2020
A. Grama, A. H. Sameh (eds.), *Parallel Algorithms in Computational Science and Engineering*, Modeling and Simulation in Science, Engineering and Technology,
https://doi.org/10.1007/978-3-030-43736-7_2

35

to parallel performance or convergence. The duality between matrices and graphs is successfully used to translate the computations on graphs into the language of linear algebra [34], where there exist several decades of effort in optimizing them. The hypergraphs generalize graphs where an edge may connect more than two vertices and they are better compared to graphs in representing multiway relations. However, the nature of the problem dictates whether it is better to utilize a graph or a hypergraph model. Although this is usually the case, another important issue is the availability of certain features in partitioners that may be necessitated by the model (such as support for fixed vertices, multiple constraints, multiple objectives, etc.).

We consider four different sparse matrix structures or forms: (1) singly-bordered block-diagonal form (SB form), (2) doubly-bordered block-diagonal form (DB form), (3) nonempty off-diagonal block minimization, and (4) block diagonal with overlap form (BDO form). The SB form is made up of a number of diagonal blocks and a single row or column border stripe, where the former is referred to as the rowwise SB form (Fig. 1a) and the latter is referred to as the columnwise SB form (Fig. 1b). The goals in the SB form are to obtain balanced-size diagonal blocks and minimize the border size. The DB form is similar to the SB form but it consists of both a row and a column border stripe (Fig. 1c). The goals in the DB form are also similar to the goals in the SB form. Another problem we investigate is the minimization of number of nonempty off-diagonal blocks for a given sparse matrix (Fig. 1d). This is not really a structured sparse matrix form in the strictest sense. Nonetheless, it enables optimizations important for parallel performance. The BDO form consists of a number of diagonal blocks where only the successive diagonal blocks overlap with each other (Fig. 1e). The two goals of this form are to minimize the overlaps between blocks and to obtain balanced-size diagonal blocks.

We give the background related to graph/hypergraph partitioning and the notations used for matrices in Sect. 2. We investigate each form on its own section. Each of these sections consists of: (1) describing the target form and the goals sought in the form, (2) the graph/hypergraph method(s) for obtaining the form, and (3) how this form is utilized for different applications and how it benefits these applications by enhancing parallel performance and/or convergence. The four forms are covered between Sects. 3–6. We give our concluding remarks in Sect. 7.

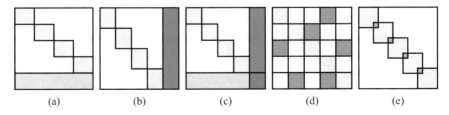

Fig. 1 The sparse matrix forms examined in this work. (**a**) Rowwise SB form. (**b**) Columnwise SB form. (**c**) DB form. (**d**) Nonempty off-diagonal block minimization. (**e**) BDO form

2   Preliminaries

In this section we introduce the terminology used in the rest of the paper. We first describe the graph and hypergraph partitioning problems, which are central to obtaining the matrix forms described in the paper. Then we give some basic definitions related to matrices.

2.1   Graph Partitioning by Vertex Separator (GPVS)

An undirected graph $G = (V, E)$ is defined by a set V of vertices and a set E of edges, where each edge $e_{i,j} \in E$ connects a pair of distinct vertices $v_i, v_j \in V$. The vertices adjacent to v_i in G are denoted by $Adj_G(v_i)$. We appropriately extend this notation to a subset $V' \subseteq V$ of vertices. Each vertex v_i is associated with a weight, which we denote by $w(v_i)$. A vertex subset V_S is a K-way *vertex separator* if the subgraph induced by the vertices in $V - V_S$ has at least K connected components. V_S is called wide if a strict subset of it also forms a separator, and narrow otherwise.

   $\Pi_{VS} = \{V_1, \ldots, V_K; V_S\}$ is a K-way vertex partition of G by vertex separator V_S if (1) the union of the vertices in all parts give V, (2) each part except the separator is nonempty, (3) the parts are pairwise disjoint, and (4) the K parts are pairwise nonadjacent. An edge is internal if the pair of vertices connected by it are in the same part, and external, otherwise. A vertex is a boundary vertex if it is connected by one or more external edges. The goal in the GPVS problem is to find a K-way vertex separator with the objective of minimizing the separator size, which is usually defined as

$$cost(\Pi_{VS}) = |V_S|, \tag{1}$$

under the constraint of maintaining the balance criterion (usually provided as a parameter) on the weights of the K parts $V_1, \ldots, V_K$. The weight $W(V_k)$ of V_k is usually defined as the summation of the weights of the vertices in it, i.e., $W(V_k) = \sum_{v_i \in V_k} w(v_i)$, for $1 \le k \le K$.

   A relevant problem is the graph partitioning by edge separator (GPES), which is similar to the GPVS problem except that its goal is to find an edge separator with minimum size instead of a vertex separator with minimum size. The multilevel approaches for GPES [9, 21] led to successful tools [20, 28, 41]. There are also tools that solve the GPVS problem directly [22, 28]. The GPVS problem can also be solved indirectly by first solving the GPES problem and forming a wide separator with all the vertices that are incident to the edges in the edge separator, and then narrowing down this separator using various algorithms based on refinement heuristics or vertex cover [43]. It is shown that the direct approaches [22] are better for the GPVS problem. The deficiency of GPVS-based approaches in multilevel frameworks is pointed out [11, 29]. Note both the GPES and GPVS problems are NP-hard [8].

A variant of the GPVS problem is the ordered GPVS problem [2]. $\Pi_{oVS} = \{\mathcal{V}_1, \ldots, \mathcal{V}_K; \mathcal{V}_{S_1}, \ldots, \mathcal{V}_{S_{K-1}}\}$ is a K-way ordered partition of $\mathcal{G}$ by the $K - 1$ vertex separators $\{\mathcal{V}_{S_1}, \ldots, \mathcal{V}_{S_{K-1}}\}$ if (1) the union of the parts and the separators give $\mathcal{V}$, (2) each separator is nonempty, (3) all parts and separators are pairwise disjoint, (4) the K parts are pairwise nonadjacent, and (5) $\mathcal{V}_{S_k}$ is only adjacent to separators $\mathcal{V}_{S_{k-1}}$ and $\mathcal{V}_{S_{k+1}}$, and the parts $\mathcal{V}_k$ and $\mathcal{V}_{k+1}$. The first four conditions are similar to those in the GPVS problem. The fifth condition imposes an order on the vertices of the parts and the separators, in which $\mathcal{V}_{S_k}$ is the vertex separator of the parts $\mathcal{V}_k$ and $\mathcal{V}_{k+1}$. The goal in the ordered GPVS problem is to find a K-way partition of $\mathcal{G}$ by an ordered set of $K - 1$ vertex separators such that the total separator size

$$cost(\Pi_{oVS}) = \sum_{k=1}^{K-1} |\mathcal{V}_{S_k}|, \tag{2}$$

is minimized and the balance criterion on the weights of the K parts is satisfied.

2.2 Hypergraph Partitioning (HP)

A hypergraph $\mathcal{H} = (\mathcal{U}, \mathcal{N})$ is defined by a set $\mathcal{U}$ of nodes and a set $\mathcal{N}$ of nets, where each net $n_i \in \mathcal{N}$ connects a subset of nodes, denoted with $Pins(n_i) \subseteq \mathcal{U}$. We refer to the vertices of a hypergraph as nodes in order to separate them from the vertices of a graph. This notation is extended to the subset of nodes $Pins(\mathcal{N}')$ connected by a subset of nets $\mathcal{N}' \subseteq \mathcal{N}$. The set of nets that connect a node v_i is denoted by $Nets(v_i) \subseteq \mathcal{N}$. This notation is similarly extended to the subset of nets $Nets(\mathcal{U}')$ that connect a subset of nodes $\mathcal{U}' \subseteq \mathcal{U}$. Each node v_i is associated with a weight $w(v_i)$.

$\Pi_{HP} = \{\mathcal{U}_1, \ldots, \mathcal{U}_K\}$ is a K-way node partition of $\mathcal{H}$ if (1) the union of nodes in all parts give $\mathcal{U}$, (2) each part is nonempty, and (3) the parts are pairwise disjoint. In Π_{HP}, a net is said to connect a part if it has at least one pin in it. The connectivity set $\Lambda(n_i)$ denotes the set of parts connected by n_i and the connectivity $\lambda(n_i) = |\Lambda(n_i)|$ of n_i denotes the number of parts connected by n_i. A net is said to be external if it connects more than one part, and internal otherwise. A K-way partition Π_{HP} on the nodes of the hypergraph induces a $(K + 1)$-way partition on $\mathcal{N}$ as well. If we denote the nets internal to part $\mathcal{U}_k$ with $\mathcal{N}_k$ and the external nets with $\mathcal{N}_S$, we can interpret this partition also as $\Pi_{HP} = \{\mathcal{N}_1, \ldots, \mathcal{N}_K; \mathcal{N}_S\}$. The goal in the hypergraph partitioning problem is to find a K-way partition with objective of minimizing the cutsize, which is defined either as

$$cost(\Pi_{HP}) = \sum_{n_i \in \mathcal{N}_S} \lambda(n_i) - 1 \text{ or as} \tag{3}$$

$$cost(\Pi_{HP}) = |\mathcal{N}_S|. \tag{4}$$

These two are commonly referred to as the *connectivity* and the *cutsize* metrics, respectively, and they are widely adopted in the scientific computing [10] and VLSI communities [32]. We are interested in two variants of constraints in the HP problem. The first variant is to maintain a balance on part weights similar to the one in the GPVS problem. The second variant is to maintain a balance on the number of internal nets.

The HP problem is NP-hard [32]. The most successful tools [12, 27] for solving the HP problem rely on the multilevel schemes.

2.3 Matrix Definitions

We give the notation related to matrices. The row i and column j of matrix A are, respectively, denoted by $a_{i,*}$ and $a_{*,j}$, and the nonzero at the intersection of these two is denoted by $a_{i,j}$. We use the functions $nnz(\cdot)$, $nr(\cdot)$, and $nc(\cdot)$ to, respectively, denote the number of nonzeros, rows, and columns in a (sub)matrix, which we may apply to a row, column, block, stripe, or the entire matrix.

The $m \times n$ sparse matrix A will often appear in blocked form with $K \times L$ blocks, in which the block at the intersection of kth row stripe and ℓth column stripe is of size $m_k \times n_\ell$. We use the capital letters B, C, D to denote the blocks and use two subscripts to denote a block. When clear from the context, we utilize a single subscript. The kth row stripe contains the blocks $B_{k,1}, \ldots, B_{k,L}$ and the ℓth column stripe contains the blocks $B_{1,\ell}, \ldots, B_{K,\ell}$. A block $B_{k,\ell}$ is said to be diagonal if $k = \ell$, and off-diagonal, otherwise. A row (column) is called a *coupling row* (*coupling column*) if it has nonzeros in at least two blocks. In our discussions there exist both symmetric and nonsymmetric matrices. Figure 2 illustrates a nonsymmetric matrix that is used for reordering purposes throughout the paper.

3 Singly-Bordered Block-Diagonal Form

Target Form The singly-bordered block-diagonal form of an $m \times n$ sparse matrix A consists of K diagonal blocks and a border stripe. In the rowwise SB form, A is permuted into A_{rSB} as

Fig. 2 A 16×18 sparse matrix used for reordering purposes throughout the paper

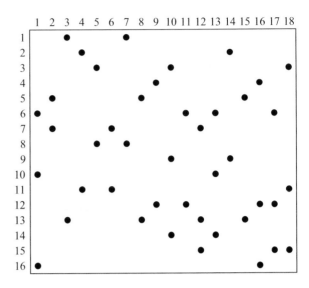

$$PAQ = \begin{bmatrix} B_1 & & \\ & \ddots & \\ & & B_K \\ R_1 & \cdots & R_K \end{bmatrix} = A_{rSB}, \tag{5}$$

where P and Q denote the row and column permutation matrices, respectively. The row border stripe $R = [R_1 \cdots R_K]$ consists of coupling rows, each of which has nonzeros in the columns of at least two diagonal blocks. In the columnwise SB form, A is permuted into A_{cSB} as

$$PAQ = \begin{bmatrix} B_1 & & C_1 \\ & \ddots & \vdots \\ & & B_K \ C_K \end{bmatrix} = A_{cSB}. \tag{6}$$

The column border stripe $C = [C_1^T \cdots C_K^T]^T$ consists of coupling columns, each of which has nonzeros in the rows of at least two diagonal blocks. The rowwise and columnwise SB forms are also referred to as primal and dual SB forms, respectively [5]. For the SB form, we do not impose a symmetric permutation on the rows and columns of the matrix, i.e., we do not enforce that $P^T = Q$.

The two goals in permuting a matrix into SB form are to reduce the border size and satisfy a balance criterion on the sizes of the blocks. The border size is given by the number of rows in R, $nr(R)$, for the rowwise SB form and number of columns in C, $nc(C)$, for the columnwise SB form. The block size is usually defined in terms of its dimensions or the number of nonzeros in it.

Methods The SB form of a given sparse matrix can be attained via different approaches [5, 15, 42]. Here, we focus on the hypergraph models, which are shown to be more effective [5]. In addition, we only describe the rowwise SB form. The columnwise SB form can be obtained using the dual of the discussed method. The reordering process can be summarized in three successive steps: (1) the modeling of the given matrix with the row-net hypergraph model, (2) partitioning of the hypergraph, and (3) interpretation of this partition to reorder the matrix. We describe each of these steps next.

The row-net hypergraph $\mathcal{H}_{RN} = (\mathcal{U}, \mathcal{N})$ used to model sparse matrix A consists of n nodes and m nets. In $\mathcal{H}_{RN}$, there exists a node $u_j \in \mathcal{U}$ for each column $a_{*,j}$ of A and there exists a net $n_i \in \mathcal{N}$ for each row $a_{i,*}$ of A. The net n_i connects u_j if and only if $a_{i,j} \neq 0$. Hence, $Pins(n_i)$ is given by the nodes that represent the columns which have a nonzero in $a_{i,*}$. In a dual manner, $Nets(u_i)$ is given by the nets that represent the rows which have a nonzero in $a_{*,j}$. The vertices can be assigned either unit weights or the number of nonzeros of the columns they represent depending on the application's need. Figure 3a displays the row-net hypergraph that models the sparse matrix in Fig. 2.

We can use a K-way partition $\Pi_{HP}(\mathcal{H}_{RN}) = \{\mathcal{U}_1, \ldots, \mathcal{U}_K\} = \{\mathcal{N}_1, \ldots, \mathcal{N}_K; \mathcal{N}_S\}$ of this hypergraph to obtain A_{rSB}. To permute A into the SB form, we order the rows associated with the internal nets in $\mathcal{N}_{k+1}$ after the rows associated with the internal nets in $\mathcal{N}_k$ for $1 \leq k \leq K - 1$, and we order the rows associated with the external nets $\mathcal{N}_S$ all to the end. We obtain the column permutation by ordering the columns associated with the nodes in $\mathcal{U}_{k+1}$ after the columns associated with the nodes in $\mathcal{U}_k$. Figure 3b illustrates the reordered matrix induced by the partition in Fig. 3a. Observe that the external nets with ids 1, 11, 14, 15 correspond to the rows in the border in A_{rSB}.

Obtaining K-way partition with the aim of minimizing the cutsize (4) minimizes the border size as the external nets in the partition corresponds to the rows in the

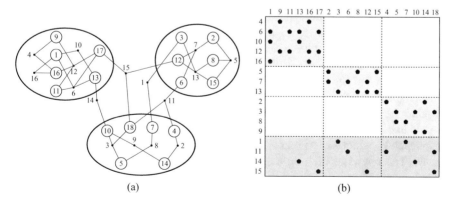

(a)   (b)

Fig. 3 Obtaining the SB form via partitioning the row-net hypergraph model. (**a**) Row-net hypergraph model of the matrix in Fig. 2 and a 3-way partition of it. (**b**) Reordered matrix in the rowwise SB form

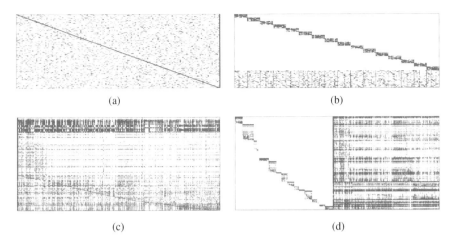

(a)   (b)

(c)   (d)

Fig. 4 A linear programming problem matrix `karted` with 47K rows, 133K columns, 1.8M nonzeros (top) and a least squares problem matrix `Maragal_8` with 33K rows, 75K columns, 1.3M nonzeros (bottom). The SB forms are obtained through the described method. (**a**) Linear programming problem matrix. (**b**) Rowwise SB form of the matrix in left for 16 processes. (**c**) Least squares problem matrix. (**d**) Columnwise SB form of the matrix in left for 16 processes

border. Maintaining a balance on part weights in $\Pi_{HP}(\mathcal{H}_{RN})$ infers balance among the diagonal blocks. A dual methodology can be adopted by using the column-net hypergraph model [5] to obtain A_{cSB}. Figure 4 displays two matrices reordered using the described methodology.

Application Areas and Parallelism Enhancement Various decomposition techniques [15, 31, 36] for the parallel solution of linear programs (LPs) exploit the coarse-grain parallelism inherent in the block-angular form [14]. These techniques solve K independent LP subproblems corresponding to the block constraints followed by a coordination phase, usually performed serially and referred to as the master problem. Such a decomposition is useful since solving a small independent subset of problems is more efficient compared to the aggregate problem due to the quadratic or cubic complexity. In terms of parallel efficiency, it is crucial to keep the sizes of the independent problems close and reduce the master problem size [15, 35], which affects the convergence and the amount of communicated data in each iteration. Hence, for efficient parallel solution of the LP problems, one would like to order the constraints and variables of the problem, which, respectively, correspond to the rows and columns of the matrix, in such a way to ensure these goals. The rowwise SB form captures these goals effectively by minimizing the row border size, which relates to minimizing the size of the master problem, and by attaining balanced blocks, which translates into attaining balance in the sizes of independent LP subproblems.

Another application is the QR factorization which is a common method for solving the linear least squares problems and linear systems. The QR factorization decomposes a given $m \times n$ coefficient matrix A into a product of an $m \times m$ orthogonal matrix Q and an $n \times n$ upper triangular matrix R, where $m \geq n$. The columnwise SB form leads to efficient parallelization of the sparse QR factorization, where A_{cSB} is utilized by enabling each process factoring its row stripe independently as the first step, for example for the last row block,

$$\begin{bmatrix} B_k & C_k \end{bmatrix} = Q_k \begin{bmatrix} R_k & S_k \\ 0 & C'_k \end{bmatrix}. \tag{7}$$

Finally, $C' = [C'_1 \cdots C'_K]^T$ [6] is factored to obtain the QR factorization of the original matrix. A small-sized column border leads to less overhead in factoring C' and attaining balanced blocks leads to load balance in factoring those blocks. Furthermore, the same reordering could be also used for the incomplete QR factorization.

Finally, the rowwise and columnwise SB forms are recently exploited to parallelize sparse matrix-vector and matrix-transpose-vector multiplication (SpMM^{T}V) on shared memory environments [26]. The range of algorithms covered is quite wide as these two operations frequently occur in many application areas. These include interior-point methods for solving LP problems [25, 37], Krylov subspace methods for nonsymmetric systems such as the Biconjugate Gradient method or the Conjugate Gradient on the Normal Equations [44], the LSQR method [40] for solving the least squares problem, the Surrogate Constraints method [51, 52] for solving the linear feasibility problem, and many more. The SB form is exploited for a number of optimizations that are important for multi-threaded programs. The objective in the SB form used for the parallelization of SpMM^{T}V is the minimization of the connectivity metric (3), which is different from minimizing the border size. Figure 5 illustrates how the SB form is used to perform $z = A_{rSB}^T x$ followed by $y = Az$ with four threads. The matrix blocks or subvectors stored by a thread are indicated with the same color, whereas the subvectors that require some sort of coordination among the threads are indicated with green color. The SB form enables four benefits for the threads: (1) the reduction of cache misses in reading the elements of x_S and updating the elements of y_S, both of which have to be performed by multiple threads, (2) the reduction of concurrent writes to y_S, (3) the reuse of A-matrix nonzeros together with their indices, and (4) balancing of computational loads of the threads, which are proportional to $nnz(B_k) + nnz(R_k)$ for the kth thread.

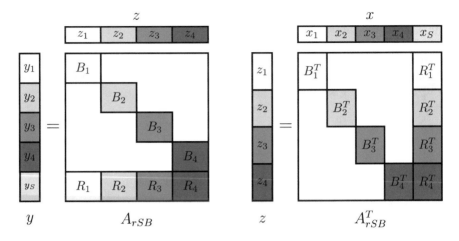

Fig. 5 Exploiting the rowwise SB form on shared memory systems (image due to [26])

4 Doubly-Bordered Block-Diagonal Form

Target Form The doubly-bordered block-diagonal form of an $m \times n$ sparse matrix A consists of K diagonal blocks and two border stripes. In the DB form, A is permuted into A_{DB} as

$$PAQ = \begin{bmatrix} B_1 & & & C_1 \\ & \ddots & & \\ & & B_K & C_K \\ R_1 & \cdots & R_K & D \end{bmatrix} = A_{DB}, \qquad (8)$$

where P and Q denote the row and column permutation matrices, respectively. The row border stripe $R = [R_1 \cdots R_K D]$ and the column border stripe $C = [C_1^T \cdots C_K^T D^T]^T$ consists of coupling rows and columns, respectively. For the DB form, we consider both symmetric permutation and nonsymmetric permutation of A in our discussions, respectively, referred to as the symmetric DB form and the nonsymmetric DB form. The former is occasionally used for the symmetric matrices and the latter for the nonsymmetric matrices.

The two main goals in permuting a matrix into DB form are to reduce the summation of sizes of the two borders, $nr(R)+nc(C)$, and satisfy a balance criterion on the block sizes.

Methods A common way of permuting a matrix into the symmetric DB form is to first represent the matrix with the standard graph model and then use a GPVS-based multilevel partitioner for reordering. The standard graph model $G = (V, \mathcal{E})$ simply contains a vertex v_i for each row/column i of A, and there exists an edge

$e_{i,j} \in \mathcal{E}$ for each nonzero of A. The deficiency of this graph model when used within a multilevel framework is that a narrow vertex separator at any level of the multilevel partitioning does not usually form a narrow separator in the finer levels. This causes overestimation of the separator size and degrades the partition quality. A remedy has been proposed based on hypergraph partitioning [11]. In order to obtain the symmetric DB form we focus on this model, which we refer to HP-based GPVS, and for the nonsymmetric DB form we focus on a bipartite graph model [5].

The HP-based GPVS for permuting a given symmetric matrix A into the symmetric DB form consists of six steps. A sample symmetric matrix used for reordering purposes is illustrated in Fig. 6a.

1. The matrix A is first represented with the standard graph model $\mathcal{G} = (\mathcal{V}, \mathcal{E})$. The graph $\mathcal{G}$ that represents the matrix in Fig. 6a is displayed in Fig. 6b.
2. In the second step of the HP-based GPVS, an edge-clique cover [30] $C = \{C_1, \ldots C_m\}$ of $\mathcal{G}$ is computed. A set of cliques cover $\mathcal{G}$ if the pair of vertices v_i and v_j for each edge $e_{i,j} \in \mathcal{E}$ are contained in at least one clique.
3. Using this set of cliques, a clique-node hypergraph (CNH) $\mathcal{H}_{CNH} = (\mathcal{U}, \mathcal{N})$ is formed. In $\mathcal{H}_{CNH}$, there exists a node $u_i \in \mathcal{U}$ for each clique $C_i \in C$ and there exists a net $n_j \in \mathcal{N}$ for each vertex in $v_j \in \mathcal{V}$. n_j connects u_i if and only if

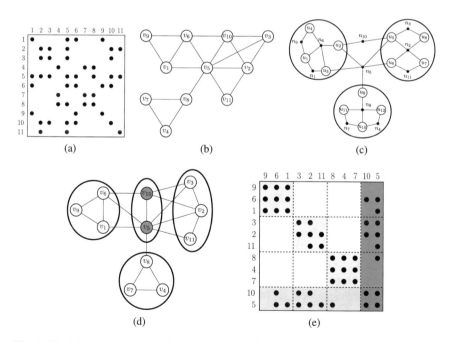

Fig. 6 Obtaining the symmetric DB form through HP-based GPVS. The example is reproduced from [11]. (**a**) A symmetric matrix A. (**b**) Standard graph representation $\mathcal{G}$. (**c**) The clique-node hypergraph $\mathcal{H}_{CNH}$ formed from the cliques in $\mathcal{G}$ and 3-way partition $\Pi_{HP}(\mathcal{H}_{CNH})$. (**d**) Obtaining $\Pi_{VS}(\mathcal{G})$ from $\Pi_{HP}(\mathcal{H}_{CNH})$. (**e**) The symmetric DB form obtained using $\Pi_{VS}(\mathcal{G})$

C_i contains v_j. Figure 6c shows the CNH formed from a 12-clique edge-clique cover of G in Fig. 6b. One can see which cliques are used to form $\mathcal{H}_{CNH}$ by $Nets(u_i)$ of a node u_i. For example, $Nets(u_5) = \{n_2, n_3, n_5, n_{10}\}$ indicates that the clique containing vertices v_2, v_3, v_5, v_{10} in G exists as node u_5 in $\mathcal{H}_{CNH}$.

4. $\mathcal{H}_{CNH}$ is partitioned to obtain $\Pi_{HP}(\mathcal{H}_{CNH}) = \{\mathcal{N}_1, \ldots, \mathcal{N}_K; \mathcal{N}_S\}$. In HP, the cutsize objective (4) and the balancing constraint of internal nets are utilized. Figure 6c illustrates a 3-way partition of the CNH described in the previous step.

5. The partition $\Pi_{VS}(G) = \{\mathcal{V}_1, \ldots, \mathcal{V}_K; \mathcal{V}_S\}$ is obtained using $\Pi_{HP}(\mathcal{H}_{CNH})$. This is done by using the net intersection graph (NIG) [3, 7, 13] representation of $\mathcal{H}_{CNH}$. Simply put, there exists a vertex in the NIG representation of a hypergraph for each net and two vertices v_i and v_j in the graph have an edge if the respective nets in the hypergraph share at least one common node. Hence, the internal net sets of $\Pi_{HP}(\mathcal{H}_{CNH})$ correspond to the vertex sets of $\Pi_{VS}(G)$ and the external net set of $\Pi_{HP}(\mathcal{H}_{CNH})$ become the separator in $\Pi_{VS}(G)$. Figure 6d shows the $\Pi_{VS}(G)$ obtained using the $\Pi_{HP}(\mathcal{H}_{CNH})$ in Fig. 6c.

6. In the last step, the partition found on the previous step is used to permute A into the symmetric DB form. To permute A into the target form, we order the rows/columns associated with the vertices in $\mathcal{V}_{k+1}$ after the rows/columns associated with the vertices in $\mathcal{V}_k$ for $1 \leq k \leq K - 1$, and we order the rows/columns associated with the vertices in the separator $\mathcal{V}_S$ all to the end. Figure 6e illustrates the ordered matrix induced by the partition in Fig. 6d.

Minimizing the cutsize in $\Pi_{HP}(\mathcal{H}_{CNH})$ in the fourth step corresponds to minimizing the number of vertices in the separator in $\Pi_{VS}(G)$, which in turn corresponds to minimizing the border size in the DB form. Maintaining a balance on the number of internal nets in $\Pi_{HP}(\mathcal{H}_{CNH})$ in the fourth step corresponds to maintaining a balance on the number of vertices in parts of $\Pi_{VS}(G)$, which in turn infers a balance among the block sizes.

The method for permuting A into the nonsymmetric DB form is less involved and achieved by formulating this problem as a GPVS problem on the bipartite graph representation of A. In the bipartite graph $\mathcal{B} = (V = \mathcal{V}^r \cup \mathcal{V}^c, \mathcal{E})$, there exists a vertex $v_i \in \mathcal{V}^r$ for each row i of A and there exists a vertex $v_j \in \mathcal{V}^c$ for each column j of A. There is an edge $e_{i,j} \in \mathcal{E}$ for each nonzero of A and it connects the vertices r_i and c_j. $Adj(r_i)$ and $Adj(c_j)$ are, respectively, given by the vertices corresponding to the columns and rows that have nonzeros in row i and column j. Figure 7a shows the bipartite graph the represents the matrix in Fig. 2.

We can use a K-way partition $\Pi_{VS}(\mathcal{B}) = \{\mathcal{V}_1^r \cup \mathcal{V}_1^c, \ldots, \mathcal{V}_K^r \cup \mathcal{V}_K^c; \mathcal{V}_S^r \cup \mathcal{V}_S^c\}$ of this bipartite graph to obtain the nonsymmetric DB form. In order to do so, for the permutation of the rows we use $\{\mathcal{V}_1^r, \ldots, \mathcal{V}_K^r; \mathcal{V}_S^r\}$ and order the rows associated with the vertices in $\mathcal{V}_{k+1}^r$ after the rows associated with the vertices in $\mathcal{V}_k^r$ for $1 \leq k \leq K-1$, and we order the rows associated with the vertices in $\mathcal{V}_S^r$ all to the end. In a similar manner, for the permutation of the columns we use $\{\mathcal{V}_1^c, \ldots, \mathcal{V}_K^c; \mathcal{V}_S^c\}$ and order the columns associated with the vertices in $\mathcal{V}_{k+1}^c$ after the columns associated with the vertices in $\mathcal{V}_k^c$ for $1 \leq k \leq K-1$, and we order the columns associated with

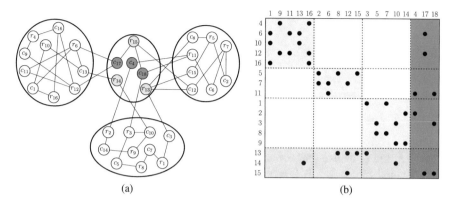

Fig. 7 Obtaining the nonsymmetric DB form via bipartite graph model. (**a**) Bipartite graph model of the matrix in Fig. 2 and a 3-way GPVS partition of it. (**b**) Reordered matrix in the DB form

the vertices in $\mathcal{V}_S^c$ all to the end. Figure 7b illustrates the ordered matrix induced by partition in Fig. 7a. In Fig. 7, the vertices that represent the row border are displayed in green and the vertices that represent the column border are displayed in blue.

Minimizing the separator size in $\Pi_{VS}(\mathcal{B})$ with respect to (1) minimizes the summation of the sizes of the row and column borders in A as the vertices in the separator correspond to the rows and columns in the border. Maintaining a balance on part weights in $\Pi_{VS}(\mathcal{B})$ infers balance among the diagonal blocks.

Application Areas and Parallelism Enhancement One of the most common areas that the DB form finds its application is the solution of sparse linear systems of equations using direct methods such as LU or Cholesky factorizations. For most problems a reordering is required in order to reduce the fill-in and enhance parallelism. The fill-in of a matrix is the set of nonzeros that are introduced in the factors. Some of the most widely used fill-in reducing reordering schemes are [4, 17, 18, 33, 47]. Furthermore, these reordering techniques are used not only for direct solvers but also for preconditioners in iterative solvers, for computing incomplete factorization based preconditioners such as incomplete LU and Cholesky.

In order to show how obtaining the DB form through the described methodology benefits the direct methods, we consider the coarse-grain parallelization of LU factorization. Given a reordered matrix A_{DB} in the DB form, first the diagonal blocks can effectively be factored independently in parallel by each process to get $B_k = L_k U_k$. In the following stage, the unfactored rows/columns and the rows/columns in the border are factored. One of the two main goals in obtaining A_{DB}, satisfying a balance criterion on the block sizes, relates balancing the computational load of the processes in the former stage in factoring diagonal blocks. The other goal, minimizing the border size, relates to reducing the work done in the latter stage, which is usually less amenable for parallelization. Hence, minimizing the border size in A_{DB} enhances parallelism in the LU factorization.

5 Nonempty Off-Diagonal Block Minimization

Target Form Consider an $m \times n$ sparse matrix A in blocked into $K \times K$ grid:

$$PAQ = \begin{bmatrix} B_{1,1} & \cdots & B_{1,K} \\ \vdots & \ddots & \vdots \\ B_{K,1} & \cdots & B_{K,K} \end{bmatrix} = A_{BL}, \tag{9}$$

where P and Q denote the row and column permutation matrices, respectively. An off-diagonal block $B_{k,\ell}$ is nonempty if it contains at least one nonzero element, i.e., $\exists a_{i,j} \in B_{k,\ell}$. Note that we actually do not require A to be permuted into some specific prior form, i.e., P and Q can be identity matrices. The requirement, however, is that there is some sort of blocking in which the rows and columns of A are grouped and these groupings may be arbitrary. It is also possible that $K = m$ or $K = n$. An example of a 4×4 blocked matrix is given in Fig. 8a and it has seven nonempty off-diagonal blocks. Note that this matrix is a permutation of the sparse matrix in Fig. 2.

The first of the two goals in this section is to minimize the number of nonempty off-diagonal blocks, which is given by

$$|\{B_{k,\ell} : \exists a_{i,j} \in B_{k,\ell} \text{ and } k \neq \ell\}|, \tag{10}$$

by permuting either rows or columns of A_{BL}. The second goal is to satisfy a balance criterion on the coupling row or column sizes in K row or column stripes. The size of a coupling row/column can be defined as unit or the number of nonempty off-diagonal blocks that this row/column is contained in (referred to as the degree weighting). In our discussions for describing the methodology, we only focus on the row permutation. Therefore, we are looking for a row permutation matrix P' to obtain $P'A_{BL} = A_{NOD}$ in which there exists as few nonempty off-diagonal blocks as possible and the coupling rows in each row stripe are balanced. A dual methodology holds for the column permutation where $A_{BL}Q' = A_{NOD}$.

Methods We focus on a hypergraph model called the communication hypergraph model [46, 49, 50] in order to minimize the nonempty off-diagonal blocks. Given a $K \times K$ blocked sparse matrix A_{BL}, the reordering process consists of four steps: (1) the formation of the communication matrix from A_{BL}, (2) modeling of the communication matrix with the communication hypergraph, (3) partitioning of the communication hypergraph, and (4) interpretation of this partition to reorder the matrix. We describe each of these steps next.

The communication matrix A_{CM} corresponding to A_{BL} contains only the coupling rows of A_{BL} and there exists a column in A_{CM} for each column stripe in A_{BL}. Hence, if there are m_c coupling rows in A_{BL}, then A_{CM} is an $m_c \times K$ matrix. There exists a nonzero $a_{i,j} \in A_{CM}$ if the coupling row $a_{i,*}$ in A_{BL} has a nonzero in the jth column stripe. A_{CM} is called communication matrix as it is

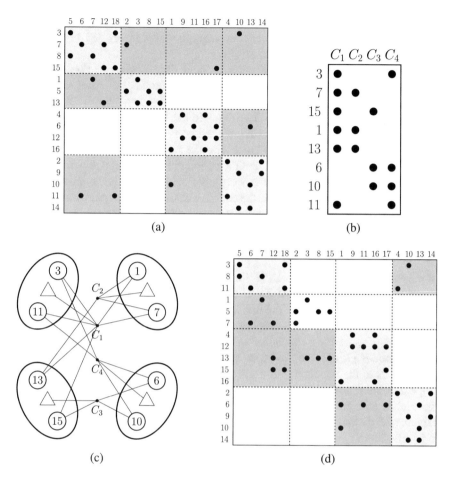

Fig. 8 Nonempty off-diagonal block minimization via communication hypergraph model. (**a**) The matrix in Fig. 2 blocked into 4×4 A_{BL} with 7 nonempty off-diagonal blocks. (**b**) Communication matrix A_{CM} of A_{BL}. (**c**) Communication hypergraph $\mathcal{H}_{CM}$ and a 4-way partition of it. (**d**) Reordered matrix A_{NOD} with 5 nonempty off-diagonal blocks

originally proposed for encapsulating the communication requirements of sparse matrix-vector multiplication on distributed memory systems. Figure 8b displays the 8×4 communication matrix A_{CM} of the blocked sparse matrix in Fig. 8a.

Using the communication matrix A_{CM}, we form the communication hypergraph $\mathcal{H}_{CM} = \{\mathcal{U}, \mathcal{N}\}$ with m_c nodes and K nets. There exists a node $u_i \in \mathcal{U}$ for each row $a_{i,*}$ of the communication matrix and there exists a net $n_j \in \mathcal{N}$ for each column $a_{*,j}$ of the communication matrix. Net n_j connects u_i if and only if $a_{i,j}$ is nonzero. In A_{BL}, this connection corresponds to column stripe j having at least one nonzero in the ith coupling row. We use unit weights for the nodes, although other variants also exist [50]. The hypergraph used to model the communication matrix is also known as the column-net hypergraph model [10]. Figure 8c displays the communication hypergraph with 8 nodes and 4 nets formed using the communication matrix in Fig. 8b.

Next, we partition $\mathcal{H}_{CM}$ into K parts to obtain $\Pi_{HP}(\mathcal{H}_{CM}) = \{\mathcal{U}_1, \ldots, \mathcal{U}_K\}$. The partitioning of communication hypergraph also includes K fixed nodes (indicated in Fig. 8 with the triangle nodes) to express the ownership of the K nets, i.e., column stripes. We use $\Pi_{HP}(\mathcal{H}_{CM})$ to reorder the coupling rows of A_{BL} and obtain A_{NOD}. To do so, we order the internal rows in the $(k + 1)$th row stripe of A_{BL} and the coupling rows associated with the nodes in $\mathcal{U}_{k+1}$ after the internal rows in the kth row stripe of A_{BL} and the coupling rows associated with the nodes in $\mathcal{U}_k$ for $1 \leq k \leq K - 1$. In other words, we only reorder the coupling rows by using $\Pi_{HP}(\mathcal{H}_{CM})$ as the nodes in the communication hypergraph, which are used for ordering purposes, represent the coupling rows. Figure 8d displays the reordered matrix using the node partition in Fig. 8c.

We use the connectivity metric (3) in partitioning $\mathcal{H}_{CM}$. Minimizing the connectivity of a net $n_j \in \mathcal{N}$, i.e., $\lambda(n_j)$, corresponds to minimizing the number of nonempty off-diagonal blocks in jth column stripe as the parts connected by n_j correspond to the nonempty off-diagonal blocks in jth column stripe in A_{NOD}. For example, in Fig. 8a, the fourth column stripe has two nonempty off-diagonal blocks and with the reordering of the coupling rows, it has a single nonempty off-diagonal block in Fig. 8d. In other words, net C_4 in Fig. 8c connects two parts, one of which contains nodes $\{3, 11\}$ and the other one contains nodes $\{6, 10\}$. The former part corresponds to the nonempty off-diagonal block at the intersection of the first row stripe and the fourth column stripe. The latter part is by default assumed to be in the connectivity set of this net due to the fixed node connected by the net (i.e., it captures the diagonal block information). Maintaining a balance on the part weights in partitioning $\mathcal{H}_{CM}$ corresponds to maintaining a balance on the sizes of coupling rows. In the reordered matrix A_{NOD} in Fig. 8d, there are five nonempty off-diagonal blocks. Figure 9 illustrates a matrix reordered using the described methodology.

Application Areas and Parallelism Enhancement Permuting a given matrix A into A_{NOD} has been utilized within the context of parallelization of conjugate gradient normal equation error and residual [44] and the standard quasi-minimal residual methods [16] for solving nonsymmetric linear systems, the linear least squares method [40] for solving the least squares problem, the Lanczos method for computing the singular value decomposition [40] and the surrogate constraint method for solving the linear feasibility problem [52]. The common theme in all of these methods is the existence of repeated sparse matrix-vector (SpMV) and sparse matrix transpose-vector multiplication (SpM^{T}V) [50] in a distributed setting. A nonsymmetric matrix utilized in these methods allows for the adoption of a nonsymmetric permutation of it. This fact is exploited for addressing communication cost metrics which are important for parallel performance: total message count and maximum communication volume.

A columnwise partitioning of A induces a rowwise partitioning of A^T. Assume that A is permuted to A_{NOD} using the methodology described in the previous

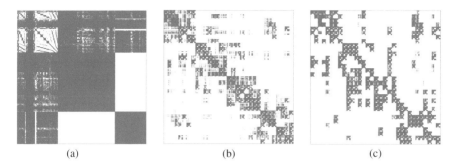

(a)   (b)   (c)

Fig. 9 An example matrix `poisson3Db` with 86K rows/columns and 2.4M nonzeros reordered using the described method for minimizing the number off-diagonal blocks for 32 processes. (**a**) A nonsymmetric matrix. (**b**) Reordered matrix with column-net hypergraph model. There are 345 off-diagonal blocks. (**c**) Reordered matrix with off-diagonal block minimization. There are 228 off-diagonal blocks

section and each process in the parallel system is responsible one of the K column stripes of A_{NOD}. In parallel SpMV, which is performed with the column-parallel algorithm, each nonempty off-diagonal block $B_{k,\ell}$ signifies a message from kth process to ℓth process and the contents of this message include the output vector elements that correspond to the nonzeros in $B_{k,\ell}$. The number of messages sent by the kth process is given by the number of nonempty off-diagonal blocks in the kth column stripe of A_{NOD}. Therefore, minimizing the nonempty off-diagonal blocks corresponds to minimizing the number of messages between processes. A dual logic applies to row-parallel SpM^{T}V.

Using degree weighting for the nodes in the communication hypergraph and then satisfying a balance constraint on the part weights relates to balancing the receive volume loads of the processes. In other words, it minimizes the maximum receive volume. One can also utilize unit weights for the nodes in order to approximate the send volume loads [50]. In SpM^{T}V, the utilization of degree weighting enables the minimization of maximum send volume. By reducing communication costs in these two multiplication operations through minimizing the nonempty off-diagonal blocks and satisfying a balance constraint on the sizes of coupling rows/columns, we can improve the parallel performance of the methods mentioned at the beginning of this section.

Note that the communication hypergraph model is also utilized for two-dimensional partitioning of the sparse matrices [46].

6 Block-Diagonal Form with Overlap

In block diagonal with overlap form, the successive blocks overlap with each other. We consider two variants of this form. In the first variant, called the general BDO form (or simply the BDO form), the diagonal blocks may overlap in both row and column dimensions. In the second variant, the diagonal blocks may overlap only along the row or column dimensions. The second variant is more restricted in the sense that the diagonal blocks are allowed to be overlap in a single dimension. For the second variant, we only consider the overlaps along the column dimension, hence the name block diagonal with column overlap form (BDCO form).

6.1 BDO Form

Target Form The block diagonal with overlap form of an $n \times n$ symmetric sparse matrix A consists of K diagonal blocks, where each diagonal block except the first and the last one overlaps with two other diagonal blocks, and the first and the last diagonal block overlaps with one diagonal block. In the BDO form, A is permuted into A_{BDO} as

$$PAP^T = \begin{bmatrix} B_{1,1} & B_{1,2} & & & & \\ B_{1,2}^T & C_{1,1} & B_{2,1} & C_{1,2} & & \\ & B_{2,1}^T & B_{2,2} & B_{2,3} & & \\ & C_{1,2}^T & B_{2,3}^T & C_{2,2} & \cdots & \\ & & & \vdots & \ddots & \\ & & & & C_{K-1,K-1} & B_{K,K-1} \\ & & & & B_{K,K-1}^T & B_{K,K} \end{bmatrix} = A_{BDO}, \qquad (11)$$

where P denotes the permutation matrix. We indicate the kth diagonal block with D_k. Each diagonal block $D_{1<k<K}$ consists of nine subblocks, and D_1 and D_K consist of four subblocks:

$$D_1 = \begin{bmatrix} B_{1,1} & B_{1,2} \\ B_{1,2}^T & C_{1,1} \end{bmatrix}, \quad D_{1<k<K} = \begin{bmatrix} C_{k-1,k-1} & B_{k,k-1} & C_{k-1,k} \\ B_{k,k-1}^T & B_{k,k} & B_{k,k+1} \\ C_{k-1,k}^T & B_{k,k+1}^T & C_{k,k} \end{bmatrix},$$

$$D_K = \begin{bmatrix} C_{K-1,K-1} & B_{K,K-1} \\ B_{K,K-1}^T & B_{K,K} \end{bmatrix}. \qquad (12)$$

The overlapping regions between diagonal blocks are referred to as coupling diagonal blocks. For two diagonal blocks D_k and D_{k+1}, where $1 \leq k < K$, the subblock $C_{k,k}$ couples these two blocks. Note that we consider a symmetric permutation of the given matrix, hence, A_{BDO} is also symmetric.

The first goal in permuting a sparse matrix into the BDO form is to minimize the summation of the number of rows/columns of the coupling diagonal blocks, i.e., to minimize $\sum_{k=1}^{K-1} nr(C_{k,k}) = \sum_{k=1}^{K-1} nc(C_{k,k})$. The second goal is to satisfy a balance criterion on the number of nonzero entries in diagonal blocks.

Methods Among the methods to obtain the BDO form [2, 23], we focus on the method that relies on recursive bipartitioning (RB) [2]. This method starts with representing the given sparse matrix A with the standard graph model G (for the standard graph model, see Methods in Sect. 4). In each bipartitioning, it makes use of a two-way GPVS with fixed vertices. To permute A into A_{BDO} with K diagonal blocks, this approach relies on the condition that G has a diameter of at least $K - 2$. We first describe the general RB framework and then review some of the issues related to fixed vertices and vertex weights.

Prior to any two-way GPVS on G' in the RB process to obtain $\Pi_{VS} = \{V_L, V_R; V_S\}$, we fix two subsets of vertices $\mathcal{F}_L$ and $\mathcal{F}_R$, $\mathcal{F}_L$ into V_L, and $\mathcal{F}_R$ into V_R, using two sets of boundary vertices $\mathcal{B}_L$ and $\mathcal{B}_R$. The function of the fixed vertices will be clear shortly. The boundary vertices are formed from the vertices that are boundary to the separator in the parent bipartitioning. In the case of the first bipartitioning in which there is no separator yet (since we are in the root level), in order to set $\mathcal{B}_L$ and $\mathcal{B}_R$, we find a pair of vertices v_i and v_j with the greatest shortest path and let $\mathcal{B}_L = \{v_i\}$ and $\mathcal{B}_R = \{v_j\}$. The bipartitioning of G' is then carried out with the fixed vertex sets $\mathcal{F}_L$ and $\mathcal{F}_R$ to obtain $\Pi_{VS} = \{\mathcal{F}_L \subseteq V_L, \mathcal{F}_R \subseteq V_R; V_S\}$. For the child bipartitionings, the separator V_S is removed and the new vertex-induced subgraphs G'_L and G'_R are, respectively, formed from the vertices in V_L and V_R. In the left bipartitioning of the two spawned child bipartitionings from G', the boundary vertex sets are formed as $\mathcal{B}_{LL} = \mathcal{B}_L$ and $\mathcal{B}_{LR} = Adj(V_S) \cap V_L$. In the right bipartitioning they are formed as $\mathcal{B}_{RR} = \mathcal{B}_R$ and $\mathcal{B}_{RL} = Adj(V_S) \cap V_R$. Then we recursively bipartition G'_L using the boundary vertex sets $\mathcal{B}_{LL}$ and $\mathcal{B}_{LR}$, and G'_R using the boundary vertex sets $\mathcal{B}_{RL}$ and $\mathcal{B}_{RR}$.

The bipartitionings in the final level of the RB tree slightly differ from those in the former levels in terms vertex fixing. In these bipartitionings, we have the flexibility of assigning the boundary vertices to the respective separators as opposed to the bipartitionings in the former levels. For that purpose, two auxiliary fixed vertices are introduced, and the boundary vertices are set free. The adjacency list of the first fixed vertex contains the vertices in the left boundary and the adjacency list of the other vertex contains the vertices in the right boundary. These fixed vertices are then removed after obtaining the final bipartitions.

The fixed vertices are central to obtaining the BDO form in the described methodology. First of all, if G_L and G_R are two graphs in the intermediate levels of the RB process, then the vertices in the right boundary set $\mathcal{B}_R$ of G_L need to be assigned to the right part in $\Pi_{VS}(G_L)$, and the vertices in the left boundary set $\mathcal{B}_L$ of G_R need to be assigned to the left part in $\Pi_{VS}(G_R)$. If G_L and G_R are two graphs in the final level, then we have the additional flexibility of assigning the vertices in $\mathcal{B}_R$ of G_L and $\mathcal{B}_L$ of G_R to the respective separators in their bipartitionings in addition to the described parts. In addition to the fixed vertices in the boundary vertex sets, for a given value of K', we fix all the vertices whose distances from the left and right boundaries smaller than $K'/2 - 1$ to the left and right parts, respectively. K' is K at the initial bipartitioning and it is halved at each level. This is the vertex fixing scheme used to obtain the fixed vertex sets $\mathcal{F}_L$ and $\mathcal{F}_R$ mentioned earlier.

This scheme ensures the existence of a valid separator in the bipartitioning, and consequently in the K-way partitioning.

At the end of this RB process, we obtain the partition $\Pi_{oVS}(\mathcal{G}) = \{\mathcal{V}_1, \ldots, \mathcal{V}_K; \mathcal{V}_{S_1}, \ldots, \mathcal{V}_{S_{k-1}}\}$. $\Pi_{oVS}(\mathcal{G})$ is used to reorder A into A_{BDO} by ordering the rows/columns corresponding to the vertices in $\mathcal{V}_{k+1}$ after the rows/columns corresponding to the vertices in $\mathcal{V}_k$, for $1 \leq k \leq K$, and the rows/columns corresponding to the vertices in separator $\mathcal{V}_{S_k}$ in between the rows/columns corresponding to the vertices in $\mathcal{V}_k$ and $\mathcal{V}_{k+1}$, for $1 \leq k \leq K - 1$. Figure 10 illustrates reordering of a given symmetric sparse matrix using $\Pi_{oVS}(\mathcal{G})$. The initial matrix is illustrated in Fig. 10a and its graph representation and a 4-way ordered GPVS of this graph are illustrated in Fig. 10b. The reordered matrix A_{BDO} is illustrated in Fig. 10c. The vertex colors in parts and separators match the color of the respective diagonal matrices and the overlapped regions.

To clarify how parts/separators relate to elements in the matrix, we present Fig. 11. Consider a part $\mathcal{V}_k$, the separators $\mathcal{V}_{S_{k-1}}$, $\mathcal{V}_{S_k}$, and the diagonal matrix D_k (12). The internal edges of $\mathcal{V}_{S_{k-1}}$, $\mathcal{V}_k$, and $\mathcal{V}_{S_k}$, respectively, correspond to the nonzeros in diagonal subblocks $C_{k-1,k-1}$, $B_{k,k}$, and $C_{k,k}$ and

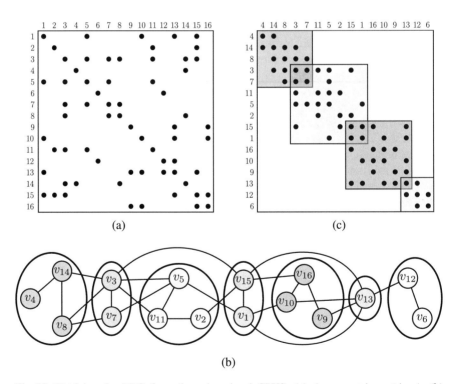

(a)

(c)

(b)

Fig. 10 Obtaining the BDO form through ordered GPVS. (**a**) A symmetric matrix A. (**b**) Reordered matrix A_{BDO}. (**c**) Π_{oVS} of standard graph representation of A

Fig. 11 The correspondence between the subblocks of the diagonal block and the edges in/between two separators and a part

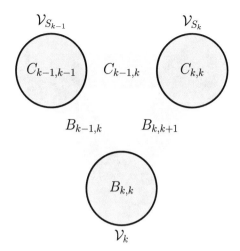

- the edges between $\mathcal{V}_{S_{k-1}}$ and $\mathcal{V}_k$ correspond to the subblocks $B_{k-1,k}$ and $B_{k-1,k}^T$,
- the edges between $\mathcal{V}_{S_{k-1}}$ and $\mathcal{V}_{S_k}$ correspond to the subblocks $C_{k-1,k}$ and $C_{k-1,k}^T$, and
- the edges between $\mathcal{V}_k$ and $\mathcal{V}_{S_k}$ correspond to the subblocks $B_{k,k+1}$ and $B_{k,k+1}^T$.

Minimizing the sizes of the $K - 1$ separators throughout the RB process corresponds to minimizing the overlap sizes of diagonal blocks in A_{BDO} as the vertices in the separators correspond to the rows/columns in the overlapped regions. Assigning each vertex a weight of number of nonzeros in the row/column representing that vertex does not correctly encapsulate the balancing the sizes of the diagonal blocks. One can exploit the flexibility of the RB framework to achieve better load balance. In the light of this direction, a weighting heuristic based on introducing two isolated fixed vertices in each bipartitioning is adopted [2]. In this way, maintaining a balance on the vertex weights of parts in $\Pi_{oVS}(\mathcal{G})$ relates to maintaining a balance on the number of elements in diagonal blocks. Figure 12 displays a matrix reordered using the described methodology.

Application Areas and Parallelism Enhancement The BDO form is utilized in the parallelization of the explicit formulation of Multiplicative Schwarz preconditioner [24] and in direct-iterative hybrid solvers [38, 39]. In these methods, typically, the kth diagonal block of the A_{BDO} is assigned to the kth process, where the computations related to this block may contain operations such as sparse matrix-vector multiplication and LU (or incomplete LU) factorization. Hence, having balanced diagonal blocks in terms of the number of nonzero elements they contain helps in balancing the computational loads. In general, minimizing the overlap size in these methods minimizes the communication overhead. A smaller overlap size means a smaller balance system [38, 39]. It also helps in speeding up the convergence [24].

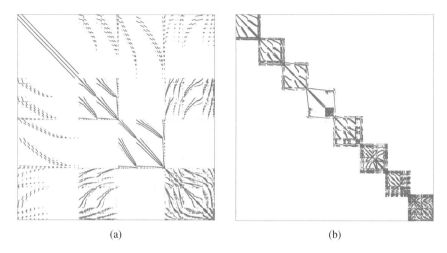

| (a) | (b) |

Fig. 12 An example problem matrix `copter2` with 56K rows/columns and 760K nonzeros. The BDO form of this matrix is obtained through the described method. (**a**) A computational fluid dynamics problem matrix. (**b**) The BDO form of the matrix in left for 8 processes

6.2 BDCO Form

Target Form The block-diagonal column-overlapped (BDCO) form of an $m \times n$ sparse matrix A consists of K diagonal blocks, where successive diagonal blocks may overlap along their columns. In the BDCO form, A is permuted into A_{BDCO} as

$$
PAQ = \begin{bmatrix}
B_1 & C_1 & & & & \\
& E_2 & B_2 & C_2 & & \\
& & E_3 & B_3 & C_3 & \\
& & & & \ddots & \\
& & & & & E_K & B_K
\end{bmatrix} = A_{BDCO}, \tag{13}
$$

where P and Q denote row and column permutation matrices, respectively. We indicate the kth diagonal block with D_k. Each diagonal block $D_{1<k<K}$ consists of three subblocks, and D_1 and D_K consist of two subblocks:

$$
D_1 = \begin{bmatrix} B_1 & C_1 \end{bmatrix}, D_{1<k<K} = \begin{bmatrix} E_k & B_k & C_k \end{bmatrix}, D_K = \begin{bmatrix} E_K & B_K \end{bmatrix}. \tag{14}
$$

The columns of the overlapping subblocks are referred to as coupling columns. Two diagonal blocks D_k and D_{k+1}, where $1 \leq k < K$, overlap along the columns of subblocks C_k and E_{k+1}.

The first goal in permuting a sparse matrix into the BDCO form is to minimize the total overlap size, i.e., the number of coupling columns. The second goal is to satisfy a balance criterion on the number of nonzero entries in diagonal blocks.

Methods Similar to the approach described in Sect. 6.1, the method [1] for obtaining the BDCO form is based on recursive bipartitioning (RB) and utilizes fixed nodes in each bipartitioning step. Here, we describe this method by comparing and contrasting it against the method described in Sect. 6.1.

The major difference of the method that obtains the BDCO form from the method that obtains the BDO form is that instead of using the standard graph representation of the given sparse matrix A, this method uses the column-net hypergraph model of A. In the column-net hypergraph $\mathcal{H}_{CN} = (\mathcal{U}, \mathcal{N})$ of A, there exists a node $u_i \in \mathcal{U}$ for each row $a_{i,*}$ of A and a net $n_j \in \mathcal{N}$ for each column $a_{*,j}$ of A [10]. Net n_j connects u_i if and only if $a_{i,j}$ is nonzero. Nets in $\mathcal{H}_{CN}$ are assigned unit cost as well as the nets in all hypergraphs formed during the RB process. Nodes in $\mathcal{H}_{CN}$ are assigned weights equal to the number of nonzeros in the corresponding rows. During the RB process, node weights are kept intact.

To permute A into A_{BDCO} form with K diagonal blocks, the diameter of $\mathcal{H}_{CN}$ should be at least $K - 1$. Here, the graph terminology for path, distance, and diameter are extended to hypergraphs. This extension relies on the consideration of two nodes as adjacent if and only if there exists at least one net connecting both of those nodes.

Similar to the vertex-fixation approach described in Sect. 6.1, each hypergraph in the RB tree has left and right boundary node sets and some nodes are fixed to the left and right parts depending on their distances to the boundary vertices. Likewise, the boundary nodes in the original hypergraph, i.e., the top-most hypergraph in the RB tree, are determined as the pseudo-peripheral nodes. However, the definition of the boundary nodes that are introduced during the RB process is different. Here, each node that is connected by at least one cut (external) net is said to be a boundary node. Instead of fixing the nodes whose distances to the boundary nodes are smaller than $K'/2 - 1$, this method fixes the nodes whose distances to the boundary nodes are smaller than $K'/2$. Furthermore, in this method, node fixing method used in each level of the RB tree is the same as opposed to the method described in Sect. 6.1.

In each RB step, after some nodes of the current hypergraph, say $\mathcal{H}$, are fixed as described above, $\mathcal{H}$ is bipartitioned and $\Pi_2 = \{\mathcal{U}_L, \mathcal{U}_R\}$ is obtained. Recall that in the method that obtains the BDO form, the separator vertices represent the rows/columns of the coupling subblocks in the reordered A_{BDO} matrix. Here in this method, cut nets represent the coupling columns of the reordered A_{BDCO} matrix. Since the overall objective here is to minimize the total number of coupling columns in A_{BDCO}, the target problem in each RB step is to minimize $cost(\Pi_{HP})$ of bipartition Π_2. Note that $cost(\Pi_{HP})$ definition in (3) becomes equal to that in (4) for a bipartition with unit net costs, hence, both objectives can be used for bipartitionings performed in this method.

After the current hypergraph $\mathcal{H}$ is bipartitioned to obtain $\Pi_2 = \{\mathcal{U}_L, \mathcal{U}_R\}$, two new hypergraphs $\mathcal{H}_L$ and $\mathcal{H}_R$ are formed out of $\mathcal{H}$ using Π_2. Node sets $\mathcal{U}_L$ and $\mathcal{U}_R$ of Π_2 correspond to the node sets of the new hypergraphs $\mathcal{H}_L$ and $\mathcal{H}_R$, respectively. Note that, in contrast to the method in Sect. 6.1, none of the nodes is removed while forming the new hypergraphs. This implies that for each level of the RB tree, the union of the node sets of the hypergraphs in that level corresponds to the node set of the original hypergraph $\mathcal{H}_{CN}$. There are two commonly-used techniques in the literature to form the net sets of $\mathcal{H}_L$ and $\mathcal{H}_R$: cut-net splitting and cut-net removal. Both can be used in this method since the existence of the split nets has no effect on the further bipartitionings. This is because they always connect boundary nodes, which are always fixed.

Recursive bipartitionings are performed on new hypergraphs $\mathcal{H}_L$ and $\mathcal{H}_R$ of each RB step until the RB tree has K leaf hypergraphs. The K-way node partition induced by the leaf hypergraphs is used to reorder the rows of A. Net sets of these hypergraphs and the external net sets in between them induce a $(2K - 1)$-way net partition of the net set of the original hypergraph $\mathcal{H}_{CN}$, which is then used to reorder the columns of A.

Figure 13 illustrates reordering of a given 10×12 sparse matrix into a 4-way BDCO form. The initial matrix A is given in Fig. 13a. A 4-way partition of the column-net hypergraph $\mathcal{H}_{CN}$ of A obtained by the above-mentioned procedure is given in Fig. 13b. The matrix A_{BDCO}, which is reordered using the node and net partitions given in Fig. 13b, is given in Fig. 10c.

Figure 14 displays a real-world matrix reordered using the described methodology.

Application Areas and Parallelism Enhancement The BCDO form is directly applicable in solution of sparse linear systems and sparse linear least squares problems. The balance scheme was first proposed as a parallel direct solver for

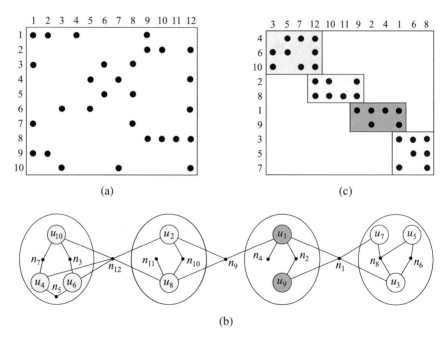

Fig. 13 Obtaining the BDCO form. (**a**) A matrix A. (**b**) Reordered matrix A_{BDCO}. (**c**) A 4-way partition of the column-net hypergraph of A

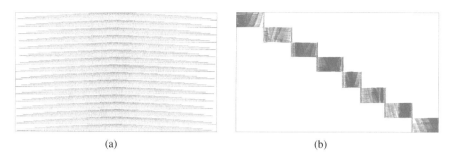

(a)   (b)

Fig. 14 An example problem matrix `mri1` with 66K rows, 115K columns, and 590K nonzeros. The BDCO form of this matrix is obtained through the described method. (**a**) A computer graphics problem matrix. (**b**) The BDCO form of the matrix in left for 8 processes

banded [19] and sparse [45] linear system of equations. Later it was extended for obtaining the minimum 2-norm solution of underdetermined linear least squares problems [48]. In this parallel algorithm, each diagonal block could be assigned to a different process and they solve independent linear systems or least squares problems using their respective diagonal blocks. This step is performed in parallel without any communication, hence, maintaining balance on the number of nonzero entries in diagonal blocks corresponds to maintaining balance on the computation loads of the processors in this step. Then a small reduced system is formed and solved either sequentially or in parallel to ensure the unique global solution vector with the same values on the overlapping parts of the vector across processes are obtained. The size of the reduced system as well as the communication amounts are determined by the total number of coupling columns. Hence, obtaining the BDCO form is a curicial step in enhancing the parallel scalability of these algorithms.

7   Conclusions

We presented a number of sparse matrix forms that are especially useful in enhancing parallel performance and/or convergence of a wide range of parallel numerical methods. We described the goals that are sought within these forms and presented partitioning-based combinatorial graph and hypergraph models to attain these goals. The partitionings obtained by the models are used to reorder the given sparse matrix into the desired form. We then described some applications that can benefit from these forms and how they benefit from them.

Acknowledgments This work was supported by the Director, Office of Science, U.S. Department of Energy under Contract No. DE-AC02-05CH11231. Sandia National Laboratories is a multi-mission laboratory managed and operated by National Technology and Engineering Solutions of Sandia LLC., a wholly owned subsidiary of Honeywell International, Inc., for the U.S. Department of Energy's National Nuclear Security Administration under contract DE-NA-0003525.

References

1. Acer, S., Aykanat, C.: Reordering sparse matrices into block-diagonal column-overlapped form. Journal of Parallel and Distributed Computing **140**, 99–109 (2020)
2. Acer, S., Kayaaslan, E., Aykanat, C.: A recursive bipartitioning algorithm for permuting sparse square matrices into block diagonal form with overlap. SIAM Journal on Scientific Computing **35**(1), C99–C121 (2013)
3. Alpert, C.J., Kahng, A.B.: Recent directions in netlist partitioning: a survey. Integr. VLSI J. **19**, 1–81 (1995)
4. Amestoy, P., Davis, T., Duff, I.: An approximate minimum degree ordering algorithm. SIAM Journal on Matrix Analysis and Applications **17**(4), 886–905 (1996)
5. Aykanat, C., Pinar, A., Çatalyürek, U.V.: Permuting sparse rectangular matrices into block-diagonal form. SIAM J. Sci. Comput. **25**, 1860–1879 (2004)
6. Bjorck, A.: Numerical Methods for Least Squares Problems. Society for Industrial and Applied Mathematics (1996)
7. Brandstädt, A., Le, V.B., Spinrad, J.P.: Graph Classes: A Survey. Society for Industrial and Applied Mathematics, Philadelphia, PA, USA (1999)
8. Bui, T.N., Jones, C.: Finding good approximate vertex and edge partitions is np-hard. Inf. Process. Lett. **42**(3), 153–159 (1992)
9. Bui, T.N., Jones, C.: A heuristic for reducing fill-in in sparse matrix factorization. In: Proceedings of the 6th SIAM Conference on Parallel Processing for Scientific Computing, pp. 445–452. Society for Industrial and Applied Mathematics (1993)
10. Catalyurek, U., Aykanat, C.: Hypergraph-partitioning-based decomposition for parallel sparse-matrix vector multiplication. IEEE Trans. Parallel Distrib. Syst. **10**, 673–693 (1999)
11. Catalyurek, U., Aykanat, C., Kayaaslan, E.: Hypergraph partitioning-based fill-reducing ordering for symmetric matrices. SIAM Journal on Scientific Computing **33**(4), 1996–2023 (2011)
12. Çatalyürek, U.V., Aykanat, C.: Patoh: partitioning tool for hypergraphs. Tech. rep., Department of Computer Engineering, Bilkent University (1999)
13. Cong, J., Lubio, W., Shivakumur, N.: Multi-way VLSI circuit partitioning based on dual net representation. In: IEEE/ACM International Conference on Computer-Aided Design, pp. 56–62 (1994)
14. Dantzig, G.B., Wolfe, P.: Decomposition principle for linear programs. Oper. Res. **8**(1), 101–111 (1960)
15. Ferris, M.C., Horn, J.D.: Partitioning mathematical programs for parallel solution. Mathematical Programming **80**(1), 35–61 (1998)
16. Freund, R., Nachtigal, N.: QMR: a quasi-minimal residual method for non-Hermitian linear systems. Numerische Mathematik **60**(1), 315–339 (1991)
17. George, A.: Nested dissection of a regular finite element mesh. SIAM Journal on Numerical Analysis **10**(2), 345–363 (1973)
18. George, A., Liu, J.W.: Computer Solution of Large Sparse Positive Definite. Prentice Hall Professional Technical Reference (1981)
19. Golub, G., Sameh, A.H., Sarin, V.: A parallel balance scheme for banded linear systems. Numerical Linear Algebra with Applications **8**(5), 285–299 (2001)
20. Hendrickson, B., Leland, R.: The Chaco user's guide version 2.0 (1995)
21. Hendrickson, B., Leland, R.: A multilevel algorithm for partitioning graphs. In: Proceedings of the 1995 ACM/IEEE conference on Supercomputing (CDROM), Supercomputing '95. ACM, New York, NY, USA (1995)
22. Hendrickson, B., Rothberg, E.: Improving the run time and quality of nested dissection ordering. SIAM J. Sci. Comput. **20**(2), 468–489 (1998)
23. Kahou, G.A.A., Grigori, L., Sosonkina, M.: A partitioning algorithm for block-diagonal matrices with overlap. Parallel Comput. **34**(6–8), 332–344 (2008)

24. Kahou, G.A.A., Kamgnia, E., Philippe, B.: An explicit formulation of the multiplicative Schwarz preconditioner. Applied Numerical Mathematics **57**(11), 1197–1213 (2007). Numerical Algorithms, Parallelism and Applications (2)

25. Karmarkar, N.: A new polynomial-time algorithm for linear programming. Combinatorica **4**(4), 373–395 (1984)

26. Karsavuran, M.O., Akbudak, K., Aykanat, C.: Locality-aware parallel sparse matrix-vector and matrix-transpose-vector multiplication on many-core processors. IEEE Transactions on Parallel and Distributed Systems **27**(6), 1713–1726 (2016)

27. Karypis, G., Aggarwal, R., Kumar, V., Shekhar, S.: Multilevel hypergraph partitioning: applications in VLSI domain. IEEE Trans. Very Large Scale Integr. Syst. **7**, 69–79 (1999)

28. Karypis, G., Kumar, V.: A fast and high quality multilevel scheme for partitioning irregular graphs. SIAM J. Sci. Comput. **20**(1), 359–392 (1998)

29. Kayaaslan, E., Pinar, A., Çatalyürek, U., Aykanat, C.: Partitioning hypergraphs in scientific computing applications through vertex separators on graphs. SIAM J. Sci. Comput. **34**(2), 970–992 (2012)

30. Kou, L.T., Stockmeyer, L.J., Wong, C.K.: Covering edges by cliques with regard to keyword conflicts and intersection graphs. Commun. ACM **21**(2), 135–139 (1978)

31. Lemaréchal, C., Nemirovskii, A., Nesterov, Y.: New variants of bundle methods. Mathematical Programming **69**(1), 111–147 (1995)

32. Lengauer, T.: Combinatorial Algorithms for Integrated Circuit Layout. John Wiley & Sons, Inc., New York, NY, USA (1990)

33. Liu, J.W.H.: Modification of the minimum-degree algorithm by multiple elimination. ACM Trans. Math. Softw. **11**(2), 141–153 (1985)

34. Mattson, T., Bader, D.A., Berry, J.W., Buluç, A., Dongarra, J.J., Faloutsos, C., Feo, J., Gilbert, J.R., Gonzalez, J., Hendrickson, B., Kepner, J., Leiserson, C.E., Lumsdaine, A., Padua, D.A., Poole, S.W., Reinhardt, S.P., Stonebraker, M., Wallach, S., Yoo, A.: Standards for graph algorithm primitives. CoRR **abs/1408.0393** (2014)

35. Medhi, D.: Parallel bundle-based decomposition for large-scale structured mathematical programming problems. Annals of Operations Research **22**(1), 101–127 (1990)

36. Medhi, D.: Bundle-based decomposition for large-scale convex optimization: Error estimate and application to block-angular linear programs. Mathematical Programming **66**(1), 79–101 (1994)

37. Mehrotra, S.: On the implementation of a primal-dual interior point method. SIAM Journal on Optimization **2**(4), 575–601 (1992)

38. Naumov, M., Manguoglu, M., Sameh, A.H.: A tearing-based hybrid parallel sparse linear system solver. J. Comput. Appl. Math. **234**(10), 3025–3038 (2010)

39. Naumov, M., Sameh, A.H.: A tearing-based hybrid parallel banded linear system solver. J. Comput. Appl. Math. **226**(2), 306–318 (2009)

40. Paige, C.C., Saunders, M.A.: LSQR: An algorithm for sparse linear equations and sparse least squares. ACM Trans. Math. Softw. **8**(1), 43–71 (1982)

41. Pellegrini, F., Roman, J.: Scotch: A software package for static mapping by dual recursive bipartitioning of process and architecture graphs. In: H. Liddell, A. Colbrook, B. Hertzberger, P. Sloot (eds.) High-Performance Computing and Networking, *Lecture Notes in Computer Science*, vol. 1067, pp. 493–498. Springer Berlin Heidelberg (1996)

42. Pinar, A., Çatalyürek, Ü.V., Aykanat, C., Pinar, M.: Decomposing linear programs for parallel solution. In: J. Dongarra, K. Madsen, J. Waśniewski (eds.) Applied Parallel Computing Computations in Physics, Chemistry and Engineering Science, pp. 473–482. Springer Berlin Heidelberg, Berlin, Heidelberg (1996)

43. Pothen, A., Simon, H.D., Liou, K.P.: Partitioning sparse matrices with eigenvectors of graphs. SIAM J. Matrix Anal. Appl. **11**(3), 430–452 (1990)

44. Saad, Y.: Iterative Methods for Sparse Linear Systems, 2nd edn. Society for Industrial and Applied Mathematics, Philadelphia, PA, USA (2003)

45. Sameh, A.H., Sarin, V.: Hybrid parallel linear system solvers. Int. J. Comp. Fluid Dyn **12**, 213–223 (1998)

46. Selvitopi, O., Aykanat, C.: Reducing latency cost in 2d sparse matrix partitioning models. Parallel Computing **57**, 1–24 (2016)
47. Tinney, W.F., Walker, J.W.: Direct solutions of sparse network equations by optimally ordered triangular factorization. Proceedings of the IEEE **55**(11), 1801–1809 (1967)
48. Torun, F.S., Manguoglu, M., Aykanat, C.: Parallel minimum norm solution of sparse block diagonal column overlapped underdetermined systems. ACM Trans. Math. Softw. **43**(4), 31:1–31:21 (2017)
49. Uçar, B., Aykanat, C.: Minimizing communication cost in fine-grain partitioning of sparse matrices. In: A. Yazıcı, C. Şener (eds.) Computer and Information Sciences - ISCIS 2003, *Lecture Notes in Computer Science*, vol. 2869, pp. 926–933. Springer Berlin Heidelberg (2003)
50. Uçar, B., Aykanat, C.: Encapsulating multiple communication-cost metrics in partitioning sparse rectangular matrices for parallel matrix-vector multiplies. SIAM J. Sci. Comput. **25**(6), 1837–1859 (2004)
51. Uçar, B., Aykanat, C., Pinar, M.c., Malas, T.: Parallel image restoration using surrogate constraint methods. J. Parallel Distrib. Comput. **67**(2), 186–204 (2007)
52. Yang, K., Murty, K.G.: New iterative methods for linear inequalities. Journal of Optimization Theory and Applications **72**(1), 163–185 (1992)

Structure-Exploiting Interior Point Methods

Juraj Kardoš, Drosos Kourounis, and Olaf Schenk

1 Introduction

Interior point (IP) methods have became a successful tool for solving the nonlinearly constrained optimal control problems. Their origin can be traced back to 1984 when Karmarkar [10] announced a polynomial time linear program that was considerably faster than the most popular simplex method to date. Furthermore, IP methods can also be applied to quadratic and other nonlinear programs, unlike the simplex method which can be applied only to linear programming. The main advantages of the IP methods lie in the convenience they offer for handling nonlinear inequality constraints using logarithmic barrier functions, so that a strictly feasible initial point is unnecessary. Another advantage of IP methods is that they are applicable to large-scale problems and allow for a variety of different direct sparse or iterative solution methods for the underlying linear systems solved at each iteration until convergence. Since different sparse system solvers can be plugged in with ease, large-scale structured problems can be solved by exploiting parallel computing infrastructures.

An example of successful application of the IP methods is the class of problems known as the optimal power flow (OPF). OPF is a nonlinear, nonconvex, large-scale optimization problem with the objective of minimizing the electricity generation cost while satisfying the physical constraints of the electric grid. The security constrained OPF (SCOPF) is an extension of the OPF problem that additionally ensures the system security with respect to a set of postulated contingencies. The SCOPF has become an essential tool for many transmission system operators for

J. Kardoš · D. Kourounis · O. Schenk (✉)
Institute of Computational Science, Faculty of Informatics, Universitá della Svizzera Italiana, Lugano, Switzerland
e-mail: juraj.kardos@usi.ch; drosos.kourounis@usi.ch; olaf.schenk@usi.ch

© Springer Nature Switzerland AG 2020
A. Grama, A. H. Sameh (eds.), *Parallel Algorithms in Computational Science and Engineering*, Modeling and Simulation in Science, Engineering and Technology,
https://doi.org/10.1007/978-3-030-43736-7_3

the planning, operational planning, and real time operation of the power system. An increase of the number of considered contingencies requires the introduction of additional variables and constraints, which in turn results in a significant problem size growth, rendering the solution computationally intractable for standard general purpose optimization tools. The structure of the SCOPF problems is appropriate for the parallel structure-exploiting IP methods, where each contingency corresponds to a separate partition on the linear level. The nonlinear IP framework leverages the bordered block-diagonal sparse structure specific to these optimal control problems by applying a Schur complement elimination on a block-per-block basis in order to exploit parallelism intrinsic to sparse block-diagonal structures by distributing the block contributions to the global Schur complement. In this way, the solution of the large-scale optimization problems can be approached more efficiently, as demonstrated in [19]. Similar structures arise also in the multistage stochastic optimal control problems [3, 16], multiperiod OPF problems (MPOPF) [11], dynamic simulations of the power grid [6], or problems such as natural gas dispatch [3].

This overview summarizes the algorithmic improvements in the recent years that have significantly advanced IP methods. The focus is on parallel implementations demonstrated on problems arising from the optimal control of the power grid. The presented primal-dual IP method is based on the IPOPT algorithm [21, 23].

1.1 Notation

Throughout we adopt the following notation. Scalar values are denoted by lowercase letters x in normal font, while vector objects are represented by bold lowercase letters $\boldsymbol{x}$. The vector $\boldsymbol{e}$ is a vector of ones with an appropriate dimension. If not specified otherwise, column vectors are assumed. Similarly, scalar functions are represented by a lowercase letter f, while vector functions are shown in bold lowercase $\boldsymbol{f}$. Concatenation of column vectors $(\boldsymbol{x}_1^\mathsf{T}, \boldsymbol{x}_2^\mathsf{T}, \ldots)^\mathsf{T}$ will be denoted by $(\boldsymbol{x}_1, \boldsymbol{x}_2, \ldots)$. The elementwise product of two vectors $\boldsymbol{x}, \boldsymbol{y}$ will be denoted by $\boldsymbol{x}\boldsymbol{y}$, while $\boldsymbol{x}^\mathsf{T}\boldsymbol{y}$ stands for the inner product of the two vectors. Matrices are represented by uppercase letters; for general (sparse) matrices we use bold fonts $\boldsymbol{X}$ while we will use normal font to distinguish diagonal matrices X. Sets will be represented by a calligraphic font $\mathcal{X}$ or uppercase Greek letters.

2 IP Algorithm

Definition 1 A general nonlinear programming (NLP) problem is formulated as a minimization problem

$$\underset{\boldsymbol{x}}{\text{minimize}} \ f(\boldsymbol{x}) \tag{1a}$$

$$\text{subject to } c_\varepsilon(x) = 0, \tag{1b}$$

$$c_I(x) \geq 0, \tag{1c}$$

$$x \geq 0, \tag{1d}$$

where $x \in \mathbb{R}^{N_x}$, the objective function f is a mapping $f : \mathbb{R}^{N_x} \to \mathbb{R}$, the constraints $c_\varepsilon : \mathbb{R}^{N_x} \to \mathbb{R}^{N_\varepsilon}$ and $c_I : \mathbb{R}^{N_x} \to \mathbb{R}^{N_I}$ are assumed to be sufficiently smooth, with continuous second-order derivatives, and $N_x > N_\varepsilon, N_I$, where N_ε, N_I are the number of equality and inequality constraints, respectively.

Definition 2 The *feasible set* Ω is a set of points x that satisfy the constraints of the NLP problem (1); that is

$$\Omega = \{x \in \mathbb{R}^{N_x} \mid c_\varepsilon(x) = 0, \; c_I(x) \geq 0, \; x \geq 0\}. \tag{2}$$

Definition 3 The *active set* at any feasible point x is a set of inequality constraints indices, for which the equality constraint holds; that is, $\mathcal{A}(x) = \{i \mid c_I^i(x) = 0\}$.

Definition 4 Given the solution of the NLP problem $x*$ and the active set $\mathcal{A}(x^*)$, the *linear independence constraint qualification* (LICQ) holds if the set of active constraint gradients $\{\nabla c_\varepsilon^i(x^*), \; i = 1 \ldots N_\varepsilon; \quad \nabla c_I^j(x^*), \; j \in \mathcal{A}(x^*)\}$ is linearly independent.

The NLP problem (1) can be transformed into the equivalent problem formulation where the inequality constraints are converted to equality constraints by introducing the slack variables $s \in \mathbb{R}^{N_I}$ with additional nonnegativity bounds $s \geq 0$. The NLP problem can be written as

$$\underset{x}{\text{minimize }} f(x) \tag{3a}$$

$$\text{subject to } c_\varepsilon(x) = 0, \tag{3b}$$

$$c_I(x) - s = 0, \tag{3c}$$

$$(x, s) \geq 0. \tag{3d}$$

Definition 5 The *Lagrangian* for the NLP problem (3) is defined as

$$\mathcal{L}(x, s, \lambda_\varepsilon, \lambda_I, \lambda_x, \lambda_s) = f(x) + \lambda_\varepsilon^\mathsf{T} c_\varepsilon(x) + \lambda_I^\mathsf{T}(c_I(x) - s) - \lambda_x^\mathsf{T} x - \lambda_s^\mathsf{T} s. \tag{4}$$

The vectors $\lambda_\varepsilon, \lambda_I, \lambda_x$, and λ_s are the Lagrange multipliers associated with the equality, original inequality, and the bound constraints on the primal and slack variables. This allows us to state the Karush–Kuhn–Tucker (KKT) first-order necessary conditions for the NLP problem (3) which characterize the solution.

Theorem 1 *Suppose that* x^* *is a local solution of the NLP problem* (3) *and that the LICQ holds at* x^*. *Then there exist Lagrange multiplier vectors* $\lambda_\varepsilon^* \in \mathbb{R}^{N_\varepsilon}, \lambda_I^* \in$

$\mathbb{R}^{N_I}$, $\lambda_x^* \in \mathbb{R}^n$ and $\lambda_s^* \in \mathbb{R}^{N_I}$, $(\lambda_x^*, \lambda_s^*) \geq 0$, such that the following conditions are satisfied at $(x^*, s^*, \lambda_\varepsilon^*, \lambda_I^*, \lambda_x^*, \lambda_s^*)$:

$$\nabla_x f(x^*) + \nabla_x c_\varepsilon(x^*)^\mathsf{T} \lambda_\varepsilon^* + \nabla_x c_I(x^*)^\mathsf{T} \lambda_I^* - \lambda_x^* = 0, \tag{5a}$$

$$-\lambda_I^* - \lambda_s^* = 0, \tag{5b}$$

$$c_\varepsilon(x^*) = 0, \tag{5c}$$

$$c_I(x^*) - s^* = 0, \tag{5d}$$

$$\lambda_x^* x^* = 0, \tag{5e}$$

$$\lambda_s^* s^* = 0, \tag{5f}$$

$$(x^*, s^*) \geq 0. \tag{5g}$$

The conditions (5a) and (5b) are referred to as dual feasibility, (5c), (5d) as primal feasibility, and (5f), (5e) as complementarity conditions. The point x^* satisfying the KKT conditions is called a stationary, or critical, point. In order to ensure that any stationary point x^* is indeed an optimal (local) solution of the NLP problem (3), the second-order sufficient conditions are needed.

Theorem 2 *Let x^* be a point at which LICQ holds, the KKT conditions are satisfied, and strict complementarity holds for the active inequality constraints. Then, the point x^* satisfies the second-order sufficient conditions for the NLP problem (3) if the Hessian of the Lagrangian $\nabla_{xx}^2 \mathcal{L}(x^*, s^*, \lambda_\varepsilon^*, \lambda_I^*, \lambda_x^*, \lambda_s^*)$ projected onto the null space of the constraint Jacobian is positive definite.*

In practice, the second-order conditions are guaranteed by monitoring the inertia of the iteration matrix, which is further elaborated in Sect. 2.3. Proofs of Theorems 1 and 2 can be found in classic optimization textbooks, e.g., [15, 25]. If the active set at the solution of the NLP problem was known, we could apply a Newton-class method directly to the linearization of the KKT conditions. However, the identification of the active set is known to be an NP-hard combinatorial problem for which, in the worst case, the computation time increases exponentially with the size of the problem. Therefore, many solution strategies adopt an IP approach, introducing a barrier subproblem where the nonnegativity bounds on the variables and slacks $(x, s) \geq 0$ are handled by the standard logarithmic barrier function, which is, in fact, a penalty term penalizing the iterates that approach the boundary of the feasible region.

Definition 6 The *barrier subproblem* (BSP) reads:

$$\underset{x,s}{\text{minimize}} \ f(x) - \mu \sum_{i=1}^{n} \log(x_i) - \mu \sum_{i=1}^{N_I} \log(s_i) \tag{6a}$$

$$\text{subject to } c_\varepsilon(x) = 0, \tag{6b}$$

$$c_I(x) - s = 0. \tag{6c}$$

Under certain conditions the solution x^* of the BSP (6) converges to the solution of the original NLP problem (1) as $\mu_j \downarrow 0$. Consequently, a strategy to solve the original NLP problem is to solve a sequence of the BSPs decreasing the barrier parameter μ_j. The solution of each iterate is not relevant for the solution of the original problem, so it can be relaxed to a certain accuracy and such an approximate solution is used as a starting point for the next BSP. The strategy for updating the μ parameter and thus switching to the next BSP is discussed later in Sect. 2.4.

The solutions of the barrier problem (6) are critical points of the Lagrangian function

$$\mathcal{L}(x, s, \lambda_\varepsilon, \lambda_I) = f(x) - \mu_j \sum_{i=1}^{N_x} \log(x_i) - \mu_j \sum_{i=1}^{N_I} \log(s_i) \tag{7}$$

$$+ \lambda_\varepsilon^\mathsf{T} c_\varepsilon(x) + \lambda_I^\mathsf{T}(c_I(x) - s).$$

Formulating and solving the optimality conditions of (7) directly would lead to singularities, since the derivatives of the barrier terms involve the fractions $\frac{\mu}{x_i}$ and $\frac{\mu}{s_i}$, which are not defined at the solution x^*, s^* of the NLP problem (1) when active bounds $x_i^* = 0$ or $s_i^* = 0$ are attained. Primal-dual IP methods [5, 9] define the dual variables z and y as

$$z_i = \frac{\mu}{x_i}, \quad i = 1, 2, \ldots, N_x, \tag{8a}$$

$$y_i = \frac{\mu}{s_i}, \quad i = 1, 2, \ldots, N_I. \tag{8b}$$

From the definition of the dual variables it follows that $z_i = \frac{\mu}{x_i} > 0$; therefore, $z_i x_i = \mu \ \forall i = 1, \ldots, N_x$. Similarly, $y_i s_i = \mu, y_i > 0 \ \forall i = 1, \ldots, N_I$. The optimality conditions of the BSP (6), considering also the dual variables (8), are written

$$\nabla_x f(x^*) + \nabla_x c_\varepsilon(x^*)^\mathsf{T} \lambda_\varepsilon^* + \nabla_x c_I(x^*)^\mathsf{T} \lambda_I^* - z^* = 0, \tag{9a}$$

$$-\lambda_I^* - y^* = 0, \tag{9b}$$

$$c_\varepsilon(x^*) = 0, \tag{9c}$$

$$c_I(x^*) - s^* = 0, \tag{9d}$$

$$z^* x^* = \mu e, \tag{9e}$$

$$y^* s^* = \mu e, \tag{9f}$$

$$(x^*, s^*) \geq 0. \tag{9g}$$

Note that the dual variables z, y correspond to the Lagrange multipliers λ_x and λ_s for the bound constraints. The KKT conditions of the BSP (9) are equivalent to the perturbed conditions (5) of the original NLP problem (3), except for the strict positivity of the dual variables $(z, y) > 0$. The primal-dual equations then become

$$l_a := \nabla_x f(x) + J_\varepsilon^\mathsf{T} \lambda_\varepsilon + J_I^\mathsf{T} \lambda_I - z = 0, \tag{10a}$$

$$l_b := -\lambda_I - y = 0, \tag{10b}$$

$$l_c := c_\varepsilon(x) = 0, \tag{10c}$$

$$l_d := c_I(x) - s = 0, \tag{10d}$$

$$l_e := Zx - \mu e = 0, \tag{10e}$$

$$l_f := Ys - \mu e = 0, \tag{10f}$$

where the Jacobian of constraints is written as $J_\varepsilon = \nabla_x c_\varepsilon(x)$ and $J_I = \nabla_x c_I(x)$. The diagonal matrices X, S, Z, Y are defined as $X = \text{diag}(x)$, $S = \text{diag}(s)$, $Z = \text{diag}(z)$, and $Y = \text{diag}(y)$.

Linearizing the primal-dual equations and solving them by applying Newton's method starting from an arbitrary value of the barrier parameter μ may result in slow convergence or poor conditioning of the associated KKT systems. Following the central path ensures that certain favorable conditions for the KKT systems and primal-dual variables are satisfied and descent directions can be obtained with reasonable accuracy.

Definition 7 The *central path* C is an arc of strictly feasible points of the BSP problem (6), $C = \{(x^\mu, s^\mu, \lambda_\varepsilon^\mu, \lambda_I^\mu, z^\mu, y^\mu) \mid \mu > 0\}$, such that $(x^\mu, s^\mu, \lambda_\varepsilon^\mu, \lambda_I^\mu, z^\mu, y^\mu)$ is a solution of the BSP problem for every value of $\mu > 0$. Points on the central path are characterized by the first-order KKT conditions (10).

Definition 8 The *duality measure* τ is an average pairwise complementarity value $x_i z_i$ and $s_i y_i$,

$$\tau = \frac{x^\mathsf{T} z + s^\mathsf{T} y}{N_x + N_I}. \tag{11}$$

The barrier parameter μ is usually chosen proportionally to the duality measure and the centering parameter $\sigma \in [0, 1]$, such that $\mu = \tau\sigma$. By choosing $\sigma = 1$ the algorithm moves toward the central path C. Such a step is biased toward the interior of the feasible region defined by the constraints $(z, x) > 0$, $(y, s) > 0$. At the other extreme, the value $\sigma = 0$ results in the standard Newton step aiming to satisy the KKT conditions (5). Many algorithms use intermediate values of σ from the open interval $(0, 1)$ to trade off between the two objectives of reducing duality measure and improving centrality. A strategy for selecting the centering parameter is discussed later in Sects. 2.4.2 and 2.4.3.

Remark 1 The treatment for general box constraints $x^{\min} \le x \le x^{\max}$ and general upper and lower bounds on the nonlinear constraints $c_I^{\min} \le s \le c_I^{\max}$ requires the addition of modified logarithmic barrier terms

$$\mathcal{B}(x, x^{\min}, x^{\max}) = -\mu_j \sum_{i=1}^{N_x} \log(x_i - x_i^{\min}) - \mu_j \sum_{i=1}^{N_x} \log(x_i^{\max} - x_i), \qquad (12)$$

$$\mathcal{B}(s, c_{\mathcal{I}}^{\min}, c_{\mathcal{I}}^{\max}) = -\mu_j \sum_{i=1}^{N_{\mathcal{I}}} \log(s_i - c_{\mathcal{I}i}^{\min}) - \mu_j \sum_{i=1}^{N_{\mathcal{I}}} \log(c_{\mathcal{I}i}^{\max} - s_i). \qquad (13)$$

The dual variables for $i = 1, 2, \ldots, N_x$ are defined by

$$z_i^L = \frac{\mu}{x_i - x_i^{\min}}, \qquad\qquad z_i^U = \frac{\mu}{x_i^{\max} - x_i}, \qquad (14)$$

while for the constraints the dual variables are defined by

$$y_i^L = \frac{\mu}{s_i - c_{\mathcal{I}i}^{\min}}, \qquad\qquad y_i^U = \frac{\mu}{c_{\mathcal{I}i}^{\max} - s_i}. \qquad (15)$$

2.1 Search Direction Computation

Since the solution of the barrier problem (6) satisfies the perturbed KKT conditions (10), Newton's method may be applied to solve the system of nonlinear equations. The search direction $(\Delta x^k, \Delta s^k, \Delta\lambda_{\mathcal{E}}^k, \Delta\lambda_{\mathcal{I}}^k, \Delta z^k, \Delta y^k)$ at the kth iteration can be obtained from the linearization of (10) at the iterate $(x^k, s^k, \lambda_{\mathcal{E}}^k, \lambda_{\mathcal{I}}^k, z^k, y^k)$, resulting in a system of linear equations

$$\begin{bmatrix} H & 0 & J_{\mathcal{E}}^{\mathsf{T}} & J_{\mathcal{I}}^{\mathsf{T}} & -I & 0 \\ 0 & 0 & 0 & -I & 0 & -I \\ J_{\mathcal{E}} & 0 & 0 & 0 & 0 & 0 \\ J_{\mathcal{I}} & -I & 0 & 0 & 0 & 0 \\ Z & 0 & 0 & 0 & X & 0 \\ 0 & Y & 0 & 0 & 0 & S \end{bmatrix}^k \begin{bmatrix} \Delta x \\ \Delta s \\ \Delta\lambda_{\mathcal{E}} \\ \Delta\lambda_{\mathcal{I}} \\ \Delta z \\ \Delta y \end{bmatrix}^k = - \begin{bmatrix} l_a \\ l_b \\ l_c \\ l_d \\ l_e \\ l_f \end{bmatrix}^k, \qquad (16)$$

where $H = \nabla_{xx}^2 \mathcal{L}$. The system (16) is clearly unsymmetric. A symmetric system can be obtained after eliminating the last two block rows:

$$\begin{bmatrix} \tilde{H} & 0 & J_{\mathcal{E}}^{\mathsf{T}} & J_{\mathcal{I}}^{\mathsf{T}} \\ 0 & L_s & 0 & -I \\ J_{\mathcal{E}} & 0 & 0 & 0 \\ J_{\mathcal{I}} & -I & 0 & 0 \end{bmatrix}^k \begin{bmatrix} \Delta x \\ \Delta s \\ \Delta\lambda_{\mathcal{E}} \\ \Delta\lambda_{\mathcal{I}} \end{bmatrix}^k = - \begin{bmatrix} l_a + X^{-1}l_e \\ l_b + S^{-1}l_f \\ l_c \\ l_d \end{bmatrix}^k, \qquad (17)$$

where $\tilde{H} = H + X^{-1}Z$ and $L_s = S^{-1}Y$. The directions Δz^k and Δy^k can then be recovered from the equations

$$\Delta z^k = -X^{-1}(l_e + Z\Delta x^k), \tag{18}$$

$$\Delta y^k = -S^{-1}(l_f + Y\Delta s^k). \tag{19}$$

For a robust algorithm it is crucial to obtain highly accurate search directions. Most of the burden is shifted to the sparse linear solver, where techniques such as fill-in minimization reordering, symmetric scaling vectors, matching, and pivoting can provide substantial improvement to the solution accuracy. Additional improvement can be achieved by performing iterative refinement using the unsymmetrical version KKT linear system of form (16). It is possible to further reduce the KKT system by eliminating the slack variables s. The system (17) can be permuted to the structure with the diagonal block L_s in the lower right corner,

$$\begin{bmatrix} \tilde{H} & J_\varepsilon^\mathsf{T} & J_\mathcal{I}^\mathsf{T} & 0 \\ J_\varepsilon & 0 & 0 & 0 \\ J_\mathcal{I} & 0 & 0 & -I \\ 0 & 0 & -I & L_s \end{bmatrix}^k \begin{bmatrix} \Delta x \\ \Delta\lambda_\varepsilon \\ \Delta\lambda_\mathcal{I} \\ \Delta s \end{bmatrix}^k = - \begin{bmatrix} l_a + X^{-1}l_e \\ l_c \\ l_d \\ l_b + S^{-1}l_f \end{bmatrix}^k. \tag{20}$$

Since the block L_s is a diagonal matrix, the reordered system (20) can be trivially reduced by computing the Schur complement with respect to the 3×3 block in the upper left corner, as illustrated in Fig. 1,

$$\begin{bmatrix} \tilde{H} & J_\varepsilon^\mathsf{T} & J_\mathcal{I}^\mathsf{T} \\ J_\varepsilon & 0 & 0 \\ J_\mathcal{I} & 0 & 0 \end{bmatrix}^k - \begin{bmatrix} 0 & 0 & -I \end{bmatrix}^\mathsf{T} (L_s^k)^{-1} \begin{bmatrix} 0 & 0 & -I \end{bmatrix}. \tag{21}$$

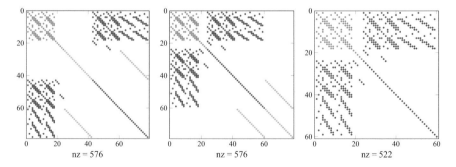

Fig. 1 Structure of the KKT system (17), reordered according to (20), and the structure of the reduced KKT with the slacks removed (22)

The additional elimination, compared to [16, 17], further reduces the memory requirements and computation time due to the smaller amount of factorization fill-in. Such an elimination, however, can be performed only for the nonzero elements of L_s^k sufficiently away from zero in order to avoid the ill-conditioning of the reduced system. The reduced linear system that needs to be solved now has the structure

$$
\begin{bmatrix} \tilde{H} & J_\varepsilon^\mathsf{T} & J_I^\mathsf{T} \\ J_\varepsilon & 0 & 0 \\ J_I & 0 & -L_s^{-1} \end{bmatrix}^k
\begin{bmatrix} \Delta x \\ \Delta \lambda_\varepsilon \\ \Delta \lambda_I \end{bmatrix}^k
= -
\begin{bmatrix} l_a + X^{-1} l_e \\ l_c \\ l_d + L_s^{-1}(l_b + S^{-1} l_f) \end{bmatrix}^k
\tag{22}
$$

and the eliminated slack variables can be recovered by solving

$$
L_s^k \Delta s^k = -l_b^k - S_k^{-1} l_f^k + \Delta \lambda_I^k.
\tag{23}
$$

2.2 Backtracking Line-Search Filter Method

After the successful computation of the search direction from (17) and (18) the step sizes $\alpha_k, \alpha_k^z \in (0, 1]$ need to be determined in order to obtain the next iterate:

$$
x^{k+1} = x^k + \alpha_k \Delta x^k,
\tag{24}
$$

$$
s^{k+1} = s^k + \alpha_k \Delta s^k,
\tag{25}
$$

$$
\lambda_\varepsilon^{k+1} = \lambda_\varepsilon^k + \alpha_k \Delta \lambda_\varepsilon^k,
\tag{26}
$$

$$
\lambda_I^{k+1} = \lambda_I^k + \alpha_k \Delta \lambda_I^k,
\tag{27}
$$

$$
z^{k+1} = z^k + \alpha_k^z \Delta z^k,
\tag{28}
$$

$$
y^{k+1} = y^k + \alpha_k^z \Delta y^k.
\tag{29}
$$

Different step sizes for the primal and dual variables is commonly employed to prevent unnecessarily small steps in either variables and delay the convergence to the optimal. A first candidate step length is chosen such that the strict positivity of x, s, and z is preserved, since it needs to hold both in the solution of the barrier problem (6) and also in every iteration, which is necessary in order to evaluate the barrier function. This is accomplished by the fraction-to-the-boundary rule, which identifies the maximum step size $\alpha_k, \alpha_k^z \in (0, 1]$, such that

$$
\alpha_k^{\max} = \max\left(\alpha \in (0, 1] : x^k + \alpha \Delta x^k \geq (1 - \tau)x^k\right),
\tag{30}
$$

$$
\alpha_k^z = \max\left(\alpha \in (0, 1] : z^k + \alpha \Delta z^k \geq (1 - \tau)z^k\right),
\tag{31}
$$

where $\tau \in (0, 1)$ is a function of the current barrier parameter μ_j. The step size for the dual variables α_k^z is used directly, but in order to ensure global convergence the step size $\alpha_k \in (0, \alpha_k^{\max})$ for the remaining variables is determined by a backtracking line-search procedure, exploring a decreasing sequence of trial step sizes $\alpha_k^i = 2^{-i} \alpha_k^{\max}$ for $i = 0, 1, 2, \ldots$.

The variant of the backtracking line-search filter method [8] used in IPOPT is based on the idea of a biobjective optimization problem with the two goals (i) minimizing the objective function

$$\varphi_{\mu_j}(x, s) := f(x) - \mu_j \sum_{i=1}^{n} \log(x_i) - \mu_j \sum_{i=1}^{N_I} \log(s_i) \tag{32}$$

and (ii) minimizing the constraint violation

$$\theta(x, s) := \| (c_{\varepsilon}(x), c_I(x) - s) \|_1. \tag{33}$$

A trial point $x^k(\alpha_k^i) := x^k + \alpha_k^i \Delta x^k$ and $s^k(\alpha_k^i) := s^k + \alpha_k^i \Delta s^k$ during the backtracking line search is considered to be acceptable, if it leads to sufficient progress toward either goal compared to the current iterate. The emphasis is put on the latter goal, until the constraint violations satisfy a certain threshold. Afterwards, the former goal is emphasized and reduction in the barrier function is required, accepting only iterates satisfying the Armijo condition.

Definition 9 The filter $\mathcal{F}$ is a set of ordered pairs containing a constraint violation value θ and the objective function value φ, such that

$$\mathcal{F} \subseteq \{(\theta, \varphi) \in \mathbb{R}^2 : \theta > 0\}. \tag{34}$$

The algorithm also maintains a filter $\mathcal{F}_j$ for each BSP j for which the μ_j is fixed. The filter $\mathcal{F}_j$ contains those combinations that are prohibited for a successful trial point in all iterations within the jth BSP. The filter is initialized so that the algorithm will never allow trial points to be accepted that have a constraint violation larger than $\theta^{\max}$. During the line search, a trial point $x^k(\alpha_k^i)$, $s^k(\alpha_k^i)$ is rejected if $(\theta(x_k(\alpha_k^i), s^k(\alpha_k^i)), \varphi_{\mu_j}(x^k(\alpha_k^i), s^k(\alpha_k^i))) \in \mathcal{F}_j$. After every iteration, in which the accepted trial step size does not satisfy the two objectives of the backtracking line search, the filter is augmented. This ensures that the iterates cannot return to the neighborhood of the unsatisfactory iterates. Overall, this procedure ensures that the algorithm cannot cycle, for example, between two points that alternate between decrease of the constraint violation and the barrier objective function.

In cases when it is not possible to identify a satisfactory trial step size, the algorithm reverts to a feasibility restoration phase. Here, the algorithm tries to find a new iterate which is acceptable to the current filter, by reducing the constraint violation with some iterative method. Note that the restoration phase algorithm might not be able to produce a new iterate for the filter line-search method, for example, when the problem is infeasible.

2.3 Inertia Correction and Curvature Detection

Definition 10 The inertia of a square matrix is defined as the ordered triplet $(n+, n-, n_0) \in \{\mathbb{N} \cup 0\}^3$, where the terms denote the number of positive, negative, and zero eigenvalues, respectively.

In order to guarantee descent properties for the line-search procedure, it is necessary to ensure that the Hessian matrix projected on the null space of the constraint Jacobian is positive definite (see Theorem 2). Also, if the constraint Jacobian does not have full rank, the iteration matrix in (17) is singular, and the solution might not exist. These conditions are satisfied if the iteration matrix has the inertia $(N_x + N_I, N_\varepsilon + N_I, 0)$. The sizes correspond to the size of the Hessian block (with respect to both primal variables x and the slack variables s) and the Jacobians of the equality and inequality constraints. If the inertia is not correct, the iteration matrix needs to be modified. In IPOPT implementation, the diagonal perturbations $\delta_w, \delta_c \geq 0$ are added to the Hessian (17), such that

$$\begin{bmatrix} \tilde{H} + \delta_w I & 0 & J_\varepsilon^\mathsf{T} & J_I^\mathsf{T} \\ 0 & L_s + \delta_w I & 0 & -I \\ J_\varepsilon & 0 & -\delta_c I & 0 \\ J_I & -I & 0 & -\delta_c I \end{bmatrix}. \tag{35}$$

The system is refactorized with different trial values of δ_w, δ_c until the inertia is correct. The inertia of the iteration matrix is readily available from several sparse indefinite linear solvers, such as PARDISO [20]. In case the correct inertia cannot be achieved, the current search direction computation is aborted and the algorithm uses a different objective function that does try to solely minimize the feasibility violation (e.g., minimizing the constraints violation), ignoring the original objective function, in the hope that the matrix has better properties close to the feasible points.

The inertia detection strategy focuses on the properties of the augmented iteration matrix (17) alone and can discard search directions that are of descent but for which the inertia of the augmented matrix is not correct. Furthermore, the inertia detection strategy might require multiple factorizations of the iteration matrix and, because the factorization is the most expensive step in the algorithm, computational performance can be greatly affected. Furthermore, the inertia estimates might vary, depending on which linear solver is used or not be available at all. To bypass the need for the inertia information, several authors suggest using the curvature test, e.g., [3, 4]:

$$d_k^\mathsf{T} W_k(\delta) d_k \geq \kappa d_k^\mathsf{T} d_k, \quad \kappa > 0, \delta \geq 0, \tag{36}$$

$$W_k(\delta) = \begin{bmatrix} \tilde{H} & 0 \\ 0 & L_s \end{bmatrix}^k + \delta I, \quad d_k = (\Delta x_k, \Delta s_k).$$

If the test is satisfied, the search direction is accepted; if it is not satisfied, the regularization parameter δ is increased and a new search direction is computed using the new regularized matrix.

Remark 2 While the curvature detection strategy usually requires more IP iterations until convergence compared with the inertia detection, it may require fewer extra factorizations. Overall, the solution time is less than that of the inertia detection because significantly fewer regularizations are needed.

2.4 Barrier Parameter Update Strategy

The strategy of the barrier parameter update is an important factor influencing the convergence properties, especially for difficult nonconvex problems. When solving nonlinear nonconvex programming problems, it is of great importance to prevent the iteration from failing. Different barrier parameter update strategies are discussed here, including the monotone Fiacco–McCormick strategy [1] and an adaptive strategy based on minimization of a quality function [14].

2.4.1 Monotone and Adaptive Strategies

Using the default monotone Fiacco–McCormick strategy, an approximate solution to the barrier problem (6) for a fixed value of μ is computed, possibly iterating over multiple primal-dual steps. Subsequently, the barrier parameter is updated and the computation continues by solution of the next barrier problem, starting from the approximate solution of the previous one. The approximate solution for the barrier problem (6), for a given value of μ_j, is required to satisfy the tolerance

$$E_\mu(x^{j+1}, s^{j+1}, \lambda_\varepsilon^{j+1}, \lambda_I^{j+1}, z^{j+1}, y^{j+1}) < \kappa_\epsilon \mu_j \tag{37}$$

for a constant $\kappa_\epsilon > 0$ before the algorithm continues with the solution of the next barrier problem. The optimality error for the barrier problem is defined by considering the individual parts of the primal-dual equations (10), that is, the dual feasibility (optimality), primal feasibility (constraint violations), and the complementarity conditions,

$$E_\mu(x, s, \lambda_\varepsilon, \lambda_I, z, y) = \max \left(\|l_a\|_\infty, \|l_b\|_\infty, \|l_c\|_\infty, \|l_d\|_\infty, \|l_e\|_\infty, \|l_f\|_\infty \right). \tag{38}$$

In the monotone barrier update strategy, the new barrier parameter is obtained from

$$\mu_{j+1} = \max \left(\frac{\epsilon_{\text{tol}}}{10}, \ \min \left(\kappa_\mu \mu_j, \ \mu_j^{\theta_\mu} \right) \right) \tag{39}$$

with constants $\kappa_\mu \in (0, 1)$ and $\theta_\mu \in (1, 2)$. In this way, the barrier parameter is eventually decreased at a superlinear rate. On the other hand, the update rule (39) does not allow μ to become smaller than necessary given the desired tolerance ϵ_{tol}, thus avoiding numerical difficulties at the end of the optimization procedure. The monotone Fiacco–McCormick strategy can be very sensitive to the choice of the initial point, the initial value of the barrier parameter, and the scaling of the problem. Furthermore, different problems might favor strategies for selecting the barrier parameter at every iteration of an IP method, that is, for every primal-dual step computation. Adaptive strategies commonly choose μ_{k+1} proportionally to the duality measure for the kth iterate,

$$\mu_{k+1} = \sigma \tau_k, \tag{40}$$

where $\sigma > 0$ is a centering parameter and τ denotes the duality measure (11). The adaptive strategies vary in how the centering parameter is determined. Two adaptive strategies implemented in IPOPT are discussed next.

2.4.2 Mehrotra's Predictor-Corrector

Mehrotra's proposed a predictor-corrector principle [12] for computing the search direction. The centering parameter is computed as the ratio between the duality measure (11) in the current iterate and the iterate updated by the predictor step, considering the longest possible step sizes that retain the nonnegativity of the variables in the barrier problem. If good progress in the duality measure is made in the predictor step, the centering parameter obtained in this way is small, $\sigma < 1$; therefore, the μ will be small in the next iteration. In other cases σ may be chosen to be greater than 1. This heuristic is based on experimentation with linear programming problems, and has proved to be effective for convex quadratic programming.

2.4.3 Quality Function

The adaptive barrier update strategy based on the quality function, as suggested in [14], is trying to determine the centering parameter by minimizing a linear approximation of the quality function. The quality function is a measure defined by the infeasibility norms in the current iterate updated by the probing search direction, which is expressed as a function of the sought parameter σ. The minimization problem is solved by a golden bisection procedure on the specified $(\sigma_{\min}, \sigma_{\max})$ interval with a maximum of 12 bisections. The evaluation of the barrier update strategies on both linear and nonlinear problems revealed superior performance of

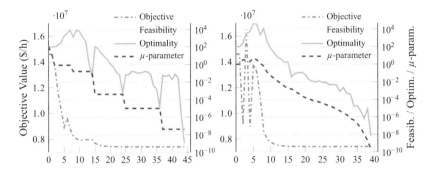

Fig. 2 Barrier parameter update strategies (left: monotone $\mu_0 = 100$; right: adaptive)

the adaptive methods over the monotone strategy, both in terms of CPU time and number of IP iterations. Although the results were more pronounced on the linear benchmarks, significant improvements can be expected by using adaptive strategies, particularly in applications where the function evaluation has the dominant cost [14]. Figure 2 depicts the convergence with different barrier parameter update strategies. The value of the barrier parameter μ over the iterations of the IP is shown for the two update strategies. Feasibility, optimality, and the objective function are shown as well. The convergence tolerance for both benchmarks was set to tol = 0.01.

2.5 Problem Scaling and Convergence Criteria

Optimal control of realistic industrial and engineering problems, such as modern power networks, multienergy carrier systems, the variables and constraints encountered, commonly involve different scales that usually differ by several orders of magnitude. Sophisticated scaling is necessary to remedy problems related to establishing accurate stopping criteria, improving convergence deteriorated by unbalanced direction vectors, and dealing with loss of accuracy of the descent direction computation due to bad conditioning of the associated KKT systems. In the ideal case, not only the variables but also the functions should be scaled so that changing a variable by a given amount has a comparable effect on any function which depends on these variables or, in other words, so that the nonzero elements of the function gradients are of the same order of magnitude. For this purpose, gradient-based scaling is commonly employed so that at the starting point the gradients are scaled close to one. The scaling factors for the gradients are defined as

$$s_f = \min(1, \ g_{\max}/\|\nabla_x f(x_0)\|_\infty), \tag{41}$$

$$s_g^{(j)} = \min(1, \ g_{\max}/\|\nabla_x c_\varepsilon^{(j)}(x_0)\|_\infty), \ j = 1 \ldots N_\varepsilon, \tag{42}$$

$$s_h^{(j)} = \min(1, \ g_{\max}/\|\nabla_x c_I^{(j)}(x_0)\|_\infty), \ j = 1 \ldots N_I, \tag{43}$$

for a given $g_{\max} > 0$. If the maximum gradient is above this value, then gradient-based scaling will be performed. Note that all gradient components in the scaled problem are at most of size $g_{\max}$ at the starting point. The scaling factors are computed only at the beginning of the optimization using the starting point and kept constant throughout the whole optimization process.

Even if the original problem is well scaled, the multipliers $\lambda_\varepsilon, \lambda_I, z$ might become very large, for example, when the gradients of the active constraints are (nearly) linearly dependent at a solution of (1). In this case, the algorithm might encounter numerical difficulties satisfying the unscaled primal-dual equations (17) to a tight tolerance. The convergence criteria in (38), therefore, need to be scaled accordingly. The scaled optimality error used to determine the convergence criteria is defined as

$$
E_0(x, s, \lambda_\varepsilon, \lambda_I, z) = \max\left(\frac{\|l_a\|_\infty}{s_1}, \frac{\|l_b\|_\infty}{s_1}, \|l_c\|_\infty, \|l_d\|_\infty, \frac{\|l_e\|_\infty}{s_2}, \frac{\|l_f\|_\infty}{s_2} \right),
$$

$$(44)$$

where the scaling factors s_1, s_2 are defined as

$$
s_1 = \frac{\max\left(s_{\max}, \frac{\|\lambda_\varepsilon\|_1 + \|\lambda_I\|_1 + \|z\|_1 + \|y\|_1}{N_\varepsilon + N_I + N_x + N_I} \right)}{s_{\max}}, \quad s_2 = \frac{\max\left(s_{\max}, \frac{\|z\|_1 + \|y\|_1}{N_x + N_I} \right)}{s_{\max}}.
$$

$$(45)$$

The overall IPOPT algorithm terminates successfully, if the NLP error for the current iterate with $\mu = 0$ in (44),

$$
E_0(x, s, \lambda_\varepsilon, \lambda_I, z, y) \leq \epsilon_{\text{tol}},
$$

$$(46)$$

becomes smaller than the user provided value $\epsilon_{\text{tol}} > 0$, and if the individual criteria according to dual, primal, and complementarity conditions in (44) are met. Each criterion uses a separate, user provided tolerance value.

3 IP Methods for OPF Problems

Recent developments in modern power grids involve widespread deployment of intermittent renewable generation, embrace installation of a wide variety of energy storage devices, as well as an increasing and widespread usage of electric vehicles. These developments will motivate fundamental changes in methods and tools for the optimal daily operation and planning of modern power grids. Operational decisions taken by power system operators on a daily basis are commonly assisted by repeatedly solving OPF problems, aiming to determine optimal operating levels for electric power plants, so that the overall electricity generation cost is minimized, while at the same time it satisfies load demands imposed throughout the transmission grid and meets safe operating limits. In actual industrial operations the entire distribution network needs to be optimized in real time, approximately every 5 min according to several independent system operators to ensure variations in load demand, renewable generation, and real-time electricity market responses to electricity prices are accurately met.

3.1 Optimal Power Flow

The OPF problem seeks a solution that minimizes the cost of the electricity generation f, while satisfying the power flow balance, maximum power flow over the transmission lines, and the bounds of the bus voltages and the generator limits. Consider a power network with N_B buses, N_G generators, and N_L transmission lines. The bus voltage vector $\underline{\mathbf{v}} \in \mathbb{C}^{N_B}$ is defined in polar notation as $\underline{\mathbf{v}} = \mathbf{v}e^{j\theta}$, where $\mathbf{v}, \theta \in \mathbb{R}^{N_B}$ specify the magnitude and phase of the complex voltage. The complex voltages $\underline{\mathbf{v}}$ determine the entire network power flow that can be computed using the Kirchhoff equations, network configuration, and properties of its components. The magnitude of the voltage components is bounded by the limits (47d), while the phase is determined relative to a single reference bus. The current injections, $\underline{\mathbf{I}} \in \mathbb{C}^{N_B}$, are defined as $\underline{\mathbf{I}} = \underline{\mathbf{Y}}^B \underline{\mathbf{v}}$, where $\underline{\mathbf{Y}}^B \in \mathbb{C}^{N_B \times N_B}$ is the bus admittance matrix. The complex power at each bus of the network, $\underline{\mathbf{S}} = \underline{\mathbf{v}}\underline{\mathbf{I}}^*$, $\underline{\mathbf{S}} \in \mathbb{C}^{N_B}$, and the power demand consumption $\underline{\mathbf{S}}^D \in \mathbb{C}^{N_B}$ are to be balanced by the net power injections from the generators $\underline{\mathbf{S}}^G \in \mathbb{C}^{N_G}$. Thus, the AC nodal power flow balance Eqs. (47b) are expressed as a function of the complex bus voltages and generator injections as $c_\varepsilon := \underline{\mathbf{S}} + \underline{\mathbf{S}}^D - C^G \underline{\mathbf{S}}^G = \mathbf{0}$, where $C^G \in \mathbb{R}^{N_B \times N_G}$ is the generator connectivity matrix.

Generator power injections $\underline{\mathbf{S}}^G = \mathbf{p} + j\mathbf{q}$ are expressed in terms of real and reactive power components $\mathbf{p}, \mathbf{q} \in \mathbb{R}^{N_G}$, respectively. The output of the generators is limited by the lower and upper bounds (47e) and (47f). Each bus has an associated complex power demand $\underline{\mathbf{S}}^D$, which is assumed to be known at all of the buses and is modeled by a static polynomial (ZIP) model [26]. If there are no loads connected to the bus i, then $\{\underline{\mathbf{S}}^D\}_i = \mathbf{0}$. Real-world transmission lines are limited by the instantaneous amount of power that can flow through the lines due to the thermal limits (47c). The apparent power flow in the transmission lines, $\underline{\mathbf{S}}^f \in \mathbb{C}^{N_L}$ and $\underline{\mathbf{S}}^t \in \mathbb{C}^{N_L}$, are therefore limited by the power injections at both ends of the lines, which cannot exceed a prescribed upper bound $\mathbf{F}_L^{max}$. The "from" and "to" ends of the line, denoted as f and t, respectively, specify the buses that are connected to the corresponding ends of the line. Squared values of the apparent power magnitude are usually used in practice, such that $c_I := \underline{\mathbf{S}}^f (\underline{\mathbf{S}}^f)^* \leq (\mathbf{F}_L^{max})^2$. Overall, the OPF problem is formulated as

$$\underset{\theta, \mathbf{v}, \mathbf{p}, \mathbf{q}}{\text{minimize}} \sum_{l=1}^{N_G} f_l(\mathbf{p}^l) \tag{47a}$$

$$\text{subject to } c_\varepsilon(\boldsymbol{\theta}, \mathbf{v}, \mathbf{p}, \mathbf{q}) = \mathbf{0}, \tag{47b}$$

$$c_I(\boldsymbol{\theta}, \mathbf{v}) \leq \mathbf{F}_L^{\max}, \tag{47c}$$

$$\mathbf{v}^{\min} \leq \mathbf{v} \leq \mathbf{v}^{\max}, \quad \boldsymbol{\theta}^{ref} = 0, \tag{47d}$$

$$\mathbf{p}^{\min} \leq \mathbf{p} \leq \mathbf{p}^{\max}, \tag{47e}$$

$$\mathbf{q}^{\min} \leq \mathbf{q} \leq \mathbf{q}^{\max}. \tag{47f}$$

The presented AC steady-state power grid model is following MATPOWER [27].

3.2 Structure-Exploiting IP Methods—Security Constrained and Multiperiod OPF

Real-world real-time implementation of OPF problems for energy systems still remains computationally intractable. This is mainly for two reasons. The real-world OPF problem is time coupled, owing to the presence of smart loads and energy storage devices such as batteries for demand shaping and deferral. Additional time couplings of the OPF problem at each time period are introduced by generator ramp rate limits. The higher the number of time periods considered, the larger the resulting optimal control problem becomes. For a significantly large number of time periods (each of 5 min length) the problem becomes notoriously difficult to solve and for this purpose several approximations and simplifications are currently employed by the industry in order to meet real-time responses. Furthermore, the system operators have to foresee possible contingency events and operate the grid in such a way that its operation will remain secure in the event of any contingencies.

Grid security is the focus of the SCOPF problem [13, 18], which seeks an optimal solution that remains feasible under any postulated contingency event, thus making the grid operation secure. It supplements the standard OPF problem with constraints for the nodal power flow balance (48a), the branch flow limits (48b), and other operational limits (48c), (48e), which have to be honored not only for the nominal case c_0, but also for every contingency event $c \in C$, $N_c = |C|$, such as a generator or a transmission line failure. An increase of the number of considered contingencies requires the introduction of additional variables and constraints that in turn result in a significant problem size growth, rendering it computationally intractable for standard general purpose optimization tools. The contingencies are modeled by the admittance matrices $\underline{\mathbf{Y}}_c^B$, which are updated accordingly for each scenario. The values of the control variables are coupled in all system scenarios, as expressed by the two nonanticipatory constraints (48g) and (48h). These declare that the voltage magnitude and real power generation at the PV buses $\mathcal{B}_{PV}$ should remain the same as in the nominal scenario c_0, regardless of which contingency they are associated with. The only generator that is allowed to change its output is the generator at the

$$
\begin{aligned}
&\underset{\boldsymbol{\theta}_c, \mathbf{v}_c, \mathbf{p}_c, \mathbf{q}_c}{\text{minimize}} \sum_{l=1}^{N_\mathrm{G}} f_l(\mathbf{p}_0^l) \\
&\text{subject to } \forall c \in \{c_0, c_1, \ldots, c_{N_c}\},
\end{aligned}
\qquad
\begin{aligned}
&\underset{\boldsymbol{\theta}_n, \mathbf{v}_n, \mathbf{p}_n, \mathbf{q}_n}{\text{minimize}} \sum_{l=1}^{N_\mathrm{G}} f_l(\mathbf{p}_0^l) \\
&\text{subject to } \forall n \in \{1, 2, \ldots, N\},
\end{aligned}
$$

(48a) $\quad c_{\varepsilon}^c(\boldsymbol{\theta}_c, \mathbf{v}_c, \mathbf{p}_c, \mathbf{q}_c) = \mathbf{0},$ $\qquad\qquad$ $c_{\varepsilon}^c(\boldsymbol{\theta}_n, \mathbf{v}_n, \mathbf{p}_n, \mathbf{q}_n) = \mathbf{0},$ (49a)

(48b) $\quad c_{\mathcal{I}}^c(\boldsymbol{\theta}_c, \mathbf{v}_c) \leq \mathbf{F}_\mathrm{L}^{\max},$ $\qquad\qquad\qquad$ $c_{\mathcal{I}}^c(\boldsymbol{\theta}_n, \mathbf{v}_n) \leq \mathbf{F}_\mathrm{L}^{\max},$ (49b)

(48c) $\quad \mathbf{v}^{\min} \leq \mathbf{v}_c \leq \mathbf{v}^{\max},$ $\qquad\qquad\qquad$ $\mathbf{v}^{\min} \leq \mathbf{v}_n \leq \mathbf{v}^{\max},$ (49c)

(48d) $\quad \boldsymbol{\theta}_c^{ref} = 0,$ $\qquad\qquad\qquad\qquad\qquad$ $\boldsymbol{\theta}_n^{ref} = 0,$ (49d)

(48e) $\quad \mathbf{p}^{\min} \leq \mathbf{p}_c \leq \mathbf{p}^{\max},$ $\qquad\qquad\qquad$ $\mathbf{p}^{\min} \leq \mathbf{p}_n \leq \mathbf{p}^{\max},$ (49e)

(48f) $\quad \mathbf{q}^{\min} \leq \mathbf{q}_c \leq \mathbf{q}^{\max},$ $\qquad\qquad\qquad$ $\mathbf{q}^{\min} \leq \mathbf{q}_n \leq \mathbf{q}^{\max},$ (49f)

(48g) $\quad \forall b \in \mathcal{B}_{PV} : \mathbf{v}_c = \mathbf{v}_{c_0},$ $\qquad\qquad$ $\boldsymbol{\epsilon}^{\min} \leq \boldsymbol{\epsilon}_n \leq \boldsymbol{\epsilon}^{\max}.$ (49g)

(48h) $\quad \forall g \in \mathcal{B}_{PV} : \mathbf{p}_c = \mathbf{p}_{c_0}.$

Fig. 3 SCOPF (left) and MPOPF (right) problem formulations

singleton reference bus $\mathcal{B}_{ref}$, as its real power generation can be modified to refill the power transmission losses occurring in each contingency c (Fig. 3).

Time-coupled formulations, such as storage scheduling, or storage placement, are collectively known as MPOPF problems (49). Similar to the SCOPF, addition of a large number of time periods results in problem size growth, rendering it computationally intractable [11]. The OPF constraints must hold in each time period, and the inter-temporal coupling is introduced by energy storage devices and generator ramp limits. For a practical MPOPF application, consider N_S energy storage units. Each storage unit in the network is modeled by two network power injections for each time period n. A positive active power injection $\mathbf{p}_n^{\mathrm{Sd},i} \in \mathbb{R}$, $\mathbf{p}_n^{\mathrm{Sd},i} \geq 0$ models the discharging of storage unit i. A negative active power injection $\mathbf{p}_n^{\mathrm{Sc},i} \in \mathbb{R}$, $\mathbf{p}_n^{\mathrm{Sc},i} \leq 0$ models the charging of storage unit i. The vector of active storage power injections $\mathbf{p}_n^\mathrm{S} \in \mathbb{R}^{2N_\mathrm{S}}$ is defined as

$$
\mathbf{p}_n^\mathrm{S} = (\mathbf{p}_n^{\mathrm{Sd},1}, \cdots, \mathbf{p}_n^{\mathrm{Sd},N_\mathrm{S}}, \mathbf{p}_n^{\mathrm{Sc},1}, \cdots, \mathbf{p}_n^{\mathrm{Sc},N_\mathrm{S}}) \tag{50}
$$

and bounded by $\mathbf{p}^{\mathrm{S,\,min}} \leq \mathbf{p}_n^\mathrm{S} \leq \mathbf{p}^{\mathrm{S,\,max}}$. Identical definitions apply for the reactive storage power injections $\mathbf{q}_n^{\mathrm{Sd},i}$, $\mathbf{q}_n^{\mathrm{Sc},i}$, $\mathbf{q}_n^\mathrm{S}$ with bounds $\mathbf{q}^{\mathrm{S,\,min}}$ and $\mathbf{q}^{\mathrm{S,\,max}}$. Together, they yield the complex storage power injections $\underline{\mathbf{S}}_n^\mathrm{S} = \mathbf{p}_n^\mathrm{S} + j\mathbf{q}_n^\mathrm{S}$. Similarly, $\underline{\mathbf{S}}_n^\mathrm{G} = \mathbf{p}_n^\mathrm{G} + j\mathbf{q}_n^\mathrm{G}$ is a vector of generator power injections. The complex power at each bus must be balanced by the power demand $\underline{\mathbf{S}}_n^D$ and the vector of free complex power injections

$$
\mathbf{S}_n = \begin{pmatrix} \underline{\mathbf{S}}_n^\mathrm{G} \\ \underline{\mathbf{S}}_n^\mathrm{S} \end{pmatrix} = \underbrace{\begin{pmatrix} \mathbf{p}_n^\mathrm{G} \\ \mathbf{p}_n^\mathrm{S} \end{pmatrix}}_{\mathbf{p}_n} + j \underbrace{\begin{pmatrix} \mathbf{q}_n^\mathrm{G} \\ \mathbf{q}_n^\mathrm{S} \end{pmatrix}}_{\mathbf{q}_n} \tag{51}
$$

in each time period, as specified by the constraint (49a). The evolution of the vector of storage levels $\epsilon_n \in \mathbb{R}^{N_S}$ follows the update equation

$$\epsilon_{n+1} = \epsilon_n + \mathbf{B}^S \, \mathbf{p}_n^S \quad n = 0, 1, \ldots, N-1, \tag{52}$$

and introduces a coupling between the individual time periods. The energy level in each period needs to honor the storage capacity, as expressed by the constraint (49 g). The initial storage level is denoted ϵ_0 and the constant matrix $\mathbf{B}^S \in \mathbb{R}^{N_S \times 2N_S}$ models discharging and charging efficiencies of the storage devices.

3.3   Impact of Slack Variables Elimination

Figure 4 illustrates the symmetric KKT structure of the SCOPF problem for a simple power grid, together with the reduced variant, where the slack variables are eliminated. Realistic power grids are significantly larger and contain proportionally more nonzero entries, but the structure remains very similar. The expected benefits of solving the reduced KKT system compared to the original system are savings both in terms of memory requirements for storing the sparse L factor of the LDL^T factorization of the symmetric indefinite system, and possibly faster factorization and solution times due to a smaller number of required floating point operations. The numerical evaluation of the benefits of solving the reduced system are summarized in Fig. 5. The elimination of the slack variables from the KKT system reduces its dimension by approximately 30% with 13% fewer nonzeros in the KKT system and up to 12% fewer nonzeros in the L factor, resulting in up to 28% memory savings, with similar reduction in solution time. Since in the neighborhood of the optimal solution some of the diagonal terms in L_s approach zero, the associated slacks variables whose coefficients in L_s are close to machine epsilon are not eliminated and are left to be treated by the direct sparse solver. This prevents the excessive ill-conditioning of the reduced system.

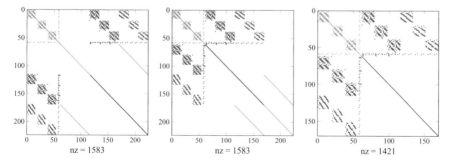

Fig. 4 Structure of the SCOPF KKT system (17) with two contingencies, reordered according to (20), and, finally, the reduced KKT with the slacks removed (22)

Fig. 5 Improvement rate
considering elimination of the
slack variables

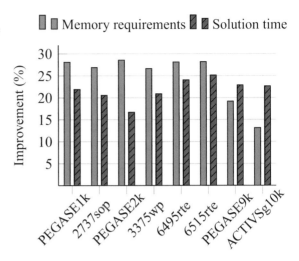

4 Structure-Exploiting Solution Strategies for IP Optimization

Computers have evolved significantly over the past decade, at an even faster pace than modern power grids. Multicore and many-core computer architectures and distributed compute clusters are ubiquitous today, while at the same time no significant performance gains are expected for sequential codes due to faster clock frequencies of modern processors. Significant performance gains, however, may be achieved by algorithmic redesign tailored to the particular application that is also able to utilize multicore and many-core architectures with deep memory hierarchies. More importantly, the practical efficiency of the IP algorithms highly depends on the linear algebra kernels used. For large-scale optimal control problems, the computation of the search direction (17) determines the overall runtime. Hence, any attempt at accelerating the solution should be focused on the efficient solution of the KKT linear system. In Fig. 6 we demonstrate how various IP method components contribute to the overall time for various OPF benchmarks. The number of IP iterations was fixed to five. Note that the solution of the linear system represents the majority of the overall time.

4.1 Revealing the Structure of SCOPF and MPOPF Problems

A widespread approach for solving KKT systems consists of employing black-box techniques such as direct sparse solvers, due to their accuracy and robustness. The direct sparse solvers obtain the solution of the linear system by factorization and subsequent forward-backward substitutions. The factorization is a computationally

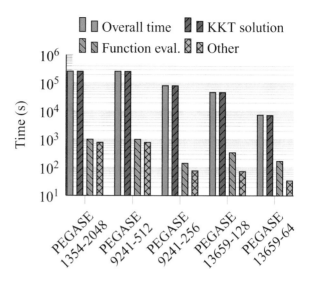

Fig. 6 Computational complexity of the IP method components

expensive operation commonly introducing significant fill-in, which may quickly exhaust available memory on shared memory machines for large-scale linear systems. Furthermore, these solvers are not aware of the underlying structural properties of the KKT systems arising from many engineering problems which make it possible to significantly decrease time to solution by employing structure-exploiting algorithms and distributed memory computers.

The appropriate structure emerges from the fact that each of the variables in the SCOPF optimization vector $(x, \lambda_\varepsilon, \lambda_I)$ or the MPOPF optimization vector $(x, \lambda_\varepsilon, \lambda_I, \lambda_A)$ correspond to some contingency scenario $c = 0, 1, \ldots, N_c$, or the time period $n = 1, 2, \ldots, N$:

$$x = (x_0, \ldots, x_{N_c}, x_g), \tag{53}$$

$$\lambda_\varepsilon = (\lambda_{\varepsilon 0}, \ldots, \lambda_{\varepsilon N_c}), \tag{54}$$

$$\lambda_I = (\lambda_{I0}, \ldots, \lambda_{I N_c}), \tag{55}$$

$$x = (x_0, \ldots, x_N), \tag{56}$$

$$\lambda_\varepsilon = (\lambda_{\varepsilon 0}, \ldots, \lambda_{\varepsilon N}), \tag{57}$$

$$\lambda_I = (\lambda_{I0}, \ldots, \lambda_{I N}). \tag{58}$$

In order to reveal the scenario-local structure of the Hessian (22), the variables corresponding to the same contingency are grouped together, i.e.,

$$u_c = (x_c, \lambda_{\varepsilon c}, \lambda_{I c}), \tag{59}$$

$$u_n = (x_n, \lambda_{\varepsilon n}, \lambda_{I n}), \tag{60}$$

and, thus, the global ordering will be

$$u = (u_0, \ldots, u_{N_c}, u_g), \tag{61}$$

$$u = (u_0, \ldots, u_N, u_g), \tag{62}$$

where the coupling variables u_g are placed at the end of the new optimization vector u. Coupling in the SCOPF problem, $u_g = x_g$, is introduced by the two nonanticipatory constraints (48g) and (48h). The coupling in a case of the MPOPF problem, $u_g = \lambda_A$, is introduced by the linear energy constraints (49g). Under the new orderings (61) and (62), the Hessian matrix of the system (22) obtains the arrowhead structure (also described as bordered block-diagonal [7] or dual block-angular [16]) structure, as illustrated in Figs. 7 and 8,

$$\begin{pmatrix} A_0 & & & & B_0^\mathsf{T} \\ & A_1 & & & B_1^\mathsf{T} \\ & & \ddots & & \vdots \\ & & & A_{N_c} & B_{N_c}^\mathsf{T} \\ B_0 & B_1 & \ldots & B_{N_c} & C \end{pmatrix} \begin{pmatrix} \Delta u_0 \\ \Delta u_1 \\ \vdots \\ \Delta u_n \\ \Delta u_g \end{pmatrix} = \begin{pmatrix} b_0 \\ b_1 \\ \vdots \\ b_n \\ b_C \end{pmatrix}, \tag{63}$$

where the block matrices A_i,

$$A_i = \begin{pmatrix} \tilde{H}_{x_i,x_i} & J_{\varepsilon_i,x_i}^\mathsf{T} & J_{\mathcal{I}_i,x_i}^\mathsf{T} \\ J_{\varepsilon_i,x_i} & 0 & 0 \\ J_{\mathcal{I}_i,x_i} & 0 & -L_{s_i}^{-1} \end{pmatrix}, \tag{64}$$

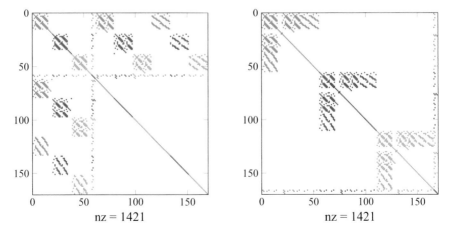

Fig. 7 Symmetrized SCOPF system (22) permuted to the arrowhead structure (63)

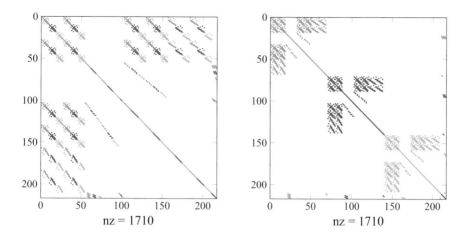

Fig. 8 Symmetrized MPOPF system (22) permuted to the arrowhead structure (63)

$$
A_i = \begin{pmatrix} \tilde{H}_{x_i,x_i} & J^{\mathsf{T}}_{\varepsilon_i,x_i} & J^{\mathsf{T}}_{I_i,x_i} & 0 \\ J_{\varepsilon_i,x_i} & 0 & 0 & 0 \\ J_{I_i,x_i} & 0 & -L^{-1}_{s_i} & 0 \\ 0 & 0 & 0 & L_{A_i} \end{pmatrix}, \tag{65}
$$

incorporate the Hessian of the Lagrangian with respect to the scenario-local variables $\tilde{H}_{x_i,x_i} = \nabla^2_{x_i x_i}\mathcal{L} + X_i^{-1}Z_i$ and the Jacobians of the constraints for the ith scenario with respect to the local variables $J_{\varepsilon_i,x_i} = \nabla_{x_i}c_{\varepsilon i}$ and $J^{\mathsf{T}}_{I_i,x_i} = \nabla_{x_i}c_{I i}$, as well as the diagonal entries corresponding to the eliminated slack variables. In the case of the SCOPF problem, the block $C = \nabla^2_{x_g x_g}\mathcal{L} + X_g^{-1}Z_g$ contains Hessian of the Lagrangian with respect to the coupling variables x_g, while in the case of the MPOPF problem it is a block of zeros. The off-diagonal blocks in the arrowhead SCOPF system are

$$
B_i = \begin{pmatrix} \tilde{H}_{x_g,x_i} \\ J^{\mathsf{T}}_{\varepsilon_i,x_g} \\ J^{\mathsf{T}}_{I_i,x_g} \end{pmatrix}^{\mathsf{T}}, \quad B_i^{\mathsf{T}} = \begin{pmatrix} \tilde{H}_{x_i,x_g} \\ J_{\varepsilon_i,x_g} \\ J_{I_i,x_g} \end{pmatrix}, \tag{66}
$$

where $\tilde{H}_{x_i,x_g} = \nabla^2_{x_i x_g}\mathcal{L}$ represents the off-diagonal blocks of the Hessian of Lagrangian with respect to the local and coupling variables and $J_{\varepsilon_i,x_g} = \nabla_{x_g}c_{\varepsilon i}$ and $J_{I_i,x_g} = \nabla_{x_g}c_{I i}$ are the Jacobians of the ith scenario with respect to the coupling variables. The MPOPF coupling matrices $B_1, B_2, \ldots, B_N \in \mathbb{R}^{N N_S \times N_A}$, where N_A is the size of the diagonal blocks in (63), contain the constant subblocks, which arise from the particular form of the linear constraints (49g)

$$B_1 = \begin{pmatrix} C_1 \\ C_0 \\ C_0 \\ \vdots \\ C_0 \end{pmatrix}, B_2 = \begin{pmatrix} 0 \\ C_1 \\ C_0 \\ \vdots \\ C_0 \end{pmatrix}, \ldots, B_N = \begin{pmatrix} 0 \\ 0 \\ \vdots \\ 0 \\ C_1 \end{pmatrix}. \tag{67}$$

4.2 Schur Complement Decomposition

The direct factorization of the full KKT system is not feasible for large-scale SCOPF problems due to their growing size with the number of contingencies and associated factorization fill-in that quickly exhausts the available memory. Instead, the solution is obtained by a sequence of partial block elimination steps, which are decoupled, aiming to form the Schur complement of the system. This way, we detour the factorization of the full KKT system, by factorizing only the smaller diagonal blocks as described in the Algorithm 1. At the first step, the Schur complement S is formed,

$$S = C - \sum_{i=0}^{N_c} B_i A_i^{-1} B_i^{\mathsf{T}}, \tag{68}$$

which in the general case becomes a dense matrix. Because the size of the coupling stays constant, independently of the number of contingency scenarios, the size of the Schur complement does not increase with an increasing number of contingencies. It can therefore be solved using dense LDL^{T} factorization and back substitution algorithms. The solution of the dense Schur system,

$$S\Delta u_g = b_C - \sum_{i=0}^{N_c} B_i A_i^{-1} b_i, \tag{69}$$

yields a part of the solution corresponding to the coupling variables Δu_g, which is used to obtain all the local solutions Δu_i by solving

$$A_i \Delta u_i = b_i - B_i^{\mathsf{T}} \Delta u_g. \tag{70}$$

Since the block contributions to the Schur $B_i A_i^{-1} B_i^{\mathsf{T}}$ complement are independent, they can be evaluated in parallel, as well as the residuals $B_i A_i^{-1} b_i$ and the solution Δu_i can be computed independently at each process. Interprocess communication occurs because the local Schur complement contributions and Schur complement residuals need to be assembled by the *master* process, and during the broadcast of the Schur complement solution to the remaining processes.

   In the description of Algorithm 1, sequential steps such as reduction and broadcast are performed only by the *master* process.

Algorithm 1 Parallel procedure for solving the linear systems based on the Schur complement decomposition (68)–(70)

Input: KKT system with arrowhead structure (63), right-hand side b
Output: Δu
1: Distribute blocks from the KKT system (63) evenly across $\mathcal{P}$ processes, where $\mathcal{N}_p$ is the set of diagonal blocks assigned to process $p \in \mathcal{P}$
2: Factorize $A_i = L_i D_i L_i^\mathsf{T}$ for each $i \in \mathcal{N}_p$
3: Compute $S_i = B_i A_i^{-1} B_i^\mathsf{T}$ for each $i \in \mathcal{N}_p$
4: Accumulate $C_p = \sum_{i \in \mathcal{N}_p} S_i$
5: **if** *master* **then**
6:     Reduce $S = C - \sum_{p \in \mathcal{P}} C_p$
7: **end if**
8: Compute $r_i = B_i A_i^{-1} b_i$ for each $i \in \mathcal{N}_p$
9: Accumulate $r_p = \sum_{p \in \mathcal{P}} r_i$
10: **if** *master* **then**
11:     Reduce $r = \sum_{p \in \mathcal{P}} r_p$
12:     Factorize $S = L_s D_s L_s^\mathsf{T}$
13:     Solve $S \Delta u_g = b_C - r$
14:     Broadcast solution u_g to all $p \in \mathcal{P}$
15: **end if**
16: Solve $A_i \Delta u_i = B_i u_g - b_i$ for each $i \in \mathcal{N}_p$

Remark 3 One should bear in mind that the computational efficiency obtained by exploiting the block-diagonal structure, such as (63), is determined by the number of the coupling variables $|u_g|$. If coupling is large, then the Schur decomposition will not be efficiently compared to the direct factorization techniques because of the cubic complexity of dense factorizations (69).

The most expensive step of the presented computational scheme is evaluation of the local contributions to the Schur complement $B_i A_i^{-1} B_i^\mathsf{T}$ in (68). The standard approach uses a direct sparse solver, such as PARDISO [16], to factorize the symmetric matrix $A_i = L_i D_i L_i^\mathsf{T}$ and perform multiple forward-backward substitutions with all right-hand side (RHS) vectors in B_i^T, followed by multiplication from the left by B_i. This approach, however, does not exploit sparsity of the problem in B_i^T blocks, since the linear solver treats the RHS vectors as being dense.

An alternative approach, implemented in PARDISO [17], addresses these limitations by performing an incomplete factorization of the augmented matrix M_i:

$$M_i = \begin{pmatrix} A_i & B_i^\mathsf{T} \\ B_i & 0 \end{pmatrix}, \tag{71}$$

exploiting also the sparsity of B_i^T. The factorization of M_i is stopped after pivoting reaches the last diagonal entry of A_i. At this point, the term $-B_i A_i^{-1} B_i^\mathsf{T}$ is computed and resides in the (2, 2) block of M_i. By exploiting the sparsity not only in A_i, but also in B_i it is possible to reduce memory traffic by using in-memory

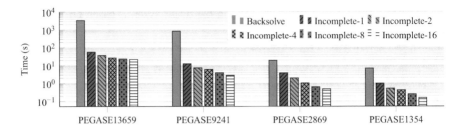

Fig. 9 Incomplete factorization of the augmented matrix

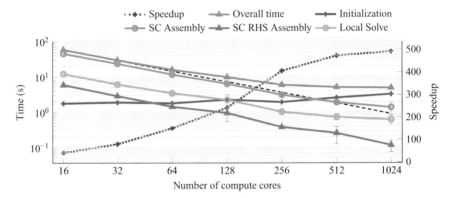

Fig. 10 Scaling of the parallel approach using the PEGASE1354-4096 benchmark and the speedup with respect to the direct sparse solver

sparse matrix compression techniques, which render this approach quite favorable for multicore parallelization.

In Fig. 9 we compare the standard, so-called backsolve, technique and the multicore incomplete factorization with increasing number of cores is shown for various benchmarks. This demonstrates that the incomplete factorization approach is orders of magnitude faster, especially for the large problems. Due to the extensive memory requirements for storing the RHS vectors in the "backsolve" approach, only its single-core execution is demonstrated.

We evaluated the strong scaling efficiency of the distributed solver on the "Piz Daint" supercomputer, using an increasing number of compute cores on the distributed compute nodes. The instance of the solved problem contained up to $1.1 \cdot 10^7$ variables and $2.7 \cdot 10^7$ constraints and the size of the KKT system is $5.48 \cdot 10^7$. Figure 10 shows the average wall time of the individual phases of Algorithm 1, indicating also the ideal strong scaling of the overall time. The algorithmic phases presented are the initialization phase, assembly of the Schur complement using the incomplete factorization of the augmented matrix in steps 2–6, RHS vector assembly and Schur complement solution in steps 12–13, and solutions of the local parts of the system in steps 14–16. Figure 10 also demonstrates the speedups of the distributed solver compared to the serial direct factorization. The benchmarks were run with a single MPI process per node and 16 threads per process.

The distributed approach using a single process outperforms the sequential direct factorization by a factor of up to 40×. With an increasing number of distributed nodes the observed speedup was up to 500×. The distributed solution time scales reasonably up to 512 cores at 32 compute nodes, which in terms of workload translates to 128 scenarios per node of PEGASE1354 benchmark. At this point, the most expensive part of the algorithm, the computation of the local contributions to the Schur complement, requires approximately the same time as the initialization phase, where the KKT system is distributed to all available compute nodes. The acceleration and efficiency of the structure-exploiting algorithm stems from the reduced complexity associated with the factorization of the smaller sparse diagonal blocks when applying the Schur decomposition scheme to the permuted KKT system with the arrowhead structure (63), as opposed to factorizing the original SCOPF KKT system (17) or its reduced variant (22). For sufficiently large power grids, however, the dense Schur complement (SC) system might become very large, and dominate the overall processing time in steps 12 and 13. Hardware accelerators such as GPUSs might be deployed to address the computational complexity of the dense linear algebra. Otherwise, the dimensions of the dense systems remain feasible for the majority of power grids, since the dimensions depend only on the power grid properties, not on the number of contingency scenarios.

4.3 Structure-Exploiting Algorithms for MPOPF

For the MPOPF problems, the size of the dense SC grows very quickly, not only with the size of the network but also proportionally to the number of installed storage devices and the number of time periods $N N_S$. As the number of time periods N or storage devices N_S increases, the solution approach based on Algorithm 1 results in a less efficient algorithm than the direct sparse approach employing PARDISO on the original KKT system (16), both with respect to computational time and memory consumption despite the benefits of the Schur decomposition. However, the MPOPF problem, unlike the SCOPF problem, can be optimized even further by exploiting the particular structure of the off-diagonal blocks B_n.

Inspecting the particular structure of the blocks B_n (67), one can see that the SC matrix computed by (71) for the nth block $S_n = -B_n A_n^{-1} B_n^{\mathsf{T}}$ has the structure

$$S_n = \begin{pmatrix} O_n & 0_n^{\mathsf{T}} & 0_n^{\mathsf{T}} & \cdots & 0_n^{\mathsf{T}} \\ 0_n & S_{11,n} & S_{10,n}^{\mathsf{T}} & \cdots & S_{10,n}^{\mathsf{T}} \\ 0_n & S_{10,n} & S_{00,n} & \cdots & S_{00,n}^{\mathsf{T}} \\ \vdots & \vdots & \vdots & \ddots & \vdots \\ 0_n & S_{10,n} & S_{00,n} & \cdots & S_{00,n} \end{pmatrix}, \tag{72}$$

where the $\mathbf{0}_n \in \mathbb{R}^{N_S \times (n-1)N_S}$, $\mathbf{O}_n \in \mathbb{R}^{(n-1)N_S \times (n-1)N_S}$, and $\mathbf{S}_{ij,n} = -\mathbf{C}_i \mathbf{A}_n^{-1} \mathbf{C}_j^\mathsf{T}$, $i, j \in \{0, 1\}$. The only blocks in $\mathbf{S}_n$ that are distinct are colored in blue and form the entries of the 2 by 2 block matrix

$$\bar{\mathbf{S}}_n = \begin{pmatrix} \mathbf{S}_{11,n} & \mathbf{S}_{10,n}^\mathsf{T} \\ \mathbf{S}_{10,n} & \mathbf{S}_{00,n} \end{pmatrix}, \tag{73}$$

where the rest of the rows and columns of $\mathbf{S}_n$ are direct replicates of the entries of the last row and column of $\bar{\mathbf{S}}_n$.

Since each one of the blocks of $\bar{\mathbf{S}}_n$ has size $N_S \times N_S$, the computation of $\mathbf{S}_n$ becomes independent of the number of time periods N and only depends on the number of storage devices N_S. It is easily verified that the global SC $\mathbf{S}_c$ obtains the form

$$\mathbf{S}_c = \begin{pmatrix} \mathbf{S}_{11} & \mathbf{S}_{12}^\mathsf{T} & \mathbf{S}_{12}^\mathsf{T} & \cdots & \mathbf{S}_{12}^\mathsf{T} \\ \mathbf{S}_{12} & \mathbf{S}_{22} & \mathbf{S}_{23}^\mathsf{T} & \cdots & \mathbf{S}_{23}^\mathsf{T} \\ \mathbf{S}_{12} & \mathbf{S}_{23} & \mathbf{S}_{33} & \cdots & \mathbf{S}_{34}^\mathsf{T} \\ \vdots & \vdots & \vdots & \ddots & \vdots \\ \mathbf{S}_{12} & \mathbf{S}_{23} & \mathbf{S}_{34} & \cdots & \mathbf{S}_{NN} \end{pmatrix}, \tag{74}$$

where each block of $\mathbf{S}_c \in \mathbb{R}^{N N_S \times N N_S}$ has dimensions $N_S \times N_S$. Storing $\mathbf{S}_c$ due to its special structure requires only two block vectors: one for all diagonal blocks $\mathbf{S}_d = [\mathbf{S}_{11}, \mathbf{S}_{22}, \cdots, \mathbf{S}_{NN}]$ of size $N_S \times N N_S$, and one for the off-diagonal blocks $\mathbf{S}_o = [\mathbf{S}_{12}, \mathbf{S}_{23}, \cdots, \mathbf{S}_{N-1N}]$ of size $N_S \times (N-1)N_S$, significantly reducing this way the storage requirements for $\mathbf{S}_c$. Furthermore, exploiting the fact that the blocks below the main diagonal of each column of $\mathbf{S}_c$ in (74) are identical, we can perform the factorization in $O(n^2)$ operations instead of $O(n^3)$, which is the case for standard dense $\mathbf{LDL}^\mathsf{T}$ factorization of $\mathbf{S}_c$ with $n = N N_S$. Similarly, the back substitution can be performed in $O(n)$ instead of $O(n^2)$. The reduction in the computational complexity and storage requirements of the SC system renders the overall approach significantly more economical in terms of overall running time and memory footprint, as demonstrated in Fig. 11.

For comparison, we also consider three alternative optimization algorithms that also adopt an IP strategy, namely IPOPT [22, 23], MIPS [24], and KNITRO [2]. The structure exploiting IP algorithm introduced in this section is referred to as BELTISTOS. The average time per iteration for $N = 3600$ up to $N = 8760$ corresponding to 1 year with a time step size corresponding to one hour is shown in Fig. 11a. For this set of benchmarks KNITRO needed more than 1 TB of memory for $N \geq 5760$ and it terminated with a related error message. For $N = 8760$ PARDISO failed due to overflow of the number of nonzero entries in the $\mathbf{L}, \mathbf{D}$ factors. It is worth noting that BELTISTOSmem (the memory saving approach of BELTISTOS that implements Algorithm 1 without storing the factors of the blocks $\mathbf{A}_i$ and computing them on the fly in steps 2, 8, and 16), although it is slightly slower than the normal mode of BELTISTOS, it is still almost four orders of magnitude faster than IPOPT and

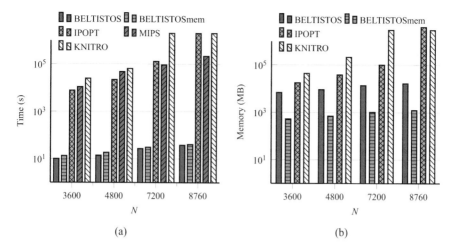

Fig. 11 Case IEEE118. Statistics for solving the KKT system. (**a**) Average time per iteration. (**b**) Memory allocated

MIPS. It also needs approximately two orders of magnitude less memory than IPOPT as it is shown in Fig. 11b, where we plot the memory (in MB) allocated by each algorithm for the solution of the KKT system. The MIPS and KNITRO solvers do not report the memory allocated and it could only be estimated for the case of KNITRO.

5 Results and Discussion

This study demonstrates that significant performance gains are possible, for specific classes of optimal control problems, not by exploiting supercomputers and parallel distributed or multithreaded programming, but through deeper understanding of the problem structure and the design of algorithms adapted to the problem structure. Orders of magnitude of faster execution time and orders of magnitude of memory savings were achieved rendering the solution of very-large-scale problems, previously intractable without a supercomputer, possible on a common laptop [11].

The Schur decomposition enables low memory SC assembly on a per-block basis, whenever a problem can be reordered to an arrowhead structure, which is the case for many real life problems composed of enumerated subproblems, such as contingency scenarios for SCOPF problems or time periods for MPOPF problems, while at the same time promoting parallel processing. Even on single-core execution for SCOPF problems, speedups from 40–270-fold were observed while further exploitation of distributed multicore and many-core computing environments for the solution of the structured KKT system drastically reduces the execution times and demonstrates significant progress towards the solution of large-scale SCOPF problems.

In contrast to SCOPF and although MPOPF problems can be reordered into an arrowhead structure, the reordering results in a dense SC that grows in size with the number of time periods and does not necessarily lead to a more efficient solution strategy. However, owing to the intrinsic structure of the linear constraints, the Schur decomposition algorithm supplemented with elimination strategies exploiting data compression, resulted in an overall solution strategy of unprecedented performance. Memory was reduced by approximately two orders of magnitude, while runtime performance still remains about three orders of magnitude higher than competitors, even on a single core.

Our findings strongly motivate further structural inspection and analysis of the present and similar problems of the same family, anticipating that adopting and extending the presented structure-exploiting techniques for other problems would result in significant acceleration of other OPF problems of interest paving the way for the next generation of OPF algorithms.

References

1. R. H. Byrd, G. Liu, and J. Nocedal. On the local behavior of an interior point method for nonlinear programming. In *Numerical Analysis 1997*, pages 37–56. Addison Wesley Longman, 1998.
2. R. H. Byrd, J. Nocedal, and R. A. Waltz. *Knitro: An Integrated Package for Nonlinear Optimization*, pages 35–59. Springer US, Boston, MA, 2006.
3. N. Chiang, C. G. Petra, and V. M. Zavala. Structured nonconvex optimization of large-scale energy systems using PIPS-NLP. In *2014 Power Systems Computation Conference*, pages 1–7, Aug 2014.
4. N.-Y. Chiang and V. M. Zavala. An inertia-free filter line-search algorithm for large-scale nonlinear programming. *Computational Optimization and Applications*, 64(2):327–354, Jun 2016.
5. A. R. Conn, N. I. M. Gould, D. Orban, and P. L. Toint. A primal-dual trust-region algorithm for non-convex nonlinear programming. *Mathematical Programming*, 87(2):215–249, Apr 2000.
6. T. Demiray and G. Andersson. Optimization of numerical integration methods for the simulation of dynamic phasor models in power systems. *International Journal of Electrical Power & Energy Systems*, 31(9):512–521, 2009. Power Systems Computation Conference (PSCC) 2008.
7. I. S. Duff and J. A. Scott. Stabilized bordered block diagonal forms for parallel sparse solvers. *Parallel Comput.*, 31(3–4):275–289, Mar. 2005.
8. R. Fletcher and S. Leyffer. Nonlinear programming without a penalty function. *Mathematical Programming*, 91(2):239–269, Jan 2002.
9. N. I. M. Gould, D. Orban, A. Sartenaer, and P. L. Toint. Superlinear convergence of primal-dual interior point algorithms for nonlinear programming. *SIAM Journal on Optimization*, 11(4):974–1002, 2001.
10. N. Karmarkar. A new polynomial-time algorithm for linear programming. *Combinatorica*, 4(4):373–395, Dec 1984.
11. D. Kourounis, A. Fuchs, and O. Schenk. Toward the next generation of multiperiod optimal power flow solvers. *IEEE Transactions on Power Systems*, 33(4):4005–4014, July 2018.
12. S. Mehrotra. On the implementation of a primal-dual interior point method. *SIAM Journal on Optimization*, 2(4):575–601, 1992.

13. A. Monticelli, M. V. F. Pereira, and S. Granville. Security-constrained optimal power flow with post-contingency corrective rescheduling. *IEEE Transactions on Power Systems*, 2(1):175–180, Feb 1987.
14. J. Nocedal, A. Wächter,, and R. A. Waltz. Adaptive barrier update strategies for nonlinear interior methods. *SIAM Journal on Optimization*, 19(4):1674–1693, 2009.
15. J. Nocedal and S. J. Wright. *Numerical Optimization*. Springer, New York, NY, USA, second edition, 2006.
16. C. G. Petra, O. Schenk, and M. Anitescu. Real-time stochastic optimization of complex energy systems on high-performance computers. *Computing in Science Engineering*, 16(5):32–42, Sept 2014.
17. C. G. Petra, O. Schenk, M. Lubin, and K. Gärtner. An augmented incomplete factorization approach for computing the Schur complement in stochastic optimization. *SIAM Journal on Scientific Computing*, 36(2):C139–C162, 2014.
18. D. T. Phan and X. A. Sun. Minimal impact corrective actions in security-constrained optimal power flow via sparsity regularization. *IEEE Transactions on Power Systems*, 30(4):1947–1956, July 2015.
19. M. Schanen, F. Gilbert, C. G. Petra, and M. Anitescu. Toward multiperiod ac-based contingency constrained optimal power flow at large scale. In *2018 Power Systems Computation Conference (PSCC)*, pages 1–7, June 2018.
20. O. Schenk, K. Gärtner, W. Fichtner, and A. Stricker. PARDISO: a high-performance serial and parallel sparse linear solver in semiconductor device simulation. *Future Generation Computer Systems*, 18(1):69–78, 2001. I. High Performance Numerical Methods and Applications. II. Performance Data Mining: Automated Diagnosis, Adaption, and Optimization.
21. A. Wächter. *An interior point algorithm for large-scale nonlinear optimization with applications in process engineering*. PhD thesis, Carnegie Mellon University, http://researcher.watson. ibm.com/researcher/files/us-andreasw/thesis.pdf, 2003.
22. A. Wächter and L. T. Biegler. Line search filter methods for nonlinear programming: motivation and global convergence. *SIAM J. Optim.*, 16(1):1–31 (electronic), 2005.
23. A. Wächter and L. T. Biegler. On the implementation of an interior-point filter line-search algorithm for large-scale nonlinear programming. *Math. Program.*, 106(1, Ser. A):25–57, 2006.
24. H. Wang, C. E. Murillo-Sanchez, R. D. Zimmerman, and R. J. Thomas. On computational issues of market-based optimal power flow. *IEEE Trans. on Power Syst.*, 22(3):1185–1193, Aug 2007.
25. S. Wright. *Primal-Dual Interior-Point Methods*. Society for Industrial and Applied Mathematics, 1997.
26. R. Zimmerman and C. Murillo-Sanchez. *Matpower 6.0 User's manual*. Power Systems Engineering Research Center, 2016.
27. R. D. Zimmerman, C. E. Murillo-Sanchez, and R. J. Thomas. MATPOWER: Steady-state operations, planning, and analysis tools for power systems research and education. *IEEE Transactions on Power Systems*, 26(1):12–19, Feb 2011.

Parallel Hybrid Sparse Linear System Solvers

Murat Manguoğlu, Eric Polizzi, and Ahmed H. Sameh

1 Introduction

Some science and engineering applications give rise to large banded linear systems in which the bandwidth is a very small percentage of the system size. Often, these systems arise in the inner-most computational loop of these applications which indicates that these systems need to be solved efficiently as fast as possible on parallel computing platforms. This motivated the development of the earliest version of the SPIKE tridiagonal linear systems in the late 1970s, e.g. see [27] followed by an investigation of the communication complexity of this solver in [12] in 1984. In both of these studies this solver was not named "SPIKE" until it was further developed in [23, 24] in 2006. In this chapter, we also present an extension of this algorithm for solving sparse linear systems. This is done through reordering schemes that bring as many of the heaviest off-diagonal elements as closer to the main diagonal, followed by extracting effective preconditioners (that encapsulate as many of these heaviest elements as possible) for outer Krylov subspace methods for

M. Manguoğlu (✉)
Department of Computer Engineering, Middle East Technical University, Ankara, Turkey
e-mail: manguoglu@ceng.metu.edu.tr

E. Polizzi (✉)
Department of Electrical and Computer Engineering, University of Massachusetts at Amherst, Amherst, MA, USA

Department of Mathematics and Statistics, University of Massachusetts at Amherst, Amherst, MA, USA
e-mail: polizzi@ecs.umass.edu

A. H. Sameh
Department of Computer Science, Purdue University, West Lafayette, IN, USA
e-mail: sameh@cs.purdue.edu

© Springer Nature Switzerland AG 2020
A. Grama, A. H. Sameh (eds.), *Parallel Algorithms in Computational Science and Engineering*, Modeling and Simulation in Science, Engineering and Technology,
https://doi.org/10.1007/978-3-030-43736-7_4

solving these sparse systems. In each outer iteration variants of the SPIKE algorithm are used for solving linear systems involving these preconditioners. An extensive survey of the SPIKE algorithm and its extensions are given in [10].

2   The SPIKE for Banded Linear Systems (Dense Within the Band)

Consider the nonsingular banded linear system

$$Ax = f \tag{1}$$

shown in Fig. 1 with $A \in \mathbb{R}^{N \times N}$ being of bandwidth $\beta = 2m + 1$. Let N be an integer multiple of p (the number of partitions). In Fig. 1, p is chosen as 4. The off-diagonal blocks are given by

$$\bar{B}_j = \begin{pmatrix} 0 & 0 \\ B_j & 0 \end{pmatrix} \text{ and } \bar{C}_j = \begin{pmatrix} 0 & C_j \\ 0 & 0 \end{pmatrix} \tag{2}$$

for $j = 1, 2, \ldots, p - 1$, where $B_j, C_j \in \mathbb{R}^{m \times m}$.

In what follows, we first describe "Spike" as a direct banded solver when A is diagonally dominant followed by the general case for which "Spike" becomes a hybrid (direct-iterative) banded solver for the linear system (1).

$$\underbrace{\begin{pmatrix} A_1 & \bar{B}_1 & & \\ \bar{C}_2 & A_2 & \bar{B}_2 & \\ & \bar{C}_3 & A_3 & \bar{B}_3 \\ & & \bar{C}_4 & A_4 \end{pmatrix}}_{A} \underbrace{\begin{pmatrix} x_1 \\ x_2 \\ x_3 \\ x_4 \end{pmatrix}}_{x} = \underbrace{\begin{pmatrix} f_1 \\ f_2 \\ f_3 \\ f_4 \end{pmatrix}}_{f}$$

First, let A be a diagonally dominant matrix. Thus, each A_j is also diagonally dominant, $j = 1, 2, \ldots, p$, with the block diagonal matrix

$$\underbrace{\begin{pmatrix} A_1 & \bar{B}_1 & & \\ \bar{C}_2 & A_2 & \bar{B}_2 & \\ & \bar{C}_3 & A_3 & \bar{B}_3 \\ & & \bar{C}_4 & A_4 \end{pmatrix}}_{A} \underbrace{\begin{pmatrix} x_1 \\ x_2 \\ x_3 \\ x_4 \end{pmatrix}}_{x} = \underbrace{\begin{pmatrix} f_1 \\ f_2 \\ f_3 \\ f_4 \end{pmatrix}}_{f}$$

Fig. 1 $A \in \mathbb{R}^{N \times N}$; bandwidth: $\beta = 2m + 1$; number of partitions: $p = 4$, $A_j \in \mathbb{R}^{n \times n}$, $j = 1, 2, \ldots, p$; $\bar{B}_j, \bar{C}_{j+1} \in \mathbb{R}^{n \times n}$, $j = 1, 2, \ldots, p - 1$; $N = 4n$; $n = 3m$

Fig. 2 $p = 4, n = 3m,$
$N = 4n$

$$\begin{pmatrix} I & \bar{V}_1 & & \\ \bar{W}_2 & I & \bar{V}_2 & \\ & \bar{W}_3 & I & \bar{V}_3 \\ & & \bar{W}_4 & I \end{pmatrix} \begin{pmatrix} x_1 \\ x_2 \\ x_3 \\ x_4 \end{pmatrix} = \begin{pmatrix} g_1 \\ g_2 \\ g_3 \\ g_4 \end{pmatrix}$$

$$\underbrace{\hspace{3cm}}_{S} \quad \underbrace{\hspace{0.8cm}}_{x} \quad \underbrace{\hspace{0.8cm}}_{g}$$

$$D = \begin{pmatrix} A_1 & & & \\ & A_2 & & \\ & & \ddots & \\ & & & A_p \end{pmatrix} \tag{3}$$

nonsingular. Premultiplying both sides of (1) by D^{-1}, we obtain the modified system

$$Sx = g, \tag{4}$$

where the matrix $S = D^{-1}A$, and the updated right-hand side is given in which the bandwidth is very small percentage of the system size. Often, these updated right-hand side are given in Fig. 2. Here, the off-diagonal blocks, $\bar{V}_j$ and $\bar{W}_j$, are given by

$$\bar{V}_j = (V_j, 0) \text{ and } \bar{W}_{j+1} = (0, W_{j+1}), j = 1, 2, \ldots, p - 1. \tag{5}$$

$$\begin{pmatrix} I & \bar{V}_1 & & \\ \bar{W}_2 & I & \bar{V}_2 & \\ & \bar{W}_3 & I & \bar{V}_3 \\ & & \bar{W}_4 & I \end{pmatrix} \begin{pmatrix} x_1 \\ x_2 \\ x_3 \\ x_4 \end{pmatrix} = \begin{pmatrix} g_1 \\ g_2 \\ g_3 \\ g_4 \end{pmatrix}$$

$$\underbrace{\hspace{3cm}}_{S} \quad \underbrace{\hspace{0.8cm}}_{x} \quad \underbrace{\hspace{0.8cm}}_{g}$$

The spikes,

$$V_j = \begin{pmatrix} V_3^{(j)} \\ V_2^{(j)} \\ V_1^{(j)} \end{pmatrix} \text{ and } W_j = \begin{pmatrix} W_1^{(j)} \\ W_2^{(j)} \\ W_3^{(j)} \end{pmatrix} \tag{6}$$

are obtained by solving the linear systems

$$A_j^{-1}[\hat{C}_j, A_j, \hat{B}_j] = [W_j, I_n, V_j] \tag{7}$$

and

$$g_j = A_j^{-1} f_j, j = 1, 2, \ldots, p \tag{8}$$

in which

$$\hat{B}_j = \begin{pmatrix} 0 \\ I_m \end{pmatrix} B_j; \hat{C}_j = \begin{pmatrix} I_m \\ 0 \end{pmatrix} C_j \tag{9}$$

with $V_j, W_j \in \mathbb{R}^{n \times m}$. Solving the linear system in (4), involving the "Spike matrix" S, reduces to solving a much smaller block-tridiagonal system of order $2m(p-1)$ of the form (See Fig. 2),

$$\begin{pmatrix} I_m & V_1^{(1)} & 0 & 0 & & & \\ W_1^{(2)} & I_m & 0 & V_3^{(2)} & & & \\ W_3^{(2)} & 0 & I_m & V_1^{(2)} & 0 & 0 & \\ 0 & 0 & W_1^{(3)} & I_m & 0 & V_3^{(3)} & \\ & & W_3^{(3)} & 0 & I_m & V_1^{(3)} & \\ & & 0 & 0 & W_1^{(4)} & I_m \end{pmatrix} \begin{pmatrix} x_1^{(b)} \\ x_2^{(t)} \\ x_2^{(b)} \\ x_3^{(t)} \\ x_3^{(b)} \\ x_4^{(t)} \end{pmatrix} = \begin{pmatrix} g_1^{(b)} \\ g_2^{(t)} \\ g_2^{(b)} \\ g_3^{(t)} \\ g_3^{(b)} \\ g_4^{(t)} \end{pmatrix} \tag{10}$$

in which $x_i^{(t)} = (I_m, 0)x_i$, and $x_i^{(b)} = (0, I_m)x_i$, and similarly for $g_i^{(t)}$ and $g_i^{(b)}$. We refer to (10) as the reduced system,

$$Ry = h, \tag{11}$$

where R results from the symmetric permutation

$$PSP^T = \begin{pmatrix} I_v & G \\ 0 & R \end{pmatrix} \tag{12}$$

in which $v = pn - 2m(p-1)$. Note that since A is nonsingular, so is S as well as R. Solving the reduced system (10) for $x_i^{(b)}$ and $x_{i+1}^{(t)}$, $i = 1, 2, \ldots, p-1$, the solution x of system (4) is retrieved directly via

$$x_1 = g_1 - V_1 x_2^{(t)}$$
$$x_i = g_i - W_i x_{i-1}^{(b)} - V_i x_{i+1}^{(t)}, i = 2, 3, \ldots, p-1 \tag{13}$$
$$x_p = g_p - W_p x_{p-1}^{(b)}.$$

In summary, the SPIKE algorithm for solving the diagonally dominant banded system $Ax = f$ consist of the D-S factorization scheme in which D is block diagonal and S is the corresponding spike matrix. Consequently, solving system (1) consists of two phases:

(i) solve $Dg = f$ followed by
(ii) solve $Sx = g$ via the reduced system approach.

In (i) each system $A_j g_j = f_j$ is solved via the classical LU-factorization as implemented in Lapack [2].

Observe that if we assign one processor (or one multicore node) to each partition, then solving $Dg = f$ realizes maximum parallelism with no interprocessor communications. Solving the reduced system (10), however, requires interprocessor communications which increases as the number of partitions p increases. The retrieval process (13) again achieves almost perfect parallelism.

Variations of this basic form of the SPIKE algorithm are given in [24]. Also note that this SPIKE algorithm requires a larger number of arithmetic operations than those required by the classical banded LU-factorization scheme. In spite of this higher arithmetic operation count, this direct form of the SPIKE algorithm realizes higher parallel performance than ScaLapack on an 8-core Intel processor, see Fig. 3 [17], due to enhanced data locality.

The reason for the superior performance of Spike is illustrated in Fig. 4 showing that the total number of off-chip data accessed (in bytes) for Spike (red color) is less than that required by ScaLapack. Also, Fig. 5 shows that the number of instructions executed by Spike is almost half that required by ScaLapack. For more details about the measurements shown in Figs. 4 and 5, see Liu et al. [13].

Second, if the banded linear system (1) is not diagonally dominant, there is no guarantee that any of the diagonal blocks A_j, $j = 1, 2, \ldots, p$, see Fig. 1, is nonsingular. In this case, the banded linear system (1) is solved via a preconditioned Krylov subspace method such as GMRES or BiCGStab, e.g. see Saad [25]. Here the preconditioner M is chosen as $M = \hat{D}\hat{S}$ where $\hat{D} = diag(\hat{A}_1, \hat{A}_2, \ldots, \hat{A}_p)$ with

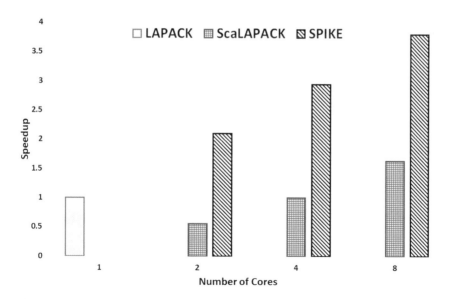

Fig. 3 Speedup of SPIKE and ScaLapack compared to the sequential Lapack for solving a linear system of size 960,000 with a bandwidth of 201

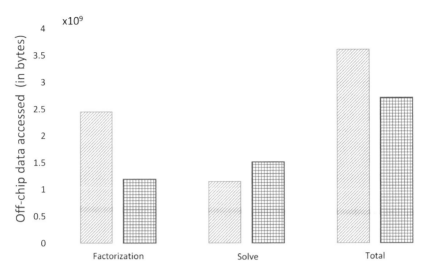

Fig. 4 Off-chip data being accessed for Spike (red) and ScaLapack (blue), using 4 cores

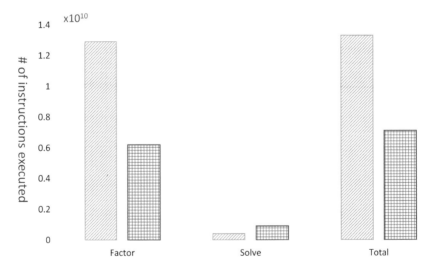

Fig. 5 Number of instructions executed by Spike (red) and ScaLapack (blue), using 4 cores

$\hat{A}_j = \hat{L}_j \hat{U}_j$, in which the factors $\hat{L}_j$ and $\hat{U}_j$ are obtained via the LU=factorization of each A_j using diagonal pivoting with a "boosting" strategy. In other words, if a diagonal element α during the factorization satisfies $|\alpha| \leq \varepsilon ||A||_1$ where ε is a multiple of the unit roundoff, then α is modified as follows:

$$\alpha := \alpha + \theta ||A_j||_1 \text{ if } \alpha \geq 0$$
$$\alpha := \alpha - \theta ||A_j||_1 \text{ if } \alpha < 0, \tag{14}$$

where $\theta \sim \sqrt{\varepsilon}$. $\hat{S}$ is then of a form identical to that of S, see Fig. 2, except that the spikes V_j and W_j are obtained as follows:

$$(\hat{L}_j\hat{U}_j)^{-1}[\hat{C}_j, \hat{B}_j] = [V_j, W_j], j = 1, 2, \ldots, p \tag{15}$$

which entails a block forward sweep followed by a block backsweep. In each iteration of GMRES, for example, one needs to solve a system of the form $Mv = r$. This is accomplished in two steps: (1) solve $\hat{D}u = r$, and (2) solve $\hat{S}v = u$. As outlined above, the first system is solved via two triangular solvers: $\hat{L}_j\dot{u}_j = r_j$, and $\hat{U}_ju_j = \dot{u}_j$, $j = 1, 2, \ldots, p$. The second system, $\hat{S}v = u$, is solved via the reduced system approach, see (10), with retrieving the rest of the solution vector v via (13).

Since the elements of the inverse of a banded matrix decay as they move away from the main diagonal, the elements of the spikes V, W decay as they move away from the main diagonal as well. Such decay becomes more pronounced as the degree of diagonal dominance increases. We define the degree of diagonal dominance of A by

$$\tau = \min_{1 \leq k \leq N} [|a_{kk}|/\sum_{k \neq j} |a_{kj}|]. \tag{16}$$

Even for system (1) for which $\tau \geq 0.25$, one can take advantage of the decay in the spikes V_j, $(\|V_q^{(j)}\| \ll \|V_1^{(j)}\|)$, and W_j, $(\|W_q^{(j)}\| \ll \|W_1^{(j)}\|)$ (In our example above $q = 3$). Taking advantage from such a property by replacing $V_q^{(j)}$ and $W_q^{(j)}$ by zero, the reduced system (10) becomes a block diagonal system that requires only obtaining $V_1^{(j)}$ and $W_1^{(j)}$, $j = 1, 2, \ldots, p$. In other words, we need only to obtain the bottom $(m \times m)$ tip of each right spike V_j, and the top $(m \times m)$ tip of each left spike W_j, $1 \leq j \leq p$. Consequently, if we assign enough processors to obtain the LU-factorization of slightly perturbed $A_1, A_2, \ldots, A_{p-1}$ using the diagonal boosting strategy, and the UL-factorization of similarly perturbed $A_2, A_3, \ldots, A_p$, we need not obtain the whole spikes V_j and W_j. The LU-factorizations will obtain the bottom tips of V_j, while the UL-factorizations will enable obtaining the top tips of W_j resulting in significant savings for computing the coefficient matrix R of the reduced system. Further, since R in this case is block diagonal, solving the reduced system achieves maximum parallelism. This "truncated" version of the SPIKE algorithm was compared with Lapack and MKL-ScaLapack (i.e., Intel's Math Kernel Library) on an Intel multicore processor for solving 8 banded linear systems with coefficient matrices obtained from Matrix-Market (see Table 1). Table 2 shows the ratios:

$$\frac{\text{Average time(MKL-2 cores*)}}{\text{Average time (MKL-1 core)}}, \tag{17}$$

and

Table 1 A Matrix-Market collection of banded systems ($n > 10,000$) where kl,ku, N, $\sim$Cond are the lower, upper bandwidths, matrix size, and the condition number estimate, respectively

Matrix name	kl	ku	N	$\sim$Cond
s3dkq4m2	614	614	90,449	N/A
s3dkt3m2	614	614	90,449	N/A
fidap035	244	247	19,716	4.3×10^{12}
e40r0000	451	451	17,281	2.2×10^{8}
e40r5000	451	451	17,281	2.2×10^{10}
bcsstk25	292	292	15,439	1.3×10^{13}
bcsstk18	1243	1243	11,948	6.5×10^{11}
bcsstk17	521	521	10,974	2.0×10^{10}

Table 2 Time ratios for Spike and MKL-ScaLapack

	MKL 1-core	MKL 2-cores	Spike 2-cores
Avg. time (ratio)	1.0	8.5	0.4
Rel. res. (norm)	$O(10^{-1})-O(10^{-10})$	$O(10^{-2})-O(10^{-10})$	$O(10^{-5})-O(10^{-11})$

$$\frac{\text{Average time(Spike-2 cores}^*)}{\text{Average time (MKL-1 core)}}, \tag{18}$$

together with the lowest and highest relative residual for each solver for the 8 benchmarks. *Note that for the 2-core entries each core belongs to a different node.

2.1 Multithreaded SPIKE

In shared memory systems, the parallelism in LAPACK LU algorithms can directly benefit from the threaded implementation of the low-level BLAS routines. In order to achieve further scalability improvement, however, it is necessary to move to a higher level of parallelism based on divide-and-conquer techniques. As a result, the OpenMP implementation of SPIKE on multithreaded systems [19, 31], is inherently better suited for parallelism than the traditional LAPACK banded LU solver. A recent stand-alone SPIKE-OpenMP solver (v1.0) [1] has been developed and released to the community.

Among the large number of variants available for SPIKE, the OpenMP solver was implemented using the recursive SPIKE algorithm [23, 24]. The latter consists of solving the reduced system (10) using SPIKE again but where the number of partitions had been divided by two. This process is repeated recursively until only two partitions are left (making the problem straightforward to solve). The SPIKE algorithm applied to two partitions is actually the kernel of recursive SPIKE, and from Fig. 3, we note the efficiency of 2×2 SPIKE which reaches a speedup of two using two partitions with two processors. The recursive SPIKE technique demonstrates parallel efficiency and is applicable to both diagonally and non-diagonally dominant systems. However, it was originally known for its lack of flexibility on distributed architectures since its application was essentially limited to a power of two number of processors. The scheme was then prone to potential waste of parallel resources when applied to shared memory systems using OpenMP [19]; for instance, if 63 cores were available, then only 32 would be effectively used by recursive SPIKE (i.e., the lowest nearest power of two). This limitation was overcome in [31] with the introduction of a new flexible threading scheme that can consider any number of threads. If the number of threads is not a power of two, some partitions are given two threads which, in turn, would benefit from the 2×2 SPIKE kernel. Load balancing is achieved by changing the size of each partition so that the computational costs of the large matrix operations on each partition are matched. This multithreaded SPIKE approach is then ideally suited for shared memory systems since optimized ratios between partition sizes can be tuned for a given system matrix and architecture, independently from user input [1]. Figure 6 demonstrates the efficiency of the scheme. The results show that the speedup performance of the new threaded recursive SPIKE is not limited to a power of two number of threads since the scalability keeps increasing with the number of threads. For example, at 30 threads the overall speed improvement increases from roughly $\times 6$ to roughly $\times 9$, as a result of the increased overall utilization of resources. The results also show that the SPIKE computation time is significantly superior to LAPACK Intel-MKL. We note that the two solvers' scaling performance are similar until 10 threads are reached, at which point SPIKE begins pulling away. Unlike SPIKE, parallelism performance of the inherently recursive serial LU approach used by MKL mainly relies on parallelism available via the BLAS which is rather poor for this matrix.

The SPIKE-openMP solver has been designed as an easy to use, "black-box" replacement to the standard LAPACK banded solver. In order to achieve near feature-parity with the standard LAPACK banded matrix solver, we add to SPIKE the feature known as transpose option, i.e. solve $A^T x = f$. Transpose solve

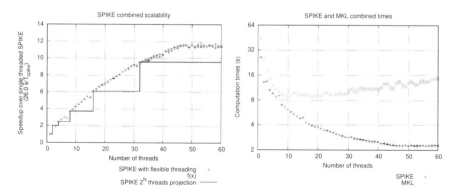

Fig. 6 SPIKE scalability and computation time compared to MKL-LAPACK for a system matrix of size N=1M, bandwidth 321, and with 160 right-hand sides

operation allows improved algorithmic flexibility and efficiency by eliminating the
need for an explicit factorization of the matrix transpose when solving:

$$A^T x = f. \tag{19}$$

As a result, if the factorization $A = DS$ is already available, it can now be used to
address the new SPIKE solve stages, which are now swapped:

1. solve $S^T y = f$ via the transposed reduced system approach, followed by
2. solve $D^T x = y$.

A transpose version of the recursive reduced system solver which has been proposed
in [31] achieves near performance parity with the non-transpose solver.

3 Hybrid Methods for General Sparse Linear Systems

In large-scale computational science and engineering application one is often faced
with solving large general sparse linear systems that cannot be reordered into a
narrow banded form. Therefore, we use nonsymmetric and symmetric reorderings
to maximize the magnitude of the product of the diagonal elements, move as many
of the largest off-diagonal elements as possible close to the main diagonal, and
extract a generalized banded preconditioner. In the next subsection we describe such
a reordering process after which an effective preconditioner can be extracted where
linear systems involving the such a preconditioner is solved using a variant of the
SPIKE algorithm.

3.1 Weighted Nonsymmetric and Symmetric Reorderings for Sparse Matrices

As the first step of the reordering scheme we apply a nonsymmetric permutation and
scaling (if needed) to make the diagonal of the coefficient matrix as large as possible.
Such nonsymmetric permutation and scaling techniques are already available in the
Harwell Subroutine Library (HSL) and is called MC64 [8] which (without scaling)
creates permutations Π_1 and Π_2 such that the magnitude of the product of the
diagonal elements of $B = \Pi_1 A \Pi_2$ is maximize, where A is the original coefficient
matrix. This is followed by obtaining a symmetric permutation of B, $C = PBP^T$,
where P is determined by the Fiedler vector [9] derived from B. The Fiedler vector
is the eigenvector corresponding to the second smallest eigenvalue of the "weighted
Laplacian" matrix based on B. This eigenvalue is sometimes called the algebraic
connectivity of the graph. Note that the smallest eigenvalue is zero. As a result, many
of the heaviest off-diagonal elements of C are much closer to the main diagonal.

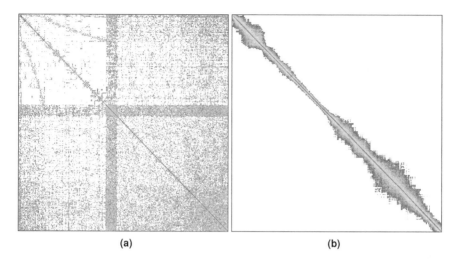

(a)   **(b)**

Fig. 7 The effect of the weighted spectral reordering using the Fiedler vector on $F2$ matrix (colors indicate the magnitude of the absolute value of the elements. Red, green, and blue are the largest, intermediate, and smallest elements, respectively. (**a**) Original matrix. (**b**) Reordered matrix

Here, PSPIKE refers to a solver that is a hybrid of the sparse direct solver Pardiso [29] and the SPIKE algorithm. In Fig. 7, we illustrate the effect of such reordering on a symmetric stiffness matrix with 71,505 rows and columns, obtained from the SuiteSparse Matrix Collection [6].

From the original sparse linear system $Ax = f$, we obtain $By = g$, where $y = \Pi_2^T x$, and $g = \Pi_1 f$. If B is symmetric, one can form the "weighted Laplacian" matrix, $L_{ij}^w = -|b_{ij}|$, and

$$L_{jj}^w = \sum_k |b_{kj}| \tag{20}$$

as follows: Note that one can obtain the unweighted Laplacian by simply replacing each nonzero element of the matrix B by 1. In this subsection, we consider the weighted case as a preprocessing tool for the PSPIKE algorithm given in Sect. 3.3.

We assume that the corresponding graph is connected since the disconnected components can be easily identified and the Fiedler vector can be computed independently for each if the graph is disconnected. The eigenvalues of L^w are $0 = \lambda_1 < \lambda_2 \leq \lambda_3 \leq \ldots \leq \lambda_n$. The Fiedler vector, x_F, is the eigenvector corresponding to smallest nontrivial eigenvalue, λ_2. Since we assume a connected graph, the trivial eigenvector x_1 is a vector of all ones. If the coefficient matrix, B, is nonsymmetric, we simply construct L^w using the elements of $(|B| + |B^T|)/2$, instead of those of $|B|$.

A Trace Minimization [26, 28] based parallel algorithm for computing the Fiedler vector, TRACEMIN-Fiedler, has been proposed in [16]. We consider the standard symmetric eigenvalue problem,

$$L^w x = \lambda x. \tag{21}$$

The trace minimization eigensolver is based on the observation,

$$\min_{X \in \mathcal{X}_p} tr(X^T L^w X) = \sum_{i=1}^{p} \lambda_i, \tag{22}$$

where $\mathcal{X}_p$ is the set of all $n \times p$ matrices, X for which $X^T X = I$. The equality holds if and only if the columns of the matrix X span the eigenspace corresponding to the smallest p eigenvalues. At each iteration of the trace minimization algorithm an approximation $X_k \in \mathcal{X}_p$ which satisfies $X_k^T L^w X_k = \Theta_k$ for some diagonal Θ_k is obtained. The approximation X_k is corrected with Δ_k obtained by

$$\text{minimizing } tr[(X_k - \Delta_k)^T L^w (X_k - \Delta_k)] \tag{23}$$
$$\text{subject to } X_k^T \Delta_k = 0.$$

The solution of the (23) can be obtained by solving the following saddle point problem:

$$\begin{pmatrix} L^w & X_k \\ X_k^T & 0 \end{pmatrix} \begin{pmatrix} \Delta_k \\ L_k \end{pmatrix} = \begin{pmatrix} L^w X_k \\ 0 \end{pmatrix}. \tag{24}$$

Once Δ_k is known, X_{k+1} is obtained by computing $(X_k - \Delta_k)$ which forms the section $X_{k+1}^T L^w X_{k+1} = \Theta_{k+1}$, $X_{k+1}^T X_{k+1} = I$. In [16], we solve those saddle point systems by computing the block LU-factorization of the coefficient matrix in (24), i.e. by forming the Schur complement matrix explicitly since we are only interested in the second smallest eigenvector and hence p is small. Then, the main computational cost is solving sparse linear systems of equations with a few right-hand side vectors where the coefficient matrix, L^w, is a large sparse and symmetric positive semi-definite matrix. The details of the TRACEMIN-Fiedler algorithm are given in [16]. This algorithm proved (see Fig. 8) to be more suitable for implementation on parallel architectures compared to the eigensolver used in HSL for (21). Table 3 shows the dimension, number of nonzeros, and symmetry properties of four large matrices obtained from the SuiteSparse Matrix Collection [7].

Table 3 Properties of matrices from the SuiteSparse matrix collection

Matrix group/name	n	nnz	Symmetry
Rajat/rajat31	4,690,002	20,316,253	No
Schenk/nlpkkt120	3,542,400	95,117,792	Yes
Freescale/freescale1	3,428,755	17,052,626	No
Zaoui/kkt_power	2,063,494	12,771,361	Yes

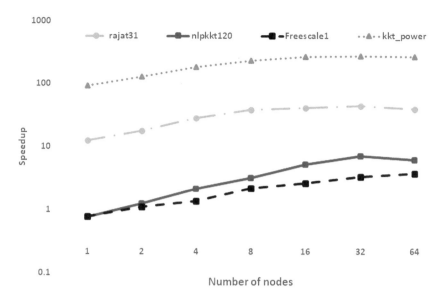

Fig. 8 Seedup of TRACEMIN-Fiedler reordering (using 8 cores per node) compared to the sequential HSL_MC73

After these two reordering steps, the resulting sparse linear system is of the form $Cz = h$, where $C = PBP^T$, $z = Py$, and $h = Pg$, with C having its heaviest off-diagonal elements as close to the main diagonal as possible. Choosing a central "band" of bandwidth $(2\beta + 1)$ as a preconditioner M of a Krylov subspace method with β chosen such that

$$||M||_F \simeq (1 - \epsilon)||C||_F. \tag{25}$$

Here $|| \cdot ||_F$ denotes the Frobenius norm, and ϵ chosen in the interval $[0.001, 0.05]$. Assuming C is of sufficiently large order n, say $n = 10^6$, then if $\beta \leq 10$, we call M a Narrow banded Preconditioner (NBP). If $\beta > 10$, we choose M as a block-tridiagonal preconditioner in which the diagonal blocks are sparse with relatively large "bandwidth," and the interconnecting off-diagonal blocks are dense square matrices of small dimensions. We call such M as Medium banded Preconditioner (MBP). When β becomes much larger than 10 in order to encapsulate as many off-diagonal as possible, we construct the preconditioner M as overlapped block diagonal sparse matrices. In this case, M is referred to as Wide banded Preconditoner (WBP). In each outer Krylov subspace iteration, one needs to solve linear systems involving M. For the cases of "MBP" and "WBP," one needs to use a sparse linear system solver. In Fig. 9 we show the classical computational loop that arises in many science and engineering applications. Solving linear systems occurs in the inner-most loop where the solution of such systems is needed to yield only modest relative residuals. For this purpose, we created a family of solvers that generalizes SPIKE for solving sparse linear systems $Ax = f$ using hybrid

Fig. 9 Target computational
loop

Loop: Integration
   | **Loop:** Nonlinear iteration
   | | **Loop:** Linear system solvers
   | | | Implemented on parallel computing platforms;
   | | **End** η_k
   | **End** ϵ_l
End Δt

schemes, i.e. a combination of the direct sparse linear system solver Pardiso [29] and
SPIKE. Even though SPIKE, rather than Pardiso is used for the case M being narrow
banded, we refer to our family of hybrid solvers as PSpike_NBP, PSpike_MBP,
and PSpike_WBP, respectively. In Fig. 10, we illustrate the structure of each of
those preconditioners obtained from the reordered matrix C. Next, we describe
and present some results illustrating the performance of each of Narrow Banded,
Medium Banded, and Wide Banded preconditioners.

3.2   PSPIKE_NBP

Certain sparse linear systems $Ax = f$ yield, after the reordering procedures
described in Sect. 3.1, effective narrow banded preconditioners to Krylov subspace
methods like GMRES or BiCGStab.

Example 1
The first system $A_1 x_1 = f_1$ considered here has the sparse coefficient matrix
$A_1 :=$ "Rajat31" from the SuiteSparse Matrix Collection[6] of order $\simeq 4.7$M, see
Fig. 11, which is in the form of an arrowhead. After reordering, and choosing
$\epsilon = 0.05$, we extract a banded preconditioner M of bandwidth $2\beta + 1 = 11$,
i.e. $\beta = 5$. Using an outer Krylov solver (BiCGStab) with a stopping criterion
of relative residual $= 10^{-5}$, Fig. 12 shows that PSPIKE_NBP consumes ~ 2.8 s
on an Intel cluster of 32 nodes (8 cores/node). Here, solving linear systems of
the form $Mz = r$ in each BiCGStab iteration is achieved by using the truncated
version of the SPIKE algorithm outlined in Sect. 2. We compare the performance of
PSPIKE_NBP with IBM's direct sparse linear system solver WSMP (implemented
on the same Intel cluster). Figure 12 shows that while the factorization stage of
WSMP is quite scalable, solving $A_1 x_1 = f_1$ using WSMP on 16 nodes of this Intel
cluster consumes ~ 27 s (approximately 9.6 times slower than PSPIKE_NBP). This
is due to solving the sparse triangular systems resulting from the LU-factorization
of A_1. Note, however, that solving $A_1 x_1 = f_1$ via WSMP yields a relative residual
of order 10^{-10}.

Example 2
Here, we consider the sparse linear system $A_2 x_2 = f_2$, where A_2 results from
a Microelectromechanical System (MEMS) simulation—a mix of structural and

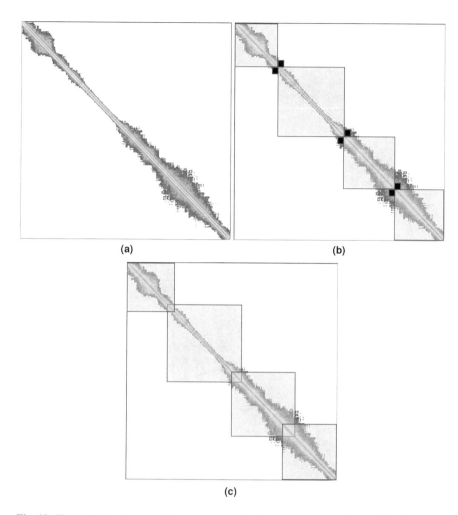

Fig. 10 Three forms of preconditioners based on the band structure and bandwidth, illustrated on $F2$ matrix after reordering. Yellow and Black colors indicate the preconditioners. (**a**) Narrow banded preconditioner. (**b**) Medium banded preconditioner. (**c**) Wide banded preconditioner

electromagnetic—with A_2 banded (sparse within the band) of order 11.0M and bandwidth of 0.3M, see Fig. 13. On an Intel cluster of 64 nodes (8 cores/node) PSPIKE_NBP with a preconditioner of bandwidth 11 consumes $\sim$2.4 s to obtain an approximation of x_2 with the required relative residual of 10^{-2}, see Fig. 14. WSMP could not be implemented on more than 32 nodes and requiring 86 s ($\sim$21.5 times slower than PSPIKE_NBP) to obtain a solution with relative residual of order 10^{-10}.

Example 3

Using the same linear system $A_2x_2 = f_2$, we compare the performance of PSPIKE_NBP and the algebraic multigrid preconditioned Krylov subspace solver

Fig. 11 Sparsity plot of
Rajat31 (the figure is
obtained from [6])

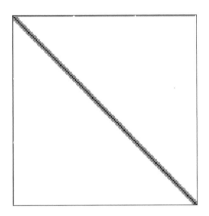

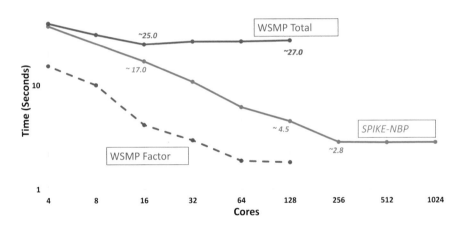

Fig. 12 SPIKE-NBP for Rajat31 system

Fig. 13 Sparsity plot of
MEMS matrix

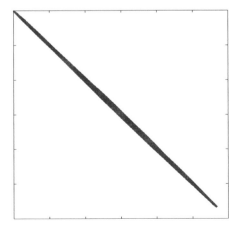

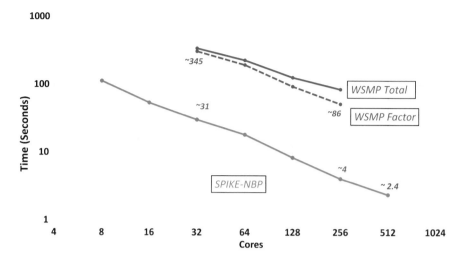

Fig. 14 SPIKE-NBP for MEMS system

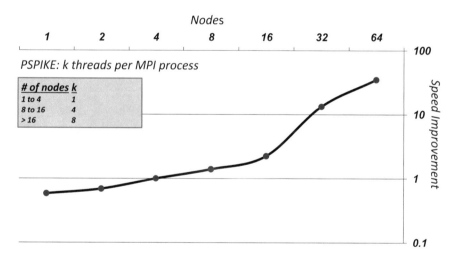

Fig. 15 Speed improvement: time (Trilinos-ML)/time (PSPIKE)

in Trilinos-ML developed at Sandia National Lab. on an Intel cluster of 64 nodes (8 cores/node). Using the Chebyshev smoother for Trilinos-ML, Fig. 15 shows the speed improvement realized by PSPIKE_NBP once we use more than 4 nodes. In PSPIKE we use a hybrid programming paradigm, OpenMP within each node (k threads per MPI process) and one node per MPI process with k depending on the number of nodes used to obtain a solution with relative residual of order 10^{-10}.

3.3 SPIKE_MBP

Here, we note how a system of the form $Mz = r$ is solved in each iteration of a Krylov subspace method, where $M \in \mathbb{R}^{n \times n}$ is of the form of a block-tridiagonal matrix

$$M = \begin{pmatrix} M_1 & \tilde{B}_1 & & & \\ \tilde{C}_2 & M_2 & \tilde{B}_2 & & \\ & \ddots & \ddots & \ddots & \\ & & \tilde{C}_{k-1} & M_{k-1} & \tilde{B}_{k-1} \\ & & & \tilde{C}_k & M_k \end{pmatrix}, \tag{26}$$

where k is the number of partitions (often chosen as the number of nodes), where each M_j is a large sparse matrix of order $m = \lceil n/k \rceil$, and

$$\tilde{B}_j = \begin{pmatrix} 0 & 0 \\ B_j & 0 \end{pmatrix}, \tilde{C}_j = \begin{pmatrix} 0 & C_j \\ 0 & 0 \end{pmatrix} \tag{27}$$

in which B_j and C_j are dense matrices of order $v << m$. Now, $Mz = r$ is solved using the SPIKE algorithm by forming only the reduced system by solving

$$M_j \begin{pmatrix} V_j & W_j \\ * & * \\ \vdots & \vdots \\ * & * \\ V'_j & W'_j \end{pmatrix} = \begin{pmatrix} C_j & 0 \\ 0 & 0 \\ \vdots & \vdots \\ 0 & 0 \\ 0 & B_j \end{pmatrix} \tag{28}$$

only for the tips of the spikes V_j, V'_j and W_j, W'_j via an interesting feature of the sparse direct solver Pardiso. Using the spike tips only, the reduced system is formed and solved via ScaLapack. Once this is achieved, the solution of $Mz = r$ is realized by employing the factors of each M_j obtained by Pardiso.

The description of PSPIKE_MBP is given in more detail with parallel scalability results for large-scale problems in [18] and its application to a PDE-constrained optimization problem in [30]. While Pardiso is primarily suitable for single node platforms, PSPIKE is scalable across multiple nodes. Furthermore, we would like to mention that PSPIKE is capable of using message passing-multithreaded hybrid parallelism. In Fig. 16, we present the required solution time of PSPIKE compared to Pardiso (on one node) for a medium size and large 3D PDE-constrained optimization problems with $75 \times 75 \times 75$ and $150 \times 150 \times 150$ meshes, respectively, using hybrid parallelism with 8 threads (cores) per node. Note that for the larger problem Pardiso runs out of memory due to fill-in. Further details of these problems and the results are given in [30].

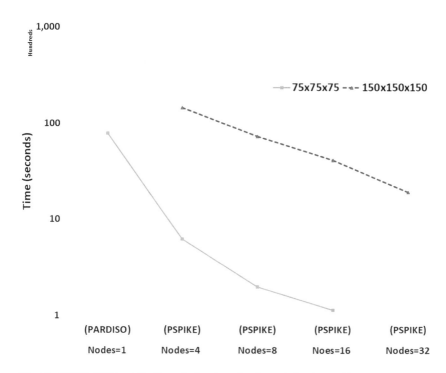

Fig. 16 SPIKE_MBP and Pardiso solution times for the optimization problem

3.4 *SPIKE_WBP*

For some applications it is not possible to obtain, after reordering, a narrow banded preconditioner, or a block-tridiagonal preconditioner in which the interconnecting off-diagonal blocks are of much smaller size than the diagonal blocks. An example of that is illustrated in Fig. 17. Note that, after reordering, the "heavy" off-diagonal elements (black color) cannot be contained in either of the two previous forms of the preconditioner $M \in \mathbb{R}^{n \times n}$. As an alternative, one way to encapsulate as many of the heavy elements in M is to create a preconditioner that consists of overlapped diagonal blocks, see Fig. 17, for M consisting of two overlapped blocks. In each outer Krylov subspace iteration we solve systems of the form $Mz = r$ via the algorithm given in [22]. Using the two overlapped blocks, $Mz = r$ becomes of the form

$$
\begin{pmatrix}
\begin{array}{|cc|}
\hline
M_{11} & M_{12} \\
\end{array} \\
\begin{array}{|cc|c}
M_{21} & M_{22} & M_{23} \\
\hline
& M_{32} & M_{33} \\
\hline
\end{array}
\end{pmatrix}
$$

$$(29)$$

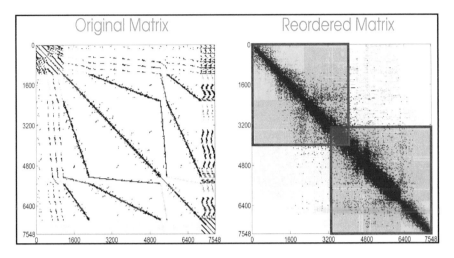

Fig. 17 WBP highlighted after reordering, see [21] for the tearing based parallel hybrid sparse solver

which can be "torn" into two linear systems

$$\begin{pmatrix} M_{11} & M_{12} \\ M_{21} & M_{22}^{(1)} \end{pmatrix} \begin{pmatrix} z_1(y) \\ z_2^{(1)}(y) \end{pmatrix} = \begin{pmatrix} r_1 \\ \alpha r_2 + y \end{pmatrix} \tag{30}$$

$$\begin{pmatrix} M_{22}^{(2)} & M_{23} \\ M_{32} & M_{33}^{(1)} \end{pmatrix} \begin{pmatrix} z_2^{(2)}(y) \\ z_2^{(3)}(y) \end{pmatrix} = \begin{pmatrix} (1-\alpha)r_2 - y \\ \alpha r_3 \end{pmatrix}, \tag{31}$$

where the overlap matrix $M_{22} = M_{22}^{(1)} + M_{22}^{(2)}$, and $0 < \alpha < 1$. Clearly, we need to choose y so that $z_2^{(1)} = z_2^{(2)}$. Enforcing $z_2^{(1)}(y) = z_2^{(2)}(y)$ results in a linear system $Gy = g$ of size equal to that of overlap matrix M_{22}, $\nu << n$. In solving $Gy = g$ for the unknown y, using a Krylov subspace method, it is shown in [22] that one needs not generate either G or g explicitly, in fact the residual $r(p) = g - G * p$ is given by $[z_2^{(2)}(p) - z_2^{(1)}(p)]$, $r(0) = g = [z_2^{(2)}(0) - z_2^{(1)}(0)]$, and the matrix-vector product $G * g = r(0) - r(g)$. The case of more than two overlapped blocks is considered in detail in [22].

3.5   The General SPIKE

Now we describe the general case where the coefficient matrix has not been subjected to the reordering process described earlier. In other words it is a general sparse matrix and also there are multiple right-hand side vectors. Given a nonsingular linear system of equations,

$$AX = F, \tag{32}$$

where $A \in \mathbb{R}^{n \times n}$ is a general sparse matrix and assume we have m right-hand side vectors F, we can still apply the General SPIKE algorithm as follows. As in the banded case, let us assume A, X, and F are partitioned conformably into k block rows and A is also partitioned into k block columns. The Spike factorization can be described as the factorization of the coefficient matrix [14],

$$A = DS, \tag{33}$$

where D is the block diagonal of A and S is the "spike" matrix. Let $A = D + R$ where R is a matrix that contains elements except diagonal blocks. Assuming D is invertible and using (33) we obtain the spike matrix,

$$S = I + \bar{S}, \tag{34}$$

where $\bar{S} = D^{-1}R$. Note that the diagonal of S consists of ones and the off-diagonals are the spikes ($\bar{S}$). Going back to the original linear system in (32), if we multiply both sides of the equality with D^{-1} from left, we have the modified system

$$SX = G, \tag{35}$$

where $G = D^{-1}F$. The modified system in (35) has the same solution vector, X, as the original system in (32). Furthermore, let idx be the nonzero column indices of R which also correspond to nonzero column indices of $D^{-1}R$. Then, there is an independent subsystem corresponding to the unknowns with row indices idx, i.e. $X(idx, :)$ in (35) such that,

$$\hat{S}\hat{X} = \hat{G}, \tag{36}$$

where $\hat{S} = S(idx, idx)$, $\hat{X} = X(idx, :)$, and $\hat{G} = G(idx, :)$. Dimensions of the reduced system in (36) are $r \times r$ where $r = length(idx)$ with $r \leq n$. After solving the reduced system we can retrieve the remaining unknowns in parallel,

$$X = G - \bar{S}X. \tag{37}$$

Note that we only need a subset of unknowns, $\hat{X}$, to evaluate the right-hand side of the equality since the other columns in $\bar{S}$ are zeros. This approach requires $\bar{S}$ to be computed explicitly. Alternatively, one can obtain the solution by solving the following system in parallel,

$$DX = F - RX. \tag{38}$$

Again, the right-hand side can be evaluated once we obtain $\hat{X}$. In contrast to (37), (38) does not require the computation of $\bar{S}$ completely, even though it still requires the solution of the reduced system involving $\hat{S}$ which may be explicitly formed via partially computing $\bar{S}$. Alternatively, the reduced system can be solved iteratively without forming $\hat{S}$ explicitly. Some of these alternatives might be preferred in practice, depending on the current availability of efficient software tools to perform those operations.

In any case, the size of the reduced system depends on r. A smaller r not only enhances parallelism by enabling a smaller reduced system and less communication requirements, but also reduces the arithmetic complexity in computing $\hat{S}$ and $\bar{S}$ (if needed) as well as the complexity of (37) and (38).

In practice, we assume $r \ll n$. Ideally, $r = 0$ and some matrices can be reordered into a block diagonal form. In this case, there is no reduced system and the block diagonal systems are solved independently in parallel. Most applications, however, give rise to sparse linear system of equations that does not contain independent blocks, then the objective is to reorder and partition those matrices in such a way that the number of the nonzero columns in $\bar{R}$ is minimized [15].

The main difference between the sparse and the banded SPIKE algorithms is the dependence of the reduced system size (r) on the sparsity structure of the matrix (and hence on the corresponding graph or hypergraph representation of the sparse matrix). Therefore, sparse graph/hypergraph partitioning methods are key ingredients for the algorithm to be scalable and to perform efficiently. METIS [11] and PaToH [4], are suitable for graph and hypergraph partitioning, respectively, and they fit well to the objective of minimizing the reduced system dimension.

To illustrate the algorithm, we give a small (9×9) coefficient matrix, A, in Fig. 18a for simplicity we ignore the numerical values. Given $k = 3$, the coefficient matrix and right-hand side are comformably partitioned,

$$A = \begin{pmatrix} D_{11} & R_{12} & R_{13} \\ R_{21} & D_{22} & R_{23} \\ R_{31} & R_{32} & D_{33} \end{pmatrix} \text{ and } F = \begin{pmatrix} F_1 \\ F_2 \\ F_3 \end{pmatrix}. \tag{39}$$

The set of indices of nonzero columns of $\bar{R}$ are $idx = \{1, 5, 8\}$. After partitioning, S and G can be computed as follows:

$$S = \begin{pmatrix} I & D_{11}^{-1} R_{12} & D_{11}^{-1} R_{13} \\ D_{22}^{-1} R_{21} & I & D_{22}^{-1} R_{23} \\ D_{33}^{-1} R_{31} & D_{33}^{-1} R_{32} & I \end{pmatrix} \text{ and } G = \begin{pmatrix} D_{11}^{-1} F_1 \\ D_{22}^{-1} F_2 \\ D_{33}^{-1} F_3 \end{pmatrix}. \tag{40}$$

If the right-hand side vector is available immediately, the computation involved is the solution of independent linear systems with multiple right-hand sides,

$$D_{11}[S_{12}, S_{13}, G_1] = [R_{12}, R_{13}, F_1], \tag{41}$$

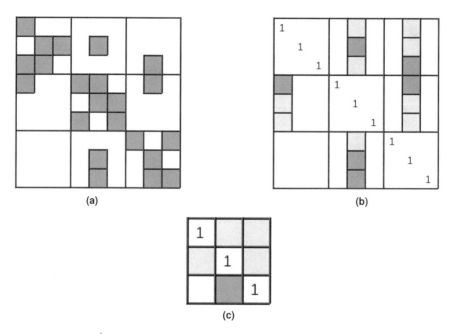

Fig. 18 A, S, and $\hat{S}$ for the small example. (**a**) Coefficient matrix (A). (**b**) Spike matrix (S). (**c**) Reduced system coefficient matrix ($\hat{S}$)

$$D_{22}[S_{21}, S_{23}, G_2] = [R_{21}, R_{23}, F_2],\tag{42}$$

$$D_{33}[S_{31}, S_{32}, G_3] = [R_{31}, R_{32}, F_3].\tag{43}$$

Note that in (41),(42), and (43) only a few columns of $R_{ij,i\neq j}$ are nonzero and the rest are zeros. We do not store or perform operations with zero columns since the corresponding solution vector is already zero. The resulting S matrix is shown in Fig. 18b. Light green elements are fill-ins and some of them can be negligible as in the banded case [20, 24] if A is diagonally dominant or near diagonally dominant.

Further savings can be obtained, if a sparse solver with sparse right-hand side vectors is available and if it is capable of solving only for a few unknowns, one can compute only those components of vectors in S_{ij} that is required for forming the reduced system (defined by idx). One of the implementation of the General SPIKE algorithm in [3] performs partial solves via the sparse right-hand side feature of PARDISO [29]. Next, we can form the reduced system explicitly by selecting $\hat{S} = S(idx, idx)$ and $\hat{G} = G(idx, :)$ and solve the reduced system, (36), to obtain $\hat{X} = X(idx, :)$. $\hat{S}$ for the small example is shown in Fig. 18c. The complete solution is obtained in parallel via either:

$$X = G - \bar{S}X\tag{44}$$

Algorithm 1: General spike algorithm

1: **procedure** GENERALSPIKE(A, X, F, k)   ▷ to solve $AX = F$ with k partitions
2:    $D + R \leftarrow A$
3:    Identify nonzero columns of R and store their indices in idx
4:    $D[S_{(:,idx)}, G] = [R_{(:,idx)}, F]$, solve for:

- $[S_{(:,idx)}, G]$ (full solve) or
- $[S_{(idx,idx)}, G_{(idx,:)}]$ (partial solve)

5:    $S_{(idx,idx)} X_{(idx,:)} = G_{(idx,:)}$ (Solve for $X_{(idx,:)}$)
6:    Retrieve the solution vector (X):

- $X \leftarrow G - S_{(:,idx)} X_{(idx)}$ if $S_{(:,idx)]}$ is available
- $DX = [F - R_{(:,idx)} x_{(idx)}]$ (Solve for X), otherwise

7: **end procedure**

---

or

$$DX = F - RX. \tag{45}$$

The former is preferred if the spikes are formed explicitly since the multiplication $\bar{S}X$ can be implemented using dense matrix-vector (BLAS Level 2) or matrix-matrix (BLAS Level 3) operations, for $m = 1$ and $m > 1$, respectively. The latter requires sparse matrix-dense matrix multiplications (RX), followed by the solution of independent sparse linear systems. It is preferred if the spikes are not explicitly available. The pseudocode of the algorithm is summarized in Algorithm 1.

Numerical results and the performance of this scheme are given in [14] in the context of a parallel solver for the preconditioned linear system and in [3] as a direct multithreaded recursive parallel sparse solver. Furthermore, a multithreaded general sparse triangular solver is proposed in [5].

4 Conclusions

The SPIKE algorithm for banded linear systems that are dense withing the band has been shown to be competitive in parallel scalability with the parallel banded solver in ScaLapack on a variety of parallel architectures. Also, the hybrid PSPIKE (Pardiso-SPIKE) algorithm for large sparse linear systems has proven to be equally competitive with: (1) direct sparse solvers such as Pardiso and WSMP if one requires only approximate solutions that correspond to modest relative residuals, and (2) black-box preconditioned Krylov subspace methods including algebraic multigrid preconditioners.

Acknowledgments The authors would like to thank Drs. Maxim Naumov and Faisal Saied for performing many of the numerical experiments reported in this chapter.

References

1. *SPIKE openMP package.* http://www.spike-solver.org/.
2. E. ANDERSON, Z. BAI, C. BISCHOF, S. BLACKFORD, J. DEMMEL, J. DONGARRA, J. DU CROZ, A. GREENBAUM, S. HAMMARLING, A. MCKENNEY, AND D. SORENSEN, *LAPACK Users' Guide*, Society for Industrial and Applied Mathematics, Philadelphia, PA, third ed., 1999.
3. E. S. BOLUKBASI AND M. MANGUOGLU, *A multithreaded recursive and nonrecursive parallel sparse direct solver*, in Advances in Computational Fluid-Structure Interaction and Flow Simulation, Springer, 2016, pp. 283–292.
4. U. V. CATALYUREK AND C. AYKANAT, *Hypergraph-partitioning-based decomposition for parallel sparse-matrix vector multiplication*, IEEE Transactions on parallel and distributed systems, 10 (1999), pp. 673–693.
5. I. CUGU AND M. MANGUOGLU, *A parallel multithreaded sparse triangular linear system solver*, Computers & Mathematics with Applications, (2019).
6. T. A. DAVIS AND Y. HU, *The University of Florida Sparse Matrix Collection*, ACM Transactions on Mathematical Software (TOMS), 38 (2011), p. 1.
7. T. A. DAVIS AND Y. HU, *The University of Florida Sparse Matrix Collection*, ACM Trans. Math. Softw., 38 (2011), pp. 1:1–1:25.
8. I. S. DUFF AND J. KOSTER, *The design and use of algorithms for permuting large entries to the diagonal of sparse matrices*, SIAM Journal on Matrix Analysis and Applications, 20 (1999), pp. 889–901.
9. M. FIEDLER, *Algebraic connectivity of graphs*, Czechoslovak Mathematical Journal, 23 (1973), pp. 298–305.
10. E. GALLOPOULOS, B. PHILIPPE, AND A. H. SAMEH, *Parallelism in matrix computations*, Springer, 2016.
11. G. KARYPIS AND V. KUMAR, *A fast and high quality multilevel scheme for partitioning irregular graphs*, SIAM Journal on scientific Computing, 20 (1998), pp. 359–392.
12. D. H. LAWRIE AND A. H. SAMEH, *The computation and communication complexity of a parallel banded system solver*, ACM Transactions on Mathematical Software (TOMS), 10 (1984), pp. 185–195.
13. L. LIU, Z. LI, AND A. H. SAMEH, *Analyzing memory access intensity in parallel programs on multicore*, in Proceedings of the 22nd annual international conference on Supercomputing, ACM, 2008, pp. 359–367.
14. M. MANGUOGLU, *A domain-decomposing parallel sparse linear system solver*, Journal of computational and applied mathematics, 236 (2011), pp. 319–325.
15. ———, *A parallel sparse solver and its relation to graphs*, in CEM'11 Computational Electromagnetics International Workshop, IEEE, 2011, pp. 91–94.
16. M. MANGUOGLU, E. COX, F. SAIED, AND A. SAMEH, *TRACEMIN-Fiedler: A Parallel Algorithm for Computing the Fiedler Vector*, High Performance Computing for Computational Science–VECPAR 2010, (2011), pp. 449–455.
17. M. MANGUOGLU, F. SAIED, A. SAMEH, AND A. GRAMA, *Performance models for the spike banded linear system solver*, Scientific Programming, 19 (2011), pp. 13–25.
18. M. MANGUOGLU, A. H. SAMEH, AND O. SCHENK, *Pspike: A parallel hybrid sparse linear system solver*, in European Conference on Parallel Processing, Springer, 2009, pp. 797–808.
19. K. MENDIRATTA AND E. POLIZZI, *A threaded spike algorithm for solving general banded systems*, Parallel Computing, 37 (2011), pp. 733–741. 6th International Workshop on Parallel Matrix Algorithms and Applications (PMAA'10).
20. C. C. K. MIKKELSEN AND M. MANGUOGLU, *Analysis of the truncated SPIKE algorithm*, SIAM Journal on Matrix Analysis and Applications, 30 (2008), pp. 1500–1519.
21. M. NAUMOV, M. MANGUOGLU, AND A. H. SAMEH, *A tearing-based hybrid parallel sparse linear system solver*, J. Computational Applied Mathematics, 234 (2010), pp. 3025–3038.

22. M. NAUMOV AND A. H. SAMEH, *A tearing-based hybrid parallel banded linear system solver*, Journal of Computational and Applied Mathematics, 226 (2009), pp. 306–318.
23. E. POLIZZI AND A. SAMEH, *SPIKE: A parallel environment for solving banded linear systems*, Computers & Fluids, 36 (2007), pp. 113–120. Challenges and Advances in Flow Simulation and Modeling.
24. E. POLIZZI AND A. H. SAMEH, *A parallel hybrid banded system solver: the SPIKE algorithm*, Parallel computing, 32 (2006), pp. 177–194.
25. Y. SAAD, *Iterative methods for sparse linear systems*, vol. 82, siam, 2003.
26. A. SAMEH AND Z. TONG, *The trace minimization method for the symmetric generalized eigenvalue problem*, J. Comput. Appl. Math., 123 (2000), pp. 155–175.
27. A. H. SAMEH AND D. J. KUCK, *On stable parallel linear system solvers*, Journal of the ACM (JACM), 25 (1978), pp. 81–91.
28. A. H. SAMEH AND J. A. WISNIEWSKI, *A trace minimization algorithm for the generalized eigenvalue problem*, SIAM Journal on Numerical Analysis, 19 (1982), pp. 1243–1259.
29. O. SCHENK AND K. GÄRTNER, *Solving unsymmetric sparse systems of linear equations with PARDISO*, Future Generation Computer Systems, 20 (2004), pp. 475–487.
30. O. SCHENK, M. MANGUOGLU, A. SAMEH, M. CHRISTEN, AND M. SATHE, *Parallel scalable PDE-constrained optimization: antenna identification in hyperthermia cancer treatment planning*, Computer Science-Research and Development, 23 (2009), pp. 177–183.
31. B. S. SPRING, E. POLIZZI, AND A. H. SAMEH, *A feature complete SPIKE banded algorithm and solver*, CoRR, abs/1811.03559 (2018).

Part II
High Performance Computational Science and Engineering Applications

Computational Materials Science and Engineering

Eric Polizzi and Yousef Saad

1 Introduction

Among the many jobs running at any given time on a high-performance computing facility today, it is likely that those related to quantum mechanical calculations will figure prominently. The numerical simulations that arise from the modeling of matter are very demanding both in terms of memory and computational power. These simulations combine ideas and techniques from a variety of disciplines including physics, chemistry, applied mathematics, numerical linear algebra, and computer science.

Determining matter's electronic structure can be a major challenge: The number of particles is large [a macroscopic amount contains $\approx 10^{23}$ electrons and nuclei] and the physical problem is intrinsically complex.

The most significant change in computational methods used in materials in the past two decades has undoubtedly been the systematic use of parallel processing. This revolution in methodology has taken some time to unravel and then mature. For example, it was not clear in the early 1990s whether massively parallel computing could be achieved with vector processors or if a message passing interface would

E. Polizzi (✉)
Department of Electrical and Computer Engineering, University of Massachusetts at Amherst, Amherst, MA, USA

Department of Mathematics and Statistics, University of Massachusetts at Amherst, Amherst, MA, USA
e-mail: polizzi@ecs.umass.edu

Y. Saad
Department of Computer Science and Engineering, University of Minnesota, Minneapolis, MN, USA
e-mail: saad@cs.umn.edu

© Springer Nature Switzerland AG 2020
A. Grama, A. H. Sameh (eds.), *Parallel Algorithms in Computational Science and Engineering*, Modeling and Simulation in Science, Engineering and Technology,
https://doi.org/10.1007/978-3-030-43736-7_5

be best. There were phases in which programming models and languages took different directions. As architectures changed over the years, the software and techniques have been in constant flux. At the same time algorithms have also evolved considerably, in part to cope with the new computing environments and the enormous power afforded by new hardware.

Most of the gains in speed combine advances from three areas: simplifications or improvements from physical models, effective numerical algorithms, and powerful hardware and software tools.

In terms of physical models, the biggest advances in nanotechnology were made in the sixties with the emergence of Density Functional Theory (DFT) which made it possible to approximate the initial problem by one which involves unknowns that are functions of only one space variables instead of N space variables, for N-particle systems in the original Schrödinger equation. Thus instead of dealing with functions in $\mathbb{R}^{3N}$ we only need to handle functions in $\mathbb{R}^3$. DFT provides (in principle) an exact method for calculating the ground state energy and electron density of a system of interacting electrons using exchange-correlation density functionals, and a set of single electron wavefunctions solution of an eigenvalue equation.

The number of atoms contained in nanostructures of technological interests usually range from few hundreds to many thousands posing a unique challenge for DFT electronic structure modeling and computation. Many modeling advances were made in designing various discretization techniques to accommodate atomistic systems with high level of accuracy. In addition, since both system size and number of needed eigenpairs to compute the electron density depend linearly on the number of atoms, progress in electronic structure calculations are tied together with advances in eigenvalue algorithm and their scalability on parallel architectures.

The goal of this paper is not to provide another exhaustive review of the state of the art in materials but rather to discuss the impact that parallel processing has had on the design of algorithms. From physics to algorithms, we will begin with a review of the basics, and then discuss the recent advances made in electronic structure calculations using appropriate discretization schemes and new parallel algorithms that can fully capitalize on modern HPC computing platforms.

2   Quantum Descriptions of Matter

Consider N nucleons of charge Z_n at positions $\{\mathbf{R}_n\}$ for $n = 1, \cdots, N$ and M electrons at positions $\{\mathbf{r}_i\}$ in space, for $i = 1, \cdots, M$. The non-relativistic, time-independent Schrödinger equation that describes the physical state of the system can be written as:

$$\mathcal{H}\,\Psi = E\,\Psi \tag{1}$$

where the many-body wave function Ψ is of the form

$$\Psi \equiv \Psi(\mathbf{R}_1, \mathbf{R}_2, \mathbf{R}_3, \cdots; \mathbf{r}_1, \mathbf{r}_2, \mathbf{r}_3, \cdots) \tag{2}$$

and E is the total electronic energy. The Hamiltonian $\mathcal{H}$ in its simplest form can be written as:

$$\mathcal{H}(\mathbf{R}_1, \mathbf{R}_2, \mathbf{R}_3, \cdots; \mathbf{r}_1, \mathbf{r}_2, \mathbf{r}_3, \cdots) = \sum_{n=1}^{N} \frac{-\hbar^2 \nabla_n^2}{2M_n} + \frac{1}{2} \sum_{\substack{n,n'=1, \\ n \neq n'}}^{N} \frac{Z_n Z_{n'} e^2}{|\mathbf{R}_n - \mathbf{R}_{n'}|}$$

$$+ \sum_{i=1}^{M} \frac{-\hbar^2 \nabla_i^2}{2m} - \sum_{n=1}^{N} \sum_{i=1}^{M} \frac{Z_n e^2}{|\mathbf{R}_n - \mathbf{r}_i|} + \frac{1}{2} \sum_{\substack{i,j=1 \\ i \neq j}}^{M} \frac{e^2}{|\mathbf{r}_i - \mathbf{r}_j|} \tag{3}$$

Here, M_n is the mass of the nucleon, $\hbar$ is Planck's constant divided by 2π, m is the mass of the electron, and e is the charge of the electron.

The above Hamiltonian includes the kinetic energies for each nucleon (first sum in $\mathcal{H}$), and each electron (3rd sum), the inter-nuclei repulsion energies (2nd sum), the nuclei-electronic (Coulomb) attraction energies (4th sum), and the electron-electron repulsion energies (5th sum). Each Laplacian ∇_n^2 involves differentiation with respect to the coordinates of the nth nucleon. Similarly the term ∇_i^2 involves differentiation with respect to the coordinates of the ith electron.

In principle, the electronic structure of any system is completely determined by (1) by finding the wave function Ψ that minimizes the energy $< \Psi|\mathcal{H}|\Psi >$ over all normalized wavefunctions Ψ. The function Ψ has a probabilistic interpretation: for the minimizing wave function Ψ,

$$|\Psi(\mathbf{R}_1, \cdots, \mathbf{R}_N; \mathbf{r}_1, \cdots, \mathbf{r}_M)|^2 d^3\mathbf{R}_1 \cdots d^3\mathbf{R}_N d^3\mathbf{r}_1 \cdots d^3\mathbf{r}_M$$

represents the probability of finding nucleon 1 in volume $|\mathbf{R}_1 + d^3\mathbf{R}_1|$, nucleon 2 in volume $|\mathbf{R}_2 + d^3\mathbf{R}_2|$, etc. However, solving (1) is not practically feasible for systems that include more than just a few atoms.

The main computational difficulty stems from the nature of the wavefunction which depends on all coordinates of all particles (nuclei and electrons) simultaneously. To give an illustration of this, imagine we have 10 atoms each with 14 electrons [e.g., Silicon]. This represents a total of $15 * 10 = 150$ particles. The wave function in its form without spin is $\Psi(R_1, \cdots, R_{10}, r_1, \cdots, r_{140})$ and it must be discretized. A simple scheme would be some finite difference method. If we use 100 points for each of the 150 coordinates, we would get a huge number of unknowns:

$$\# \text{Unknowns} = \underbrace{100}_{part.1} \times \underbrace{100}_{part.2} \times \cdots \times \underbrace{100}_{part.150} = 100^{150}$$

The original Schrödinger equation (1) can be viewed as an eigenvalue problem: we need to compute the smallest eigenvalue and associated eigenvector of the

Hamiltonian. It can also be viewed from the point of view of optimization since finding the smallest eigenpair is known to be equivalent to finding the wavefunction Ψ that minimizes the Rayleigh quotient:

$$E = <\Psi|\mathcal{H}|\Psi> \equiv \frac{\int \Psi^*\mathcal{H}\Psi \, d^3\mathbf{R}_1 \, d^3\mathbf{R}_2 \, d^3\mathbf{R}_3 \cdots . d^3\mathbf{r}_1 \, d^3\mathbf{r}_2 \, d^3\mathbf{r}_3 \cdots}{\int \Psi^*\Psi \, d^3\mathbf{R}_1 \, d^3\mathbf{R}_2 \, d^3\mathbf{R}_3 \cdots . d^3\mathbf{r}_1 \, d^3\mathbf{r}_2 \, d^3\mathbf{r}_3 \cdots} \quad (4)$$

The symbols *bra* (for $< |$) and *ket* (for $| >$) are common in chemistry and physics. When applying the Hamiltonian to a state function Ψ the result is another state function: $\Phi = |\mathcal{H}|\Psi >$. The inner product of this function with another function Θ is $< \Theta|\Phi >$ which is a scalar.

The first, and basic, approximation made to reduce complexity is the *Born-Oppenheimer* or adiabatic approximation. This approximation separates the nuclear and electronic degrees of freedom: exploiting the fact that the nuclei have a much bigger mass than the electrons, it can be assumed that the electrons will respond "instantaneously" to the nuclear coordinates. This allows one to treat the nuclear coordinates as classical parameters. For most condensed matter systems, this assumption is highly accurate [29, 79]. Under this approximation the first term in (3) vanishes and the second becomes a constant, so we end up with the simplified Hamiltonian:

$$\mathcal{H}(\mathbf{r}_1, \mathbf{r}_2, \mathbf{r}_3, \cdots) = \sum_{i=1}^{M} \frac{-\hbar^2 \nabla_i^2}{2m} - \sum_{n=1}^{N}\sum_{i=1}^{M} \frac{Z_n e^2}{|\mathbf{R}_n - \mathbf{r}_i|} + \frac{1}{2} \sum_{\substack{i,j=1 \\ i\neq j}}^{M} \frac{e^2}{|\mathbf{r}_i - \mathbf{r}_j|} \quad (5)$$

This simplified Hamiltonian is often taken as a practical replacement of the original problem.

Its eigenfunctions determine the *states*. There are infinitely many states, labeled $1, 2, \cdots$ by increasing eigenvalue. Each eigenvalue represents an "energy" level of the state. The state with lowest energy (smallest eigenvalue) is the *ground state*. It determines stable structures, mechanical deformations, phase transitions, and phonons. States above the ground state are known as "excited states." They are used to study many body effects, quasi-particles, electronic band gaps, optical properties, etc.

A direct numerical treatment of the Schrödinger equation using the simplified many-body Hamiltonian (5) leads to a deceptively simple linear eigenvalue problem which is still intractable because of its exponential growing dimension with the number of electrons. This limitation has historically motivated the need for lower levels of sophistication in the description of the electronic structure using a single electron picture approximation where the size of the Hamiltonian operator scales linearly with the number of electrons. It is within the single electron picture that first-principle electronic structure calculations are usually performed [49] using either (post) Hartree–Fock type methods widely used in quantum chemistry, or as an alternative to wave function based methods, the Density Functional Theory (DFT) associated with the Kohn–Sham equations [31, 36].

3 Density Functional Theory and the Kohn–Sham Equation

A breakthrough in the solution of the Schrödinger equation came with the discovery of Density Functional Theory. In a series of papers, Hohenberg, Kohn, and Sham established a theory in which the many-body wave function was replaced by one-electron orbitals [31, 36, 48]. The basic idea is that the state of the system will now be expressed in terms of the the charge density ρ, which is a distribution of probability, i.e., $\rho(\mathbf{r}_1)d^3\mathbf{r}_1$ represents—in a probabilistic sense—the number of electrons (all electrons) in the infinitesimal volume $d^3\mathbf{r}_1$. It is easy to calculate the charge density from a given wavefunction. The fundamental theorem which these authors were able to state is that this mapping is one-to-one, i.e., given the charge density it should be possible to obtain the ground state wavefunction. In essence there is a certain Hamiltonian—as defined by a certain potential (that depends on ρ) whose minimum energy is reached for the ground state Ψ. Kohn and Sham wrote this Hamiltonian as

$$H_{KS} = \frac{\hbar^2}{2m}\nabla^2 + V_N(\rho) + V_H(\rho) + V_{xc}(\rho) \qquad (6)$$

where $V_N(\rho)$ is the external potential, $V_H(\rho)$ is the Hartree potential, and $V_{xc}(\rho)$ is the exchange-correlation potential. Note the dependence on the charge density ρ which is itself implicitly defined from the set of occupied eigenstates $\phi_i, i = 1, \cdots, N$ of (6) by:

$$\rho(\mathbf{r}) = 2 \sum_{j=1}^{occup} |\phi_j(\mathbf{r})|^2, \qquad (7)$$

where N is the number of occupied states (i.e., number of electrons) and the factor 2 accounts for the electron spin.

3.1 The Kohn–Sham Equation

We can now write the *Kohn–Sham equation* [36] for the electronic structure of matter as

$$\left(\frac{-\hbar^2\nabla^2}{2m} + V_N(\mathbf{r}) + V_H(\mathbf{r}) + V_{xc}[\rho(\mathbf{r})] \right) \phi_i(\mathbf{r}) = E_i\phi_i(\mathbf{r}) \qquad (8)$$

As stated above the charge density is defined in terms of the orbitals ϕ_i given by (7).

Given a charge density ρ the Hartree potential V_H is the solution of Poisson equation:

$$\nabla^2 V_H = -4\pi\rho(r) \tag{9}$$

The exchange and correlation potential V_{xc} is unknown in theory but it is approximated by a potential in different ways, the simplest of which is the Local Density Approximation (LDA).

Therefore, this equation is usually solved "self-consistently" in the sense that if a given ρ^{in}, as obtained from a set of occupied states $\phi_i(r)$, $i = 1, \cdots, N$ is utilized to compute new occupied states from (6), and a new charge density ρ^{out} is then computed according to (7) then ρ and ρ^{out} should be the same. The SCF procedure takes some initial approximate charge to estimate the exchange-correlation potential and this charge is used to determine the Hartree potential from (9). These approximate potentials are inserted in the Kohn–Sham equation and the total charge density determined as in (7). The "output" charge density is used to construct new exchange-correlation and Hartree potentials. The process is repeated until the input and output charge densities (or potentials) are close enough. This process is illustrated in Fig. 1.

DFT has been widely used in computational material science and quantum chemistry over the past few decades, since it provides (in principle) an exact method for calculating the ground state density and energy of a system of interacting electrons using a nonlinear single electron equation associated with exchange-correlation (XC) functionals. In practice, the reliability of DFT depends on the

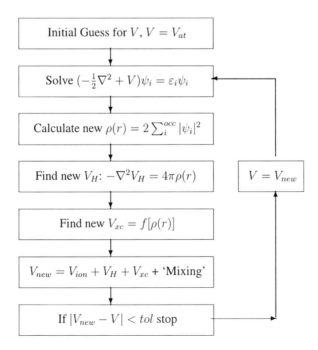

Fig. 1 The self-consistent field iteration

numerical approximations used for the XC terms that range from the simplest local density approximation (LDA) to more advanced schemes which are still the subject of active research efforts [14, 65, 70]. Solutions of the DFT/Kohn–Sham problem are routinely used in the calculations of many ground state properties including: total energy and ionization potential and, via perturbation: crystal-atomic structure, ionic forces, vibrational frequencies, and phonon bandstructure.

3.2 Pseudopotentials

When discretizing the KS equation, we run into a major difficulty which arises from the different scales of the lengths involved. The inner (core) electrons are highly localized and tightly bound compared to the outer (valence electrons). Another major advance in the solid-state physics field was the advent of pseudopotential techniques which remove the core states from the problem and replacing the all electron potential by one that replicates only the chemically active, valence electron states[16]. This is possible because the physical properties of solids depend much more on the valence electrons than on the core electrons. The whole art is then to construct pseudopotentials that reproduce the valence state properties such as the eigenvalue spectrum and the charge density outside the ion core.

3.3 Discretization

One can identify three main discretization techniques that have been widely used over the past four decades by both the quantum chemistry and the solid-state physics communities [49]: (1) the plane wave expansion scheme, (2) the linear combination of atomic orbitals (LCAO) (along with the dominant use of Gaussian local basis sets), and (3) the real-space mesh techniques (also loosely called "numerical grids") based on the finite difference method (FDM), finite element method (FEM), spectral element, or wavelets methods. Each of these approaches has advantages and disadvantages.

3.3.1 Plane waves

Plane wave bases have been very popular in materials science and solid-state physics for performing bandstructure calculations. For example, in the context of pseudopotentials methods, plane wave bases can be quite effective in representing the orbitals for crystalline periodic matter, requiring a small number of plane waves. This leads to a compact representation of the Schrödinger operator. The resulting matrix is dense in Fourier (plane wave) space, but it is not formed explicitly. Instead, matrix-vector product operations are performed with the help of fast Fourier transforms. This plane wave approach is akin to spectral techniques used in solving certain types of partial differential equations [24]. The plane wave basis used is of the form:

$$\psi_{\mathbf{k}}(\mathbf{r}) = \sum_{\mathbf{G}} \alpha(\mathbf{k}, \mathbf{G}) \exp\left(i(\mathbf{k} + \mathbf{G}) \cdot \mathbf{r}\right) \tag{10}$$

where $\mathbf{k}$ is the wave vector, $\mathbf{G}$ is a reciprocal lattice vector, and $\alpha(\mathbf{k}, \mathbf{G})$ represent the coefficients of the basis. Thus, each plane wave is labeled by a wave vector which is a triplet of 3 integers, i.e., $\mathbf{k} = (\mathbf{k}_1, \mathbf{k}_2, \mathbf{k}_3)$. The vector parameter $\mathbf{G}$ translates the periodicity of the wave function with respect to a lattice which attempts to describe a crystalline structure of the atoms.

3.3.2 Linear Combination of Atomic Orbitals (LCAO)

An appealing approach uses a basis set of orbitals localized around the atoms. This is the approach, for example, taken in the SIESTA code [68] where with each atom a is associated with a basis set of functions which combine radial functions around a with spherical harmonics:

$$\phi_{lmn}^{a}(\mathbf{r}) = \phi_{ln}^{a}(\mathbf{r}_a) Y_{lm}(\hat{\mathbf{r}}_a)$$

where $\mathbf{r}_a = \mathbf{r} - \mathbf{R}_a$.

In contrast to plane wave methods, LCAO techniques cannot be universally and systematically improved towards convergence. On the positive side, LCAO benefits from a large collection of local basis sets that has been refined over the years by the quantum chemistry community to obtain high level of accuracy in simulations. Atomic orbital basis also yields much smaller matrices and requires less memory than plane wave methods. The sparsity of the matrices depends on how many neighboring atoms are accounted for in the linear combination.

A popular basis employed with pseudopotentials is that of Gaussian orbitals[13, 17, 32, 33]. Gaussian bases have the advantage of yielding analytical matrix elements provided the potentials are also expanded in Gaussians. However, the implementation of a Gaussian basis is not as straightforward as with plane waves. For example, numerous indices must be employed to label the state, the atomic site, and the Gaussian orbitals used.

3.3.3 Real-Space Methods

When applied to electronic structure calculations, real-space mesh techniques exhibit the following significant advantages: (1) they avoid deriving global basis sets for a specific problem by employing universal mathematical approximations at local regions in the physical space; (2) they can easily handle the treatment of various boundary conditions such as Dirichlet, Neumann, or mixed (such as self-energy functions useful in transport problems [59]); (3) they produce very sparse matrices and are cast as linear scaling electronic structure discretization methods; (4) they allow solving the Poisson equation for electrostatics using the same numerical grid; (5) they can benefit from the recent advances made in mathematical modeling techniques and numerical algorithm design including multigrids, domain decomposition, or direct and Krylov-subspace iterative techniques. All of these properties motivated the development of real-space mesh software packages for electronic structure calculations such as Octopus [5, 8], MIKA [3], PARSEC [6, 37], and NESSIE [4].

Finite Differences An appealing discretization alternative is to avoid traditional explicit bases altogether and work instead in real space, by discretizing the space variable. This can be achieved with Finite Difference Methods (FDM), see, e.g., [9, 22, 27, 30, 37, 41, 53, 72]. FDM is the simplest real-space method which utilizes finite difference discretization on a cubic grid. One of the most popular schemes is to use regular grids with high-order discretizations [25] for the Laplacian which represents the kinetic energy operator. Such high order schemes significantly improve convergence of the eigenvalue problem when compared with standard, low order, finite difference methods. With a uniform grid where the points are described in a finite domain by (x_i, y_j, z_k), $\frac{\partial^2 \psi}{\partial x^2}$ at (x_i, y_j, z_k) is approximated by

$$\frac{\partial^2 \psi}{\partial x^2} = \sum_{n=-M}^{M} C_n \psi(x_i + nh, y_j, z_k) + O(h^{2M}) \tag{11}$$

where h is the grid spacing. Thus using a total of $2M + 1$ points in each direction yields an error of order $O(h^{2M})$. Algorithms are available to compute the coefficients C_n for arbitrary order in h [25].

With the kinetic energy operator expanded as in (11), one can set up a one-electron Schrödinger equation over a grid. One may assume a uniform grid, but this is not a necessary requirement. Once the Kohn–Sham equation is discretized using high order finite differences, we obtain a standard matrix eigenvalue problem of the form:

$$A\psi = \lambda\psi \tag{12}$$

in which A is a real sparse symmetric matrix. Note that the discretization (11) for the kinetic energy term will lead to $2M$ nonzero entries for each of the 3 directions, plus the diagonal entry, so we end up with a total of $6M + 1$ nonzero entries, to which we need to add the nonzero entries that correspond to the other terms of the Hamiltonian. The Hartree and exchange correlation terms usually lead to a diagonal matrix, while the external potential is non-local and leads to a sort of low-rank matrix centered around each atom. An example of such a matrix is shown in Fig. 2.

Fig. 2 Matrix resulting from a 12-th order ($M = 6$) FD discretization of the Kohn–Sham equation. The matrix is obtained from a Parsec simulation of a small silicon cluster passivated by hydrogen atoms (Si10H16). A spherical domain is used which explains the curved diagonals

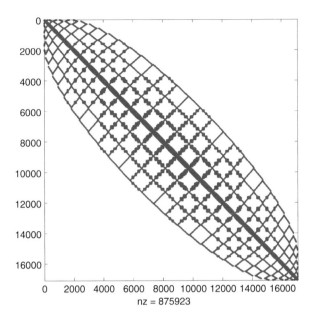

nz = 875923

A grid based on points uniformly spaced in a three- dimensional cube is typically used. Many points in the cube are far from any atoms in the system and the wave function on these points may be replaced by zero. Special data structures may be used to discard these points and keep only those having a nonzero value for the wave function. The size of the Hamiltonian matrix is usually reduced by a factor between two and three with this strategy, which is quite important considering the large number of eigenvectors which must be saved. Further, since the Laplacian can be represented by a simple stencil, and since all local potentials sum up to a simple diagonal matrix, the Hamiltonian need not be stored explicitly as a sparse matrix. Handling the ionic pseudopotential is complex as it consists of a local and a non-local term. In the discrete form, the non-local term becomes a sum over all atoms, a, and quantum numbers, (l, m) of rank-one updates:

$$V_{ion} = \sum_a V_{loc,a} + \sum_{a,l,m} c_{a,l,m} \mathbf{U}_{a,l,m} \mathbf{U}_{a,l,m}^T \tag{13}$$

where $\mathbf{U}_{a,l,m}$ are sparse vectors which are only nonzero in a localized region around each atom, and $c_{a,l,m}$ are normalization coefficients.

Finite Elements One of the main advantages of the finite element method (FEM) is its flexibility to be used with non-uniform meshes and include local refinement by adding more nodes in various regions of interests. In electronic structure calculations, local refinement is important to capture the strong variations of potential and electron density in the vicinity of the atom center regions. Consequently, FEM has been employed in some electronic structure codes [42, 43] as a way to bypass the

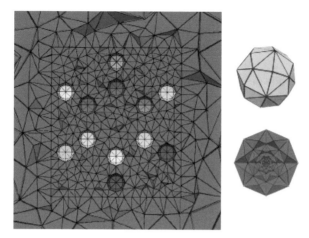

Fig. 3 The figures represent a 2D cross-section of 3D finite element mesh using a coarser interstitial mesh (left) connecting all of the atoms of a benzene molecule, and a much finer mesh (right) for the atom-centered regions suitable to capture the highly localized core states around the nuclei

pseudopotential approach and consider the full core potential. These calculations are called all-electron calculations since both core and valence electrons are included.

As illustrated in Fig. 3 with the example of a benzene molecule, the 3D finite-element mesh can be built in two steps: (1) a 3D atom-centered mesh which is highly refined around the nucleus to capture the core states and (2) a much coarser 3D interstitial mesh that connects all the atom-centered regions. For the atom-centered mesh, successive layers of polyhedra as proposed in [42], along with cubic finite element, do provide high level of accuracy for solving single atom systems. Not only, the distance between layers can be systematically refined while approaching the nucleus, the outer-layer is consistently providing the same (relatively small) number of connectivity nodes that will be used by the coarser interstitial mesh at the surface with the atoms. This approach, used in the NESSIE code [4], is ideally suited for domain-decomposition techniques and parallel computing [34].

3.4   Comparison of Discretization Approaches

Real-space approaches have a number of advantages and have become popular in recent years, see [11, 12, 18–20, 23, 24, 27, 35, 41, 54, 74, 80]. It is worth mentioning that the Gordon Prize in 2011 was awarded to a team that relied on finite difference discretization [28] a testimony of the capability of this approach. One of the attractions of space approaches relative to plane waves is that they bypass many of the difficulties involved with non-periodic systems. Although the resulting matrices are usually (much) larger than with plane waves, they are sparse and the methods are easy to implement on parallel computers. However, even on sequential machines, real-space methods can be faster than methods based on traditional approaches.

Comparing finite difference with finite element discretization methods, one can state that finite element methods yield a smaller number of variables but are more difficult to implement.

3.5 Computing the Electron Density

Within the SCF-DFT procedure, solving the linear and symmetric eigenvalue problem at each given iteration step becomes a very challenging part of the calculations.

The most challenging aspect of electronic structure calculations is the high computational cost of calculating the electron density (7) at each step of the DFT/Kohn–Sham self-consistent iterations (see Fig. 1). The electron density is traditionally calculated using all the wave functions (eigenvectors) solution of the Kohn–Sham eigenvalue problem over all occupied energy states. In order to characterize complex systems and nanostructures of current technological interests, many thousands of eigenpairs may indeed be needed. Indeed, all valence electrons (and core electrons if applicable) need to be included in the calculation.

An alternative approach to the wave function formalism consists of performing a contour integration of the Green's function matrix $G(z) = (zB - A)^{-1}$ over the complex energy space [71, 76]. We note that A is the Hamiltonian matrix, and B represents the basis function overlap matrix (i.e., or mass matrix) which is obtained after discretization ($S = I$ when using FDM). At zero temperature, the resulting expression for the electron density in real space is

$$\rho(\mathbf{r}) = -\frac{1}{\pi \iota} \int_{\mathcal{C}} \mathrm{diag}(G(z))dz = 2 \sum_{j=1}^{occup} |\phi_j|^2 \tag{14}$$

where the clockwise complex contour $\mathcal{C}$ includes all the occupied eigenvalues. The contour integration technique represents apriori an attractive alternative approach to the traditional eigenvalue problem since the number of Green's function to be calculated (typically of order $\sim O(10)$ using Gaussian quadrature) is independent of the size of the system. In addition, only the diagonal elements of the Green's function need to be computed (independently) along the integration points. This problem has motivated the development of new algorithms that are able to directly and economically obtain the diagonal elements of the inverse of sparse matrices. For 1D physical structures such as long nanowires which give rise to banded matrices after discretization, it is possible to perform efficient $O(N)$ calculations for obtaining the diagonal elements of the Green's function [10, 47, 77]. For arbitrary

3D systems (i.e., beyond nanowire structures), however, the numerical complexity of a direct solver such as PEXSI is $O(N^2)$ [7].

The Green's function-based alternative to the wave function formalism for computing electron density gives rise to difficulties in algorithmic complexity, parallel scalability, and accuracy. In that regard, it is difficult to bypass the wave function formalism, and progress in large-scale electronic structure calculations are dependent on advances in numerical algorithms for addressing the eigenvalue problem. This is discussed in the next section.

4 Solution of the Eigenvalue Problem

One significant characteristic of the eigenvalue problem that arises from the Kohn–Sham equation is that the number of required eigenvectors is proportional to the number of atoms in the system, and can grow up to thousands or possibly many more depending on the compound being studied. This means that we will have to store an eigenbasis consisting of a large number of vectors. In addition, the vectors of this basis need to be orthogonal. In fact, the biggest part of the cost of existing eigenvalue codes is related to orthogonalization.

In this Section, we will briefly review various diagonalization methods ranging from Lanczos and Davidson to polynomial and rational filtering, and introduce the notion of "slicing." One of the main motivations of filtering is to allow "slices" of the spectrum to be computed independently of one another and orthogonalization between eigenvectors in different slices is no longer necessary.

4.1 Traditional Methods: Subspace Iteration, Lanczos, and Davidson

Large computations based on DFT approaches started in the 1970s after the breakthrough results of Kohn, Hohenberg, and Sham. The use of plane wave bases dominated the arena of electronic structure from that period onward—starting with the trend-setting Car and Parrinello [15] article which was the catalyst in the development of computational codes using plane waves and pseudopotentials. Most computations in the mid-1980s to the 1990s, and still today, rely on plane wave bases. Since the matrices involved were dense and memory was expensive, this was a major limiting factor at the beginning. However, it was soon realized that it was not necessary to store the dense matrix if a code that accesses the matrix only to perform matrix-vector products ("matvecs" thereafter) is employed [50], see also [51]. This is achieved by working in Fourier space and using FFT to go back and forth from real to Fourier space to perform the operations needed for the matvec. An early code based on subspace iteration for eigenvalue problems and called Ritzit, initially written by Rutishauser in Algol [60], became a de facto standard.

The Lanczos algorithm [38] discovered in 1950 re-emerged in the early 1980s in the linear algebra community as a contender to subspace iteration due mainly to its superior effectiveness when computing a small number of eigenvalues at one end of the spectrum. In exact arithmetic, the Lanczos algorithm generates an orthonormal basis $\mathbf{v}_1, \mathbf{v}_2, \ldots, \mathbf{v}_m$, of the Krylov subspace Span$\{\mathbf{v}, \mathcal{A}\mathbf{v}, \mathcal{A}^2\mathbf{v}, \cdots, \mathcal{A}^{m-1}\mathbf{v}\}$ via an inexpensive 3-term recurrence of the form :

$$\beta_{j+1}\mathbf{v}_{j+1} = \mathcal{A}\mathbf{v}_j - \alpha_j\mathbf{v}_j - \beta_j\mathbf{v}_{j-1}$$

In the above sequence, $\alpha_j = \mathbf{v}_j^H \mathcal{A}\mathbf{v}_j$ and $\beta_{j+1} = \|\mathcal{A}\mathbf{v}_j - \alpha_j\mathbf{v}_j - \beta_j\mathbf{v}_{j-1}\|_2$. So the jth step of the algorithm starts by computing α_j and then proceeds to form the vector $\hat{\mathbf{v}}_{j+1} = \mathcal{A}\mathbf{v}_j - \alpha_j\mathbf{v}_j - \beta_j\mathbf{v}_{j-1}$ and then $v_{j+1} = \hat{\mathbf{v}}_{j+1}/\beta_{j+1}$. Note that for $j = 1$, the formula for $\hat{\mathbf{v}}_2$ changes to $\hat{\mathbf{v}}_2 = \mathcal{A}\mathbf{v}_2 - \alpha_2\mathbf{v}_2$.

Suppose that m steps of the recurrence are carried out and consider the tridiagonal matrix,

$$\mathcal{T}_m = \begin{pmatrix} \alpha_1 & \beta_2 & & \\ \beta_2 & \alpha_2 & \beta_3 & \\ & \ddots & \ddots & \ddots \\ & & \beta_m & \alpha_m \end{pmatrix}$$

Further, denote by $\mathcal{V}_m$ the $n \times m$ matrix $\mathcal{V}_m = [\mathbf{v}_1, \ldots, \mathbf{v}_m]$ and by $\mathbf{e}_m$ the mth column of the $m \times m$ identity matrix. After m steps of the algorithm, the following relation holds:

$$\mathcal{A}\mathcal{V}_m = \mathcal{V}_m\mathcal{T}_m + \beta_{m+1}\mathbf{v}_{m+1}\mathbf{e}_m^T$$

It is observed, and can be theoretically shown, that some of the eigenvalues of the tridiagonal matrix $\mathcal{T}_m$ will start approximating corresponding eigenvalues of $\mathcal{A}$ when m becomes large enough. An eigenvalue $\tilde{\lambda}$ of $\mathcal{T}_m$ is called a Ritz value, and if $\mathbf{y}$ is an associated eigenvector, then the vector $\mathcal{V}_m\mathbf{y}$ is, by definition, the Ritz vector, i.e., the approximate eigenvector of $\mathcal{A}$ associated with $\tilde{\lambda}$. If m is large enough, the process may yield good approximations to the desired eigenvalues $\lambda_1, \ldots, \lambda_s$ of $\mathcal{A}$, corresponding to the occupied states, i.e., all occupied eigenstates.

In practice, orthogonality of the Lanczos vectors, which is guaranteed in theory, is lost and this phenomenon takes place as soon as one of the eigenvectors starts to converge [55, 56]. Orthogonality can be reinstated in a number of ways, see [39, 40, 66, 67, 75].

The Davidson [52] method is a sort of preconditioned version of the Lanczos algorithm, in which the preconditioner is the diagonal of $\mathcal{A}$. We refer to the generalized Davidson algorithm as a Davidson approach in which the preconditioner is not restricted to being a diagonal matrix (A detailed description can be found in [62].)

The Davidson algorithm differs from the Lanczos method in the way in which it defines new vectors to add to the projection subspace. Instead of adding just $\mathcal{A}\mathbf{v}_j$, it preconditions a given residual vector $\mathbf{r}_i = (\mathcal{A} - \mu_i\mathcal{I})\mathbf{u}_i$ and adds it to the subspace (after orthogonalizing it against current basis vectors). The algorithm consists of an "eigenvalue loop" which computes the desired eigenvalues one by one (or a few at a time), and a "basis" loop which gradually computes the subspace on which to perform the projection. Consider the eigenvalue loop which computes the ith eigenvalue and eigenvector of $\mathcal{A}$. If $\mathcal{M}$ is the current preconditioner, and $\mathcal{V} = [\mathbf{v}_1, \cdots, \mathbf{v}_k]$ is the current basis, the main steps of the main loop are as follows:

1. Compute the ith eigenpair $(\mu_k, \mathbf{y}_k)$ of $\mathcal{C}_k = \mathcal{V}_k^T \mathcal{A}\mathcal{V}_k$
2. Compute the residual vector $\mathbf{r}_k = (\mathcal{A} - \mu_k\mathcal{I})\mathcal{V}_k\mathbf{y}_k$
3. Precondition $\mathbf{r}_k$, i.e., compute $\mathbf{t}_k = \mathcal{M}^{-1}\mathbf{r}_k$
4. Orthonormalize $\mathbf{t}_k$ against $\mathbf{v}_1, \cdots, \mathbf{v}_k$ and call $\mathbf{v}_{k+1}$ the resulting vector, so $\mathcal{V}_{k+1} = [\mathcal{V}_k, \mathbf{v}_{k+1}]$
5. Compute last column-row of $\mathcal{C}_{k+1} = \mathcal{V}_{k+1}^T \mathcal{A}\mathcal{V}_{k+1}$

The original Davidson approach used the diagonal of the matrix as a preconditioner but this works only for limited cases. For plane wave bases, it is possible to construct fairly effective preconditioners by exploiting the lower order bases. By this we mean that if $\mathcal{H}_k$ is the matrix representation obtained by using k plane waves, we can construct a good approximation to $\mathcal{H}_k$ from $\mathcal{H}_m$ with $m \ll k$, by completing it with a diagonal matrix representing the larger (undesirable) modes. Note that these matrices are not explicitly computed as they are dense. This possibility of building lower dimensional approximations to the Hamiltonian which can be used to precondition the original matrix constitutes an advantage of plane wave-based methods.

4.2 Nonlinear Chebyshev Filtered Subspace Iteration

A big disadvantage of the Lanczos and Davidson iterations is that they do not allow to exploit previous bases that have been calculated from earlier SCF iterations. A look at Fig. 1 indicates that what matters for convergence is how well the procedure is approximating the basis of the subspace corresponding to the n occupied states. At the next SCF iteration, the Lanczos algorithm starts with one vector only. This means that we cannot fully take advantage of the basis that has been computed previously. In contrast, the subspace iteration algorithm is ideal in this context. All we need to do at the next SCF iteration is update the Hamiltonian—and use whatever subspace we had from the previous SCF iteration. This constitutes a major attraction of subspace iteration. Another attraction is clearly its added parallelism.

Algorithm 1: $[Y] = \texttt{Chebyshev\_filter}(X, m, a, b, g)$

---

1 $e = (b - a)/2$; $\ c = (a + b)/2$; $\ \sigma = e/(c - g)$; $\ \tau = 2/\sigma$;
2 $Y = (A * X - c * X) * (\sigma/e)$;
3 **for** $i = 2$ *to* m **do**
4 $\quad$ $\sigma_{new} = 1/(\tau - \sigma)$;
5 $\quad$ $Y_t = (A * Y - c * Y) * (2 * \sigma_{new}/e) - (\sigma * \sigma_{new}) * X$;
6 $\quad$ $X = Y$; $\ Y = Y_t$; $\ \sigma = \sigma_{new}$;

The main ingredient of a subspace iteration procedure is the Chebyshev filtering. Given a basis $[v_1, \ldots, v_m]$, each vector is "filtered" as $\hat{v}_i = P_k(A)v_i$, where p_k is a low degree polynomial whose goal is to enhance the wanted components of these vectors in the desired eigenvectors of A. The most common filters used are shifted and scaled Chebyshev polynomials. If $[a, \ b]$ is the interval containing unwanted eigenvalues, those that must be dampened, then we use the polynomial

$$p_k(t) = \frac{C_k(l(t))}{C_k(l(g))}; \quad \text{with} \quad l(t) = \frac{2t - b - a}{b - a}$$

where C_k is the Chebyshev polynomial of degree k of the first kind and g is some approximation of the eigenvalue that is farthest from the center $(a + b)/2$ of the interval—which is used for scaling. One such polynomial of degree 7 is shown in Fig. 4. The 3-term recurrence of Chebyshev polynomial is exploited to compute $p_k(A)v$. If $B = l(A)$, then $C_{k+1}(t) = 2tC_k(t) - C_{k-1}(t) \rightarrow w_{k+1} = 2Bw_k - w_{k-1}$. Algorithm 1 provides an illustration of Chebyshev filtering.

What was discussed above is what might be termed a standard SCF approach in which a filtered subspace iteration is used to compute the eigenvalues at each

Fig. 4 Degree 8 Chebyshev filter

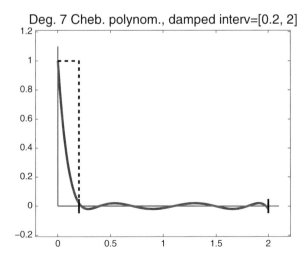

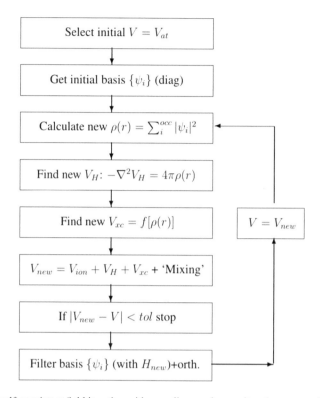

Fig. 5 The self-consistent field iteration with a nonlinear subspace iteration approach

SCF iteration. The subspace iteration can also be used in a *nonlinear* way. In the nonlinear subspace iteration, the filtering step is not used to compute eigenvectors accurately. Instead, the basis is filtered and the Hamiltonian is updated immediately using these vectors. In essence the process amounts to removing one loop from the algorithm in that the SCF and the diagonalization loops merged. The new SCF iteration is illustrated in Fig. 5. Experiments reported in [78] reported that this procedure can yield a factor of 10 speed-up over the more traditional one in which the inner eigenvalue loop is kept

4.3 EVSL: Filtering and Spectrum Slicing

As mentioned earlier, a big part of the cost of computing a large number of eigenvectors is related to the process of maintaining orthogonality between these vectors. The number of vectors to orthogonalize is typically of the order of the number of states which is itself proportional to the number of particles, and so the cost increases quadratically with the number of particles. This was observed early on and a number of articles sought inexpensive alternatives. One of the main ideas proposed was one based on *filtering*, i.e., transforming the Hamiltonian so as to enhance or magnify the desired part of the spectrum by a polynomial of rational transformation to enable a projection method like subspace iteration, to extract the desired eigenvalues easily. An early contribution along these lines is the article by Zunger [73] which discusses a scheme, whereby the Hamiltonian $\mathcal{H}$ is replaced by $B = (\mathcal{H} - \sigma I)^2$. Extracting the smallest eigenpairs of B will yield the eigenvectors associated with the eigenvalues closest to the shift σ. A similarly simple technique is one that is based on shift-and-invert [56] which uses a rational filter.

The essence of a filtering technique is to replace the original matrix $\mathcal{A}$ by $B = \phi(\mathcal{A})$, where the filter ϕ is either a polynomial or rational function. The main advantage of filtering is that it allows to compute different parts of the spectrum independently. A *spectrum slicing* method refers to a technique that computes the desired spectrum by sub-intervals or "slices." The recently developed package named EVSL (for Eigenvalues Slicing Library) relies entirely on this strategy [1, 44, 45]. Figure 6 illustrates the main motivation for this strategy, namely that eigenvectors belonging to slices that are far apart need not be orthogonalized against each other.

The gain in computational cost that comes from avoiding limiting or orthogonalization can be significant both in terms of computational time and in terms of memory. For example, Fig. 7 illustrates a calculation with EVSL in which all eigenvalues in the interval [0, 1] of a Laplacian discretized on a $49 \times 49 \times 49$ centered finite difference grid. A spectrum slicing strategy is exploited and the total cost is shown as the number of intervals varies from 1 to 6. Note that in EVSL the degree of the polynomial filter is computed automatically. One can observe that orthogonalization costs are drastically reduced along with costs related to the projection process. At the same time the cost of matvecs increases but it remains insignificant relative to the rest. This calculation is performed without fully taking advantage of parallelism. If a fully parallel computation was to be implemented, each of the total times would have been divided by the number of intervals used.

Fig. 6 Two filters to compute two slices of the spectrum that are far apart. Note that eigenvectors associated with two distinct slices need not be orthogonalized against each other

Fig. 7 Cost of calculating all eigenvalues of a Laplacian matrix in the interval [0, 1] by a polynomial filtered non-restart Lanczos method. There are 1971 eigenvalues in the interval and they are computed by slicing the spectrum into 1, 2, ⋯ , 6 sub-intervals

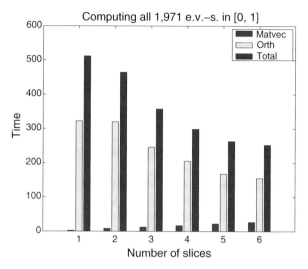

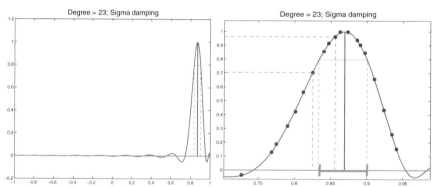

Fig. 8 A filter polynomial of degree 23 (left) and a zoom of the same polynomial near the interval of interest (right)

To illustrate how polynomial filtering is combined with a non-restarted version of the Lanczos algorithm we show in Fig. 8 a polynomial filter of the type used in EVSL. In the figure, an eigenvalue λ_i located inside the interval of desired eigenvalues is transformed to $\phi(\lambda_i)$. The filter is designed so that any eigenvalue λ_i located inside the interval of desired eigenvalues is transformed into an eigenvalue $\phi(\lambda_i)$ that is larger than or equal to a certain value (called the "bar") which is $\beta = 0.8$ in the figure. This makes it easy to distinguish between wanted eigenvalues ($\phi(\lambda_i \geq \beta$) and unwanted ones ($\phi(\lambda_i < \beta$). Figure 9 shows the filtered eigenvalues for the same problem. As is highlighted in the figure, all wanted eigenvalues of the original problem are now eigenvalues that are not smaller than $\beta = 0.8$ for the filtered matrix. It is therefore possible to devise a strategy, whereby these eigenvalues are all computed from a Lanczos algorithm with full

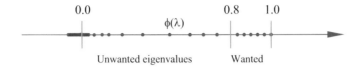

Fig. 9 Eigenvalues of the filtered matrix with the filter of Fig. 8

reorthogonalization and no restarts. If the degree of the polynomial is well selected and the sub-interval contains a reasonable number of eigenvalues, this strategy works quite well in practice.

EVSL solves large sparse real symmetric standard and generalized eigenvalue problems. In order to enable a spectrum slicing strategy, the methods in EVSL rely on a quick calculation of the spectral density of a given matrix, or a matrix pair. Once this is done the driver will then cut the interval into slices so that each slice will have approximately the same number of eigenvalues. What distinguishes EVSL from other currently available packages is that EVSL relies entirely on filtering techniques. While much effort has been devoted to develop effective polynomial filtering the package also implements rational filters. The projection methods developed in the package are the Lanczos methods without restart, or with thick restart, as well as the subspace iteration method. Various interfaces are available for various scenarios, including matrix-free modes, whereby the user can supply his/her own functions to perform matrix-vector operations or to solve sparse linear systems. A fully parallel version is currently being developed.

4.4 FEAST: Rational Filtering and Spectrum Slicing

Equation (14) indicates that the contour integration technique does not provide a natural route for obtaining the individual occupied wave functions but rather the summation of their amplitudes square. The FEAST algorithm was originally proposed to reconcile both wave function and Green's function formalism and provide an efficient and scalable new approach for solving the eigenvalue problem [58]. FEAST can be applied for solving both standard and generalized form of the Hermitian or non-Hermitian problem, and it belongs to the family of contour integration eigensolvers along with the Sakurai and Sugiura (SS) method [63, 64]. In contrast to the Krylov-based SS method, FEAST is a subspace iteration method that uses the Rayleigh–Ritz projection and an approximate spectral projector as a filter [57]. Given a Hermitian generalized eigenvalue problem $AX = BX\Lambda$ of size n, the algorithm in Fig. 10 outlines the main steps of a generic Rayleigh-Ritz subspace iteration procedure for computing m eigenpairs. At convergence, the algorithm yields the B-orthonormal eigensubspace $Y_m \equiv X_m = \{x_1, x_2, \ldots, x_m\}_{n \times m}$ and associated eigenvalues $\Lambda_{Q_m} \equiv \Lambda_m$. Taking $\rho(B^{-1}A) = B^{-1}A$ yields the bare-bone subspace iteration (generalization of the power method) which converges

0. Start: Select random subspace $Y_{m_0} \equiv \{y_1, y_2, \ldots, y_{m_0}\}_{n \times m_0}$ $(n \gg m_0 \geq m)$
1. Repeat until convergence
2.   Compute $Q_{m_0} = \rho(B^{-1}A)Y_{m_0}$
3.   Orthogonalize Q_{m_0}
4.   Compute $A_Q = Q_{m_0}^H A Q_{m_0}$ and $B_Q = Q_{m_0}^H B Q_{m_0}$
5.   Solve $A_Q W = B_Q W \Lambda_Q$ with $W^H B_Q W = I_{m_0 \times m_0}$
6.   Compute $Y_{m_0} = Q_{m_0} W$
7.   Check convergence of Y_{m_0} and $\Lambda_{Q_{m_0}}$ for the m wanted eigenvalues
8. End

Fig. 10 Subspace iteration method with Rayleigh–Ritz projection

towards the m dominant eigenvectors with the linear rate $|\lambda_{m_0+1}/\lambda_i|_{i=1,\ldots,m}$ [56, 61]. This standard approach is never used in practice. Instead, it is combined with filtering using the function ρ which aims at improving the convergence rate (i.e., $|\rho(\lambda_{m_0+1})/\rho(\lambda_i)|_{i=1,\ldots,m}$) by increasing the gap between wanted and unwanted eigenvalues. An ideal filter for the interior eigenvalue problem which maps all m wanted eigenvalues to one and all unwanted ones to zero can be derived from the Cauchy (or Dunford) integral formula:

$$\rho(\lambda) = \frac{1}{2\pi\iota} \oint_C dz (z - \lambda)^{-1} \tag{15}$$

where the wanted eigenvalues are located inside a complex contour C. The filter then becomes a spectral projector, with $\rho(B^{-1}A) = X_m X_m^H B$, for the eigenvector subspace X_m (i.e., $\rho(B^{-1}A)X_m = X_m$) and can be written as:

$$\rho(B^{-1}A) = \frac{1}{2\pi\iota} \oint_C dz (zB - A)^{-1} B. \tag{16}$$

FEAST uses a numerical quadrature to approximately compute the action of this filter onto a set of m_0 vectors along the subspace iterations. The resulting rational function ρ_a that approximates the filter (15) is given by

$$\rho_a(z) = \sum_{j=1}^{n_e} \frac{\omega_j}{z_j - z} \tag{17}$$

where $\{z_j, \omega_j\}_{1 \leq j \leq n_e}$ are the nodes and related weights of the quadrature. We obtain for the subspace Q_{m_0} in step 2 of the algorithm in Fig. 10:

$$Q_{m_0} = \rho_a(B^{-1}A)Y_{m_0} = \sum_{j=1}^{n_e} \omega_j (z_j B - A)^{-1} B Y_{m_0} \equiv X \rho_a(\Lambda) X^H B Y_{m_0} \tag{18}$$

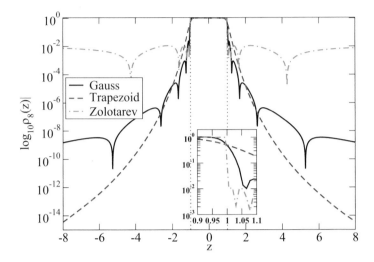

Fig. 11 Variations of the rational functions $\rho_8(\lambda)$ ($n_e = 8$ contour points) associated with Gauss, Trapezoidal, and Zolotarev quadrature rules. While Trapezoidal presents a more regular decay than Gauss, the latter produces smaller values for the rational function just outside the edges of the search interval $|z| > 1$. From the caption, we note that Zolotarev presents a dramatic drop in the rational function at $z = 1$ (i.e., fastest possible decay), but this value quickly saturates

In practice, Q_{m_0} can be computed by solving a small number of (independent) shifted linear systems over a complex contour.

$$Q_{m_0} = \sum_{j=1}^{n_e} \omega_j Q_{m_0}^{(j)}, \quad \text{with } Q_{m_0}^{(j)} \text{ solution of} \quad (z_j B - A) Q_{m_0}^{(j)} = B Y_{m_0} \quad (19)$$

As shown in Fig. 11, a relatively small number of quadrature nodes (using Gauss, Trapezoidal, or Zolotarev [26] rules) on a circular contour suffices to produce a rapid decay of the function ρ_a from ≈ 1 within the search contour to ≈ 0 outside. In comparison with more standard polynomial filtering [61, 69], the rational filter (17) can lead to a very fast convergence of the subspace iteration procedure. In addition, all the m desired eigenvalues are expected to converge at the same rate (since $\rho_a(\lambda_i) \simeq 1$ if λ_i is located within the search interval). The convergence rate of FEAST does not only depend upon the decay properties of the rational function ρ_a, but also on the size of the search subspace m_0 which must not be chosen smaller than the number of eigenvalues inside the search contour (i.e., $m_0 \geq m$). Users of the FEAST eigensolver[2] are then responsible for specifying an interval to search for the eigenvalues and a subspace size m_0 that overestimate the number of the wanted eigenvalues. Once these conditions are satisfied, FEAST offers the following set of appealing features:

(i) high robustness with well-defined convergence rate $|\rho_a(\lambda_{m_0+1})/\rho_a(\lambda_i)|_{i=1,\ldots,m}$;

(ii) all multiplicities naturally captured;

(iii) no explicit orthogonalization procedure on long vectors required in practice (i.e., step-3 in Fig. 10 is unnecessary as long as B_Q is positive definite). We note in (18) that Q_{m_0} is naturally spanned by the eigenvector subspace;

(iv) reusable subspace capable of generating suitable initial guess when solving a series of eigenvalue problems such as the ones that appear in DFT-SCF iterations;

(v) can exploit natural parallelism at three different levels: search intervals can be treated separately (no overlap) while maintaining orthogonality—linear systems can be solved independently across the quadrature nodes of the complex contour—each complex linear system with m_0 multiple right-hand sides can be solved in parallel. Consequently, in a parallel environment, the algorithm complexity depends on solving a single linear system using a direct or an iterative method.

Using FEAST, the total number of processes n_{pp} can be distributed over three levels of parallelism: (i) eigenvalue level parallelism using i filters (i.e., i slices); (ii) block level parallelism where all the k linear systems are solved independently; (iii) domain level parallelism which handles the system matrices and the multiple right-hand sides using the remaining p processes available since $n_{pp} = i \times k \times p$. Achieving a good balance in suitable distribution of the parallel resources among all slices would require that the number of eigenvalues in each slice be roughly the same. Obviously, it can be quite challenging for a user to perform a customized slicing by first guessing the distribution of the eigenvalue spectrum. Recent work on stochastic estimates can be helpful in this regard [21, 46]. One possible estimate on the eigenvalue count in an interval consists of approximating the trace of the spectral projector by exploiting the rational function expansion (19), i.e.,

$$\text{tr}(\tilde{P}) \approx \frac{n}{n_v} \sum_{j=1}^{k} \omega_j \sum_{i=1}^{n_v} v_i^T (\sigma_j B - A)^{-1} B v_i \qquad (20)$$

The cost of this estimation can remain relatively small since the linear systems can be solved with low accuracy and with a very small number of right-hand sides n_v. Furthermore, if the factorizations can already be computed at each complex shift σ_j, they can be reused in the subsequent subspace iteration.

5   Conclusion

Atom-by-atom large-scale first-principle calculations have become critical for supplementing the experimental investigations and obtaining detailed electronic structure properties and reliable characterization of emerging nanostructures. First-principle calculations most often rely on a succession of modeling trade-offs between accuracy and performances, which can be broadly divided into four major steps: (1) physical, (2) mathematical, (3) discretization, and (4) computing. These modeling steps contain different layers of numerical approximations which are most often tightly tied together. In order to improve on current software implementation by fully capitalizing on modern HPC computing platforms, it is essential to revisit not one, but all the various stages of the electronic structure modeling process which have been summarized in this chapter.

Solutions of the DFT/Kohn–Sham problem are routinely used in the calculations of many ground state properties of small molecular systems or crystal unit-cells containing a handful of atoms. In order to characterize large-scale complex systems and nanostructures of current technological interest, the SCF-DFT procedure would require repeated computations of many tens of thousands of eigenvectors, for eigenvalue systems that can have sizes in the tens of millions. In this case, a divide-and-conquer approach that can compute wanted eigenpairs by parts, becomes mandatory, since windows or slices of the spectrum can be computed independently of one another and orthogonalization between eigenvectors in different slices is no longer necessary. All these issues have originally motivated the development of the EVSL and FEAST approaches that were discussed here.

Acknowledgments The work of the author "Eric Polizzi" was supported by NSF awards CCF-1510010 and SI2-SSE-1739423. The work of the author "Yousef Saad" was supported by NSF award CCF-1505970.

References

1. *Evsl.* http://www.cs.umn.edu/~saad/software/EVSL.
2. *Feast.* http://www.feast-solver.org.
3. *Mika.* http://www.csc.fi/physics/mika.
4. *Nessie.* http://www.nessie-code.org.
5. *Octopus.* http://www.tddft.org/programs/octopus/.
6. *Parsec.* http://www.ices.utexas.edu/parsec/index.html.
7. *Pexsi.* https://math.berkeley.edu/~linlin/pexsi/download/doc_v0.6.0/.
8. X. ANDRADE, J. ALBERDI-RODRIGUEZ, D. A. STRUBBE, M. J. T. OLIVEIRA, F. NOGUEIRA, A. CASTRO, J. MUGUERZA, A. ARRUABARRENA, S. G. LOUIE, A. ASPURU-GUZIK, A. RUBIO, AND M. A. L. MARQUES, *Time-dependent density-functional theory in massively parallel computer architectures: the octopus project*, Journal of Physics: Condensed Matter, 24 (2012), p. 233202.
9. X. ANDRADE, D. A. STRUBBE, U. D. GIOVANNINI, A. H. LARSEN, M. J. T. OLIVEIRA, J. ALBERDI-RODRIGUEZ, A. VARAS, I. THEOPHILOU, N. HELBIG, M. VERSTRAETE, L. STELLA, F. NOGUEIRA, A. ASPURU-GUZIK, A. CASTRO, M. A. L. MARQUES, AND A. RUBIO, *Real-space grids and the octopus code as tools for the development of new simulation approaches for electronic systems*, Physical Chemistry Chemical Physics, 17 (2015), pp. 31371–31396.
10. S. BARONI AND P. GIANNOZZI, *Towards very large-scale electronic-structure calculations*, Europhysics Letters (EPL), 17 (1992), pp. 547–552.
11. T. L. BECK, *Real-space mesh techniques in density functional theory*, Rev. Mod. Phys., 74 (2000), pp. 1041–1080.

12. E. L. Briggs, D. J. Sullivan, and J. Bernholc, *Large-scale electronic-structure calculations with multigrid acceleration*, Phys. Rev. B, 52 (1995), pp. R5471–R5474.
13. A. Briley, M. R. Pederson, K. A. Jackson, D. C. Patton, and D. V. Porezag, *Vibrational frequencies and intensities of small molecules: All-electron, pseudopotential, and mixed-potential methodologies*, Phys. Rev. B, 58 (1997), pp. 1786–1793.
14. K. Burke, *Perspective on density functional theory.*, The Journal of chemical physics, 136 15 (2012), p. 150901.
15. R. Car and M. Parrinello, *Unified approach for molecular dynamics and density functional theory*, Phys. Rev. Lett., 55 (1985), pp. 2471–2474.
16. J. R. Chelikowsky and M. L. Cohen, *Pseudopotentials for semiconductors*, in Handbook of Semiconductors, T. S. Moss and P. T. Landsberg, eds., Elsevier, Amsterdam, 2nd edition, 1992.
17. J. R. Chelikowsky and S. G. Louie, *First-principles linear combination of atomic orbitals method for the cohesive and structural properties of solids: Application to diamond*, Phys. Rev. B, 29 (1984), pp. 3470–3481.
18. J. R. Chelikowsky, N. Troullier, X. Jing, D. Dean, N. Biggeli, K. Wu, and Y. Saad, *Algorithms for the structural properties of clusters*, Comp. Phys. Comm., 85 (1995), pp. 325–335.
19. J. R. Chelikowsky, N. Troullier, and Y. Saad, *The finite-difference-pseudopotential method: Electronic structure calculations without a basis*, Phys. Rev. Lett., 72 (1994), pp. 1240–1243.
20. J. R. Chelikowsky, N. Troullier, K. Wu, and Y. Saad, *Higher order finite difference pseudopotential method: An application to diatomic molecules*, Phys. Rev. B, 50 (1994), pp. 11355–11364.
21. E. Di Napoli, E. Polizzi, and Y. Saad, *Efficient estimation of eigenvalue counts in an interval*, Numerical Linear Algebra with Applications, 23 (2016), pp. 674–692.
22. J.-L. Fattebert and J. Bernholc, *Towards grid-based O(N) density-functional theory methods: Optimized nonorthogonal orbitals and multigrid acceleration*, Phys. Rev. B, 62 (2000), pp. 1713–1722.
23. ——, *Towards grid-based O(N) density-functional theory methods: Optimized nonorthogonal orbitals and multigrid acceleration*, Phys. Rev. B, 62 (2000), pp. 1713–1722.
24. C. Y. Fong, *Topics in Computational Materials Science*, World Scientific, 1998.
25. B. Fornberg and D. M. Sloan, *A review of pseudospectral methods for solving partial differential equations*, Acta Numer., 94 (1994), pp. 203–268.
26. S. Güttel, E. Polizzi, P. Tang, and G. Viaud, *Zolotarev quadrature rules and load balancing for the feast eigensolver*, SIAM Journal on Scientific Computing, 37 (2015), pp. A2100–A2122.
27. F. Gygi and G. Galli, *Real-space adaptive-coordinate electronic-structure calculations*, Phys. Rev. B, 52 (1995), pp. R2229–R2232.
28. Y. Hasegawa, J. Iwata, M. Tsuji, D. Takahashi, A. Oshiyama, K. Minami, T. Boku, F. Shoji, A. Uno, M. Kurokawa, H. Inoue, I. Miyoshi, and M. Yokokawa, *First-principles calculations of electron states of a silicon nanowire with 100,000 atoms on the k computer*, in Proceedings of 2011 International Conference for High Performance Computing, Networking, Storage and Analysis, SC '11, New York, NY, USA, 2011, ACM, pp. 1:1–1:11.
29. A. Haug, *Theoretical Solid State Physics*, Pergamon Press, 1972.
30. M. Heikanen, T. Torsti, M. J. Puska, and R. M. Nieminen, *Multigrid method for electronic structure calculations*, Phys. Rev. B, 63 (2001), pp. 245106–245113.
31. P. Hohenberg and W. Kohn, *Inhomogeneous electron gas*, Phys. Rev., 136 (1964), pp. B864–B871.
32. K. A. Jackson, M. R. Pederson, D. V. Porezag, Z. Hajnal, and T. Fraunheim, *Density-functional-based predictions of Raman and IR spectra for small Si clusters*, Phys. Rev. B, 55 (1997), pp. 2549–2555.

33. R. W. JANSEN AND O. F. SANKEY, *Ab initio linear combination of pseudo-atomic-orbital scheme for the electronic properties of semiconductors: Results for ten materials*, Phys. Rev. B, 36 (1987), pp. 6520–6531.

34. J. KESTYN, V. KALANTZIS, E. POLIZZI, AND Y. SAAD, *PFEAST: A high performance sparse eigenvalue solver using distributed-memory linear solvers*, in Proceedings of the International Conference for High Performance Computing, Networking, Storage and Analysis, SC '16, Piscataway, NJ, USA, 2016, IEEE Press, pp. 16:1–16:12.

35. Y. H. KIM, I. H. LEE, AND R. M. MARTIN, *Object-oriented construction of a multigrid electronic structure code with Fortran*, Comp. Phys. Comm., 131 (2000), pp. 10–25.

36. W. KOHN AND L. J. SHAM, *Self-consistent equations including exchange and correlation effects*, Phys. Rev., 140 (1965), pp. A1133–A1138.

37. L. KRONIK, A. MAKMAL, M. L. TIAGO, M. M. G. ALEMANY, M. JAIN, X. HUANG, Y. SAAD, AND J. R. CHELIKOWSKY, *PARSEC – the pseudopotential algorithm for real-space electronic structure calculations: recent advances and novel applications to nano-structure*, Phys. Stat. Sol. (B), 243 (2006), p. 1063–1079.

38. C. LANCZOS, *An iteration method for the solution of the eigenvalue problem of linear differential and integral operators*, J. Research Nat. Bur. Standards, 45 (1950), pp. 255–282.

39. R. M. LARSEN, *PROPACK: A software package for the symmetric eigenvalue problem and singular value problems on Lanczos and Lanczos bidiagonalization with partial reorthogonalization, SCCM, Stanford University URL:* http://sun.stanford.edu/~rmunk/PROPACK/.

40. ——, *Efficient Algorithms for Helioseismic Inversion*, PhD thesis, Dept. Computer Science, University of Aarhus, DK-8000 Aarhus C, Denmark, October 1998.

41. I. H. LEE, Y. H. KIM, AND R. M. MARTIN, *One-way multigrid method in electronic-structure calculations*, Phys. Rev. B, 61 (2000), p. 4397.

42. L. LEHTOVAARA, V. HAVU, AND M. PUSKA, *All-electron density functional theory and time-dependent density functional theory with high-order finite elements*, The Journal of Chemical Physics, 131 (2009), p. 054103.

43. A. R. LEVIN, D. ZHANG, AND E. POLIZZI, *Feast fundamental framework for electronic structure calculations: Reformulation and solution of the muffin-tin problem*, Computer Physics Communications, 183 (2012), pp. 2370 – 2375.

44. R. LI, Y. XI, L. ERLANDSON, AND Y. SAAD, *The EigenValues Slicing Library (EVSL): Algorithms, implementation, and software*, Tech. Report ys-2018-02, Dept. Computer Science and Engineering, University of Minnesota, Minneapolis, MN, 2018. Submitted. ArXiv: https://arxiv.org/abs/1802.05215.

45. R. LI, Y. XI, E. VECHARYNSKI, C. YANG, AND Y. SAAD, *A Thick-Restart Lanczos algorithm with polynomial filtering for Hermitian eigenvalue problems*, SIAM Journal on Scientific Computing, 38 (2016), pp. A2512–A2534.

46. L. LIN, Y. SAAD, AND C. YANG, *Approximating spectral densities of large matrices*, SIAM Review, 58 (2016), pp. 34–65.

47. L. LIN, C. YANG, J. C. MEZA, J. LU, L. YING, AND E. WEINAN, *Selinv - an algorithm for selected inversion of a sparse symmetric matrix*, ACM Trans. Math. Softw., 37 (2011), pp. 40:1–40:19.

48. S. LUNDQVIST AND N. H. MARCH, eds., *Theory of the Inhomogeneous Electron Gas*, Plenum, 1983.

49. R. MARTIN, R. MARTIN, AND C. U. PRESS, *Electronic Structure: Basic Theory and Practical Methods*, Cambridge University Press, 2004.

50. J. L. MARTINS AND M. COHEN, *Diagonalization of large matrices in pseudopotential band-structure calculations: Dual-space formalism*, Phys. Rev. B, 37 (1988), pp. 6134–6138.

51. J. L. MARTINS, N. TROULLIER, AND S.-H. WEI, *Pseudopotential plane-wave calculations for ZnS*, Phys. Rev. B, 43 (1991), pp. 2213–2217.

52. R. B. MORGAN AND D. S. SCOTT, *Generalizations of Davidson's method for computing eigenvalues of sparse symmetric matrices*, SIAM J. Sci. Comput., 7 (1986), pp. 817–825.

53. T. ONO AND K. HIROSE, *Timesaving double-grid method for real-space electronic structure calculations*, Phys. Rev. Lett., 82 (1999), pp. 5016–5019.

54. ——, *Timesaving double-grid method for real-space electronic structure calculations*, Phys. Rev. Lett., 82 (1999), pp. 5016–5019.
55. C. C. PAIGE, *The Computation of Eigenvalues and Eigenvectors of Very Large Sparse Matrices*, PhD thesis, University of London, 1971.
56. B. N. PARLETT, *The Symmetric Eigenvalue Problem*, no. 20 in Classics in Applied Mathematics, SIAM, Philadelphia, 1998.
57. P. PETER TANG AND E. POLIZZI, *Feast as a subspace iteration eigensolver accelerated by approximate spectral projection*, SIAM Journal on Matrix Analysis and Applications, 35 (2014), pp. 354–390.
58. E. POLIZZI, *A density matrix-based algorithm for solving eigenvalue problems*, phys. rev. B, 79 (2009).
59. E. POLIZZI AND S. DATTA, *Multidimensional nanoscale device modeling: the finite element method applied to the non-equilibrium Green's function formalism*, in 2003 Third IEEE Conference on Nanotechnology, 2003. IEEE-NANO 2003., vol. 1, Aug 2003, pp. 40–43 vol.2.
60. H. RUTISHAUSER, *Simultaneous iteration for symmetric matrices*, in Handbook for automatic computations (linear algebra), J. Wilkinson and C. Reinsch, eds., New York, 1971, Springer Verlag, pp. 202–211.
61. Y. SAAD, *Numerical Methods for Large Eigenvalue Problems*, John Wiley, New York, 1992.
62. Y. SAAD, A. STATHOPOULOS, J. R. CHELIKOWSKY, K. WU, AND S. OGUT, *Solution of large eigenvalue problems in electronic structure calculations*, BIT, 36 (1996), pp. 563–578.
63. T. SAKURAI AND H. SUGIURA, *A projection method for generalized eigenvalue problems using numerical integration*, Journal of Computational and Applied Mathematics, 159 (2003), pp. 119–128. Japan-China Joint Seminar on Numerical Mathematics; In Search for the Frontier of Computational and Applied Mathematics toward the 21st Century.
64. T. SAKURAI AND H. TADANO, *Cirr: a Rayleigh-Ritz type method with contour integral for generalized eigenvalue problems*, Hokkaido Mathematical Journal, 36 (2007), p. 745–757.
65. L. J. SHAM, *Theoretical and computational development some efforts beyond the local density approximation*, International Journal of Quantum Chemistry, Vol. 56, 4 , pp 345–350, (2004).
66. H. D. SIMON, *Analysis of the symmetric Lanczos algorithm with reorthogonalization methods*, Linear Algebra Appl., 61 (1984), pp. 101–132.
67. ——, *The Lanczos algorithm with partial reorthogonalization*, Math. Comp., 42 (1984), pp. 115–142.
68. J. M. SOLER, E. ARTACHO, J. D. GALE, A. GARCIA, J. JUNQUERA, P. ORDEJÓN, AND D. SÁNCHEZ-PORTAL, *The SIESTA method for ab-initio order-N materials simulation*, J. Phys.: Condens. Matter, 14 (2002), pp. 2745–2779.
69. D. C. SORENSEN, *Implicit application of polynomial filters in a k-step Arnoldi method*, SIAM J. Matrix Anal. Appl., 13 (1992), pp. 357–385.
70. E. M. STOUDENMIRE, L. O. WAGNER, S. R. WHITE, AND K. BURKE, *One-dimensional continuum electronic structure with the density matrix renormalization group and its implications for density functional theory*, Phys. Rev. Lett. 109, I5, 056402, (2012).
71. J. TAYLOR, H. GUO, AND J. WANG, *Ab initio modeling of quantum transport properties of molecular electronic devices*, Phys. Rev. B, 63 (2001), p. 245407.
72. T. TORSTI, M. HEISKANEN, M. PUSKA, AND R. NIEMINEN, *MIKA: A multigrid-based program package for electronic structure calculations*, Int. J. Quantum Chem., 91 (2003), pp. 171–176.
73. L.-W. WANG AND A. ZUNGER, *Electronic structure pseudopotential calculations of large (1000 atoms) SI quantum dots*, J. Phys. Chem., 98 (1994), pp. 2158–2165.
74. S. R. WHITE, J. W. WILKINS, AND M. P. TETER, *Finite-element method for electronic structure*, Phys. Rev. B, 39 (1989), pp. 5819–5833.
75. K. WU AND H. SIMON, *A parallel Lanczos method for symmetric generalized eigenvalue problems*, Tech. Report 41284, Lawrence Berkeley National Laboratory, 1997. Available on line at http://www.nersc.gov/research/SIMON/planso.html.

76. R. ZELLER, J. DEUTZ, AND P. DEDERICHS, *Application of complex energy integration to self-consistent electronic structure calculations*, Solid State Communications, 44 (1982), pp. 993–997.

77. D. ZHANG AND E. POLIZZI, *Linear scaling techniques for first-principle calculations of large nanowire devices*, 2008 NSTI Nanotechnology Conference and Trade Show. Technical Proceedings, Vol. 1 pp12–15, (2008).

78. Y. ZHOU, Y. SAAD, M. L. TIAGO, AND J. R. CHELIKOWSKY, *Parallel self-consistent-field calculations via Chebyshev-filtered subspace acceleration*, Phy. rev. E, 74 (2006), p. 066704.

79. J. M. ZIMAN, *Electrons and Phonons*, Oxford University Press, 1960.

80. G. ZUMBACH, N. A. MODINE, AND E. KAXIRAS, *Adaptive coordinate, real-space electronic structure calculations on parallel computers*, Solid State Commun., 99 (1996), pp. 57–61.

Computational Cardiovascular Analysis with the Variational Multiscale Methods and Isogeometric Discretization

Thomas J. R. Hughes, Kenji Takizawa, Yuri Bazilevs, Tayfun E. Tezduyar, and Ming-Chen Hsu

1 Introduction

In this article we review general computational fluid dynamics (CFD) methods that we have developed and used over an almost five-decade period on a variety of applications in science, engineering, and medicine. However, our focal application area herein is computational medicine and in particular computational cardiovascular analysis. This area has a long history, in fact the senior author (TJRH) did his PhD thesis in it in 1974, and there was even earlier work than this, but the area took on a new direction in the mid-1990s when the first patient-specific calculations were performed with models created from medical imaging

T. J. R. Hughes (✉)
Institute for Computational Engineering and Sciences, The University of Texas at Austin, Austin, TX, USA
e-mail: hughes@ices.utexas.edu

K. Takizawa
Department of Modern Mechanical Engineering, Waseda University, Shinjuku-ku, Tokyo, Japan
e-mail: Kenji.Takizawa@tafsm.org

Y. Bazilevs
School of Engineering, Brown University, Providence, RI, USA
e-mail: yuri_bazilevs@brown.edu

T. E. Tezduyar
Mechanical Engineering, Rice University, Houston, TX, USA

Faculty of Science and Engineering, Waseda University, Shinjuku-ku, Tokyo, Japan
e-mail: tezduyar@tafsm.org

M.-C. Hsu
Department of Mechanical Engineering, Iowa State University, Ames, IA, USA
e-mail: jmchsu@iastate.edu

© Springer Nature Switzerland AG 2020
A. Grama, A. H. Sameh (eds.), *Parallel Algorithms in Computational Science and Engineering*, Modeling and Simulation in Science, Engineering and Technology, https://doi.org/10.1007/978-3-030-43736-7_6

151

data, such as MRI and CT. The archival journal paper that began this trend was [1]. Up to that time computational cardiovascular analysis was focused on very simple two-dimensional geometries such as straight and circular channels, and thus had almost no clinical significance. After [1], the subject began to dramatically transform to where it is today, in which detailed analyses of a wide variety of patient-specific configurations are routinely analyzed to diagnose disease, plan surgeries, and interventions, such as stenting and bypass grafting, and to virtually evaluate medical devices, such as left ventricular assist devices (LVADs), implanted in individual patients. Our purpose here is not to describe the array of medical applications of computational cardiovascular analysis (for these we would refer in particular to the works of Charles A. Taylor, Alison Marsden, and Alberto Figueroa, among others), but rather to describe the main technologies that support these applications. This started with the seminal work of the senior author [2] and the algorithm which has become known by the acronym SUPG, which was extracted from the name given by the authors, the "Streamline-Upwind Petrov-Galerkin" method. Reference [2] was the first archival journal publication of the basic ideas, but earlier, starting in 1979, there were several now obscure, conference proceedings papers that preceded it. We have to acknowledge that the name is not great. However, the ideas embodied therein were important and have had significant subsequent impact. The basic problem of computational fluid dynamics (CFD) at the time was achieving a combination of good stability and high accuracy in one algorithm. Many investigators viewed stability and accuracy as competing attributes. Reference [2] proved otherwise computationally, and mathematical analyses justified what was observed subsequently, the first being [3]. The fundamental concept employed was "residual-based stabilization," which added weighted residuals of the numerical solution to basic Galerkin formulations. Residual-based methods are a priori consistent and thus capable of preserving the underlying accuracy of Galerkin methods, while at the same time appropriate weighting enhanced their stability. Numerous "Stabilized Methods," as they have been commonly referred to subsequently, were then developed over the years based on this paradigm. The success of Stabilized Methods, another somewhat unfortunate name in our opinion, cannot be overestimated. The number of citations these works have garnered is staggering, e.g., [2] alone has received approximately 6000 citations. Although the mathematical analysis of Stabilized Methods developed as a field in its own right shortly after the initial publications, the creation of new Stabilized Methods technologies, such as, for example, residual-based discontinuity-capturing operators, was essentially based largely on intuition. The breakthrough concept that derived Stabilized Methods from the fundamental governing equations was the Variational Multiscale Method [4–7]. This provided an approach to derive consistent Stabilized Methods directly from any system of linear or nonlinear equations in fluid dynamics, or any scientific discipline, and it has been perhaps the most powerful development tool in the arsenal of CFD technologies.

Stabilized Methods and the Variational Multiscale Method are fundamental to all our works in computational cardiovascular analysis. Many other technologies have been developed that further extend these basic building blocks to specific classes of problems and phenomena. This article describes the use of these methods in computational cardiovascular analysis, with a focus on two specific areas, namely, aortic flow phenomena [8] and patient-specific and bioprosthetic heart-valve fluid–structure interaction [9, 10]. We wish to also emphasize that these applications are only a small sample of activity in this rapidly growing field. There are many formidable challenges posed by problems of these types, including highly unsteady flows, complex diseased geometries, moving boundaries and interfaces (e.g., motion of heart valve leaflets), contact between moving solid surfaces within a flow (e.g., contact between heart valve leaflets), and the fluid–structure interaction of blood flow with cardiovascular structures, such as arteries, heart valves, etc. Many of these challenges have been or are being addressed by the Space–Time Variational Multiscale (ST-VMS) method [11], Arbitrary Lagrangian–Eulerian VMS (ALE-VMS) method [12], and the VMS-based immersogeometric analysis (IMGA-VMS) [9], which serve as the core computational methods. The special methods used in combination with the ST-VMS include the Space–Time Slip Interface (ST-SI) method [13], Space–Time Topology Change (ST-TC) [14] method, Space–Time Isogeometric Analysis (ST-IGA) [15, 16], integration of these methods, and a general-purpose NURBS mesh generation method for complex geometries [17]. The special methods used in combination with ALE-VMS include weak enforcement of no-slip boundary conditions [18], "sliding interfaces" [19] (the acronym "SI" will also indicate that), and backflow stabilization [20].

Despite the focus of this article on problems of computational cardiovascular analysis, the methods described herein are general CFD and fluid–structure interaction technologies that have wide applicability to diverse scientific and engineering applications, and therefore we also take the opportunity to draw attention to many such applications that the authors of this chapter have been actively involved with.

1.1 Space–Time Stabilized and VMS Methods

The Deforming-Spatial-Domain/Stabilized Space–Time (DSD/SST) method [21] was introduced for computation of flows with moving boundaries and interfaces (MBI), including fluid–structure interaction (FSI). In MBI computations the DSD/SST functions as a moving-mesh method. Moving the fluid mechanics mesh to follow an interface enables mesh-resolution control near the interface and, consequently, high-resolution boundary-layer representation near fluid–solid interfaces. The stabilization components of the original DSD/SST are the Streamline-Upwind/Petrov-Galerkin (SUPG) [2] and Pressure-Stabilizing/Petrov-Galerkin (PSPG) [21] stabilizations, which are used widely. Because of the SUPG and PSPG components, the original DSD/SST is now called "ST-SUPS." The ST-VMS is the VMS version of the DSD/SST. The VMS components of the ST-VMS are from the residual-based VMS (RBVMS) method [4, 7]. The ST-VMS has two more stabilization terms beyond those in the ST-SUPS, and the additional terms give the method better turbulence modeling features. The ST-SUPS and ST-VMS, because of the higher-order accuracy of the Space–Time (ST) framework (see [11]), are desirable also in computations without MBI.

The ST-SUPS and ST-VMS have been applied to many classes of FSI, MBI, and fluid mechanics problems (see [22] for a comprehensive summary). The classes of problems include spacecraft parachute analysis for the landing-stage parachutes [23], cover-separation parachutes [24] and drogue parachutes [25], wind-turbine aerodynamics for horizontal-axis wind-turbine rotors [26], full horizontal-axis wind turbines [27] and vertical-axis wind turbines [13], flapping-wing aerodynamics for an actual locust [28], bioinspired MAVs [29] and wing-clapping [30], blood flow analysis of cerebral aneurysms [31], stent-treated aneurysms [32], aortas [8] and heart valves [10], spacecraft aerodynamics [24], thermo-fluid analysis of ground vehicles and their tires [33], thermo-fluid analysis of disk brakes [34], flow-driven filament dynamics in turbomachinery [35], flow analysis of turbocharger turbines [36], flow around tires with road contact and deformation [37], fluid films [38], ram-air parachutes [39], and compressible-flow spacecraft parachute aerodynamics [40].

The space–time computational methods have a relatively long track record in arterial FSI analysis, starting with computations reported in [41, 42]. These were among the earliest arterial FSI computations, and the core method was the ST-SUPS. Many space–time computations were also reported in the last 15 years. In the first 8 years of that period the space–time computations were performed for FSI of the abdominal aorta [43], carotid artery [43], and cerebral aneurysms [44]. In the last 7 years, the space–time computations focused on even more challenging aspects of cardiovascular fluid mechanics and FSI, including comparative studies of cerebral aneurysms [31], stent treatment of cerebral aneurysms [45], heart valve flow computation [10], aortic flow analysis [8], and coronary arterial dynamics [46].

In the flow analyses presented here, the space–time framework provides higher-order accuracy. The VMS feature of the ST-VMS addresses the computational challenges associated with the multiscale nature of the unsteady flow. The moving-mesh feature of the space–time framework enables high-resolution computation near the moving heart valve leaflets.

1.2   ALE Stabilized and VMS Methods

The ALE-VMS method [12] is the VMS version of ALE [47]. It succeeded the ST-SUPS [21] and ALE-SUPS [48] and preceded the ST-VMS. The VMS components are from the RBVMS [4, 7]. The ALE-VMS originated from the RBVMS formulation of incompressible turbulent flows proposed in [7] for non-moving meshes, and may be thought of as an extension of the RBVMS to moving meshes. As such, it was presented for the first time in [12] in the context of FSI. To increase their scope and accuracy, the ALE-VMS and RBVMS are often supplemented with special methods, such as those for weakly enforced no-slip boundary conditions [18], "sliding interfaces" [19] and backflow stabilization [20]. The ALE-SUPS, RBVMS, and ALE-VMS have been applied to many classes of

FSI, MBI, and fluid mechanics problems including ram-air parachute FSI [48], wind-turbine aerodynamics and FSI [49, 50], vertical-axis wind turbines [50], floating wind turbines [51], wind turbines in atmospheric boundary layers [50], fatigue damage in wind-turbine blades [52], patient-specific cardiovascular fluid mechanics and FSI [53, 54], biomedical-device FSI [55, 56], ship hydrodynamics with free-surface flow and fluid–object interaction [57], hydrodynamics and FSI of hydraulic arresting gear [58], hydrodynamics of tidal-stream turbines with free-surface flow [59], passive-morphing FSI in turbomachinery [60], bioinspired FSI for marine propulsion [61], and bridge aerodynamics and fluid–object interaction [62]. Recent advances in stabilized and multiscale methods may be found for stratified incompressible flows [63], divergence-conforming discretizations of incompressible flows [64], and compressible flows with emphasis on gas-turbine modeling [65].

In the flow analyses presented here, the VMS feature of ALE-VMS addresses the computational challenges associated with the multiscale nature of the unsteady flow. The moving-mesh feature of the ALE framework enables high-resolution computation near the moving wall of a thoracic aorta.

1.3 Slip Interface Space–Time Method

The Space–Time version of the slip interface (ST-SI) method was introduced in [13] in the context of incompressible-flow equations to retain the desirable moving-mesh features of the ST-VMS and ST-SUPS when there are spinning solid surfaces, such as for a turbine rotor. The mesh covering the spinning surface spins with it, retaining the high-resolution representation of boundary layers. The starting point in the development of ST-SI was the version of ALE-VMS for computations with sliding interfaces [19]. Interface terms similar to those in the ALE-VMS version are added to ST-VMS to account for the compatibility conditions for velocity and stress at the slip interface. That accurately connects the two sides of the solution. An ST-SI version where the slip interface is between fluid and solid domains was also presented in [13]. The slip interface in this case is a "fluid–solid" interface rather than a standard "fluid–fluid" interface, and enables weak enforcement of the Dirichlet boundary conditions for the fluid. The ST-SI introduced in [34] for the coupled incompressible-flow and thermal-transport equations retains the high-resolution representation of the thermo-fluid boundary layers near spinning solid surfaces. These ST-SI methods have been applied to aerodynamic analysis of vertical-axis wind turbines [13], thermo-fluid analysis of disk brakes [34], flow-driven filament dynamics in turbomachinery [35], flow analysis of turbocharger turbines [36], flow around tires with road contact and deformation [37], fluid films [38], aerodynamic analysis of ram-air parachutes [39], and flow analysis of heart valves [10].

In the ST-SI version presented in [13] the slip interface is between a thin porous structure and the fluid on its two sides. This enables dealing with the porosity in a fashion consistent with how the standard fluid–fluid slip interfaces are dealt with and how the Dirichlet conditions are enforced weakly with fluid–solid slip interfaces. This version also enables handling thin structures that have T-junctions. This method has been applied to incompressible-flow aerodynamic analysis of ram-air parachutes with fabric porosity [39]. The compressible-flow ST-SI methods were introduced in [40], including the version where the slip interface is between a thin porous structure and the fluid on both its sides. Compressible-flow porosity models were also introduced in [40]. These, together with the compressible-flow space–time SUPG method [66], extended the space–time computational analysis range to compressible-flow aerodynamics of parachutes with fabric and geometric porosities. That enabled space–time computational flow analysis of the Orion spacecraft drogue parachute in the compressible-flow regime [67].

1.4 Immersogeometric VMS Analysis

The immersogeometric analysis (IMGA) was introduced in [56] as a geometrically flexible technique for solving FSI problems involving large, complex structural deformations and change of fluid-domain topology (e.g., structural contact). The motivating application is the simulation of heart valve function over a complete cardiac cycle. The method directly analyzes a spline representation of a thin structure by immersing it into a non-body-fitted discretization of the background fluid domain, and focuses on accurately capturing the immersed design geometry within non-body-fitted analysis meshes. A new semi-implicit numerical method, which we now refer to as the Dynamic Augmented Lagrangian (DAL) approach [68], was introduced in [56] for weakly enforcing constraints in time-dependent immersogeometric FSI problems. A mixed ALE-VMS/IMGA-VMS (ALE-IMGA-VMS) method was developed in [9] in the framework of the Fluid–Solid Interface-Tracking/Interface-Capturing Technique [69]; a single computation combines a body-fitted, moving-mesh treatment of some fluid–structure interfaces, with a non-body-fitted treatment of others. This approach enables us to simulate the FSI of a bioprosthetic heart valve (BHV) in a deforming artery over the entire cardiac cycle under physiological conditions, and study the effect of arterial-wall elasticity on valve dynamics [9]. The DAL-based ALE-IMGA-VMS was integrated with Computer-Aided Design (CAD) for heart valve analysis in [55] with a thorough comparison between pressure-driven only and full FSI computations. An anisotropic constitutive modeling of BHV leaflets for immersogeometric FSI, based on the Kirchhoff–Love shell formulation for general hyperelastic materials [70], is proposed in [71]. A divergence-conforming formulation of incompressible flow, which gives a pointwise divergence-free velocity field everywhere in the domain, completely eliminates mass loss error across the valve interface in [72]. Stable coupling strategies and suitable definition of Lagrange multipliers for the DAL numerical approach were proposed and analyzed in [73]. The FSI framework of ALE-IMGA-VMS was employed in patient-specific valve design in [74]. The DAL-based IMGA has also been combined with surrogate modeling in [58] for an efficient and effective use of FSI to optimize the design of a hydraulic arresting gear.

1.5   Stabilization Parameters

The methods discussed in this chapter all have some embedded stabilization parameters that play a significant role (see [13, 75]). There are many ways of defining these stabilization parameters (for examples, see [33, 37, 76–80]). The stabilization-parameter definitions used in the computations reported in this article can be found from the references cited in the sections where those computations are described.

1.6   Topology Change Space–Time Method

The Topology Change Space–Time method (ST-TC) [14] was introduced for moving-mesh computation of flow problems with topology change, such as contact between solid surfaces. Even before the ST-TC, the ST-SUPS and ST-VMS, when used with robust mesh update methods, have proven effective in flow computations where the solid surfaces are in near contact or create other near topology change. Many classes of problems can be solved that way with sufficient accuracy by approximating actual contact with a small gap between the solid surfaces. For examples of such computations, see the references mentioned in [14]. The ST-TC made moving-mesh computations possible even when there is an actual contact between solid surfaces or other topology change. By collapsing elements as needed, without changing the connectivity of the "parent" mesh, the ST-TC can handle an actual topology change while maintaining high-resolution boundary layer representation near solid surfaces. This enabled successful moving-mesh computation of heart valve flows [10], wing clapping [30], and flow around a rotating tire with road contact and prescribed deformation [37].

For more on the ST-TC, see [14]. In the computational analyses here, the ST-TC enables moving-mesh computation even with the topology change created by the actual contact between the valve leaflets. It deals with the contact while maintaining high-resolution flow representation near the leaflet.

1.7   Topology Change Slip Interface Space–Time Method

The Topology Change Slip Interface Space–Time Method (ST-SI-TC) is the integration of the ST-SI and ST-TC. A fluid–fluid slip interface requires elements on both sides of the interface. When part of a slip interface needs to coincide with a solid surface, which happens, for example, when the solid surfaces on two sides of the interface come into contact or when the interface reaches a solid surface, the elements between the coinciding slip interface part and the solid surface need to collapse with the ST-TC mechanism. The collapse switches the slip interface from the fluid–fluid type to the fluid–solid type. With that, a slip interface can be a mixture of the fluid–fluid and fluid–solid types. With the ST-SI-TC, the elements collapse and are reborn independent of the nodes representing a solid surface. The ST-SI-TC enables high-resolution flow representation even when parts of the slip interface are coinciding with a solid surface. It also enables dealing with contact location change and contact sliding. This was applied to heart valve flow analysis [10] and tire aerodynamics with road contact and deformation [37].

For more on the ST-SI-TC, see [81]. In the computational analyses presented here, the ST-SI-TC enables high-resolution representation of the boundary layers even when the contact is between leaflets that are in mesh sectors connected by slip interfaces. It enables contact location change and contact sliding between the leaflets.

1.8   Space–Time IGA

The ST-IGA, introduced in [11], is the integration of the space–time framework with isogeometric discretization, motivated by the success of NURBS meshes in spatial discretization [12, 19, 53, 82]. Computations with the ST-VMS and ST-IGA were first reported in [11] in a 2D context, with IGA basis functions in space for flow past an airfoil, and in both space and time for the advection equation. Using higher-order basis functions in time enables getting full benefit out of using higher-order basis functions in space. This was demonstrated with the stability and accuracy analysis given in [11] for the advection equation.

The ST-IGA with IGA basis functions in time enables a more accurate representation of the motion of the solid surfaces and a mesh motion consistent with that. This was pointed out in [11] and demonstrated in [15]. It also enables more efficient temporal representation of the motion and deformation of the volume meshes, and more efficient remeshing. These motivated the development of the ST/NURBS Mesh Update Method (STNMUM) [15, 79]. The STNMUM has a wide scope that includes spinning solid surfaces. With the spinning motion represented by quadratic NURBS in time, and with sufficient number of temporal patches for a full rotation, the circular paths are represented exactly. A "secondary mapping" [11] enables also specifying a constant angular velocity for invariant speeds along the circular paths. The space–time framework and NURBS in time also enable, with the "ST-C" method, extracting a continuous representation from the computed data and, in large-scale computations, efficient data compression [83]. The STNMUM and the ST-IGA with IGA basis functions in time have been used in many 3D computations. The classes of problems solved are flapping-wing aerodynamics for an actual locust [28], bioinspired MAVs [29] and wing-clapping [30], separation aerodynamics of spacecraft [24], aerodynamics of horizontal-axis [31] and vertical-axis [13] wind turbines, thermo-fluid analysis of ground vehicles and their tires [33], thermo-fluid analysis of disk brakes [34], flow-driven string dynamics in turbomachinery [35], and flow analysis of turbocharger turbines [36].

The ST-IGA with IGA basis functions in space enables more accurate representation of the geometry and increased accuracy in the flow solution. It accomplishes that with fewer control points, and consequently with larger effective element sizes. That in turn enables using larger time-step sizes while keeping the Courant number at a desirable level for good accuracy. It has been used in space–time computational flow analysis of turbocharger turbines [36], flow-driven string dynamics in turbomachinery [35], ram-air parachutes [39], spacecraft parachutes [67], aortas [8], heart valves [10], tires with road contact and deformation [37], and fluid films [38]. Using IGA basis functions in space is now a key part of some of the newest Zero Stress State (ZSS) estimation methods [84] and related shell analysis [85].

For more on the ST-IGA, see [16]. In the computational flow analyses presented here, the ST-IGA enables more accurate representation of the cardiovascular geometries, increased accuracy in the flow solution, and using larger time-step sizes.

1.9 Space–Time IGA with Slip Interface and Topology Change

The turbocharger turbine analysis [36] and flow-driven string dynamics in turbomachinery [35] were based on the integration of the ST-SI and ST-IGA. The IGA basis functions were used in the spatial discretization of the fluid mechanics equations and also in the temporal representation of the rotor and spinning-mesh motion. That enabled accurate representation of the turbine geometry and rotor motion and increased accuracy in the flow solution. The IGA basis functions were used also in the spatial discretization of the string structural dynamics equations. That enabled increased accuracy in the structural dynamics solution, as well as smoothness in the string shape and fluid dynamics forces computed on the string.

The ram-air parachute analysis [39] and spacecraft parachute compressible-flow analysis [67] were based on the integration of the ST-IGA, the ST-SI version that weakly enforces the Dirichlet conditions, and the ST-SI version that accounts for the porosity of a thin structure. The ST-IGA with IGA basis functions in space enabled, with relatively few number of unknowns, accurate representation of the parafoil and parachute geometries and increased accuracy in the flow solution. The volume mesh needed to be generated both inside and outside the parafoil. Mesh generation inside was challenging near the trailing edge because of the narrowing space. The spacecraft parachute has a very complex geometry, including gores and gaps. Using IGA basis functions addressed those challenges and still kept the element density near the trailing edge of the parafoil and around the spacecraft parachute at a reasonable level.

The heart valve analysis [10] was based on the integration of the ST-SI, ST-TC, and ST-IGA, which we refer to as ST-SI-TC-IGA. The ST-SI-TC-IGA, beyond enabling a more accurate representation of the geometry and increased accuracy in the flow solution, kept the element density in the narrow spaces near the contact areas at a reasonable level. When solid surfaces come into contact, the elements between the surface and the slip interface collapse. Before the elements collapse, the boundaries could be curved and rather complex, and the narrow spaces might have high-aspect-ratio elements. With NURBS elements, it was possible to deal with such adverse conditions rather effectively.

In computational analysis of flow around tires with road contact and deformation [37], the ST-SI-TC-IGA enables a more accurate representation of the geometry and motion of the tire surfaces, a mesh motion consistent with that, and increased accuracy in the flow solution. It also keeps the element density in the tire grooves and in the narrow spaces near the contact areas at a reasonable level. In addition, we benefit from the mesh generation flexibility provided by using SIs.

An SI provides mesh generation flexibility in a general context by accurately connecting the two sides of the solution computed over nonmatching meshes. This type of mesh generation flexibility is especially valuable in complex-geometry flow computations with isogeometric discretization, removing the matching requirement between the NURBS patches without loss of accuracy. This feature was used in the flow analysis of heart valves [10], turbocharger turbines [36], and spacecraft parachute compressible-flow analysis [67].

For more on the ST-SI-TC-IGA, see [10]. In the computations presented here, the ST-SI-TC-IGA is used in the heart valve flow analysis, for the reasons given and as described in an earlier paragraph of this section.

1.10 General-Purpose NURBS Mesh Generation Method

To make the ST-IGA use, and in a wider context the IGA use, even more practical in computational flow analysis with complex geometries, NURBS volume mesh generation needs to be easier and more automated. To that end, a general-purpose NURBS mesh generation method was introduced in [17]. The method is based on multi-block-structured mesh generation with existing techniques, projection of that mesh to a NURBS mesh made of patches that correspond to the blocks, and recovery of the original model surfaces. The method is expected to retain the refinement distribution and element quality of the multi-block-structured mesh that we start with. Because there are ample good techniques and software for generating multi-block-structured meshes, the method makes general-purpose mesh generation relatively easy.

Mesh-quality performance studies for 2D and 3D meshes, including those for complex models, were presented in [86]. A test computation for a turbocharger turbine and exhaust manifold was also presented in [86], with a more detailed computation in [36]. The mesh generation method was used also in the pump-flow analysis part of the flow-driven string dynamics presented in [35] and in the aorta flow analysis presented in [8]. The performance studies, test computations, and actual computations demonstrated that the general-purpose NURBS mesh generation method makes the IGA use in fluid mechanics computations even more practical.

For more on the general-purpose NURBS mesh generation method, see [17, 86]. In the computations presented here, the method used in the aorta flow analysis.

1.11 Outline of the Remaining Sections

We provide the governing equations in Sect. 2. The ST-VMS and ST-SI are described in Sect. 3, and the ALE-VMS and IMGA-VMS in Sect. 4. In Sect. 5 we provide some brief comments on the parallel computations. In Sects. 6 and 7, as examples of space–time computations, we present an aortic-valve flow analysis and a patient-specific aorta flow analysis. In Sect. 8, as an example of IMGA computations, we present a patient-specific heart valve design and analysis. The concluding remarks are given in Sect. 9.

2 Governing Equations

2.1 Incompressible Flow

Let $\Omega_t \subset \mathbb{R}^{n_{\mathrm{sd}}}$ be the spatial domain with boundary Γ_t at time $t \in (0, T)$, where n_{sd} is the number of space dimensions. The subscript t indicates the time-dependence of the domain. The Navier–Stokes equations of incompressible flows are written on Ω_t and $\forall t \in (0, T)$ as

$$\rho \left(\frac{\partial \mathbf{u}}{\partial t} + \mathbf{u} \cdot \nabla \mathbf{u} - \mathbf{f} \right) - \nabla \cdot \boldsymbol{\sigma} = \mathbf{0}, \tag{1}$$

$$\nabla \cdot \mathbf{u} = 0, \tag{2}$$

where ρ, $\mathbf{u}$, and $\mathbf{f}$ are the density, velocity, and body force. The stress tensor $\boldsymbol{\sigma}(\mathbf{u}, p) = -p\mathbf{I} + 2\mu\boldsymbol{\varepsilon}(\mathbf{u})$, where p is the pressure, $\mathbf{I}$ is the identity tensor, $\mu = \rho\nu$ is the viscosity, ν is the kinematic viscosity, and the strain rate $\boldsymbol{\varepsilon}(\mathbf{u}) = \left(\nabla \mathbf{u} + (\nabla \mathbf{u})^T \right)/2$. The essential and natural boundary conditions for Eq. (1) are represented as $\mathbf{u} = \mathbf{g}$ on $(\Gamma_t)_{\mathrm{g}}$ and $\mathbf{n} \cdot \boldsymbol{\sigma} = \mathbf{h}$ on $(\Gamma_t)_{\mathrm{h}}$, where $\mathbf{n}$ is the unit normal vector and $\mathbf{g}$ and $\mathbf{h}$ are given functions. A divergence-free velocity field $\mathbf{u}_0(\mathbf{x})$ is specified as the initial condition.

2.2 Structural Mechanics

In this article we will not provide any of our formulations requiring fluid and
structure definitions simultaneously; we will instead give reference to earlier journal
articles where the formulations were presented. Therefore, for notation simplicity,
we will reuse many of the symbols used in the fluid mechanics equations to
represent their counterparts in the structural mechanics equations. To begin with,
$\Omega_t \subset \mathbb{R}^{n_{sd}}$ and Γ_t will represent the structure domain and its boundary. The
structural mechanics equations are then written, on Ω_t and $\forall t \in (0, T)$, as

$$\rho \left(\frac{d^2 \mathbf{y}}{dt^2} - \mathbf{f} \right) - \nabla \cdot \boldsymbol{\sigma} = \mathbf{0}, \tag{3}$$

where $\mathbf{y}$ and $\boldsymbol{\sigma}$ are the displacement and Cauchy stress tensor. The essential and
natural boundary conditions for Eq. (3) are represented as $\mathbf{y} = \mathbf{g}$ on $(\Gamma_t)_g$ and $\mathbf{n} \cdot \boldsymbol{\sigma} = \mathbf{h}$ on $(\Gamma_t)_h$. The Cauchy stress tensor can be obtained from

$$\boldsymbol{\sigma} = J^{-1} \mathbf{F} \mathbf{S} \mathbf{F}^T, \tag{4}$$

where $\mathbf{F}$ and J are the deformation gradient tensor and its determinant, and $\mathbf{S}$ is the
second Piola–Kirchhoff stress tensor. It is obtained from the strain-energy density
function φ as follows:

$$\mathbf{S} \equiv \frac{\partial \varphi}{\partial \mathbf{E}}, \tag{5}$$

where $\mathbf{E}$ is the Green–Lagrange strain tensor:

$$\mathbf{E} = \frac{1}{2} (\mathbf{C} - \mathbf{I}), \tag{6}$$

and $\mathbf{C}$ is the Cauchy–Green deformation tensor:

$$\mathbf{C} \equiv \mathbf{F}^T \cdot \mathbf{F}. \tag{7}$$

From Eqs. (5) and (6),

$$\mathbf{S} = 2 \frac{\partial \varphi}{\partial \mathbf{C}}. \tag{8}$$

2.3 Fluid–Structure Interface

In an FSI problem, at the fluid–structure interface, we will have the velocity and
stress compatibility conditions between the fluid and structure parts. The details on
those conditions can be found in Section 5.1 of [75].

3   ST-VMS and ST-SI

We include from [13, 81] the ST-VMS and ST-SI methods.
  The ST-VMS is given as

$$
\int_{Q_n} \mathbf{w}^h \cdot \rho \left(\frac{\partial \mathbf{u}^h}{\partial t} + \mathbf{u}^h \cdot \nabla \mathbf{u}^h - \mathbf{f}^h \right) \mathrm{d}Q + \int_{Q_n} \boldsymbol{\varepsilon}(\mathbf{w}^h) : \boldsymbol{\sigma}(\mathbf{u}^h, p^h) \mathrm{d}Q
$$

$$
- \int_{(P_n)_\mathrm{h}} \mathbf{w}^h \cdot \mathbf{h}^h \mathrm{d}P + \int_{Q_n} q^h \nabla \cdot \mathbf{u}^h \mathrm{d}Q + \int_{\Omega_n} (\mathbf{w}^h)_n^+ \cdot \rho \left((\mathbf{u}^h)_n^+ - (\mathbf{u}^h)_n^- \right) \mathrm{d}\Omega
$$

$$
+ \sum_{e=1}^{(n_\mathrm{el})_n} \int_{Q_n^e} \frac{\tau_\mathrm{SUPS}}{\rho} \left[\rho \left(\frac{\partial \mathbf{w}^h}{\partial t} + \mathbf{u}^h \cdot \nabla \mathbf{w}^h \right) + \nabla q^h \right] \cdot \mathbf{r}_\mathrm{M}(\mathbf{u}^h, p^h) \mathrm{d}Q
$$

$$
+ \sum_{e=1}^{(n_\mathrm{el})_n} \int_{Q_n^e} \nu_\mathrm{LSIC} \nabla \cdot \mathbf{w}^h \rho r_\mathrm{C}(\mathbf{u}^h) \mathrm{d}Q
$$

$$
- \sum_{e=1}^{(n_\mathrm{el})_n} \int_{Q_n^e} \tau_\mathrm{SUPS} \mathbf{w}^h \cdot \left(\mathbf{r}_\mathrm{M}(\mathbf{u}^h, p^h) \cdot \nabla \mathbf{u}^h \right) \mathrm{d}Q
$$

$$
- \sum_{e=1}^{(n_\mathrm{el})_n} \int_{Q_n^e} \frac{\tau_\mathrm{SUPS}^2}{\rho} \mathbf{r}_\mathrm{M}(\mathbf{u}^h, p^h) \cdot \left(\nabla \mathbf{w}^h \right) \cdot \mathbf{r}_\mathrm{M}(\mathbf{u}^h, p^h) \mathrm{d}Q = 0, \tag{9}
$$

where

$$
\mathbf{r}_\mathrm{M}(\mathbf{u}^h, p^h) = \rho \left(\frac{\partial \mathbf{u}^h}{\partial t} + \mathbf{u}^h \cdot \nabla \mathbf{u}^h - \mathbf{f}^h \right) - \nabla \cdot \boldsymbol{\sigma}(\mathbf{u}^h, p^h), \tag{10}
$$

$$
r_\mathrm{C}(\mathbf{u}^h) = \nabla \cdot \mathbf{u}^h \tag{11}
$$

are the residuals of the momentum equation and incompressibility constraint. The
test functions associated with the velocity and pressure are $\mathbf{w}$ and q. A superscript
"h" indicates that the function is coming from a finite-dimensional space. The
symbol Q_n represents the ST slice between time levels n and $n + 1$, $(P_n)_\mathrm{h}$ is the part
of the lateral boundary of that slice associated with the traction boundary condition
$\mathbf{h}$, and Ω_n is the spatial domain at time level n. The superscript "e" is the ST element
counter, and n_el is the number of ST elements. The functions are discontinuous in
time at each time level, and the superscripts "$-$" and "$+$" indicate the values of
the functions just below and just above the time level. See [13, 33, 76, 77, 79] for
the definitions used here for the stabilization parameters τ_SUPS and ν_LSIC. For more
ways of calculating the stabilization parameters in finite element computation of
flow problems, see [37, 78, 80]).

Remark 1 The ST-SUPS method can be obtained from the ST-VMS method by dropping the eighth and ninth integrations.

In the ST-SI, labels "Side A" and "Side B" represent the two sides of the SI. We add boundary terms to Eq. (9). The boundary terms are first added separately for the two sides, using test functions $\mathbf{w}_A^h$ and q_A^h and $\mathbf{w}_B^h$ and q_B^h. Putting them together, the complete set of terms added becomes

$$
- \int_{(P_n)_{\text{SI}}} \left(q_B^h \mathbf{n}_B - q_A^h \mathbf{n}_A \right) \cdot \frac{1}{2} \left(\mathbf{u}_B^h - \mathbf{u}_A^h \right) \mathrm{d}P
$$

$$
- \int_{(P_n)_{\text{SI}}} \rho \mathbf{w}_B^h \cdot \frac{1}{2} \left(\left(\mathcal{F}_B^h - \left| \mathcal{F}_B^h \right| \right) \mathbf{u}_B^h - \left(\mathcal{F}_B^h - \left| \mathcal{F}_B^h \right| \right) \mathbf{u}_A^h \right) \mathrm{d}P
$$

$$
- \int_{(P_n)_{\text{SI}}} \rho \mathbf{w}_A^h \cdot \frac{1}{2} \left(\left(\mathcal{F}_A^h - \left| \mathcal{F}_A^h \right| \right) \mathbf{u}_A^h - \left(\mathcal{F}_A^h - \left| \mathcal{F}_A^h \right| \right) \mathbf{u}_B^h \right) \mathrm{d}P
$$

$$
+ \int_{(P_n)_{\text{SI}}} \left(\mathbf{n}_B \cdot \mathbf{w}_B^h + \mathbf{n}_A \cdot \mathbf{w}_A^h \right) \frac{1}{2} \left(p_B^h + p_A^h \right) \mathrm{d}P
$$

$$
- \int_{(P_n)_{\text{SI}}} \left(\mathbf{w}_B^h - \mathbf{w}_A^h \right) \cdot \left(\hat{\mathbf{n}}_B \cdot \mu \left(\boldsymbol{\varepsilon}(\mathbf{u}_B^h) + \boldsymbol{\varepsilon}(\mathbf{u}_A^h) \right) \right) \mathrm{d}P
$$

$$
- \gamma_{\text{ACI}} \int_{(P_n)_{\text{SI}}} \hat{\mathbf{n}}_B \cdot \mu \left(\boldsymbol{\varepsilon} \left(\mathbf{w}_B^h \right) + \boldsymbol{\varepsilon} \left(\mathbf{w}_A^h \right) \right) \cdot \left(\mathbf{u}_B^h - \mathbf{u}_A^h \right) \mathrm{d}P
$$

$$
+ \int_{(P_n)_{\text{SI}}} \frac{\mu C}{h} \left(\mathbf{w}_B^h - \mathbf{w}_A^h \right) \cdot \left(\mathbf{u}_B^h - \mathbf{u}_A^h \right) \mathrm{d}P, \tag{12}
$$

where

$$
\mathcal{F}_B^h = \mathbf{n}_B \cdot \left(\mathbf{u}_B^h - \mathbf{v}_B^h \right), \tag{13}
$$

$$
\mathcal{F}_A^h = \mathbf{n}_A \cdot \left(\mathbf{u}_A^h - \mathbf{v}_A^h \right), \tag{14}
$$

$$
h = \frac{h_B + h_A}{2}, \tag{15}
$$

$$
h_B = 2 \left(\sum_{\alpha=1}^{n_{\text{ent}}} \sum_{a=1}^{n_{\text{ens}}} \left| \mathbf{n}_B \cdot \nabla N_a^\alpha \right| \right)^{-1} \quad \text{(for Side B)}, \tag{16}
$$

$$
h_A = 2 \left(\sum_{\alpha=1}^{n_{\text{ent}}} \sum_{a=1}^{n_{\text{ens}}} \left| \mathbf{n}_A \cdot \nabla N_a^\alpha \right| \right)^{-1} \quad \text{(for Side A)}, \tag{17}
$$

$$
\hat{\mathbf{n}}_B = \frac{\mathbf{n}_B - \mathbf{n}_A}{\| \mathbf{n}_B - \mathbf{n}_A \|}. \tag{18}
$$

Here, $(P_n)_{\text{SI}}$ is the SI in the ST domain, $\mathbf{v}$ is the mesh velocity, n_{ens} and n_{ent} are the number of spatial and temporal element nodes, N_a^α is the basis function associated with spatial and temporal nodes a and α, $\gamma_{\text{ACI}} = 1$, and C is a nondimensional constant. For our element length definition, we typically set $C = 1$.

A number of remarks were provided in [13] to explain the added terms and to comment on related interpretations. We refer the reader interested in those details to [13].

Remark 2 A coefficient γ_{ACI} was added in [81] to the sixth integration so that we have the option of using $\gamma_{\text{ACI}} = -1$. This option was added, in [40], also in the context of compressible flows. Using $\gamma_{\text{ACI}} = 1$ in a discontinuous Galerkin method was introduced in the symmetric interior penalty Galerkin method [87], and using $\gamma_{\text{ACI}} = -1$ was introduced in the nonsymmetric interior penalty Galerkin method [88]. Stabilized methods based on both $\gamma_{\text{ACI}} = 1$ and -1 were reported in [18] in the context of the advection–diffusion equation. In the computations reported in this article, we set $\gamma_{\text{ACI}} = 1$.

4 ALE-VMS and ALE-IMGA-VMS

The ALE-VMS formulation is posed on a spatial domain Ω that is discretized into elements Ω^e. While $\{\Omega^e\}$, Ω, and its boundary Γ are time-dependent, when there is no risk of confusion, we drop the subscript t to simplify notation. The superscript h indicates association with discrete function spaces defined over Ω, which moves with the velocity $\hat{\mathbf{u}}^h$, which is the same as the mesh velocity $\mathbf{v}^h$ in Sect. 3. The semi-discrete formulation is given as

$$
\int_\Omega \mathbf{w}^h \cdot \rho \left(\left.\frac{\partial \mathbf{u}^h}{\partial t}\right|_{\hat{\mathbf{x}}} + (\mathbf{u}^h - \hat{\mathbf{u}}^h) \cdot \nabla \mathbf{u}^h - \mathbf{f}^h \right) \mathrm{d}\Omega + \int_\Omega \boldsymbol{\varepsilon}(\mathbf{w}^h) : \boldsymbol{\sigma}(\mathbf{u}^h, p^h) \, \mathrm{d}\Omega
$$

$$
- \int_\Gamma \mathbf{w}^h \cdot \mathbf{h}^h \mathrm{d}\Gamma + \int_\Omega q^h \nabla \cdot \mathbf{u}^h \, \mathrm{d}\Omega
$$

$$
- \beta \int_\Gamma \mathbf{w}^h \cdot \rho \left\{ \left(\mathbf{u}^h - \hat{\mathbf{u}}^h \right) \cdot \mathbf{n} \right\}_{-} \mathbf{u}^h \mathrm{d}\Gamma
$$

$$
+ \sum_e \int_{\Omega^e} \tau_{\text{SUPS}} \left((\mathbf{u}^h - \hat{\mathbf{u}}^h) \cdot \nabla \mathbf{w}^h + \frac{1}{\rho}\nabla q^h \right) \cdot \mathbf{r}_{\text{M}}(\mathbf{u}^h, p^h) \, \mathrm{d}\Omega
$$

$$
+ \sum_e \int_{\Omega^e} \nu_{\text{LSIC}} \nabla \cdot \mathbf{w}^h \rho r_{\text{C}}(\mathbf{u}^h) \, \mathrm{d}\Omega
$$

$$
- \sum_e \int_{\Omega^e} \tau_{\text{SUPS}} \mathbf{w}^h \cdot \left(\mathbf{r}_{\text{M}}(\mathbf{u}^h, p^h) \cdot \nabla \mathbf{u}^h \right) \mathrm{d}\Omega
$$

$$- \sum_e \int_{\Omega^e} \frac{\tau_{\text{SUPS}}^2}{\rho} \mathbf{r}_{\text{M}}(\mathbf{u}^h, p^h) \cdot \left(\nabla \mathbf{w}^h \right) \cdot \mathbf{r}_{\text{M}}(\mathbf{u}^h, p^h) \, d\Omega$$

$$+ \sum_e \int_{\Omega^e} \left(\tau_{\text{SUPS}} \mathbf{r}_{\text{M}}(\mathbf{u}^h, p^h) \cdot \nabla \mathbf{w}^h \right) \overline{\tau} \cdot \left(\tau_{\text{SUPS}} \mathbf{r}_{\text{M}}(\mathbf{u}^h, p^h) \cdot \nabla \mathbf{u}^h \right) \, d\Omega = 0 \,,$$

(19)

where $\left. \frac{\partial(\cdot)}{\partial t} \right|_{\hat{\mathbf{x}}}$ is the time derivative taken with respect to the fixed reference coordinates $\hat{\mathbf{x}}$ of the spatial configuration, β (≥ 0) is associated with the backflow stabilization (see Remark 4), and $\{\cdot\}_-$ isolates the negative part of its argument. The additional stabilization parameter $\overline{\tau}$ is defined as

$$\overline{\tau} = \left(\tau_{\text{SUPS}} \mathbf{r}_{\text{M}}(\mathbf{u}^h, p^h) \cdot (\mathbf{G}) \cdot \tau_{\text{SUPS}} \mathbf{r}_{\text{M}}(\mathbf{u}^h, p^h) \right)^{-1/2} \,,$$

(20)

where $\mathbf{G}$ generalizes element size to physical elements mapped through $\mathbf{x}(\boldsymbol{\xi})$ from a parametric parent element: $G_{ij} = \xi_{k,i} \xi_{k,j}$.

The ALE-VMS formulation can be combined with the immersogeometric analysis (IMGA) [56], which we refer to as the ALE-IMGA-VMS method [9, 55, 74]. In the IMGA problem, the kinematic and traction compatibility conditions at the immersed fluid–structure interface are imposed weakly using the DAL. The details of this method can be found in [56, 68].

Remark 3 To improve mass conservation of the ALE-IMGA-VMS technique near immersed boundaries, the following modification to τ_{SUPS} is introduced in [56]:

$$\tau_{\text{SUPS}} = \left(s \left(\frac{4}{\Delta t^2} + (\mathbf{u}^h - \hat{\mathbf{u}}^h) \cdot \mathbf{G}(\mathbf{u}^h - \hat{\mathbf{u}}^h) + C_I \left(\frac{\mu}{\rho} \right)^2 \mathbf{G} : \mathbf{G} \right) \right)^{-1/2} \,.$$

(21)

Almost everywhere in Ω we set $s = 1$, which yields a traditional definition of τ_{SUPS}. However, in an $O(h)$ neighborhood of the immersed fluid–structure interface we set $s \geq 1$, which effectively reduces the size of τ_{SUPS} in that region. A theoretical motivation for this scaling is given in [72], and a numerical investigation of its effect is given in [73].

Remark 4 Unsteady flow computations may sometimes diverge due to significant inflow through the Neumann boundary $\Gamma_{\text{f}}^{\text{h}}$; this is known as backflow divergence and is frequently encountered in cardiovascular simulations. In order to preclude backflow divergence, a backflow stabilization method (the β term in Eq. (19)) originally proposed in [89] and further studied in [90] is employed in our ALE-VMS and ALE-IMGA-VMS formulations.

Remark 5 The $\overline{\tau}$ term of Eq. (19) is not derived from VMS analysis; it is an additional residual-based stabilization term that is included to provided extra stabilizing dissipation near steep solution gradients while maintaining consistency with the exact solution. It was introduced in [1] and bears resemblance to the DCDD [76] and YZβ [91, 92] discontinuity-capturing methods.

5 Parallel Computations

Parallel computations with space–time methods go as far back as 1992 [93], with the 3D computations reported as early as 1993 [94]. All computations reported in this chapter were carried out on parallel computing platforms. The number of cores used in a typical computation ranges from 96 to 576. Because the computations were mostly for the purpose of testing a new computational method, parallel efficiency was not a high priority. Still the efficiencies we see are high enough to justify the use of the maximum number of cores available in the computer resources we have.

6 ST Computation: Aortic-Valve Flow Analysis

This section is from [10].

6.1 Geometry and Leaflet Motion

We have a typical aortic-valve model, such as the one in [30]. The model, shown in Fig. 1, has three leaflets and one main outlet, corresponding to the beginning of the aorta. The leaflet motion is prescribed. They move in an asymmetric fashion. We identify the individual leaflets as shown in Fig. 2. The leaflet positions are defined by means of a pseudo-time parameter θ, with the values 0 and 1 corresponding to the fully open and fully closed positions. The prescribed motion is given through θ as shown in Fig. 3.

6.2 Mesh, Flow Conditions and Computational Conditions

We create the mesh with five SIs, with three of them connecting the mesh sectors containing the leaflets in the valve region of the aorta (see Fig. 4). The other two SIs, which are the top and bottom circular planes in Fig. 4, connect the meshes in the inlet and outlet regions to the valve region. They are for independent meshing in the inlet and outlet regions. The volume mesh is made of quadratic NURBS elements. The number of control points is 84,534, and the number of elements is 54,000. We prescribe the motion of the interior control points, and specify in each domain

Fig. 1 Aortic-valve flow
analysis. Model geometry.
Aorta, leaflets, and sinuses.
The *left* picture shows the
entire computational domain,
and the *right* picture is the
zoomed view of the valve

Fig. 2 Aortic-valve flow
analysis. Leaflet
identification. Leaflet 1 (*red*),
2 (*green*), and 3 (*blue*)

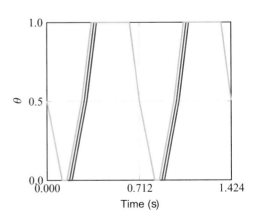

Fig. 3 Aortic-valve flow
analysis. Leaflet motion.
Pseudo-time parameter θ as a
function of time for each of
the three leaflets

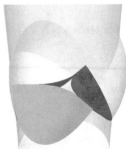

Fig. 4 Aortic-valve flow
analysis. Aortic valve and the
five SIs

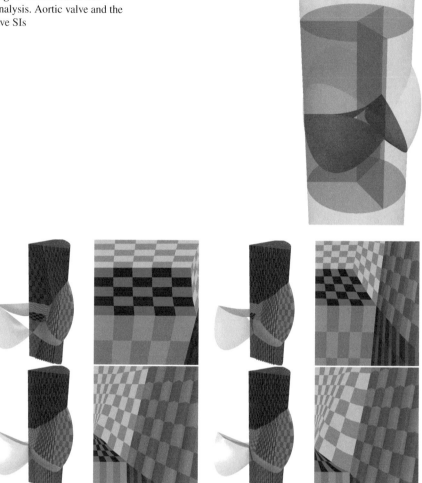

Fig. 5 Aortic-valve flow analysis. A set of selected NURBS elements, from when the valve is
fully open (*top-left*) to when it is fully closed (*bottom-right*). The corresponding θ values are 0.0,
0.42, 0.97, and 1.0. The *right* pictures are the zoomed views around the leaflet

the master–slave mapping for all leaflet positions. Figure 5 shows a set of selected
NURBS elements to illustrate how elements collapse.

The density and kinematic viscosity of the blood are 1050 kg/m^3 and
4.2×10^{-6} m^2/s. The boundary conditions are no-slip on the arterial walls and
the leaflets, traction-free at the outflow boundary, and uniform velocity at the inflow
boundary, with a temporal profile as shown in Fig. 6. The cycle period is 0.712 s.
The no-slip condition on the arterial walls is enforced weakly.

Fig. 6 Aortic-valve flow
analysis. Inflow velocity (two
cycles)

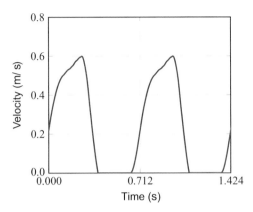

We use the ST-SUPS method. The time-step size is 4.00×10^{-3} s. There are three nonlinear iterations at each time step. The number of GMRES iterations per nonlinear iteration is 300.

6.3 Results

Figure 7 shows the isosurfaces corresponding to a positive value of the second invariant of the velocity gradient tensor, colored by the velocity magnitude. The viewing angle is as we see the leaflets in Fig. 2. We have a biased flow jet due to the asymmetric leaflet closing. This can be seen from the third, fourth, and fifth pair of pictures in Fig. 7. We also report the wall shear stress (WSS) on the leaflet surfaces. The viewing angle is as we see the leaflets in Fig. 8. Figure 9 shows the magnitude of the WSS on the upper and lower surfaces of the leaflets.

7 ST Computation: Patient-Specific Aorta Flow Analysis

This section is from [8].

We start with a geometry obtained from medical images and then use cubic T-splines to represent the surface. The density and kinematic viscosity of the blood are 1050 kg/m^3 and 4.2×10^{-6} m^2/s.

7.1 Conditions

The computational domain and boundary conditions are shown in Fig. 10. The diameters are given in Table 1. The inflow flow rate, plug flow, is in Fig. 11. The

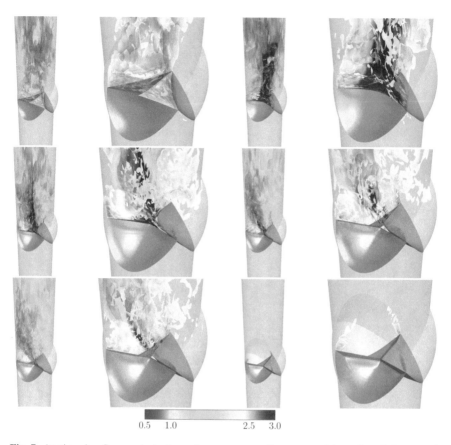

0.5 1.0   2.5 3.0

Fig. 7 Aortic-valve flow analysis. Isosurfaces corresponding to a positive value of the second invariant of the velocity gradient tensor, colored by the velocity magnitude (m/s). The frames are for $t = 0.804, 0.984, 1.028, 1.072, 1.080$, and 1.252 s

Fig. 8 Aortic-valve flow analysis. Viewing angle for reporting the WSS. The leaflet identification is same as in Fig. 2

peak value of the average inflow velocity is 0.709 m/s. We estimate the outflows as distributed by Murray's law [95]:

$$Q_o \propto D_o^3, \tag{22}$$

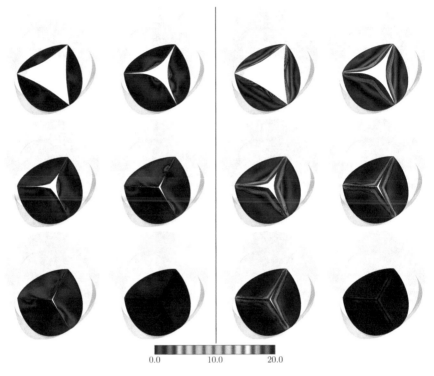

Fig. 9 Aortic-valve flow analysis. Magnitude of the WSS (Pa). Upper surface (*left*) and lower surface (*right*). The frames are for $t = 0.804$, 0.984, 1.028, 1.072, 1.080, and 1.252 s

Fig. 10 Patient-specific aorta flow analysis. Geometry and boundary conditions

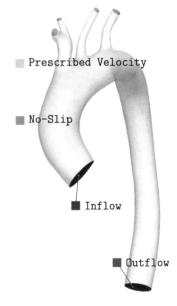

Table 1 Patient-specific aorta flow analysis

	Inlet	Outlet 1	Outlet 2	Outlet 3	Outlet 4	Outlet 5
Diameter	25.6	5.81	3.90	4.41	6.43	19.9

Diameter (mm) of the inlet and outlets. The outlets are listed in the order of closeness to the inlet

Fig. 11 Patient-specific aorta flow analysis. Volumetric flow rate at the inlet

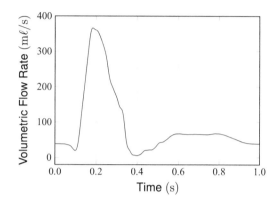

where Q_o is the volumetric outflow rate, and the outlet diameter D_o is defined based on the outlet area A_o:

$$D_o = 2\sqrt{\frac{A_o}{\pi}}. \tag{23}$$

We form a plug flow profile at the smaller outlets, and the main outlet is set to traction free.

7.2 Mesh

We create a quadratic NURBS mesh from the T-spline surface, using the technique introduced in [17, 86]. Figure 12 shows one of the NURBS patches and five of the patches together to illustrate the block-structured nature of the NURBS mesh. The function space has only C^0 continuity between the patches. Figure 13 shows the base mesh. Figure 14 shows the base and refined meshes at the inlet. The meshes are refined by knot insertion, therefore the geometry is unchanged, and the basis functions for the coarser meshes are subsets of the basis functions for the finer meshes. The refinement is in the normal direction, and at each refinement, the element thickness is halved in half of the most refined layers. For the base mesh, the element thickness in the normal direction is approximately 1% of the local diameter. There is no refinement in the tangential directions. During the refinement, the original plug flow profiles of the base mesh are retained. Table 2 shows the number of elements and control points.

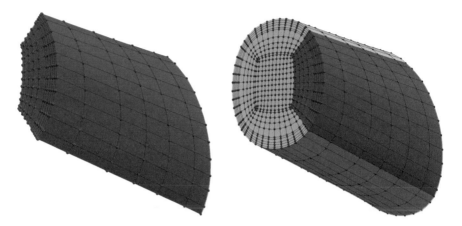

Fig. 12 Patient-specific aorta flow analysis. NURBS control mesh. One of the patches (*top*) and five of the patches together (*bottom*)

Fig. 13 Patient-specific aorta
flow analysis. Base mesh.
Control mesh and surface
(*green*). *Red* points are
control points

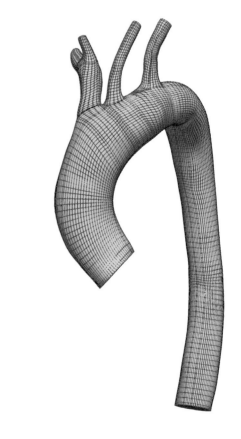

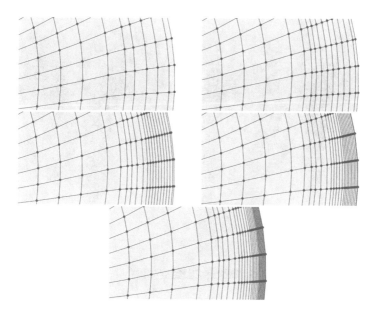

Fig. 14 Patient-specific aorta flow analysis. Control mesh at the inlet. Base Mesh, Refinement Mesh 1, Refinement Mesh 2, Refinement Mesh 3, and Refinement Mesh 4

Table 2 Patient-specific aorta flow analysis

	nc	ne
Base Mesh	202,497	151,513
Refinement Mesh 1	266,437	205,733
Refinement Mesh 2	330,377	259,953
Refinement Mesh 3	394,317	314,173
Refinement Mesh 4	458,257	368,393

Number of control points (nc) and element (ne) for the quadratic NURBS meshes used in the computations

7.3 Mesh Refinement Study

We compute with the 5 meshes in Table 2. The time-step sizes are $\Delta t = 0.0025$ s for Base Mesh and Refinement Mesh 1 and 2, and $\Delta t = 0.00125$ s for Refinement Mesh 3 and 4. The number of nonlinear iterations per time step is 3, and the number of GMRES iterations per nonlinear iteration is 800 for Base Mesh and Refinement Mesh 1, and 1200, 1400, and 1600 for Refinement Mesh 2, 3, and 4, respectively. The ST-SUPS method is used and the stabilization parameters are those given by Eqs. (2.4)–(2.6), (2.8), and (2.10) in [13].

We first compute 9 cycles with Base Mesh, and the initial condition for the refined meshes is obtained by knot insertion. The solution reported here is for the 10th cycle. Figure 15 shows the solution computed with Refinement Mesh 4. At the peak flow rate a complex flow pattern is formed, and the vortex structure breaks down into

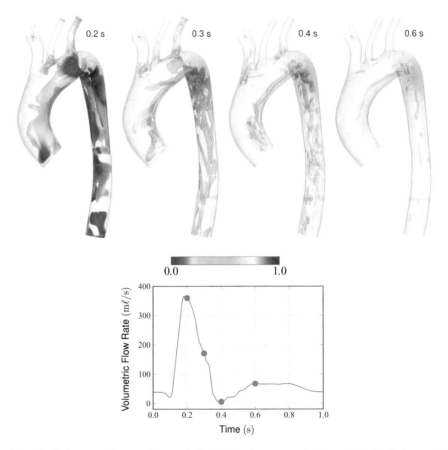

Fig. 15 Patient-specific aorta flow analysis. Mesh refinement study. Computed with Refinement Mesh 4. Isosurfaces corresponding to a positive value of the second invariant of the velocity gradient tensor, colored by the velocity magnitude (m/s) (*top*). The time instants are shown with circles (*bottom*)

smaller structures during the deceleration. The magnitude of the WSS ($\mathbf{h}_v$) at the peak flow rate is shown for each mesh in Fig. 16. Qualitatively, all results are in good agreement, and the convergence can be seen with refinement. To quantify the mesh refinement level, we calculate the y^+ value for the first-element thickness h as

$$y^+ = \frac{u^* h}{\nu}, \tag{24}$$

where the friction velocity u^* is based on the computed value of the WSS as follows:

$$u^* = \sqrt{\frac{\|\mathbf{h}_v^h\|}{\rho}}. \tag{25}$$

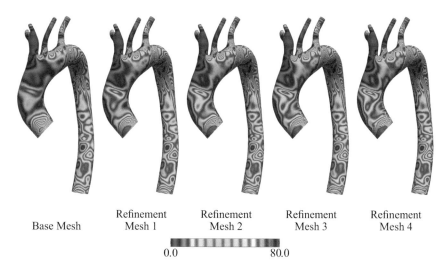

Fig. 16 Patient-specific aorta flow analysis. Mesh refinement study. WSS (dyn/cm^2) at the peak flow rate

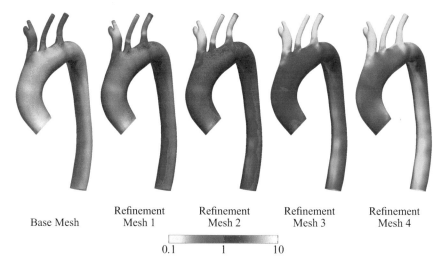

Fig. 17 Patient-specific aorta flow analysis. Mesh refinement study. y^+ value for the first-element thickness, based on the WSS computed at the peak flow rate

Figure 17 shows the spatial distribution of y^+ at the peak flow rate. It shows that for the meshes used here, y^+ range is from approximate maximum 10 to less than 1. Comparing Figs. 16 and 17, we see that the WSS values computed over different meshes are in agreement where $y^+ \leq 1$.

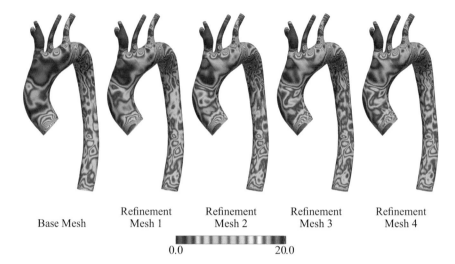

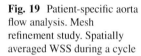

Fig. 18 Patient-specific aorta flow analysis. Mesh refinement study. TAWSS (dyn/cm^2)

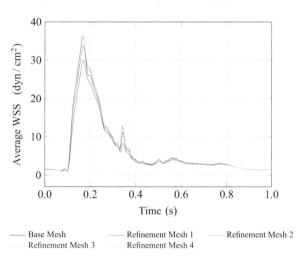

Fig. 19 Patient-specific aorta flow analysis. Mesh refinement study. Spatially averaged WSS during a cycle

The time-averaged WSS magnitude (TAWSS) is shown in Figs. 18, and 19 shows the spatially averaged WSS magnitude in a cycle. Figure 20 shows the oscillatory shear index (OSI), defined as

$$\text{OSI} = \frac{1}{2}\left(1 - \frac{\left\|\int_0^T \mathbf{h}_v^h \, dt\right\|}{\int_0^T \left\|\mathbf{h}_v^h\right\| \, dt}\right). \qquad (26)$$

Overall for OSI, even Base Mesh is in a good agreement with others. However, if we compare details such as branches, we see some difference even where y^+ value

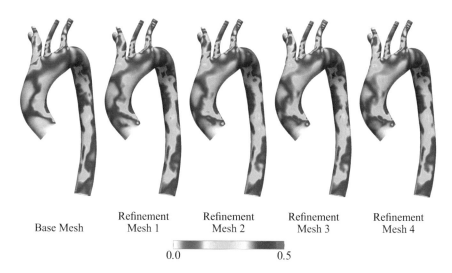

| Base Mesh | Refinement Mesh 1 | Refinement Mesh 2 | Refinement Mesh 3 | Refinement Mesh 4 |

0.0   0.5

Fig. 20 Patient-specific aorta flow analysis. Mesh refinement study. OSI

is small. To see the flow differences, using the solution from Refinement Mesh 4 as the reference solution, we inspect the velocity difference $\left\| \mathbf{u}_k^h - \mathbf{u}_4^h \right\|$, where the subscripts indicate Base Mesh and Refinement Mesh k.

Remark 6 To calculate the velocity difference, all meshes and corresponding solutions are refined by using the knot-insertion technique, and the control variables are obtained based on Refinement Mesh 4. The visualization is done after taking the difference between the control variables, interpolating the vector, and taking its magnitude.

The spatial average of the difference is maximum at around 0.5 s. This indicates that the vortex breakdown, due to the small-scale flow behavior that needs to be dealt with, would not be easy to resolve. Figure 21 shows the velocity difference at 0.5 s.

In summary, good accuracy in the WSS magnitude can be obtained with locally good representation, and the OSI requires a good flow representation overall, including the vortex breakdown.

8 IMGA Computation: Patient-Specific Heart Valve Design and Analysis

This section is from [74], where more details can be found.

Here we present a novel framework for designing personalized prosthetic heart valves using IMGA-VMS. We parameterize the leaflet geometry using several key

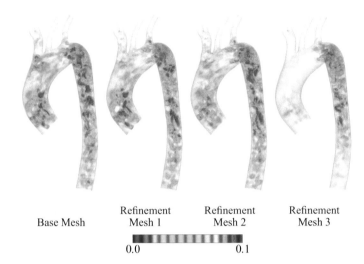

Base Mesh   Refinement   Refinement   Refinement
              Mesh 1   Mesh 2   Mesh 3

0.0   0.1

Fig. 21 Patient-specific aorta flow analysis. Mesh refinement study. Velocity difference $\left\| \mathbf{u}_k^h - \mathbf{u}_4^h \right\|$ (m/s) at 0.5 s, where the subscripts indicate Base Mesh and Refinement Mesh k

design parameters. This allows for generating various perturbations of the leaflet design for the patient-specific aortic root reconstructed from the medical image data. Each design is analyzed using the IMGA-VMS FSI methodology, which allows us to efficiently simulate the coupling of the deforming aortic root, the parametrically designed prosthetic valves, and the surrounding blood flow under physiological conditions. A parametric study is carried out to investigate the influence of the geometry on heart valve performance, indicated by the effective orifice area (EOA) and the coaptation area (CA). Finally, the FSI simulation results of a design that reasonably well balances the EOA and CA are presented.

8.1 Trivariate NURBS Parameterization of the Ascending Aorta

To obtain a volumetric parameterization of the artery and lumen, we first construct a trivariate multi-patch NURBS in a regular shape, e.g., a tubular domain, then solve a linear elastostatic, mesh moving problem [94] for the displacement from this regular domain to a deformed configuration that represents the artery and lumen. However, solving a linear elastostatic problem to obtain the deformed interior mesh is only effective for relatively mild, translation-dominant deformations. For scenarios that involve large deformations, such as the deformation of a straight tubular domain into a curved shape of a patient-specific ascending aorta, the interior elements can become severely distorted. To avoid this, we first obtain a centerline along the axial

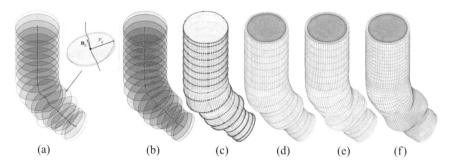

Fig. 22 The construction of the volumetric NURBS discretization of the blood and the artery wall domains. (**a**) Cross sections of the artery wall surface. (**b**) Circular cross sections. (**c**) NURBS tubular surface and corresponding control points. (**d**) Primitive volume mesh. (**e**) Deformed volume mesh. (**f**) h-refined volume mesh

direction of a patient-specific artery wall surface. Along this centerline, we define a number of cross sections corresponding to the control points of the NURBS artery wall surface in the axial direction. (These cross sections are shown as blue curves in Fig. 22a.) At each cross section, we calculate its unit normal vector $\mathbf{n}_c$ and the effective radius r_c, which is determined such that the area of a circle calculated using this radius matches the area of the cross section. (A circle corresponding to one of the cross sections is shown in the red curve in Fig. 22a.) Finally, using this information, we construct a tubular NURBS surface that has the same control-point and knot-vector topology as the target patient-specific artery wall surface, as shown in Fig. 22b, c. Another tubular surface corresponding to the lumenal surface is also constructed, using the same cross sections but smaller effective radii coming from the lumenal NURBS surface.

These two tubular NURBS surfaces are used to construct a primitive trivariate multi-patch NURBS that includes the solid and fluid subdomains, shown in gray and red, respectively, in Fig. 22d. Basis functions are made C^0-continuous at the fluid–solid interface, so that velocity functions defined using the resulting spline space conform to standard fluid–structure kinematic constraints while retaining the ability to represent non-smooth behavior across the material interface. The resulting volumetric NURBS can then be morphed to match the patient-specific geometry with minimal rotation, so an elastostatic problem can provide an analysis-suitable parameterization. Displacements at the ends of the tube are constrained to remain within their respective cross sections. Finally, we refine the deformed trivariate NURBS for analysis purposes, by inserting knots at desired locations, such as around the sinuses and the flow boundary layers. The final volumetric NURBS discretization of the patient-specific ascending aorta is shown in Fig. 22f.

8.2 Parametric BHV Design

To design effective prosthetic valves for specific patients, we focus specifically on the leaflet geometry and assume that non-leaflet components of stentless valves move with the aortic root and do not affect aortic deformation or flow. Starting from the NURBS surface of a patient-specific root, valve leaflets are parametrically designed as follows. We first pick nine "key points" located on the ends of commissure lines and the bottom of the sinuses. The positions of these points are indicated by blue spheres in Fig. 23. These define how the leaflets attach to the sinuses. The key points solely depend on the geometry of the patient-specific aortic root and will remain unchanged for different valve designs. We then parameterize families of univariate B-splines defining the free edges and radial "belly curves" of the leaflets. These curves are shown in red and green in Fig. 23. The attachment edges, free edges, and belly curves are then interpolated to obtain smooth bivariate B-spline representations of the leaflets.

Figure 24 shows the details of parameterizing the free-edge curve (red) and the belly-region curve (green). In Fig. 24, $\mathbf{p}_1$, $\mathbf{p}_2$, and $\mathbf{p}_3$ are the key points on the top of the commissure lines and $\mathbf{p}_4$ is the key point on the sinus bottom, as labeled in Fig. 23. Points $\mathbf{p}_1$-$\mathbf{p}_3$ define a triangle $\Delta\mathbf{p}_{1-3}$, with $\mathbf{p}_c$ being its geometric center. The unit vector pointing from $\mathbf{p}_c$ to $\mathbf{p}_n$ is denoted by $\mathbf{t}_p$, and the unit normal vector of $\Delta\mathbf{p}_{1-3}$ pointing downwards is $\mathbf{n}_p$. We first construct the free edge curve as a univariate quadratic B-spline curve determined by three control points, $\mathbf{p}_1$, $\mathbf{p}_f$, and $\mathbf{p}_2$. The location of $\mathbf{p}_f$ is defined by $\mathbf{p}_f = \mathbf{p}_c + x_1\mathbf{t}_p + x_2\mathbf{n}_p$. By changing x_1 and x_2 to control the location of $\mathbf{p}_f$, the curvature and the height of the free edge can be parametrically changed. We then take $\mathbf{p}_m$ as the midpoint of the free edge, the point $\mathbf{p}_b$, and the key point $\mathbf{p}_4$ to construct a univariate quadratic B-spline curve (green). The point $\mathbf{p}_b$ is defined by $\mathbf{p}_b = \mathbf{p}_o + x_3\mathbf{n}_p$, where $\mathbf{p}_o$ is the projection of $\mathbf{p}_m$ onto $\Delta\mathbf{p}_{1-3}$ along the direction of $\mathbf{n}_p$. Finally, the fixed attachment edges and the parametrically controlled free edge and belly curve are used to construct a

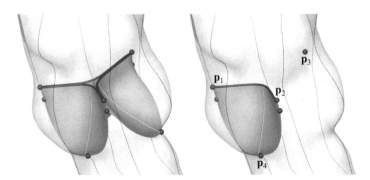

Fig. 23 The key geometric features used to parametrically control the valve designs. The *blue* key points define the attachment of the valve to the root. The *red* and *green* curves are parametrically controlled for valve design

Fig. 24 The parametric control of the valve designs. The key points (*blue* spheres) are identical to those in the *right* plot of Fig. 23. x_1, x_2, and x_3 control the location of $\mathbf{P}_f$ and $\mathbf{P}_b$ and thus control the curvature and height of the *red* free edge, and the curvature of the *green* belly curve

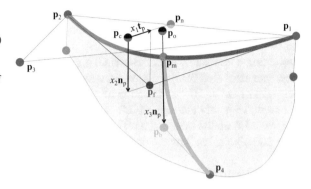

cubic B-spline surface with desired parameterization. By choosing x_1, x_2, and x_3 as design variables, we can parametrically change the free edge and belly curve, and therefore change the valve design. This procedure is implemented in an interactive geometry modeling and parametric design platform [96].

8.3 Application to BHV Design

To determine an effective BHV design, we first need to identify quantitative measures of its performance. We focus on two quantities of clinical interest: to measure the systolic performance, we evaluate the EOA, which indicates how well the valve permits flow in the forward direction. For a quantitative evaluation of the diastolic performance, we measure the CA, which indicates how well the valve seals and prevents flow in the reverse direction. In this section, we study the impact of the design variables x_1, x_2, and x_3 on our two quantities of interest.

Constitutive parameters in the governing equations are held constant over the design space. Fluid, solid, and shell structure mass densities are set to 1.0 g/cm^3. The parameters of the Fung-type material model for the shell structure are $c_0 = 2.0 \times 10^6$ dyn/cm^2, $c_1 = 2.0 \times 10^5$ dyn/cm^2, and $c_2 = 100$. The thickness of the leaflet is set to 0.0386 cm. The bulk and shear moduli for the arterial wall are selected to give a Young's modulus of 10^7 dyn/cm^2 and Poisson's ratio of 0.45 in the small strain limit. The inlet and outlet cross sections are free to slide in their tangential planes and deform radially, but constrained not to move in the orthogonal directions [97]. Mass-proportional damping with constant $C_{\text{damp}} = 10^4$ Hz is used to model the interaction of the artery with the surrounding tissue. The dynamic viscosity of the blood is set to $\mu_f = 3 \times 10^{-2}$ g/(cm s).

We apply a physiologically realistic left ventricular pressure time history as a traction boundary condition at the inflow. The applied pressure signal is periodic, with a period of 0.86 s for one cardiac cycle. The traction $-(p_0 + RQ)\mathbf{n}_f$ is applied at the outflow for the resistance boundary condition, where p_0 is a constant physiological pressure level, $R > 0$ is a resistance coefficient, and Q is the volumetric flow rate through the outflow. In the present computation, we set $p_0 = 80$ mmHg and $R = 200$ (dyn s)/cm^5. These values ensure a realistic transvalvular pressure difference of 80 mmHg across a closed valve when $Q = 0$, while permitting a flow rate within the normal physiological range and consistent with the flow rate estimated from the medical data (about 310 ml/s) during systole. A time-step size of $\Delta t = 10^{-4}$ s is used in all simulations. To obtain the artery wall tissue prestress, we apply the highest left ventricular pressure during systole (127 mmHg at $t = 0.25$ s) on the inlet and a resistance boundary condition ($p_0 = 80$ mmHg and $R = 200$ (dyn s)/cm^5) on the outlet for the calculation of $\tilde{\mathbf{h}}_f$ in the prestress problem [54].

We perform FSI simulations of each of $(x_1, x_2, x_3) \in (\{0.05, 0.25, 0.45\}$ cm, $\{0.1, 0.3, 0.5\}$ cm, $\{0.5, 0.8, 1.1, 1.4\}$ cm), then calculate the EOA at peak systole and the maximum CA occurring during ventricular diastole. The simulation results and quantities of interest for each case are reported in [74]. An ideal valve would have both a large EOA and a large CA. However, these two quantities tend to compete with each other: valves that close easily can be more difficult to open and vice versa. In general, the results show that increasing x_1, which corresponds to decreasing the length of the free edge, decreases EOA and CA at the same time. Increasing x_2, which decreases the height of the free edge, may increase EOA slightly but reduces CA significantly. The reduction of CA due to increasing x_2 reduces CA and causes many designs cannot seal completely. Increasing x_3, which increases the surface curvature in the leaflet belly region, improves CA but decreases EOA. Finally, the combination of $x_1 = 0.05$ cm, $x_2 = 0.1$ or 0.3 cm, and $x_3 = 0.5$ or 0.8 cm reliably yields a high EOA between 3.92 and 4.05 cm^2, near the upper end of the physiological range of 3.0–4.0 cm^2 in healthy adults, and a CA between 3.49 and 4.54 cm^2. Among these four cases, $\mathbf{x}^* = (x_1, x_2, x_3) = (0.05$ cm, 0.1 cm, 0.8 cm), which has a CA of 4.54 cm^2 and EOA of 3.92 cm^2, strikes the best compromise between EOA and CA. The valve geometry of this best-performing design and its EOA and CA from the FSI simulation are shown in Fig. 25.

Figure 26 shows several snapshots of the valve deformation and the details of the flow field at several points during the cardiac cycle. The color indicates the fluid velocity magnitude. The visualizations clearly show the instantaneous valve response to the left ventricular pressure. The valve opens with the rising left ventricular pressure in early systole (0.0–0.20 s), and then stays fully open near peak systole (0.25–0.27 s), allowing sufficient blood flow to enter the ascending aorta. A quick valve closure is then observed in early diastole (0.32–0.38 s). This

Fig. 25 The best-performing prosthetic valve design and its EOA and CA from the FSI simulation

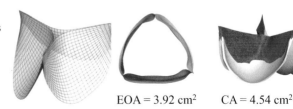

EOA = 3.92 cm^2 CA = 4.54 cm^2

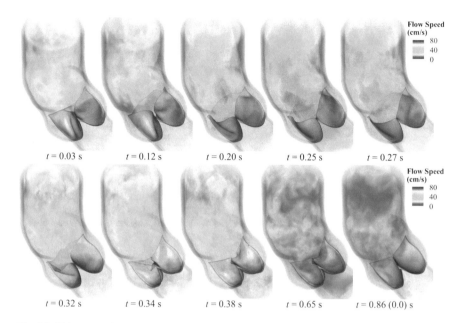

Fig. 26 Volume rendering visualization of the velocity field from our FSI simulation at several points during a cardiac cycle

quick closure of the valve minimizes the reverse flow into the left ventricle, as the left ventricular pressure drops rapidly in this period. After that, the valve properly seals, and the flow reaches a near-hydrostatic state (0.65 s). These features observed during the cardiac cycle characterize a well-functioning valve within the objectives considered in this study: a large EOA during systole and a proper CA during diastole. In Fig. 27, the models are superposed in the configurations corresponding to the fully open and fully closed phases for better visualization of the leaflet–wall coupling results. The deformation of the attachment edges can be clearly seen. The expansion and contraction of the arterial wall, as well as its sliding motion between systole and diastole can also be observed. The maximum in-plane principal Green–Lagrange strain (MIPE) evaluated on the aortic side of the leaflet is shown in Fig. 28. The figure shows that during opening the strain is concentrated in the belly region of the leaflet, while during closing the highest strain happens near the valve commissure.

9   Concluding Remarks

In this chapter we have reviewed various technologies that have been developed by us and our colleagues and used to solve general classes of problems in computational cardiovascular analysis, with focus herein on aortic flows and patient-specific and

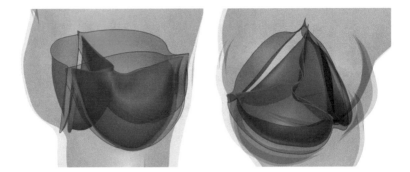

Fig. 27 Relative displacement between fully open (*red*) and fully closed (*blue*) configurations, showing the effect of leaflet–wall coupling. The deformation of the attachment edges can be clearly seen. The expansion and contraction of the arterial wall as well as its sliding motion between systole (*red*) and diastole (*blue*) can also be observed

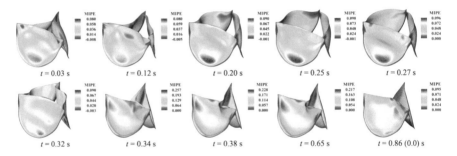

Fig. 28 Deformed valve configuration, colored by the maximum in-plane principal Green–Lagrange strain (MIPE) evaluated on the aortic side of the leaflet. Note the different scale for each time instant

bioprosthetic heart-valve FSI. Our work on these problems, and in other more general areas of engineering, science, and medicine, is based on Stabilized Methods and the Variational Multiscale Method (VMS), which have enjoyed enormous attention in the research literature and are used widely in industry and national laboratories. Stabilized Methods and the Variational Multiscale Method are at the center of development of core technologies such as Space–Time VMS, Arbitrary Lagrangian–Eulerian VMS, and Immersogeometric VMS, which we emphasized herein. They are in turn enhanced by many other special technologies that are used to deal with specific features of the applications, many of which we also described.

Computational cardiovascular analysis is now used routinely in medical device design, diagnosis of cardiovascular disease, surgical planning, virtual stent placement, and numerous other areas. It is only part of the more general field of Computational Medicine, which is rapidly growing. Just as the capacity of the underlying computational methods described in this article depend on the growing power of computers, Computational Medicine depends upon the increasing fidelity of medical imaging technologies and devices. Like computers, these are also advancing rapidly, which portends a bright future for the further development of Computational Medicine and its enormous potential impact on health and the human condition.

Acknowledgments This work was supported (second author) in part by JST-CREST; Grant-in-Aid for Scientific Research (S) 26220002 from the Ministry of Education, Culture, Sports, Science and Technology of Japan (MEXT); Grant-in-Aid for Scientific Research (A) 18H04100 from Japan Society for the Promotion of Science; and Rice–Waseda research agreement. The mathematical model and computational method parts of the work were also supported (fourth author) in part by ARO Grant W911NF-17-1-0046, ARO DURIP Grant W911NF-18-1-0234, and Top Global University Project of Waseda University. The third author was partially supported by NSF Grant 1854436, and the fifth author was partially supported by NIH/NHLBI Grants R01HL129077 and R01HL142504.

References

1. C.A. Taylor, T.J.R. Hughes, and C.K. Zarins, "Finite element modeling of blood flow in arteries", *Computer Methods in Applied Mechanics and Engineering*, **158** (1998) 155–196.
2. A.N. Brooks and T.J.R. Hughes, "Streamline upwind/Petrov-Galerkin formulations for convection dominated flows with particular emphasis on the incompressible Navier-Stokes equations", *Computer Methods in Applied Mechanics and Engineering*, **32** (1982) 199–259.
3. C. Johnson, U. Navert, and J. Pitkäranta, "Finite element methods for linear hyperbolic problems", *Computer Methods in Applied Mechanics and Engineering*, **45** (1984) 285–312.
4. T.J.R. Hughes, "Multiscale phenomena: Green's functions, the Dirichlet-to-Neumann formulation, subgrid scale models, bubbles, and the origins of stabilized methods", *Computer Methods in Applied Mechanics and Engineering*, **127** (1995) 387–401.
5. T.J.R. Hughes, G. Feijóo., L. Mazzei, and J.B. Quincy, "The variational multiscale method–A paradigm for computational mechanics", *Computer Methods in Applied Mechanics and Engineering*, **166** (1998) 3–24.
6. T.J.R. Hughes and G. Sangalli, "Variational multiscale analysis: the fine-scale Green's function, projection, optimization, localization, and stabilized methods", *SIAM Journal of Numerical Analysis*, **45** (2007) 539–557.
7. Y. Bazilevs, V.M. Calo, J.A. Cottrell, T.J.R. Hughes, A. Reali, and G. Scovazzi, "Variational multiscale residual-based turbulence modeling for large eddy simulation of incompressible flows", *Computer Methods in Applied Mechanics and Engineering*, **197** (2007) 173–201.
8. K. Takizawa, T.E. Tezduyar, H. Uchikawa, T. Terahara, T. Sasaki, and A. Yoshida, "Mesh refinement influence and cardiac-cycle flow periodicity in aorta flow analysis with isogeometric discretization", *Computers & Fluids*, **179** (2019) 790–798, https://doi.org/10.1016/j.compfluid.2018.05.025.
9. M.-C. Hsu, D. Kamensky, Y. Bazilevs, M.S. Sacks, and T.J.R. Hughes, "Fluid–structure interaction analysis of bioprosthetic heart valves: significance of arterial wall deformation", *Computational Mechanics*, **54** (2014) 1055–1071, https://doi.org/10.1007/s00466-014-1059-4.
10. K. Takizawa, T.E. Tezduyar, T. Terahara, and T. Sasaki, "Heart valve flow computation with the integrated Space–Time VMS, Slip Interface, Topology Change and Isogeometric Discretization methods", *Computers & Fluids*, **158** (2017) 176–188, https://doi.org/10.1016/j.compfluid.2016.11.012.
11. K. Takizawa and T.E. Tezduyar, "Multiscale space–time fluid–structure interaction techniques", *Computational Mechanics*, **48** (2011) 247–267, https://doi.org/10.1007/s00466-011-0571-z.

12. Y. Bazilevs, V.M. Calo, T.J.R. Hughes, and Y. Zhang, "Isogeometric fluid–structure interaction: theory, algorithms, and computations", *Computational Mechanics*, **43** (2008) 3–37.
13. K. Takizawa, T.E. Tezduyar, H. Mochizuki, H. Hattori, S. Mei, L. Pan, and K. Montel, "Space–time VMS method for flow computations with slip interfaces (ST-SI)", *Mathematical Models and Methods in Applied Sciences*, **25** (2015) 2377–2406, https://doi.org/10.1142/S0218202515400126.
14. K. Takizawa, T.E. Tezduyar, A. Buscher, and S. Asada, "Space–time interface-tracking with topology change (ST-TC)", *Computational Mechanics*, **54** (2014) 955–971, https://doi.org/10.1007/s00466-013-0935-7.
15. K. Takizawa, B. Henicke, A. Puntel, T. Spielman, and T.E. Tezduyar, "Space–time computational techniques for the aerodynamics of flapping wings", *Journal of Applied Mechanics*, **79** (2012) 010903, https://doi.org/10.1115/1.4005073.
16. K. Takizawa, T.E. Tezduyar, Y. Otoguro, T. Terahara, T. Kuraishi, and H. Hattori, "Turbocharger flow computations with the Space–Time Isogeometric Analysis (ST-IGA)", *Computers & Fluids*, **142** (2017) 15–20, https://doi.org/10.1016/j.compfluid.2016.02.021.
17. Y. Otoguro, K. Takizawa, and T.E. Tezduyar, "Space–time VMS computational flow analysis with isogeometric discretization and a general-purpose NURBS mesh generation method", *Computers & Fluids*, **158** (2017) 189–200, https://doi.org/10.1016/j.compfluid.2017.04.017.
18. Y. Bazilevs and T.J.R. Hughes, "Weak imposition of Dirichlet boundary conditions in fluid mechanics", *Computers and Fluids*, **36** (2007) 12–26.
19. Y. Bazilevs and T.J.R. Hughes, "NURBS-based isogeometric analysis for the computation of flows about rotating components", *Computational Mechanics*, **43** (2008) 143–150.
20. M.E. Moghadam, Y. Bazilevs, T.-Y. Hsia, I.E. Vignon-Clementel, A.L. Marsden, and M. of Congenital Hearts Alliance (MOCHA), "A comparison of outlet boundary treatments for prevention of backflow divergence with relevance to blood flow simulations", *Computational Mechanics*, **48** (2011) 277–291, https://doi.org/10.1007/s00466-011-0599-0.
21. T.E. Tezduyar, "Stabilized finite element formulations for incompressible flow computations", *Advances in Applied Mechanics*, **28** (1992) 1–44, https://doi.org/10.1016/S0065-2156(08)70153-4.
22. T.E. Tezduyar and K. Takizawa, "Space–time computations in practical engineering applications: A summary of the 25-year history", *Computational Mechanics*, **63** (2019) 747–753, https://doi.org/10.1007/s00466-018-1620-7.
23. K. Takizawa and T.E. Tezduyar, "Computational methods for parachute fluid–structure interactions", *Archives of Computational Methods in Engineering*, **19** (2012) 125–169, https://doi.org/10.1007/s11831-012-9070-4.
24. K. Takizawa, D. Montes, M. Fritze, S. McIntyre, J. Boben, and T.E. Tezduyar, "Methods for FSI modeling of spacecraft parachute dynamics and cover separation", *Mathematical Models and Methods in Applied Sciences*, **23** (2013) 307–338, https://doi.org/10.1142/S0218202513400058.
25. K. Takizawa, T.E. Tezduyar, and R. Kolesar, "FSI modeling of the Orion spacecraft drogue parachutes", *Computational Mechanics*, **55** (2015) 1167–1179, https://doi.org/10.1007/s00466-014-1108-z.
26. K. Takizawa, B. Henicke, D. Montes, T.E. Tezduyar, M.-C. Hsu, and Y. Bazilevs, "Numerical-performance studies for the stabilized space–time computation of wind-turbine rotor aerodynamics", *Computational Mechanics*, **48** (2011) 647–657, https://doi.org/10.1007/s00466-011-0614-5.
27. K. Takizawa, "Computational engineering analysis with the new-generation space–time methods", *Computational Mechanics*, **54** (2014) 193–211, https://doi.org/10.1007/s00466-014-0999-z.
28. K. Takizawa, B. Henicke, A. Puntel, N. Kostov, and T.E. Tezduyar, "Space–time techniques for computational aerodynamics modeling of flapping wings of an actual locust", *Computational Mechanics*, **50** (2012) 743–760, https://doi.org/10.1007/s00466-012-0759-x.

29. K. Takizawa, T.E. Tezduyar, and N. Kostov, "Sequentially-coupled space–time FSI analysis of bio-inspired flapping-wing aerodynamics of an MAV", *Computational Mechanics*, **54** (2014) 213–233, https://doi.org/10.1007/s00466-014-0980-x.
30. K. Takizawa, T.E. Tezduyar, and A. Buscher, "Space–time computational analysis of MAV flapping-wing aerodynamics with wing clapping", *Computational Mechanics*, **55** (2015) 1131–1141, https://doi.org/10.1007/s00466-014-1095-0.
31. K. Takizawa, Y. Bazilevs, T.E. Tezduyar, M.-C. Hsu, O. Øiseth, K.M. Mathisen, N. Kostov, and S. McIntyre, "Engineering analysis and design with ALE-VMS and space–time methods", *Archives of Computational Methods in Engineering*, **21** (2014) 481–508, https://doi.org/10.1007/s11831-014-9113-0.
32. K. Takizawa, K. Schjodt, A. Puntel, N. Kostov, and T.E. Tezduyar, "Patient-specific computational analysis of the influence of a stent on the unsteady flow in cerebral aneurysms", *Computational Mechanics*, **51** (2013) 1061–1073, https://doi.org/10.1007/s00466-012-0790-y.
33. K. Takizawa, T.E. Tezduyar, and T. Kuraishi, "Multiscale ST methods for thermo-fluid analysis of a ground vehicle and its tires", *Mathematical Models and Methods in Applied Sciences*, **25** (2015) 2227–2255, https://doi.org/10.1142/S0218202515400072.
34. K. Takizawa, T.E. Tezduyar, T. Kuraishi, S. Tabata, and H. Takagi, "Computational thermo-fluid analysis of a disk brake", *Computational Mechanics*, **57** (2016) 965–977, https://doi.org/10.1007/s00466-016-1272-4.
35. T. Kanai, K. Takizawa, T.E. Tezduyar, K. Komiya, M. Kaneko, K. Hirota, M. Nohmi, T. Tsuneda, M. Kawai, and M. Isono, "Methods for computation of flow-driven string dynamics in a pump and residence time", *Mathematical Models and Methods in Applied Sciences*, **29** (2019) 839–870, https://doi.org/10.1142/S021820251941001X.
36. Y. Otoguro, K. Takizawa, T.E. Tezduyar, K. Nagaoka, and S. Mei, "Turbocharger turbine and exhaust manifold flow computation with the Space–Time Variational Multiscale Method and Isogeometric Analysis", *Computers & Fluids*, **179** (2019) 764–776, https://doi.org/10.1016/j.compfluid.2018.05.019.
37. T. Kuraishi, K. Takizawa, and T.E. Tezduyar, "Tire aerodynamics with actual tire geometry, road contact and tire deformation", *Computational Mechanics*, **63** (2019) 1165–1185, https://doi.org/10.1007/s00466-018-1642-1.
38. T. Kuraishi, K. Takizawa, and T.E. Tezduyar, "Space–Time Isogeometric flow analysis with built-in Reynolds-equation limit", *Mathematical Models and Methods in Applied Sciences*, **29** (2019) 871–904, https://doi.org/10.1142/S0218202519410021.
39. K. Takizawa, T.E. Tezduyar, and T. Terahara, "Ram-air parachute structural and fluid mechanics computations with the space–time isogeometric analysis (ST-IGA)", *Computers & Fluids*, **141** (2016) 191–200, https://doi.org/10.1016/j.compfluid.2016.05.027.
40. K. Takizawa, T.E. Tezduyar, and T. Kanai, "Porosity models and computational methods for compressible-flow aerodynamics of parachutes with geometric porosity", *Mathematical Models and Methods in Applied Sciences*, **27** (2017) 771–806, https://doi.org/10.1142/S0218202517500166.
41. R. Torii, M. Oshima, T. Kobayashi, K. Takagi, and T.E. Tezduyar, "Computation of cardiovascular fluid–structure interactions with the DSD/SST method", in *Proceedings of the 6th World Congress on Computational Mechanics (CD-ROM)*, Beijing, China, (2004).
42. R. Torii, M. Oshima, T. Kobayashi, K. Takagi, and T.E. Tezduyar, "Computer modeling of cardiovascular fluid–structure interactions with the Deforming-Spatial-Domain/Stabilized Space–Time formulation", *Computer Methods in Applied Mechanics and Engineering*, **195** (2006) 1885–1895, https://doi.org/10.1016/j.cma.2005.05.050.
43. T.E. Tezduyar, S. Sathe, T. Cragin, B. Nanna, B.S. Conklin, J. Pausewang, and M. Schwaab, "Modeling of fluid–structure interactions with the space–time finite elements: Arterial fluid mechanics", *International Journal for Numerical Methods in Fluids*, **54** (2007) 901–922, https://doi.org/10.1002/fld.1443.

44. T.E. Tezduyar, K. Takizawa, T. Brummer, and P.R. Chen, "Space–time fluid–structure inter-action modeling of patient-specific cerebral aneurysms", *International Journal for Numerical Methods in Biomedical Engineering*, **27** (2011) 1665–1710, https://doi.org/10.1002/cnm.1433.

45. K. Takizawa, Y. Bazilevs, T.E. Tezduyar, C.C. Long, A.L. Marsden, and K. Schjodt, "ST and ALE-VMS methods for patient-specific cardiovascular fluid mechanics modeling", *Mathematical Models and Methods in Applied Sciences*, **24** (2014) 2437–2486, https://doi.org/10.1142/S0218202514500250.

46. K. Takizawa, R. Torii, H. Takagi, T.E. Tezduyar, and X.Y. Xu, "Coronary arterial dynamics computation with medical-image-based time-dependent anatomical models and element-based zero-stress state estimates", *Computational Mechanics*, **54** (2014) 1047–1053, https://doi.org/10.1007/s00466-014-1049-6.

47. T.J.R. Hughes, W.K. Liu, and T.K. Zimmermann, "Lagrangian–Eulerian finite element for-mulation for incompressible viscous flows", *Computer Methods in Applied Mechanics and Engineering*, **29** (1981) 329–349.

48. V. Kalro and T.E. Tezduyar, "A parallel 3D computational method for fluid–structure inter-actions in parachute systems", *Computer Methods in Applied Mechanics and Engineering*, **190** (2000) 321–332, https://doi.org/10.1016/S0045-7825(00)00204-8.

49. Y. Bazilevs, M.-C. Hsu, I. Akkerman, S. Wright, K. Takizawa, B. Henicke, T. Spielman, and T.E. Tezduyar, "3D simulation of wind turbine rotors at full scale. Part I: Geometry modeling and aerodynamics", *International Journal for Numerical Methods in Fluids*, **65** (2011) 207–235, https://doi.org/10.1002/fld.2400.

50. A. Korobenko, Y. Bazilevs, K. Takizawa, and T.E. Tezduyar, "Computer modeling of wind tur-bines: 1. ALE-VMS and ST-VMS aerodynamic and FSI analysis", *Archives of Computational Methods in Engineering*, published online, DOI: 10.1007/s11831-018-9292-1, September 2018, https://doi.org/10.1007/s11831-018-9292-1.

51. J. Yan, A. Korobenko, X. Deng, and Y. Bazilevs, "Computational free-surface fluid–structure interaction with application to floating offshore wind turbines", *Computers and Fluids*, **141** (2016) 155–174, https://doi.org/10.1016/j.compfluid.2016.03.008.

52. Y. Bazilevs, A. Korobenko, X. Deng, and J. Yan, "FSI modeling for fatigue-damage prediction in full-scale wind-turbine blades", *Journal of Applied Mechanics*, **83** (6) (2016) 061010.

53. Y. Bazilevs, V.M. Calo, Y. Zhang, and T.J.R. Hughes, "Isogeometric fluid–structure interaction analysis with applications to arterial blood flow", *Computational Mechanics*, **38** (2006) 310–322.

54. M.-C. Hsu and Y. Bazilevs, "Blood vessel tissue prestress modeling for vascular fluid–structure interaction simulations", *Finite Elements in Analysis and Design*, **47** (2011) 593–599.

55. M.-C. Hsu, D. Kamensky, F. Xu, J. Kiendl, C. Wang, M.C.H. Wu, J. Mineroff, A. Reali, Y. Bazilevs, and M.S. Sacks, "Dynamic and fluid–structure interaction simulations of biopros-thetic heart valves using parametric design with T-splines and Fung-type material models", *Computational Mechanics*, **55** (2015) 1211–1225, https://doi.org/10.1007/s00466-015-1166-x.

56. D. Kamensky, M.-C. Hsu, D. Schillinger, J.A. Evans, A. Aggarwal, Y. Bazilevs, M.S. Sacks, and T.J.R. Hughes, "An immersogeometric variational framework for fluid-structure interac-tion: Application to bioprosthetic heart valves", *Computer Methods in Applied Mechanics and Engineering*, **284** (2015) 1005–1053.

57. I. Akkerman, J. Dunaway, J. Kvandal, J. Spinks, and Y. Bazilevs, "Toward free-surface modeling of planing vessels: simulation of the Fridsma hull using ALE-VMS", *Computational Mechanics*, **50** (2012) 719–727.

58. M.C.H. Wu, D. Kamensky, C. Wang, A.J. Herrema, F. Xu, M.S. Pigazzini, A. Verma, A.L. Marsden, Y. Bazilevs, and M.-C. Hsu, "Optimizing fluid–structure interaction systems with immersogeometric analysis and surrogate modeling: Application to a hydraulic arresting gear", *Computer Methods in Applied Mechanics and Engineering*, **316** (2017) 668–693.

59. J. Yan, X. Deng, A. Korobenko, and Y. Bazilevs, "Free-surface flow modeling and simulation of horizontal-axis tidal-stream turbines", *Computers and Fluids*, **158** (2017) 157–166, https://doi.org/10.1016/j.compfluid.2016.06.016.

60. A. Castorrini, A. Corsini, F. Rispoli, K. Takizawa, and T.E. Tezduyar, "A stabilized ALE method for computational fluid–structure interaction analysis of passive morphing in turbomachinery", *Mathematical Models and Methods in Applied Sciences*, **29** (2019) 967–994, https://doi.org/10.1142/S0218202519410057.

61. J. Yan, B. Augier, A. Korobenko, J. Czarnowski, G. Ketterman, and Y. Bazilevs, "FSI modeling of a propulsion system based on compliant hydrofoils in a tandem configuration", *Computers and Fluids*, **141** (2016) 201–211, https://doi.org/10.1016/j.compfluid.2015.07.013.

62. T.A. Helgedagsrud, I. Akkerman, Y. Bazilevs, K.M. Mathisen, and O.A. Oiseth, "Isogeometric modeling and experimental investigation of moving-domain bridge aerodynamics", *ASCE Journal of Engineering Mechanics*, Accepted for publication.

63. J. Yan, A. Korobenko, A.E. Tejada-Martinez, R. Golshan, and Y. Bazilevs, "A new variational multiscale formulation for stratified incompressible turbulent flows", *Computers & Fluids*, **158** (2017) 150–156, https://doi.org/10.1016/j.compfluid.2016.12.004.

64. T.M. van Opstal, J. Yan, C. Coley, J.A. Evans, T. Kvamsdal, and Y. Bazilevs, "Isogeometric divergence-conforming variational multiscale formulation of incompressible turbulent flows", *Computer Methods in Applied Mechanics and Engineering*, **316** (2017) 859–879, https://doi.org/10.1016/j.cma.2016.10.015.

65. F. Xu, G. Moutsanidis, D. Kamensky, M.-C. Hsu, M. Murugan, A. Ghoshal, and Y. Bazilevs, "Compressible flows on moving domains: Stabilized methods, weakly enforced essential boundary conditions, sliding interfaces, and application to gas-turbine modeling", *Computers & Fluids*, **158** (2017) 201–220, https://doi.org/10.1016/j.compfluid.2017.02.006.

66. T.E. Tezduyar, S.K. Aliabadi, M. Behr, and S. Mittal, "Massively parallel finite element simulation of compressible and incompressible flows", *Computer Methods in Applied Mechanics and Engineering*, **119** (1994) 157–177, https://doi.org/10.1016/0045-7825(94)00082-4.

67. T. Kanai, K. Takizawa, T.E. Tezduyar, T. Tanaka, and A. Hartmann, "Compressible-flow geometric-porosity modeling and spacecraft parachute computation with isogeometric discretization", *Computational Mechanics*, **63** (2019) 301–321, https://doi.org/10.1007/s00466-018-1595-4.

68. M.-C. Hsu and D. Kamensky, "Immersogeometric Analysis of Bioprosthetic Heart Valves, Using the Dynamic Augmented Lagrangian Method", in T.E. Tezduyar, editor, *Frontiers in Computational Fluid–Structure Interaction and Flow Simulation*, 167–212, Springer International Publishing, Cham, 2018.

69. T.E. Tezduyar, K. Takizawa, C. Moorman, S. Wright, and J. Christopher, "Space–time finite element computation of complex fluid–structure interactions", *International Journal for Numerical Methods in Fluids*, **64** (2010) 1201–1218, https://doi.org/10.1002/fld.2221.

70. J. Kiendl, M.-C. Hsu, M.C.H. Wu, and A. Reali, "Isogeometric Kirchhoff–Love shell formulations for general hyperelastic materials", *Computer Methods in Applied Mechanics and Engineering*, **291** (2015) 280–303.

71. M.C.H. Wu, R. Zakerzadeh, D. Kamensky, J. Kiendl, M.S. Sacks, and M.-C. Hsu, "An anisotropic constitutive model for immersogeometric fluid–structure interaction analysis of bioprosthetic heart valves", *Journal of Biomechanics*, **74** (2018) 23–31.

72. D. Kamensky, M.-C. Hsu, Y. Yu, J.A. Evans, M.S. Sacks, and T.J.R. Hughes, "Immersogeometric cardiovascular fluid–structure interaction analysis with divergence-conforming B-splines", *Computer Methods in Applied Mechanics and Engineering*, **314** (2017) 408–472.

73. D. Kamensky, J.A. Evans, and M.-C. Hsu, "Stability and Conservation Properties of Collocated Constraints in Immersogeometric Fluid–Thin Structure Interaction Analysis", *Communications in Computational Physics*, **18** (04) (2015) 1147–1180.

74. F. Xu, S. Morganti, R. Zakerzadeh, D. Kamensky, F. Auricchio, A. Reali, T.J.R. Hughes, M.S. Sacks, and M.-C. Hsu, "A framework for designing patient-specific bioprosthetic heart valves using immersogeometric fluid–structure interaction analysis", *International Journal for Numerical Methods in Biomedical Engineering*, **34** (4) (2018) e2938.

75. Y. Bazilevs, K. Takizawa, and T.E. Tezduyar, *Computational Fluid–Structure Interaction: Methods and Applications*. Wiley, February 2013, ISBN 978-0470978771.

76. T.E. Tezduyar, "Computation of moving boundaries and interfaces and stabilization parameters", *International Journal for Numerical Methods in Fluids*, **43** (2003) 555–575, https://doi.org/10.1002/fld.505.
77. T.E. Tezduyar and S. Sathe, "Modeling of fluid–structure interactions with the space–time finite elements: Solution techniques", *International Journal for Numerical Methods in Fluids*, **54** (2007) 855–900, https://doi.org/10.1002/fld.1430.
78. M.-C. Hsu, Y. Bazilevs, V.M. Calo, T.E. Tezduyar, and T.J.R. Hughes, "Improving stability of stabilized and multiscale formulations in flow simulations at small time steps", *Computer Methods in Applied Mechanics and Engineering*, **199** (2010) 828–840, https://doi.org/10.1016/j.cma.2009.06.019.
79. K. Takizawa, T.E. Tezduyar, S. McIntyre, N. Kostov, R. Kolesar, and C. Habluetzel, "Space–time VMS computation of wind-turbine rotor and tower aerodynamics", *Computational Mechanics*, **53** (2014) 1–15, https://doi.org/10.1007/s00466-013-0888-x.
80. K. Takizawa, T.E. Tezduyar, and Y. Otoguro, "Stabilization and discontinuity-capturing parameters for space–time flow computations with finite element and isogeometric discretizations", *Computational Mechanics*, **62** (2018) 1169–1186, https://doi.org/10.1007/s00466-018-1557-x.
81. K. Takizawa, T.E. Tezduyar, S. Asada, and T. Kuraishi, "Space–time method for flow computations with slip interfaces and topology changes (ST-SI-TC)", *Computers & Fluids*, **141** (2016) 124–134, https://doi.org/10.1016/j.compfluid.2016.05.006.
82. T.J.R. Hughes, J.A. Cottrell, and Y. Bazilevs, "Isogeometric analysis: CAD, finite elements, NURBS, exact geometry, and mesh refinement", *Computer Methods in Applied Mechanics and Engineering*, **194** (2005) 4135–4195.
83. K. Takizawa and T.E. Tezduyar, "Space–time computation techniques with continuous representation in time (ST-C)", *Computational Mechanics*, **53** (2014) 91–99, https://doi.org/10.1007/s00466-013-0895-y.
84. T. Sasaki, K. Takizawa, and T.E. Tezduyar, "Medical-image-based aorta modeling with zero-stress-state estimation", *Computational Mechanics*, **64** (2019) 249–271, https://doi.org/10.1007/s00466-019-01669-4.
85. K. Takizawa, T.E. Tezduyar, and T. Sasaki, "Isogeometric hyperelastic shell analysis with out-of-plane deformation mapping", *Computational Mechanics*, **63** (2019) 681–700, https://doi.org/10.1007/s00466-018-1616-3.
86. Y. Otoguro, K. Takizawa, and T.E. Tezduyar, "A general-purpose NURBS mesh generation method for complex geometries", in T.E. Tezduyar, editor, *Frontiers in Computational Fluid–Structure Interaction and Flow Simulation: Research from Lead Investigators under Forty – 2018*, Modeling and Simulation in Science, Engineering and Technology, 399–434, Springer, 2018, ISBN 978-3-319-96468-3, https://doi.org/10.1007/978-3-319-96469-0_10.
87. M.F. Wheeler, "An elliptic collocation-finite element method with interior penalties", *SIAM Journal on Numerical Analysis*, **15** (1978) 152–161.
88. P. Houston, C. Schwab, and E. Suli, "Discontinuous hp-finite element methods for advection-diffusion reaction problems", *SIAM Journal on Numerical Analysis*, **39** (2002) 2133–2163.
89. Y. Bazilevs, J.R. Gohean, T.J.R. Hughes, R.D. Moser, and Y. Zhang, "Patient-specific isogeometric fluid–structure interaction analysis of thoracic aortic blood flow due to implantation of the Jarvik 2000 left ventricular assist device", *Computer Methods in Applied Mechanics and Engineering*, **198** (2009) 3534–3550.
90. M. Esmaily-Moghadam, Y. Bazilevs, T.-Y. Hsia, I.E. Vignon-Clementel, and A.L. Marsden, "A comparison of outlet boundary treatments for prevention of backflow divergence with relevance to blood flow simulations", *Computational Mechanics*, **48** (2011) 277–291.
91. T.E. Tezduyar and M. Senga, "Stabilization and shock-capturing parameters in SUPG formulation of compressible flows", *Computer Methods in Applied Mechanics and Engineering*, **195** (2006) 1621–1632, https://doi.org/10.1016/j.cma.2005.05.032.
92. Y. Bazilevs, V.M. Calo, T.E. Tezduyar, and T.J.R. Hughes, "$YZ\beta$ discontinuity-capturing for advection-dominated processes with application to arterial drug delivery", *International Journal for Numerical Methods in Fluids*, **54** (2007) 593–608, https://doi.org/10.1002/fld.1484.

93. T.E. Tezduyar, M. Behr, S. Mittal, and A.A. Johnson, "Computation of unsteady incompressible flows with the finite element methods: Space–time formulations, iterative strategies and massively parallel implementations", in *New Methods in Transient Analysis*, PVP-Vol.246/AMD-Vol.143, ASME, New York, (1992) 7–24.
94. T. Tezduyar, S. Aliabadi, M. Behr, A. Johnson, and S. Mittal, "Parallel finite-element computation of 3D flows", *Computer*, **26** (10) (1993) 27–36, https://doi.org/10.1109/2.237441.
95. C.D. Murray, "The physiological principle of minimum work: I. the vascular system and the cost of blood volume", *Proceedings of the National Academy of Sciences of the United States of America*, **12** (1926) 207–214.
96. M.-C. Hsu, C. Wang, A.J. Herrema, D. Schillinger, A. Ghoshal, and Y. Bazilevs, "An interactive geometry modeling and parametric design platform for isogeometric analysis", *Computers and Mathematics with Applications*, **70** (2015) 1481–1500.
97. Y. Bazilevs, M.-C. Hsu, Y. Zhang, W. Wang, T. Kvamsdal, S. Hentschel, and J. Isaksen, "Computational fluid–structure interaction: Methods and application to cerebral aneurysms", *Biomechanics and Modeling in Mechanobiology*, **9** (2010) 481–498.

ALE and Space–Time Variational Multiscale Isogeometric Analysis of Wind Turbines and Turbomachinery

Yuri Bazilevs, Kenji Takizawa, Tayfun E. Tezduyar, Ming-Chen Hsu,
Yuto Otoguro, Hiroki Mochizuki, and Michael C. H. Wu

1 Introduction

Sophistication level of computational analysis of wind turbines and turbomachinery defines the practical value of the computations. The Arbitrary Lagrangian–Eulerian (ALE) and Space–Time (ST) Variational Multiscale (VMS) methods and isogeometric discretization are now enabling sophisticated wind turbine and turbomachinery computational analysis (see, for example, [1–4]). The computational challenges encountered in this class of problems include turbulent rotational flows, complex geometries, moving boundaries and interfaces, such as the rotor motion, and the fluid–structure interaction (FSI), such as the FSI between the wind turbine blade and the air. As examples of the challenging computations performed, we present computational analysis of horizontal- and vertical-axis wind turbines (HAWTs and VAWTs) and flow-driven string dynamics in pumps.

Y. Bazilevs (✉) · M. C. H. Wu
School of Engineering, Brown University, Providence, RI, USA
e-mail: yuri_bazilevs@brown.edu; Michael_CH_Wu@brown.edu

K. Takizawa · Y. Otoguro · H. Mochizuki
Department of Modern Mechanical Engineering, Waseda University, Tokyo, Japan
e-mail: Kenji.Takizawa@tafsm.org; yuto.otoguro@tafsm.org

T. E. Tezduyar
Mechanical Engineering, Rice University, Houston, TX, USA

Faculty of Science and Engineering, Waseda University, Shinjuku-ku, Tokyo, Japan
e-mail: tezduyar@tafsm.org

M.-C. Hsu
Department of Mechanical Engineering, Iowa State University, Ames, IA, USA
e-mail: jmchsu@iastate.edu

© Springer Nature Switzerland AG 2020
A. Grama, A. H. Sameh (eds.), *Parallel Algorithms in Computational Science and Engineering*, Modeling and Simulation in Science, Engineering and Technology,
https://doi.org/10.1007/978-3-030-43736-7_7

195

Our core methods in addressing the computational challenges are the ALE-VMS [5] and ST-VMS [6]. We have a number special methods used in combination with them. The special methods used in combination with the ST-VMS include the ST Slip Interface (ST-SI) method [1], ST Isogeometric Analysis (ST-IGA) [7, 8], ST/NURBS Mesh Update Method (STNMUM) [7], a general-purpose NURBS mesh generation method for complex geometries [9], and a one-way-dependence model for the string dynamics [10]. The special methods used in combination with the ALE-VMS include weak enforcement of no-slip boundary conditions [11] and "sliding interfaces" [12] (the acronym "SI" will also indicate that).

1.1 ST-VMS and ST-SUPS

The ST-VMS and ST-SUPS are versions of the Deforming-Spatial-Domain/Stabilized ST (DSD/SST) method [13], which was introduced for computation of flows with moving boundaries and interfaces (MBI), including FSI. The ST-SUPS is a new name for the original version of the DSD/SST, with "SUPS" reflecting its stabilization components, the Streamline-Upwind/Petrov-Galerkin (*SU*PG) [14] and Pressure-Stabilizing/Petrov-Galerkin (*PS*PG) [13] stabilizations. The ST-VMS is the VMS version of the DSD/SST. The VMS components of the ST-VMS are from the residual-based VMS (RBVMS) method [15, 16]. The five stabilization terms of the ST-VMS include the three that the ST-SUPS has, and therefore the ST-VMS subsumes the ST-SUPS. In MBI computations the ST-VMS and ST-SUPS function as a moving-mesh methods. Moving the fluid mechanics mesh to follow an interface enables mesh-resolution control near the interface and, consequently, high-resolution boundary-layer representation near fluid–solid interfaces. Because of the higher-order accuracy of the ST framework (see [6]), the ST-SUPS and ST-VMS are desirable also in computations without MBI.

The ST-SUPS and ST-VMS have been applied to many classes of challenging FSI, MBI, and fluid mechanics problems (see [17] for a comprehensive summary of the computations prior to July 2018). The classes of problems include space-craft parachute analysis for the landing-stage parachutes [10], cover-separation parachutes [18] and the drogue parachutes [19], wind-turbine aerodynamics for HAWT rotors [20], full HAWTs [21] and VAWTs [1], flapping-wing aerodynamics for an actual locust [22], bioinspired MAVs [23] and wing-clapping [24], blood flow analysis of cerebral aneurysms [25], stent-treated aneurysms [26], aortas [27] and heart valves [28], spacecraft aerodynamics [18], thermo-fluid analysis of ground vehicles and their tires [29], thermo-fluid analysis of disk brakes [30], flow-driven string dynamics in turbomachinery [3], flow analysis of turbocharger turbines [31], flow around tires with road contact and deformation [32], fluid films [33], ram-air parachutes [34], and compressible-flow spacecraft parachute aerodynamics [35].

In the flow analyses presented here, the ST framework provides higher-order accuracy in a general context. The VMS feature of the ST-VMS addresses the computational challenges associated with the multiscale nature of the unsteady flow.

The moving-mesh feature of the ST framework enables high-resolution computation near the rotor surface. The advection equation involved in the residence time computation associated with flow-driven string dynamics in pumps is solved with the ST-SUPG method.

1.2 ALE-VMS, RBVMS, and ALE-SUPS

The ALE-VMS [5] is the VMS version of the ALE [36]. It succeeded the ST-SUPS [13] and ALE-SUPS [37] and preceded the ST-VMS. The VMS components are from the RBVMS [15, 16]. It is the moving-mesh extension of the RBVMS formulation of incompressible turbulent flows proposed in [16], and as such, it was first presented in [5] in the FSI context. The ALE-SUPS, RBVMS, and ALE-VMS have also been applied to many classes of challenging FSI, MBI, and fluid mechanics problems. The classes of problems include ram-air parachute FSI [37], wind-turbine aerodynamics and FSI [4, 38], more specifically, VAWTs [4], floating wind turbines [39], wind turbines in atmospheric boundary layers [4], and fatigue damage in wind-turbine blades [2], patient-specific cardiovascular fluid mechanics and FSI [40, 41], biomedical-device FSI [42, 43], ship hydrodynamics with free-surface flow and fluid–object interaction [44], hydrodynamics and FSI of a hydraulic arresting gear [45], hydrodynamics of tidal-stream turbines with free-surface flow [46], passive-morphing FSI in turbomachinery [47], bioinspired FSI for marine propulsion [48], bridge aerodynamics and fluid–object interaction [49], stratified incompressible flows [50], and compressible-flow gas-turbine analysis [51]. Recent advances in stabilized and multiscale methods may be found for stratified incompressible flows in [50], for divergence-conforming discretizations of incompressible flows in [52], and for compressible flows with emphasis on gas-turbine modeling in [51].

In the flow analyses presented here, the VMS feature of the ALE-VMS addresses the computational challenges associated with the multiscale nature of the unsteady flow. The moving-mesh feature of the ALE framework enables high-resolution computation near the rotor surface.

1.3 ALE-SI and ST-SI

The ALE-SI was introduced in [12] to retain the desirable moving-mesh features of the ALE-VMS in computations with spinning solid surfaces, such as a turbine rotor. The mesh covering the spinning surface spins with it, retaining the high-resolution representation of the boundary layers. The method was in the context of incompressible-flow equations. Interface terms added to the ALE-VMS to account for the compatibility conditions for the velocity and stress at the SI accurately connect the two sides of the solution. The ST-SI was introduced in [1], also in the context of incompressible-flow equations, to retain the desirable moving-mesh features of the ST-VMS and ST-SUPS in computations with spinning solid surfaces. The starting point in its development was the ALE-SI. Interface terms similar to those in the ALE-SI are added to the ST-VMS to accurately connect the two sides of the solution. An ST-SI version where the SI is between fluid and solid domains was also presented in [1]. The SI in this case is a "fluid–solid SI" rather than a standard "fluid–fluid SI" and enables weak enforcement of the Dirichlet boundary conditions for the fluid. The ST-SI introduced in [30] for the coupled incompressible-flow and thermal-transport equations retains the high-resolution representation of the thermo-fluid boundary layers near spinning solid surfaces. These ST-SI methods have been applied to aerodynamic analysis of VAWTs [1], thermo-fluid analysis of disk brakes [30], flow-driven string dynamics in turbomachinery [3], flow analysis of turbocharger turbines [31], flow around tires with road contact and deformation [32], fluid films [33], aerodynamic analysis of ram-air parachutes [34], and flow analysis of heart valves [28].

In the computations here, with the ALE-SI and ST-SI the mesh covering the rotor spins with it and we retain the high-resolution representation of the boundary layers.

1.4   Stabilization Parameters

The ST-SUPS, ALE-SUPS, RBVMS, ALE-VMS, ST-VMS, ALE-SI, and ST-SI all have some embedded stabilization parameters that play a significant role (see [1, 53]). There are many ways of defining these stabilization parameters (for examples, see [29, 32, 54–58]). The stabilization-parameter definitions used in the computations reported in this article can be found from the references cited in the sections where those computations are described.

1.5   ST-IGA

The ST-IGA is the integration of the ST framework with isogeometric discretization, motivated by the success of NURBS meshes in spatial discretization [5, 12, 40, 59]. It was introduced in [6]. Computations with the ST-VMS and ST-IGA were first reported in [6] in a 2D context, with IGA basis functions in space for flow past an airfoil, and in both space and time for the advection equation. Using higher-order basis functions in time enables getting full benefit out of using higher-order basis functions in space (see the stability and accuracy analysis given in [6] for the advection equation).

The ST-IGA with IGA basis functions in time enables, as pointed out and demonstrated in [6, 7], a more accurate representation of the motion of the solid surfaces and a mesh motion consistent with that. It also enables more efficient temporal representation of the motion and deformation of the volume meshes, and more efficient remeshing. These motivated the development of the STNMUM [7, 57]. The STNMUM has a wide scope that includes spinning solid surfaces. With the spinning motion represented by quadratic NURBS in time, and with sufficient number of temporal patches for a full rotation, the circular paths are represented exactly. A "secondary mapping" [6] enables also specifying a constant angular velocity for invariant speeds along the circular paths. The ST framework and NURBS in time also enable, with the "ST-C" method, extracting a continuous representation from the computed data and, in large-scale computations, efficient data compression [60]. The STNMUM and the ST-IGA with IGA basis functions in time have been used in many 3D computations. The classes of problems solved are flapping-wing aerodynamics for an actual locust [22], bioinspired MAVs [23] and wing-clapping [24], separation aerodynamics of spacecraft [18], aerodynamics of HAWTs [25] and VAWTs [1], thermo-fluid analysis of ground vehicles and their tires [29], thermo-fluid analysis of disk brakes [30], flow-driven string dynamics in turbomachinery [3], and flow analysis of turbocharger turbines [31].

The ST-IGA with IGA basis functions in space enables more accurate represen-tation of the geometry and increased accuracy in the flow solution. It accomplishes that with fewer control points, and consequently with larger effective element sizes. That in turn enables using larger time-step sizes while keeping the Courant number at a desirable level for good accuracy. It has been used in ST computational flow analysis of turbocharger turbines [31], flow-driven string dynamics in turbomachin-ery [3], ram-air parachutes [34], spacecraft parachutes [61], aortas [27], heart valves [28], tires with road contact and deformation [32], and fluid films [33]. Using IGA basis functions in space is now also a key part of some of the newest ZSS estimation methods [62] and related shell analysis [63].

For more on the ST-IGA, see [8]. In the computational flow analyses presented here, the ST-IGA enables more accurate representation of the turbine and turbo-machinery geometries, increased accuracy in the flow solution, and using larger time-step sizes. Integration of the ST-SI with the ST-IGA enables a more accurate representation of the rotor motion and a mesh motion consistent with that, and we will describe the ST-SI-IGA in Sect. 1.6.

1.6   ST-SI-IGA

The turbocharger turbine analysis [31] and flow-driven string dynamics in turboma-chinery [3] were based on the integration of the ST-SI and ST-IGA. The IGA basis functions were used in the spatial discretization of the fluid mechanics equations and also in the temporal representation of the rotor and spinning-mesh motion. That enabled accurate representation of the turbine geometry and rotor motion and increased accuracy in the flow solution. The IGA basis functions were used also in the spatial discretization of the string structural dynamics equations. That enabled increased accuracy in the structural dynamics solution, as well as smoothness in the string shape and fluid dynamics forces computed on the string.

The ram-air parachute analysis [34] and spacecraft parachute compressible-flow analysis [61] were based on the integration of the ST-IGA, the ST-SI version that weakly enforces the Dirichlet conditions, and the ST-SI version that accounts for the porosity of a thin structure. The ST-IGA with IGA basis functions in space enabled, with relatively few number of unknowns, accurate representation of the parafoil and parachute geometries and increased accuracy in the flow solution. The volume mesh needed to be generated both inside and outside the parafoil. Mesh generation inside was challenging near the trailing edge because of the narrowing space. The spacecraft parachute has a very complex geometry, including gores and gaps. Using IGA basis functions addressed those challenges and still kept the element density near the trailing edge of the parafoil and around the spacecraft parachute at a reasonable level. In the heart valve analysis [28], the ST-SI-IGA, beyond enabling a more accurate representation of the geometry and increased accuracy in the flow solution, kept the element density in the narrow spaces near the leaflet contact areas at a reasonable level. In computational analysis of flow around tires with road contact and deformation [32], the ST-SI-IGA enables a more accurate representation of the geometry and motion of the tire surfaces, a mesh motion consistent with that, and increased accuracy in the flow solution. It also keeps the element density in the tire grooves and in the narrow spaces near the contact areas at a reasonable level. In addition, we benefit from the mesh generation flexibility provided by using SIs.

An SI provides mesh generation flexibility by accurately connecting the two sides of the solution computed over nonmatching meshes. This type of mesh generation flexibility is especially valuable in complex-geometry flow computations with isogeometric discretization, removing the matching requirement between the NURBS patches without loss of accuracy. This feature was used in the flow analysis of heart valves [28], turbocharger turbines [31], and spacecraft parachute compressible-flow analysis [61].

For more on the ST-SI-IGA, see [34]. In the computations presented here, the ST-SI-IGA is used for the reasons given and as described in the first paragraph of this section.

1.7 General-Purpose NURBS Mesh Generation Method

While the IGA provides superior accuracy and high-fidelity solutions, to make its use even more practical in computational flow analysis with complex geometries, NURBS volume mesh generation needs to be easier and more automated. The general-purpose NURBS mesh generation method introduced in [9] serves that purpose. The method is based on multi-block-structured mesh generation with established techniques, projection of that mesh to a NURBS mesh made of patches that correspond to the blocks, and recovery of the original model surfaces. The recovery of the original surfaces is to the extent they are suitable for accurate and robust computations. The method targets retaining the refinement distribution and element quality of the multi-block-structured mesh that we start with. Because good techniques and software for generating multi-block-structured meshes are easy to find, the method makes general-purpose NURBS mesh generation relatively easy.

Mesh-quality performance studies for 2D and 3D meshes, including those for complex models, were presented in [64]. A test computation for a turbocharger turbine and exhaust manifold was also presented in [64], with a more detailed computation in [31]. The mesh generation method was used also in the pump-flow analysis part of the flow-driven string dynamics presented in [3] and in the aorta flow analysis presented in [27]. The performance studies, test computations, and actual computations demonstrated that the general-purpose NURBS mesh generation method makes the IGA use in fluid mechanics computations even more practical.

For more on the general-purpose NURBS mesh generation method, see [9, 64]. In the computations presented here, the method is used for the vertical-axis wind turbine and for the pump-flow part of the flow-driven string dynamics.

1.8 Outline of the Remaining Sections

We provide the governing equations in Sect. 2. The ST-VMS and ST-SI are described in Sect. 3, and the ALE-VMS in Sect. 4. Some of the other computational methods used are described in Sect. 5. In Sect. 6 we provide some brief comments on the parallel computations. In Sects. 7 and 8, as examples of the ST computations, we present flow-driven string dynamics in a pump and aerodynamics of a VAWT. In Sect. 9, as an example of the ALE computations, we present FSI of a HAWT with rotor–tower coupling. The concluding remarks are given in Sect. 10.

2 Governing Equations

2.1 Incompressible Flow

Let $\Omega_t \subset \mathbb{R}^{n_{sd}}$ be the spatial domain with boundary Γ_t at time $t \in (0, T)$, where n_{sd} is the number of space dimensions. The subscript t indicates the time-dependence of the domain. The Navier–Stokes equations of incompressible flows are written on Ω_t and $\forall t \in (0, T)$ as

$$\rho \left(\frac{\partial \mathbf{u}}{\partial t} + \mathbf{u} \cdot \nabla \mathbf{u} - \mathbf{f} \right) - \nabla \cdot \boldsymbol{\sigma} = \mathbf{0}, \tag{1}$$

$$\nabla \cdot \mathbf{u} = 0, \tag{2}$$

where ρ, $\mathbf{u}$ and $\mathbf{f}$ are the density, velocity, and body force. The stress tensor $\boldsymbol{\sigma}(\mathbf{u}, p) = -p\mathbf{I} + 2\mu\boldsymbol{\varepsilon}(\mathbf{u})$, where p is the pressure, $\mathbf{I}$ is the identity tensor, $\mu = \rho\nu$ is the viscosity, ν is the kinematic viscosity, and the strain rate $\boldsymbol{\varepsilon}(\mathbf{u}) =$

$\left(\nabla \mathbf{u} + (\nabla \mathbf{u})^T \right) / 2$. The essential and natural boundary conditions for Eq. (1) are represented as $\mathbf{u} = \mathbf{g}$ on $(\Gamma_t)_g$ and $\mathbf{n} \cdot \boldsymbol{\sigma} = \mathbf{h}$ on $(\Gamma_t)_h$, where $\mathbf{n}$ is the unit normal vector and $\mathbf{g}$ and $\mathbf{h}$ are given functions. A divergence-free velocity field $\mathbf{u}_0(\mathbf{x})$ is specified as the initial condition.

2.2 Structural Mechanics

In this article we will not provide any of our formulations requiring fluid and structure definitions simultaneously; we will instead give reference to earlier journal articles where the formulations were presented. Therefore, for notation simplicity, we will reuse many of the symbols used in the fluid mechanics equations to represent their counterparts in the structural mechanics equations. To begin with, $\Omega_t \subset \mathbb{R}^{n_{sd}}$ and Γ_t will represent the structure domain and its boundary. The structural mechanics equations are then written, on Ω_t and $\forall t \in (0, T)$, as

$$\rho \left(\frac{d^2 \mathbf{y}}{dt^2} - \mathbf{f} \right) - \nabla \cdot \boldsymbol{\sigma} = \mathbf{0}, \tag{3}$$

where $\mathbf{y}$ and $\boldsymbol{\sigma}$ are the displacement and Cauchy stress tensor. The essential and natural boundary conditions for Eq. (3) are represented as $\mathbf{y} = \mathbf{g}$ on $(\Gamma_t)_g$ and $\mathbf{n} \cdot \boldsymbol{\sigma} = \mathbf{h}$ on $(\Gamma_t)_h$. The Cauchy stress tensor can be obtained from

$$\boldsymbol{\sigma} = J^{-1} \mathbf{F} \mathbf{S} \mathbf{F}^T, \tag{4}$$

where $\mathbf{F}$ and J are the deformation gradient tensor and its determinant, and $\mathbf{S}$ is the second Piola–Kirchhoff stress tensor. It is obtained from the strain-energy density function φ as follows:

$$\mathbf{S} \equiv \frac{\partial \varphi}{\partial \mathbf{E}}, \tag{5}$$

where $\mathbf{E}$ is the Green–Lagrange strain tensor:

$$\mathbf{E} = \frac{1}{2} (\mathbf{C} - \mathbf{I}), \tag{6}$$

and $\mathbf{C}$ is the Cauchy–Green deformation tensor:

$$\mathbf{C} \equiv \mathbf{F}^T \cdot \mathbf{F}. \tag{7}$$

From Eqs. (5) and (6),

$$\mathbf{S} = 2 \frac{\partial \varphi}{\partial \mathbf{C}}. \tag{8}$$

2.3 Fluid–Structure Interface

In an FSI problem, at the fluid–structure interface, we will have the velocity and stress compatibility conditions between the fluid and structure parts. The details on those conditions can be found in Section 5.1 of [53].

3 ST-VMS and ST-SI

We include from [1, 65] the ST-VMS and ST-SI methods.
    The ST-VMS is given as

$$
\int_{Q_n} \mathbf{w}^h \cdot \rho \left(\frac{\partial \mathbf{u}^h}{\partial t} + \mathbf{u}^h \cdot \nabla \mathbf{u}^h - \mathbf{f}^h \right) dQ + \int_{Q_n} \boldsymbol{\varepsilon}(\mathbf{w}^h) : \boldsymbol{\sigma}(\mathbf{u}^h, p^h) dQ
$$

$$
- \int_{(P_n)_h} \mathbf{w}^h \cdot \mathbf{h}^h dP + \int_{Q_n} q^h \nabla \cdot \mathbf{u}^h dQ + \int_{\Omega_n} (\mathbf{w}^h)_n^+ \cdot \rho \left((\mathbf{u}^h)_n^+ - (\mathbf{u}^h)_n^- \right) d\Omega
$$

$$
+ \sum_{e=1}^{(n_{el})_n} \int_{Q_n^e} \frac{\tau_{\mathrm{SUPS}}}{\rho} \left[\rho \left(\frac{\partial \mathbf{w}^h}{\partial t} + \mathbf{u}^h \cdot \nabla \mathbf{w}^h \right) + \nabla q^h \right] \cdot \mathbf{r}_{\mathrm{M}}(\mathbf{u}^h, p^h) dQ
$$

$$
+ \sum_{e=1}^{(n_{el})_n} \int_{Q_n^e} \nu_{\mathrm{LSIC}} \nabla \cdot \mathbf{w}^h \rho r_{\mathrm{C}}(\mathbf{u}^h) dQ
$$

$$
- \sum_{e=1}^{(n_{el})_n} \int_{Q_n^e} \tau_{\mathrm{SUPS}} \mathbf{w}^h \cdot \left(\mathbf{r}_{\mathrm{M}}(\mathbf{u}^h, p^h) \cdot \nabla \mathbf{u}^h \right) dQ
$$

$$
- \sum_{e=1}^{(n_{el})_n} \int_{Q_n^e} \frac{\tau_{\mathrm{SUPS}}^2}{\rho} \mathbf{r}_{\mathrm{M}}(\mathbf{u}^h, p^h) \cdot \left(\nabla \mathbf{w}^h \right) \cdot \mathbf{r}_{\mathrm{M}}(\mathbf{u}^h, p^h) dQ = 0, \tag{9}
$$

where

$$
\mathbf{r}_{\mathrm{M}}(\mathbf{u}^h, p^h) = \rho \left(\frac{\partial \mathbf{u}^h}{\partial t} + \mathbf{u}^h \cdot \nabla \mathbf{u}^h - \mathbf{f}^h \right) - \nabla \cdot \boldsymbol{\sigma}(\mathbf{u}^h, p^h), \tag{10}
$$

$$
r_{\mathrm{C}}(\mathbf{u}^h) = \nabla \cdot \mathbf{u}^h \tag{11}
$$

are the residuals of the momentum equation and incompressibility constraint. The test functions associated with the velocity and pressure are $\mathbf{w}$ and q. A superscript "h" indicates that the function is coming from a finite-dimensional space. The symbol Q_n represents the ST slice between time levels n and $n+1$, $(P_n)_h$ is the part of the lateral boundary of that slice associated with the traction boundary condition $\mathbf{h}$, and Ω_n is the spatial domain at time level n. The superscript "e" is the ST element counter, and n_{el} is the number of ST elements. The functions are discontinuous in time at each time level, and the superscripts "−" and "+" indicate the values of the functions just below and just above the time level. See [1, 29, 54, 55, 57] for the definitions used here for the stabilization parameters τ_{SUPS} and ν_{LSIC}. For more ways of calculating the stabilization parameters in finite element computation of flow problems, see [32, 56, 58].

Remark 1 The ST-SUPS method can be obtained from the ST-VMS method by dropping the eighth and ninth integrations.

In the ST-SI, labels "Side A" and "Side B" represent the two sides of the SI. We add boundary terms to Eq. (9). The boundary terms are first added separately for the two sides, using test functions $\mathbf{w}_{\mathrm{A}}^h$ and q_{A}^h and $\mathbf{w}_{\mathrm{B}}^h$ and q_{B}^h. Putting them together, the complete set of terms added becomes

$$
-\int_{(P_n)_{\mathrm{SI}}} \left(q_{\mathrm{B}}^h \mathbf{n}_{\mathrm{B}} - q_{\mathrm{A}}^h \mathbf{n}_{\mathrm{A}} \right) \cdot \frac{1}{2} \left(\mathbf{u}_{\mathrm{B}}^h - \mathbf{u}_{\mathrm{A}}^h \right) \mathrm{d}P
$$

$$
-\int_{(P_n)_{\mathrm{SI}}} \rho \mathbf{w}_{\mathrm{B}}^h \cdot \frac{1}{2} \left(\left(\mathcal{F}_{\mathrm{B}}^h - \left| \mathcal{F}_{\mathrm{B}}^h \right| \right) \mathbf{u}_{\mathrm{B}}^h - \left(\mathcal{F}_{\mathrm{B}}^h - \left| \mathcal{F}_{\mathrm{B}}^h \right| \right) \mathbf{u}_{\mathrm{A}}^h \right) \mathrm{d}P
$$

$$
-\int_{(P_n)_{\mathrm{SI}}} \rho \mathbf{w}_{\mathrm{A}}^h \cdot \frac{1}{2} \left(\left(\mathcal{F}_{\mathrm{A}}^h - \left| \mathcal{F}_{\mathrm{A}}^h \right| \right) \mathbf{u}_{\mathrm{A}}^h - \left(\mathcal{F}_{\mathrm{A}}^h - \left| \mathcal{F}_{\mathrm{A}}^h \right| \right) \mathbf{u}_{\mathrm{B}}^h \right) \mathrm{d}P
$$

$$
+\int_{(P_n)_{\mathrm{SI}}} \left(\mathbf{n}_{\mathrm{B}} \cdot \mathbf{w}_{\mathrm{B}}^h + \mathbf{n}_{\mathrm{A}} \cdot \mathbf{w}_{\mathrm{A}}^h \right) \frac{1}{2} \left(p_{\mathrm{B}}^h + p_{\mathrm{A}}^h \right) \mathrm{d}P
$$

$$
-\int_{(P_n)_{\mathrm{SI}}} \left(\mathbf{w}_{\mathrm{B}}^h - \mathbf{w}_{\mathrm{A}}^h \right) \cdot \left(\hat{\mathbf{n}}_{\mathrm{B}} \cdot \mu \left(\boldsymbol{\varepsilon}(\mathbf{u}_{\mathrm{B}}^h) + \boldsymbol{\varepsilon}(\mathbf{u}_{\mathrm{A}}^h) \right) \right) \mathrm{d}P
$$

$$
-\gamma_{\mathrm{ACI}} \int_{(P_n)_{\mathrm{SI}}} \hat{\mathbf{n}}_{\mathrm{B}} \cdot \mu \left(\boldsymbol{\varepsilon} \left(\mathbf{w}_{\mathrm{B}}^h \right) + \boldsymbol{\varepsilon} \left(\mathbf{w}_{\mathrm{A}}^h \right) \right) \cdot \left(\mathbf{u}_{\mathrm{B}}^h - \mathbf{u}_{\mathrm{A}}^h \right) \mathrm{d}P
$$

$$
+\int_{(P_n)_{\mathrm{SI}}} \frac{\mu C}{h} \left(\mathbf{w}_{\mathrm{B}}^h - \mathbf{w}_{\mathrm{A}}^h \right) \cdot \left(\mathbf{u}_{\mathrm{B}}^h - \mathbf{u}_{\mathrm{A}}^h \right) \mathrm{d}P, \tag{12}
$$

where

$$
\mathcal{F}_{\mathrm{B}}^h = \mathbf{n}_{\mathrm{B}} \cdot \left(\mathbf{u}_{\mathrm{B}}^h - \mathbf{v}_{\mathrm{B}}^h \right), \tag{13}
$$

$$
\mathcal{F}_{\mathrm{A}}^h = \mathbf{n}_{\mathrm{A}} \cdot \left(\mathbf{u}_{\mathrm{A}}^h - \mathbf{v}_{\mathrm{A}}^h \right), \tag{14}
$$

$$
h = \left(\frac{h_{\mathrm{B}}^{-1} + h_{\mathrm{A}}^{-1}}{2} \right)^{-1}, \tag{15}
$$

$$
h_{\mathrm{B}} = 2 \left(\mathbf{n}_{\mathrm{B}} \mathbf{n}_{\mathrm{B}} : \mathbf{G} \right)^{-\frac{1}{2}} \qquad \text{(for Side B)}, \tag{16}
$$

$$
h_{\mathrm{A}} = 2 \left(\mathbf{n}_{\mathrm{A}} \mathbf{n}_{\mathrm{A}} : \mathbf{G} \right)^{-\frac{1}{2}} \qquad \text{(for Side A)}, \tag{17}
$$

$$\hat{\mathbf{n}}_{\mathrm{B}} = \frac{\mathbf{n}_{\mathrm{B}} - \mathbf{n}_{\mathrm{A}}}{\|\mathbf{n}_{\mathrm{B}} - \mathbf{n}_{\mathrm{A}}\|}. \tag{18}$$

Here, $\mathbf{G}$ is a kind of metric tensor (given in [58]), which is different than the metric tensor in Sect. 4, $(P_n)_{\mathrm{SI}}$ is the SI in the ST domain, $\mathbf{v}$ is the mesh velocity, $\gamma_{\mathrm{ACI}} = 1$, and C is a nondimensional constant. We note that the expressions given by Eqs. (15)–(17) were introduced in published form in [31]. At the same time we note that the element lengths given by Eqs. (16) and (17) are straightforward extensions of the one in [58]. For explanation of the added SI terms, see [1].

4 ALE-VMS

The ALE-VMS formulation is posed on a spatial domain Ω that is discretized into elements Ω^e. While $\{\Omega^e\}$, Ω, and its boundary Γ are time-dependent, when there is no risk of confusion, we drop the subscript t to simplify notation. The superscript h indicates association with discrete function spaces defined over Ω, which moves with the velocity $\hat{\mathbf{u}}^h$, which is the same as the mesh velocity $\mathbf{v}^h$ in Sect. 3. The semidiscrete formulation is given as

$$
\int_{\Omega} \mathbf{w}^h \cdot \rho \left(\left. \frac{\partial \mathbf{u}^h}{\partial t} \right|_{\hat{\mathbf{x}}} + (\mathbf{u}^h - \hat{\mathbf{u}}^h) \cdot \nabla \mathbf{u}^h - \mathbf{f}^h \right) \mathrm{d}\Omega + \int_{\Omega} \boldsymbol{\varepsilon}(\mathbf{w}^h) : \boldsymbol{\sigma}(\mathbf{u}^h, p^h) \, \mathrm{d}\Omega
$$

$$
- \int_{\Gamma} \mathbf{w}^h \cdot \mathbf{h}^h \mathrm{d}\Gamma + \int_{\Omega} q^h \nabla \cdot \mathbf{u}^h \, \mathrm{d}\Omega
$$

$$
- \beta \int_{\Gamma} \mathbf{w}^h \cdot \rho \left\{ \left(\mathbf{u}^h - \hat{\mathbf{u}}^h \right) \cdot \mathbf{n} \right\}_{-} \mathbf{u}^h \mathrm{d}\Gamma
$$

$$
+ \sum_{e} \int_{\Omega^e} \tau_{\mathrm{SUPS}} \left((\mathbf{u}^h - \hat{\mathbf{u}}^h) \cdot \nabla \mathbf{w}^h + \frac{1}{\rho} \nabla q^h \right) \cdot \mathbf{r}_{\mathrm{M}}(\mathbf{u}^h, p^h) \, \mathrm{d}\Omega
$$

$$
+ \sum_{e} \int_{\Omega^e} \nu_{\mathrm{LSIC}} \nabla \cdot \mathbf{w}^h \rho r_{\mathrm{C}}(\mathbf{u}^h) \, \mathrm{d}\Omega
$$

$$
- \sum_{e} \int_{\Omega^e} \tau_{\mathrm{SUPS}} \mathbf{w}^h \cdot \left(\mathbf{r}_{\mathrm{M}}(\mathbf{u}^h, p^h) \cdot \nabla \mathbf{u}^h \right) \mathrm{d}\Omega
$$

$$
- \sum_{e} \int_{\Omega^e} \frac{\tau_{\mathrm{SUPS}}^2}{\rho} \mathbf{r}_{\mathrm{M}}(\mathbf{u}^h, p^h) \cdot \left(\nabla \mathbf{w}^h \right) \cdot \mathbf{r}_{\mathrm{M}}(\mathbf{u}^h, p^h) \, \mathrm{d}\Omega
$$

$$
+ \sum_{e} \int_{\Omega^e} \left(\tau_{\mathrm{SUPS}} \mathbf{r}_{\mathrm{M}}(\mathbf{u}^h, p^h) \cdot \nabla \mathbf{w}^h \right) \overline{\tau} \cdot \left(\tau_{\mathrm{SUPS}} \mathbf{r}_{\mathrm{M}}(\mathbf{u}^h, p^h) \cdot \nabla \mathbf{u}^h \right) \mathrm{d}\Omega = 0, \tag{19}
$$

where $\left.\frac{\partial(\cdot)}{\partial t}\right|_{\hat{\mathbf{x}}}$ is the time derivative taken with respect to the fixed reference coordinates $\hat{\mathbf{x}}$ of the spatial configuration, β (≥ 0) is associated with the backflow stabilization (see Remark 2), and $\{\cdot\}_-$ isolates the negative part of its argument. The additional stabilization parameter $\bar{\tau}$ is defined as

$$\bar{\tau} = \left(\tau_{\mathrm{SUPS}} \mathbf{r}_{\mathrm{M}}(\mathbf{u}^h, p^h) \cdot (\mathbf{G}) \cdot \tau_{\mathrm{SUPS}} \mathbf{r}_{\mathrm{M}}(\mathbf{u}^h, p^h) \right)^{-1/2}, \tag{20}$$

where $\mathbf{G}$ generalizes element size to physical elements mapped through $\mathbf{x}(\boldsymbol{\xi})$ from a parametric parent element: $G_{ij} = \xi_{k,i}\xi_{k,j}$.

Remark 2 Unsteady flow computations may sometimes diverge due to significant inflow through the Neumann boundary $\Gamma_{\mathrm{f}}^{\mathrm{h}}$; this is known as backflow divergence. In order to preclude backflow divergence, a backflow stabilization method (the β term in Eq. (19)) originally proposed in [66] and further studied in [67] is employed in our ALE-VMS formulation.

Remark 3 The $\bar{\tau}$ term of Eq. (19) is not derived from VMS analysis; it is an additional residual-based stabilization term that is included to provided extra stabilizing dissipation near steep solution gradients while maintaining consistency with the exact solution. It was introduced in [68] and bears resemblance to the DCDD [54] and YZβ [69, 70] discontinuity-capturing methods.

5   Other Computational Methods

5.1   String Dynamics

The string in the flow-driven string dynamics is modeled with bending-stabilized cable elements [71], using the IGA with cubic NURBS basis functions. This gives us a higher-order method, and smoothness in the structure shape. It also gives us smoothness in the fluid forces acting on the string. Because a string is a very thin structure, its influence on the flow will be very small. In the one-way-dependence model, we compute the influence of the flow on the string dynamics, while avoiding the formidable task of computing the influence of the string on the flow. The fluid mechanics forces acting on the string are calculated with the method described in [10] for computing the aerodynamic forces acting on the suspension lines of spacecraft parachutes. Contact between the string and solid surfaces is handled with the Surface-Edge-Node Contact Tracking (SENCT-FC) method [72], which is a later version of the SENCT introduced in [55].

5.2 Particle Residence Time

In flow-driven string dynamics in pumps, the residence time computations help us to have a simplified but quick understanding of the string behavior. The residence time in domain $\Omega_s \subset \Omega$ can be written as

$$\frac{dR}{dt} = s(\mathbf{x}), \tag{21}$$

where $s(\mathbf{x}) = 1$ on Ω_s and $s(\mathbf{x}) = 0$ on $\Omega \setminus \Omega_s$. The Eulerian form of the equation is

$$\frac{\partial R}{\partial t} + \mathbf{u} \cdot \nabla R = s, \tag{22}$$

and we solve that with the ST-SUPG supplemented with the YZβ discontinuity capturing [69]. Integration of Eq. (22) over Ω_s gives

$$\int_{\Omega_s} \left(\frac{\partial R}{\partial t} + \mathbf{u} \cdot \nabla R \right) d\Omega = \int_{\Omega_s} s d\Omega. \tag{23}$$

We assume $\nabla \cdot \mathbf{u} = 0$ and obtain

$$\frac{d}{dt} \left(\int_{\Omega_s} R d\Omega \right) + \int_{\Gamma_s} \mathbf{n} \cdot (\mathbf{u} - \mathbf{v}) R d\Gamma = V, \tag{24}$$

where Γ_s is the boundary of Ω_s and

$$V = \int_{\Omega_s} d\Omega. \tag{25}$$

We define the flow-rate-averaged residence time as

$$\overline{R}_{\text{out}} = \frac{1}{Q} \int_{(\Gamma_s)_{\text{out}}} \mathbf{n} \cdot \mathbf{u} R d\Gamma, \tag{26}$$

$$Q = \int_{(\Gamma_s)_{\text{out}}} \mathbf{n} \cdot \mathbf{u} d\Gamma, \tag{27}$$

where subscript "out" indicates the outlet.

In a typical setting, there is no flow coming back to Ω_s, $\mathbf{u} = \mathbf{v}$ on the part of Γ_s corresponding to the rotor, and $\mathbf{v} = \mathbf{0}$ at the inlet and outlet. If we assume that first term in Eq. (24) is zero, $\overline{R}_{\text{out}} = V/Q$. If any part of Ω_s is enclosed by a closed surface with zero normal velocity, the first term cannot be zero.

Remark 4 More explanation on the residence time computation and related concepts can be found in [73] in the ALE framework. Here we do the computation in the ST framework.

5.3 Rotation Representation with Constant Angular Velocity

We use quadratic NURBS functions, as described in [7], to represent a circular-arc trajectory. We discretize time and position as follows:

$$t = \sum_{\alpha=1}^{n_{\text{ent}}} T^{\alpha}(\Theta_t(\theta))t^{\alpha}, \tag{28}$$

$$\mathbf{x} = \sum_{\alpha=1}^{n_{\text{ent}}} T^{\alpha}(\Theta_x(\theta))\mathbf{x}^{\alpha}. \tag{29}$$

Here n_{ent} is the number of temporal element nodes, T^{α} is the basis function, $\Theta_t(\theta)$ and $\Theta_x(\theta)$ are the secondary mappings for time and position, and t^{α} and $\mathbf{x}^{\alpha}$ are the time and position values corresponding to the basis function T^{α}. The basis functions could be finite element or NURBS basis functions. For the circular arc, $n_{\text{ent}} = 3$ and they are quadratic NURBS. The secondary mapping concept above was introduced in [6], and the velocity can be expressed as follows:

$$\frac{d\mathbf{x}}{dt} = \left(\sum_{\alpha=1}^{n_{\text{ent}}} \frac{dT^{\alpha}}{d\Theta_x}\frac{d\Theta_x}{d\theta}\mathbf{x}^{\alpha}\right)\left(\sum_{\alpha=1}^{n_{\text{ent}}} \frac{dT^{\alpha}}{d\Theta_t}\frac{d\Theta_t}{d\theta}t^{\alpha}\right)^{-1}, \tag{30}$$

leading to

$$\frac{d\mathbf{x}}{dt} = \left(\sum_{\alpha=1}^{n_{\text{ent}}} \frac{dT^{\alpha}}{d\Theta_x}\mathbf{x}^{\alpha}\right)\left(\sum_{\alpha=1}^{n_{\text{ent}}} \frac{dT^{\alpha}}{d\Theta_t}t^{\alpha}\right)^{-1}\left(\frac{d\Theta_x}{d\theta}\frac{d\theta}{d\Theta_t}\right). \tag{31}$$

Thus, the speed along the path can be specified only by modifying the secondary mapping. For a circular arc, two methods were introduced in [7]; one is modifying the secondary mapping for position and the other one is modifying both such that $\frac{dt}{d\theta}$ is constant. We note that, in theory, the secondary mapping selections do not make any difference as long as the relationship $\frac{d\Theta_x}{d\Theta_t}$ is the same.

   In our implementation, to keep the process general, we search for the parametric coordinate θ by using an iterative solution method [7]. We use the latter set of the secondary mappings, having constant $\frac{dt}{d\theta}$. We first calculate the time corresponding to each integration point, and then calculate Θ_x and Θ_t to interpolate the position and velocity from Eqs. (29) and (31).

6 Parallel Computations

Parallel computations with the ST methods go as far back as 1992 [74], with the 3D computations reported as early as 1993 [75]. All computations reported in this chapter were carried out on parallel computing platforms. The number of cores used in a typical computation ranges from 96 to 576. Because the computations were mostly for the purpose of testing a new computational method, parallel efficiency was not a high priority. Still the efficiencies we see are high enough to justify the use of the maximum number of cores available in the computer resources we have.

7 ST Computation: Flow-Driven String Dynamics in a Pump

7.1 Flow Analysis of the Pump

We use a vortex pump with six blades, including two higher-height blades. The rotor diameter is roughly 150 mm. We are unable to provide more details due to the industrial-partner restrictions. The quadratic NURBS mesh used in the computation is shown in Fig. 1. The number of control points and elements are 838,222 and 544,466. The pump is used for water, the density is 998.2 kg/m^3, and the kinematic viscosity 8.7×10^{-7} m^2/s. The rotation speed is 2544 rpm. The boundary conditions are shown in Fig. 2. At the inlet, $Q = 5.46 \times 10^{-3}$ m^3/s. The time-step size is 9.8×10^{-5} s. The number of nonlinear iterations per time step is 3, and the number of GMRES iterations per nonlinear iteration is 100. Stabilization parameters of the ST-VMS are those given by Eqs. (2.4)–(2.6), (2.8), and (2.10) in [1].

Figure 3 shows the second invariant of the velocity gradient tensor. The turbulent nature of the flow is well represented. The solution is compared to the experimental data from Professor Kazuyoshi Miyagawa's group (Waseda University). The conditions here are close to those corresponding to the best-efficiency operating point, and the relative error in the efficiency compared to the experimental data is less than 1.5%. The computed flow field from rotations 17 through 21 is stored with the ST-C

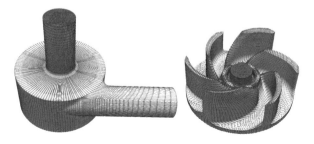

Fig. 1 Control mesh. *Red* circles represent the control points

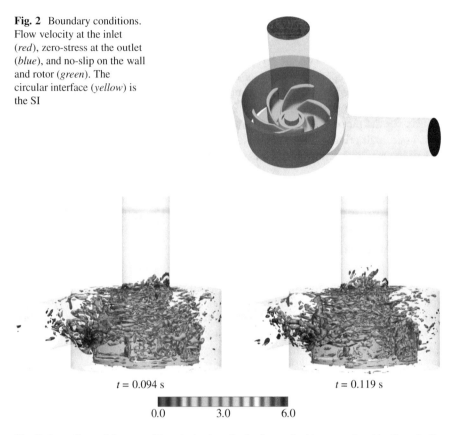

Fig. 2 Boundary conditions. Flow velocity at the inlet (*red*), zero-stress at the outlet (*blue*), and no-slip on the wall and rotor (*green*). The circular interface (*yellow*) is the SI

$t = 0.094$ s $t = 0.119$ s

0.0   3.0   6.0

Fig. 3 Isosurfaces of the second invariant value of velocity gradient tensor, colored by the velocity magnitude (m/s)

[60] as the data compression method and is used repeatedly in the string dynamics and residence time computations.

7.2 String Dynamics in the Pump

The string has 1.5 mm diameter and circular-shape cross-section. We compute with three different string lengths, 10, 50, and 70 mm. The Young's modulus and density are 5.0 MPa and 960 kg/m^3. We use a cubic NURBS mesh, with 19 control points and 16 elements. There are 17 different initial positions, shown in Fig. 4. The initial string velocity is 2.0 m/s, in the flow direction. The time-step size is 9.8×10^{-4} s, which is ten times smaller than the time-step size used in the flow computation. The number of nonlinear iterations per time step is 3, with full GMRES (i.e. until no more Krylov vectors can be found).

Fig. 4 The initial positions
of the strings at the inlet plane

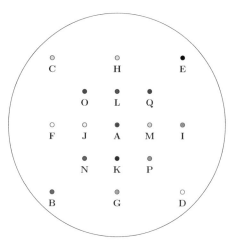

Figures 5, 6, and 7 show, for the three different string lengths, the string with the initial position at A (see Fig. 4). In all three cases the string first hits the top of the blade, and then moves to the edge of the pump casing.

7.3 Residence Time for the Pump

We set the entire pump domain as Ω_s for the residence time. The computation is carried out with a time-step size of 4.9×10^{-4} s, which is five times larger than the time-step size used in the flow computation. The number of nonlinear iterations per time step is 2, and the number of GMRES iterations per nonlinear iteration is 30.

The flow-rate-averaged residence time over the outlet is shown in Fig. 8. After 1.2 s it reaches the maximum value. Figure 9 shows the spatial distribution of the residence time at the end of the computation. The residence time under the rotor is much higher than the residence time at the outlet, which is around 0.4 s. This means that this region is not connected to the main flow.

7.4 Discussion

We discuss the relationship between the string dynamics and the residence time. Figure 10 show, for the string with length 70 mm, the time histories of the string centroid positions in radius and height. We see some strings moving in circles along the bottom edges of the casing. These strings tend to stay there and cannot rise up. Therefore they stay in the pump forever. This can be correlated with the high residence time at the bottom of the pump (Fig. 9).

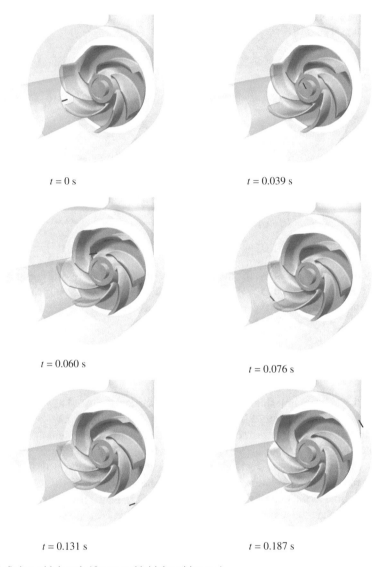

$t = 0$ s $t = 0.039$ s

$t = 0.060$ s

$t = 0.076$ s

$t = 0.131$ s $t = 0.187$ s

Fig. 5 String with length 10 mm and initial position at A

8 ST Computation: Aerodynamics of a VAWT

We present test computations with 2D and 3D models of the aerodynamics of a VAWT. The wind turbine has four support columns at the periphery. Figure 11 shows the wind turbine. The design is modeled after the wind turbine in [76]. The rotor diameter is 16 m, and the machine height is 45 m. The three blades are based on the NACA0015 airfoil, and the cord length and the blade height are 1.5 m and

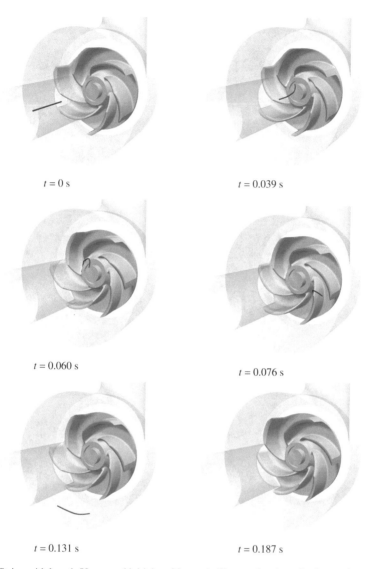

$t = 0$ s

$t = 0.039$ s

$t = 0.060$ s

$t = 0.076$ s

$t = 0.131$ s

$t = 0.187$ s

Fig. 6 String with length 50 mm and initial position at A. We note that the string leaves the casing before the sixth picture

18 m, respectively. There are two connecting rods from the hub to each blade, and the blades are supported without any tilt with respect to the tangent of the rotation path. The four support columns are cylindrical with circular cross-section, and they provide enough strength to support the rotor, which is estimated to weigh 3 tons.

We carry out the computations at a constant free-stream velocity U_∞ and with prescribed rotor motion at constant angular velocity. The rotation is clockwise viewed from the top. The air density and kinematic viscosity are 1.205 kg/m^3

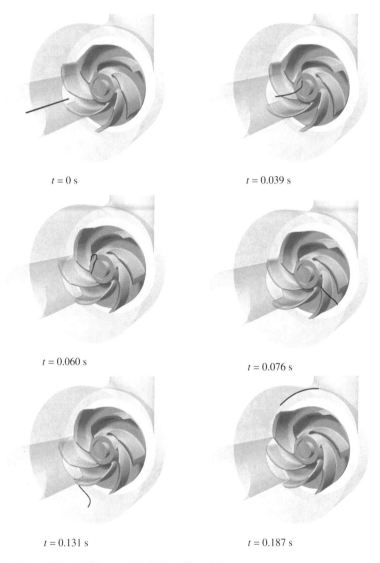

$t = 0$ s $t = 0.039$ s

$t = 0.060$ s $t = 0.076$ s

$t = 0.131$ s $t = 0.187$ s

Fig. 7 String with length 70 mm and initial position at A

and 1.511×10^{-5} m^2/s. We extract from the computations the instantaneous power coefficient C_{POW}, defined as

$$C_{\text{POW}} = \frac{P}{\frac{1}{2}\rho U_\infty^3 A},$$ (32)

where A and P are the projected area of the wind turbine and the power generated. We report the power coefficient as a function of the blade orientation as represented

Fig. 8 Flow-rate-averaged residence time over the outlet. Computed value $\left(\overline{R}_\text{out}\right)$ and theoretical value (V/Q)

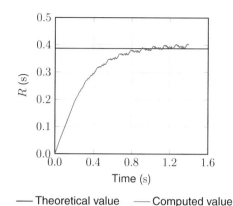

— Theoretical value   — Computed value

Fig. 9 Residence time (s) on a cut plane at $t = 1.297$ s, end of the computation

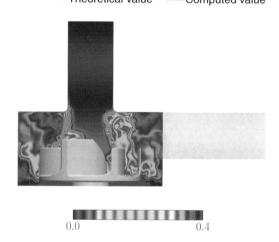

0.0   0.4

by the angle ϕ seen in Fig. 12. With that orientation, the flow speed seen by a blade can be calculated as

$$V = U_\infty\sqrt{1 - 2\lambda \sin \phi + \lambda^2}, \tag{33}$$

where λ is the tip-speed ratio (TSR). The symbol T will denote the rotation cycle.

The computational-domain size is 62.5 times the rotor diameter in the wind direction, with a distance of 18.75 times the rotor diameter between the upstream boundary and the center of the rotor. In the cross-wind direction, the domain size is 37.5 times the rotor diameter. In the 3D case, the domain height is ten times the rotor diameter. The mesh position is represented by quadratic NURBS in time. There are three patches that are 120° each, and the secondary mapping introduced in [7] is used to achieve the constant angular velocity. The free-stream velocity is 12.56 m/s.

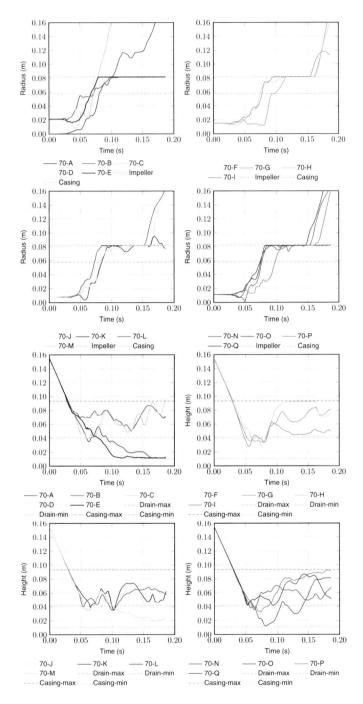

Fig. 10 String with length 70 mm. Time histories of the string centroid positions in radius and height

Fig. 11 A VAWT

Fig. 12 Blade orientation as represented by the angle ϕ

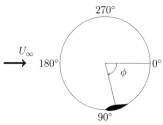

Fig. 13 2D VAWT. Model geometry and SI

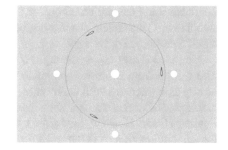

8.1 2D Computations

We compute with TSR = 4. The model geometry and the SI are shown in Fig. 13. The boundary conditions are U_∞ at the inflow, zero stress at the outflow, slip at the lateral boundaries, and no-slip on the rotor and support column surfaces. The prescribed velocity is evaluated at the integration points, with the values extracted from the NURBS representation of the rotor surface velocity.

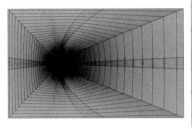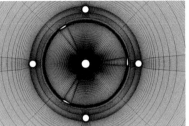

Fig. 14 2D VAWT. Mesh 3 (control mesh)

Table 1 2D VAWT

Mesh	nc	ne
Mesh 1	7510	5756
Mesh 2	26,432	23,024
Mesh 3	98,812	92,096

Number of control points (nc) and elements (ne)

We use three different meshes. We start with Mesh 1, and obtain the other two meshes by knot insertion. We halve the knot spacing to get Mesh 2, and halve it again to get Mesh 3. Figure 14 shows Mesh 3. The number of control points and elements are shown in Table 1. We compute for ten rotations, with two different time-step sizes. The two time-step sizes selected translate to $\Delta\phi = 2°$ and $\Delta\phi = 1°$ per time step. The number of nonlinear iterations per time step is 5, and the number of GMRES iterations per nonlinear iteration is 300. The first three nonlinear iterations are based on the ST-SUPS, and the last two the ST-VMS. The stabilization parameters are those given by Eqs. (4)–(8), and (10) in [31]. In the ST-SI (see Eq. (12)), we set $C = 2$.

Figure 15 shows C_{POW}, averaged over the three blades in the last three rotations. The results from different combinations of spatial and temporal resolutions are mostly in agreement with each other. The cases with the lower spatial resolution and highest Courant number show some differences in parts of the rotation cycle. Figures 16 and 17 show, for Mesh 1 with $\Delta\phi = 1°$ and Mesh 2 with $\Delta\phi = 2°$, the velocity magnitude in the wake of the support columns located at $\phi = 180°$ and $\phi = 90°$. Overall, the wakes are captured better with smaller Courant numbers, and a reasonable level of mesh refinement is needed to obtain good values for C_{POW}.

8.2 3D Computation

We compute with TSR = 3. The boundary conditions are no-slip on all turbine surfaces, U_∞ at the inflow, zero stress at the outflow, and slip at the lateral boundaries. We do not try to resolve the boundary layer near the ground since the blades are positioned relatively high (see Fig. 11). The no-slip condition on the blade

Fig. 15 2D VAWT. C_{POW}, averaged over the three blades in the last three rotations

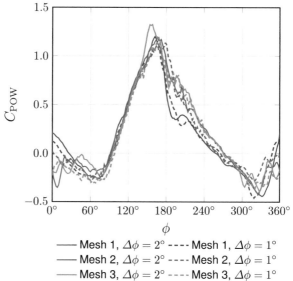

Mesh 1, $\Delta\phi = 2°$ ---- Mesh 1, $\Delta\phi = 1°$
Mesh 2, $\Delta\phi = 2°$ ---- Mesh 2, $\Delta\phi = 1°$
Mesh 3, $\Delta\phi = 2°$ ---- Mesh 3, $\Delta\phi = 1°$

and arm surfaces are enforced weakly. All prescribed velocities are evaluated at the integration points with the values extracted from the NURBS representation of the prescribed motion.

Figure 18 shows the mesh. There are 1,544,460 control points and 955,477 quadratic NURBS elements. We compute for one rotation, with a single time-step size. The time-step size selected translates to $\Delta\phi = 1°$ per time step. The number of nonlinear iterations per time step is 4, and the number of GMRES iterations per nonlinear iteration is 300. The first two nonlinear iterations are based on the ST-SUPS, and the last two the ST-VMS. The stabilization parameters are those given by Eqs. (2.4)–(2.6), (2.8) and (2.10) in [1]. In the ST-SI (see Eq. (12)), we set $C = 4$, and for the weakly enforced no-slip condition, we set $C = 2$.

Figure 19 shows the second invariant of the velocity gradient tensor near a blade, at different positions of the blade. Figure 20 shows the instantaneous power coefficient. The total C_{POW} averaged over the rotation cycle is about 0.16.

9   ALE Computation: HAWT FSI with Rotor–Tower Coupling

Dynamic coupling of a spinning rotor with flexible blades to a deformable tower presents a challenge for standalone structural and coupled FSI simulations. In this section we address this challenge by using a penalty-based approach that allows load transfer between the spinning rotor and tower (see Fig. 21). This approach presents

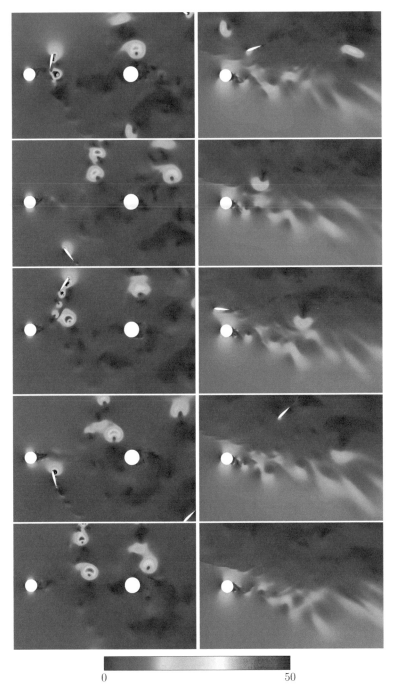

Fig. 16 2D VAWT. Velocity magnitude for Mesh 1 with $\Delta\phi = 1°$ in the wake of the support columns located at $\phi = 180°$ (*left*) and $\phi = 90°$ (*right*), for t/T ranging from 0.2 to 1

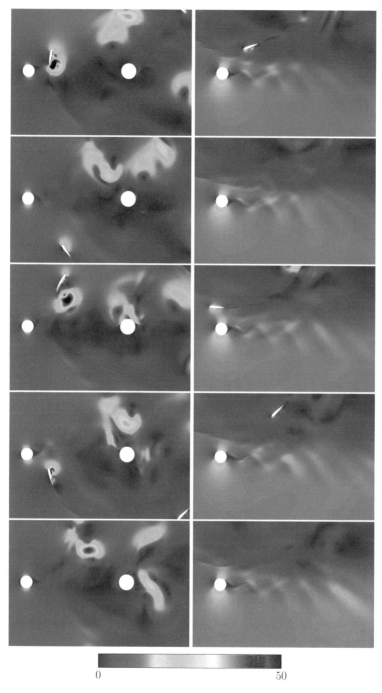

Fig. 17 2D VAWT. 2D VAWT. Velocity magnitude for Mesh 2 with $\Delta\phi = 2°$ in the wake of the support columns located at $\phi = 180°$ (*left*) and $\phi = 90°$ (*right*), for t/T ranging from 0.2 to 1

Fig. 18 3D VAWT. Control mesh

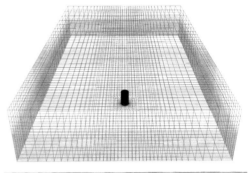

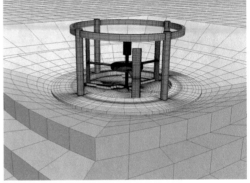

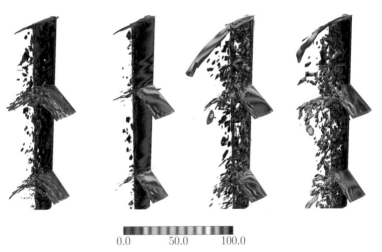

Fig. 19 3D VAWT. Isosurfaces corresponding to a positive value of the second invariant of the velocity gradient tensor, colored by the velocity magnitude (m/s), at $\phi = 0°$, $90°$, $180°$, and $270°$

an alternative technique to that proposed in [77], and naturally accommodates coupling of distinct structural models (e.g., shells and solids) and discretizations (e.g., finite elements and IGA).

Fig. 20 3D VAWT. C_{POW} for the three blades (*red*, *blue*, *green*) and the total C_{POW}, at instants defined by the position of the *red* blade

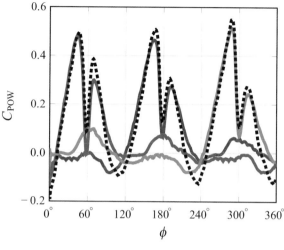

Fig. 21 Illustration of rotor and tower structural domains Ω_R and Ω_T and combined deformation accounting for the interaction of Ω_R and Ω_T

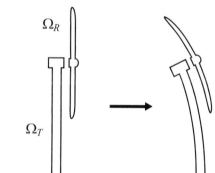

9.1 Formulation of the Rotor–Tower Penalty Coupling

In a wind turbine, the rotor hub is connected to the nacelle by the main shaft that transfers the rotational motion of the rotor hub to the gearbox. Since we do not wish to model the drivetrain operation directly, a simplified rotor–tower coupling strategy is required. We develop such a strategy by exploiting a penalty-based technique. For this, we first define the regions on both the rotor and nacelle surfaces that interact with one other, and denote them by Γ_1 (rotor side) and Γ_2 (nacelle side). These regions, which are assumed to have a circular shape, are highlighted using distinct colors in Fig. 22. We then design the penalty operator, which precludes all relative motion between Γ_1 and Γ_2 except for relative rotation about the rotor axis. This is achieved, conceptually, by using an overconstrained truss-like system to link the two interaction surfaces. More specifically, the change of distance between a point on one surface and every point on the opposing surface, as shown in Fig. 22a, is penalized. Figure 22b illustrates all penalized distances between the two surfaces. If the set of current distances (see Fig. 22c) is not the same as the set of reference

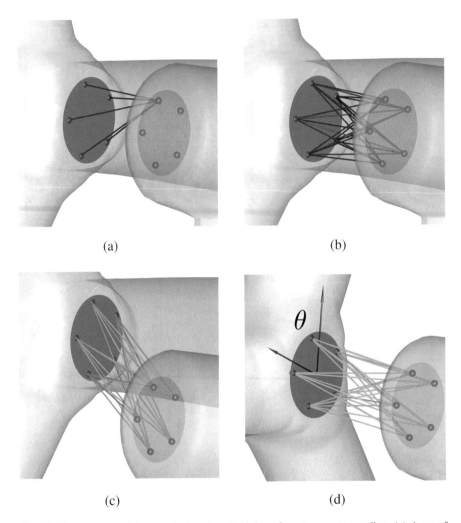

Fig. 22 Key concepts of the penalty-based methodology for rotor–tower coupling. (**a**) A set of distances between a point on a surface and points on another surface. (**b**) A set of distances in the reference configuration. (**c**) A set of distances in the current configuration. (**d**) Total rotation angle

distances (see Fig. 22b), the penalty term will produce forces to keep the current distances the same as the reference distances. The remaining challenge is to remove the forces associated with the relative spinning motion. For this, *the distances in the reference configuration are computed from the rotated configuration of the rotor.* The latter requires calculation of the total rotation angle θ (see Fig. 22d).

With these considerations, the potential form of the penalty term becomes

$$\Pi_p \equiv \frac{\beta}{2} \int_{\Gamma_1} \int_{\Gamma_2} \left(\|\mathbf{x}_1 - \mathbf{x}_2\| - \|\mathbf{X}_1^r - \mathbf{X}_2^r\| \right)^2 \mathrm{d}\Gamma_2 \mathrm{d}\Gamma_1, \tag{34}$$

where β is the penalty constant, $\mathbf{x}_1$ and $\mathbf{x}_2$ are the current positions of the two interaction surfaces, and $\mathbf{X}_1^r$ and $\mathbf{X}_2^r$ are the reference positions of the two interaction surfaces after taking their relative rotation into account. To arrive at the contribution of the penalty term to the weak form of the structural mechanics problem, we take a variation of Π_p with respect to $\mathbf{x}_1$ and $\mathbf{x}_2$ to obtain

$$
\begin{aligned}
\delta \Pi_p &= \frac{\partial \Pi_p}{\partial \mathbf{x}_1} \cdot \delta \mathbf{x}_1 + \frac{\partial \Pi_p}{\partial \mathbf{x}_2} \cdot \delta \mathbf{x}_2 \\
&= \beta \int_{\Gamma_1} \int_{\Gamma_2} (\delta \mathbf{x}_1 - \delta \mathbf{x}_2) \cdot \left(\|\mathbf{x}_1 - \mathbf{x}_2\| - \|\mathbf{X}_1^r - \mathbf{X}_2^r\| \right) \frac{\mathbf{x}_1 - \mathbf{x}_2}{\|\mathbf{x}_1 - \mathbf{x}_2\|} \mathrm{d}\Gamma_2 \mathrm{d}\Gamma_1.
\end{aligned}
\tag{35}
$$

In the discrete setting, the above integrals are approximated using numerical quadrature. Because only quadrature-point locations and weights are needed to formulate the method, it is well suited for coupling of distinct models and discretizations for the different structural components, which we do in this work.

9.2   Rotor and Tower Models and Meshes

A 3D model of the Hexcrete tower is constructed parametrically using the computer-aided design (CAD) software Rhinoceros 3D and the Grasshopper algorithmic modeling plugin for Rhinoceros (see [78] for details of the parametric modeling methodology). The profile of the tower is hexagonal with smaller hexagonal columns at each corner (see Fig. 23). The tower is comprised of two prismatic sections, located at the top and bottom of the structure, and two intermediate sections with unique rates of taper (see Fig. 23). The cylindrical nacelle is also modeled as part of the tower and approximated considered as a solid block. The tower is discretized using 295,332 linear tetrahedral elements. The columns have a Young's modulus 51.36 GPa, whereas the panels have a Young's modulus 47.23 GPa. The density and Poisson's ratio of both are assumed to be 2392 kg/m^3 and 0.2, respectively. The nacelle has a Young's modulus 500 GPa, Poisson's ratio 0.2, and density 741 kg/m^3 to produce a realistically stiff structure with a mass of 82 metric tons. Given these design characteristics, the combined tower and nacelle structure has a mass of approximately 1662 metric tons.

For the NREL 5 MW rotor design, we use the geometry definition provided in [79] to generate an initial blade model using the Grasshopper algorithmic modeling plugin for Rhinoceros. We then scale the blade by a factor appropriate to achieve a 108 m rotor and convert the model to a T-spline geometry description. Three such blades are then attached to a hub with a precone angle of 2.5° to produce the final rotor model. A simplified blade structural model is considered in this work. Internal shear webs are not modeled, and an isotropic material with an assumed thickness distribution is used (more details can be found in [80]). The Young's modulus and Poisson's ratio are set to 55.2 GPa and 0.2, respectively. The density is

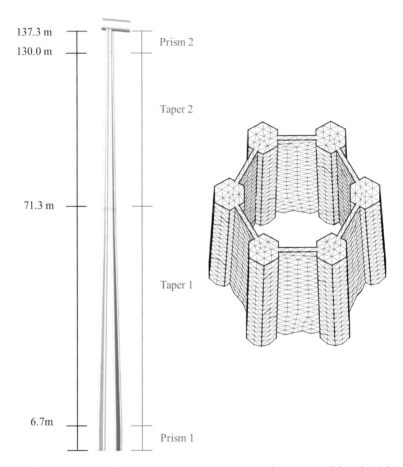

Fig. 23 CAD model of the Hexcrete tower (*left*) and a section of the tower solid mesh (*right*)

Fig. 24 T-spline mesh of the rotor surface

set to 2500 kg/m^3. Material properties and shell thickness distribution are selected such that the rotor has a mass of 60,000 kg, and such that the blade undergoes reasonable deflection and has a natural frequency of 0.705 Hz. This frequency was calculated using a simple proportional scaling law [81] applied to the original NREL 5 MW blade natural frequency of 0.870 Hz. Figure 24 shows the rotor model, where the T-spline mesh consists of 23,244 C^1-continuous cubic elements and 25,151 control points.

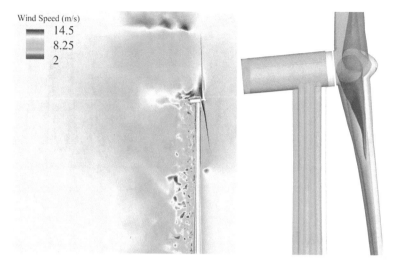

Fig. 25 Air speed contours at a planar cut (*left*) and wind-turbine deflected shape (*right*). The undeformed structure is shown in *gray* and the deformed structure is shown in *light green*

9.3   Results

The FSI simulation is performed at the rated wind speed of 11.4 m/s. Figure 25 shows the flow visualization of the full wind turbine configuration, and the deflection of the tower and blades. The figure clearly demonstrates that the rotor and tower displacements are coupled while the rotor is spinning. To assess the penalty-coupling error E_{int} we define it as

$$E_{int} \equiv \frac{\int_{\Gamma_1} \int_{\Gamma_2} \left(\|\mathbf{x}_1 - \mathbf{x}_2\| - \|\mathbf{X}_1^r - \mathbf{X}_2^r\| \right)^2 \, d\Gamma_2 d\Gamma_1}{\int_{\Gamma_1} \int_{\Gamma_2} \|\mathbf{X}_1^r - \mathbf{X}_2^r\|^2 \, d\Gamma_2 d\Gamma_1}, \tag{36}$$

and plot it a function of time in Fig. 26. The figure clearly shows that the coupling error, defined as a relative, dimensionless quantity, is very small.

Fig. 26 Penalty coupling error as a function of time

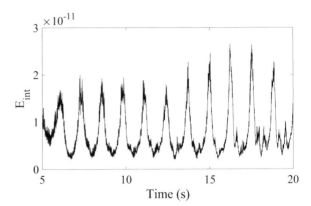

10 Concluding Remarks

We have described how the challenges encountered in computational analysis of wind turbines and turbomachinery are being addressed by the ALE-VMS and ST-VMS methods and isogeometric discretization. The computational challenges include turbulent rotational flows, complex geometries, MBI, such as the rotor motion, and the FSI, such as the FSI between the wind turbine blade and the air. The ALE-VMS and ST-VMS serve as the core computational methods. They are supplemented with special methods like the ST-ALE and ST-SI, weak enforcement of the no-slip boundary conditions, and ST-IGA with NURBS basis functions in time. We described the core methods and some of the special methods. We presented, as examples of challenging computations performed, computational analysis of a HAWT, a VAWT, and flow-driven string dynamics in pumps. The examples demonstrate the power and scope of the core and special methods in computational analysis of wind turbines and turbomachinery.

Acknowledgments This work was supported (second author) in part by Grant-in-Aid for Challenging Exploratory Research 16K13779 from Japan Society for the Promotion of Science; Grant-in-Aid for Scientific Research (S) 26220002 from the Ministry of Education, Culture, Sports, Science and Technology of Japan (MEXT); Council for Science, Technology and Innovation (CSTI), Cross-Ministerial Strategic Innovation Promotion Program (SIP), "Innovative Combustion Technology" (Funding agency: JST); and Rice–Waseda research agreement. It was also supported in part by Grant-in-Aid for Early-Career Scientists 19K20287 (fifth author). The mathematical model and computational method parts of the work were also supported (third author) in part by ARO Grant W911NF-17-1-0046, ARO DURIP Grant W911NF-18-1-0234, and Top Global University Project of Waseda University. The first author was partially supported by NSF Grant 1854436, and the fourth author was partially supported by NIH/NHLBI Grants R01HL129077 and R01HL142504.

References

1. K. Takizawa, T.E. Tezduyar, H. Mochizuki, H. Hattori, S. Mei, L. Pan, and K. Montel, "Space–time VMS method for flow computations with slip interfaces (ST-SI)", *Mathematical Models and Methods in Applied Sciences*, **25** (2015) 2377–2406, https://doi.org/10.1142/S0218202515400126.
2. Y. Bazilevs, A. Korobenko, X. Deng, and J. Yan, "FSI modeling for fatigue-damage prediction in full-scale wind-turbine blades", *Journal of Applied Mechanics*, **83** (6) (2016) 061010.
3. T. Kanai, K. Takizawa, T.E. Tezduyar, K. Komiya, M. Kaneko, K. Hirota, M. Nohmi, T. Tsuneda, M. Kawai, and M. Isono, "Methods for computation of flow-driven string dynamics in a pump and residence time", *Mathematical Models and Methods in Applied Sciences*, **29** (2019) 839–870, https://doi.org/10.1142/S021820251941001X.
4. A. Korobenko, Y. Bazilevs, K. Takizawa, and T.E. Tezduyar, "Computer modeling of wind turbines: 1. ALE-VMS and ST-VMS aerodynamic and FSI analysis", *Archives of Computational Methods in Engineering*, published online, https://doi.org/10.1007/s11831-018-9292-1, September 2018.
5. Y. Bazilevs, V.M. Calo, T.J.R. Hughes, and Y. Zhang, "Isogeometric fluid–structure interaction: theory, algorithms, and computations", *Computational Mechanics*, **43** (2008) 3–37.
6. K. Takizawa and T.E. Tezduyar, "Multiscale space–time fluid–structure interaction techniques", *Computational Mechanics*, **48** (2011) 247–267, https://doi.org/10.1007/s00466-011-0571-z.
7. K. Takizawa, B. Henicke, A. Puntel, T. Spielman, and T.E. Tezduyar, "Space–time computational techniques for the aerodynamics of flapping wings", *Journal of Applied Mechanics*, **79** (2012) 010903, https://doi.org/10.1115/1.4005073.
8. K. Takizawa, T.E. Tezduyar, Y. Otoguro, T. Terahara, T. Kuraishi, and H. Hattori, "Turbocharger flow computations with the Space–Time Isogeometric Analysis (ST-IGA)", *Computers & Fluids*, **142** (2017) 15–20, https://doi.org/10.1016/j.compfluid.2016.02.021.
9. Y. Otoguro, K. Takizawa, and T.E. Tezduyar, "Space–time VMS computational flow analysis with isogeometric discretization and a general-purpose NURBS mesh generation method", *Computers & Fluids*, **158** (2017) 189–200, https://doi.org/10.1016/j.compfluid.2017.04.017.
10. K. Takizawa and T.E. Tezduyar, "Computational methods for parachute fluid–structure interactions", *Archives of Computational Methods in Engineering*, **19** (2012) 125–169, https://doi.org/10.1007/s11831-012-9070-4.
11. Y. Bazilevs and T.J.R. Hughes, "Weak imposition of Dirichlet boundary conditions in fluid mechanics", *Computers and Fluids*, **36** (2007) 12–26.
12. Y. Bazilevs and T.J.R. Hughes, "NURBS-based isogeometric analysis for the computation of flows about rotating components", *Computational Mechanics*, **43** (2008) 143–150.
13. T.E. Tezduyar, "Stabilized finite element formulations for incompressible flow computations", *Advances in Applied Mechanics*, **28** (1992) 1–44, https://doi.org/10.1016/S0065-2156(08)70153-4.
14. A.N. Brooks and T.J.R. Hughes, "Streamline upwind/Petrov-Galerkin formulations for convection dominated flows with particular emphasis on the incompressible Navier-Stokes equations", *Computer Methods in Applied Mechanics and Engineering*, **32** (1982) 199–259.
15. T.J.R. Hughes, "Multiscale phenomena: Green's functions, the Dirichlet-to-Neumann formulation, subgrid scale models, bubbles, and the origins of stabilized methods", *Computer Methods in Applied Mechanics and Engineering*, **127** (1995) 387–401.
16. Y. Bazilevs, V.M. Calo, J.A. Cottrell, T.J.R. Hughes, A. Reali, and G. Scovazzi, "Variational multiscale residual-based turbulence modeling for large eddy simulation of incompressible flows", *Computer Methods in Applied Mechanics and Engineering*, **197** (2007) 173–201.
17. T.E. Tezduyar and K. Takizawa, "Space–time computations in practical engineering applications: A summary of the 25-year history", *Computational Mechanics*, **63** (2019) 747–753, https://doi.org/10.1007/s00466-018-1620-7.

18. K. Takizawa, D. Montes, M. Fritze, S. McIntyre, J. Boben, and T.E. Tezduyar, "Methods for FSI modeling of spacecraft parachute dynamics and cover separation", *Mathematical Models and Methods in Applied Sciences*, **23** (2013) 307–338, https://doi.org/10.1142/S0218202513400058.

19. K. Takizawa, T.E. Tezduyar, and R. Kolesar, "FSI modeling of the Orion spacecraft drogue parachutes", *Computational Mechanics*, **55** (2015) 1167–1179, https://doi.org/10.1007/s00466-014-1108-z.

20. K. Takizawa, B. Henicke, D. Montes, T.E. Tezduyar, M.-C. Hsu, and Y. Bazilevs, "Numerical-performance studies for the stabilized space–time computation of wind-turbine rotor aerodynamics", *Computational Mechanics*, **48** (2011) 647–657, https://doi.org/10.1007/s00466-011-0614-5.

21. K. Takizawa, "Computational engineering analysis with the new-generation space–time methods", *Computational Mechanics*, **54** (2014) 193–211, https://doi.org/10.1007/s00466-014-0999-z.

22. K. Takizawa, B. Henicke, A. Puntel, N. Kostov, and T.E. Tezduyar, "Space–time techniques for computational aerodynamics modeling of flapping wings of an actual locust", *Computational Mechanics*, **50** (2012) 743–760, https://doi.org/10.1007/s00466-012-0759-x.

23. K. Takizawa, T.E. Tezduyar, and N. Kostov, "Sequentially-coupled space–time FSI analysis of bio-inspired flapping-wing aerodynamics of an MAV", *Computational Mechanics*, **54** (2014) 213–233, https://doi.org/10.1007/s00466-014-0980-x.

24. K. Takizawa, T.E. Tezduyar, and A. Buscher, "Space–time computational analysis of MAV flapping-wing aerodynamics with wing clapping", *Computational Mechanics*, **55** (2015) 1131–1141, https://doi.org/10.1007/s00466-014-1095-0.

25. K. Takizawa, Y. Bazilevs, T.E. Tezduyar, M.-C. Hsu, O. Øiseth, K.M. Mathisen, N. Kostov, and S. McIntyre, "Engineering analysis and design with ALE-VMS and space–time methods", *Archives of Computational Methods in Engineering*, **21** (2014) 481–508, https://doi.org/10.1007/s11831-014-9113-0.

26. K. Takizawa, K. Schjodt, A. Puntel, N. Kostov, and T.E. Tezduyar, "Patient-specific computational analysis of the influence of a stent on the unsteady flow in cerebral aneurysms", *Computational Mechanics*, **51** (2013) 1061–1073, https://doi.org/10.1007/s00466-012-0790-y.

27. K. Takizawa, T.E. Tezduyar, H. Uchikawa, T. Terahara, T. Sasaki, and A. Yoshida, "Mesh refinement influence and cardiac-cycle flow periodicity in aorta flow analysis with isogeometric discretization", *Computers & Fluids*, **179** (2019) 790–798, https://doi.org/10.1016/j.compfluid.2018.05.025.

28. K. Takizawa, T.E. Tezduyar, T. Terahara, and T. Sasaki, "Heart valve flow computation with the integrated Space–Time VMS, Slip Interface, Topology Change and Isogeometric Discretization methods", *Computers & Fluids*, **158** (2017) 176–188, https://doi.org/10.1016/j.compfluid.2016.11.012.

29. K. Takizawa, T.E. Tezduyar, and T. Kuraishi, "Multiscale ST methods for thermo-fluid analysis of a ground vehicle and its tires", *Mathematical Models and Methods in Applied Sciences*, **25** (2015) 2227–2255, https://doi.org/10.1142/S0218202515400072.

30. K. Takizawa, T.E. Tezduyar, T. Kuraishi, S. Tabata, and H. Takagi, "Computational thermo-fluid analysis of a disk brake", *Computational Mechanics*, **57** (2016) 965–977, https://doi.org/10.1007/s00466-016-1272-4.

31. Y. Otoguro, K. Takizawa, T.E. Tezduyar, K. Nagaoka, and S. Mei, "Turbocharger turbine and exhaust manifold flow computation with the Space–Time Variational Multiscale Method and Isogeometric Analysis", *Computers & Fluids*, **179** (2019) 764–776, https://doi.org/10.1016/j.compfluid.2018.05.019.

32. T. Kuraishi, K. Takizawa, and T.E. Tezduyar, "Tire aerodynamics with actual tire geometry, road contact and tire deformation", *Computational Mechanics*, **63** (2019) 1165–1185, https://doi.org/10.1007/s00466-018-1642-1.

33. T. Kuraishi, K. Takizawa, and T.E. Tezduyar, "Space–Time Isogeometric flow analysis with built-in Reynolds-equation limit", *Mathematical Models and Methods in Applied Sciences*, **29** (2019) 871–904, https://doi.org/10.1142/S0218202519410021.
34. K. Takizawa, T.E. Tezduyar, and T. Terahara, "Ram-air parachute structural and fluid mechanics computations with the space–time isogeometric analysis (ST-IGA)", *Computers & Fluids*, **141** (2016) 191–200, https://doi.org/10.1016/j.compfluid.2016.05.027.
35. K. Takizawa, T.E. Tezduyar, and T. Kanai, "Porosity models and computational methods for compressible-flow aerodynamics of parachutes with geometric porosity", *Mathematical Models and Methods in Applied Sciences*, **27** (2017) 771–806, https://doi.org/10.1142/S0218202517500166.
36. T.J.R. Hughes, W.K. Liu, and T.K. Zimmermann, "Lagrangian–Eulerian finite element formulation for incompressible viscous flows", *Computer Methods in Applied Mechanics and Engineering*, **29** (1981) 329–349.
37. V. Kalro and T.E. Tezduyar, "A parallel 3D computational method for fluid–structure interactions in parachute systems", *Computer Methods in Applied Mechanics and Engineering*, **190** (2000) 321–332, https://doi.org/10.1016/S0045-7825(00)00204-8.
38. Y. Bazilevs, M.-C. Hsu, I. Akkerman, S. Wright, K. Takizawa, B. Henicke, T. Spielman, and T.E. Tezduyar, "3D simulation of wind turbine rotors at full scale. Part I: Geometry modeling and aerodynamics", *International Journal for Numerical Methods in Fluids*, **65** (2011) 207–235, https://doi.org/10.1002/fld.2400.
39. J. Yan, A. Korobenko, X. Deng, and Y. Bazilevs, "Computational free-surface fluid–structure interaction with application to floating offshore wind turbines", *Computers and Fluids*, **141** (2016) 155–174, https://doi.org/10.1016/j.compfluid.2016.03.008.
40. Y. Bazilevs, V.M. Calo, Y. Zhang, and T.J.R. Hughes, "Isogeometric fluid–structure interaction analysis with applications to arterial blood flow", *Computational Mechanics*, **38** (2006) 310–322.
41. M.-C. Hsu and Y. Bazilevs, "Blood vessel tissue prestress modeling for vascular fluid–structure interaction simulations", *Finite Elements in Analysis and Design*, **47** (2011) 593–599.
42. M.-C. Hsu, D. Kamensky, F. Xu, J. Kiendl, C. Wang, M.C.H. Wu, J. Mineroff, A. Reali, Y. Bazilevs, and M.S. Sacks, "Dynamic and fluid–structure interaction simulations of bioprosthetic heart valves using parametric design with T-splines and Fung-type material models", *Computational Mechanics*, **55** (2015) 1211–1225, https://doi.org/10.1007/s00466-015-1166-x.
43. D. Kamensky, M.-C. Hsu, D. Schillinger, J.A. Evans, A. Aggarwal, Y. Bazilevs, M.S. Sacks, and T.J.R. Hughes, "An immersogeometric variational framework for fluid-structure interaction: Application to bioprosthetic heart valves", *Computer Methods in Applied Mechanics and Engineering*, **284** (2015) 1005–1053.
44. I. Akkerman, J. Dunaway, J. Kvandal, J. Spinks, and Y. Bazilevs, "Toward free-surface modeling of planing vessels: simulation of the Fridsma hull using ALE-VMS", *Computational Mechanics*, **50** (2012) 719–727.
45. M.C.H. Wu, D. Kamensky, C. Wang, A.J. Herrema, F. Xu, M.S. Pigazzini, A. Verma, A.L. Marsden, Y. Bazilevs, and M.-C. Hsu, "Optimizing fluid–structure interaction systems with immersogeometric analysis and surrogate modeling: Application to a hydraulic arresting gear", *Computer Methods in Applied Mechanics and Engineering*, **316** (2017) 668–693.
46. J. Yan, X. Deng, A. Korobenko, and Y. Bazilevs, "Free-surface flow modeling and simulation of horizontal-axis tidal-stream turbines", *Computers and Fluids*, **158** (2017) 157–166, https://doi.org/10.1016/j.compfluid.2016.06.016.
47. A. Castorrini, A. Corsini, F. Rispoli, K. Takizawa, and T.E. Tezduyar, "A stabilized ALE method for computational fluid–structure interaction analysis of passive morphing in turbomachinery", *Mathematical Models and Methods in Applied Sciences*, **29** (2019) 967–994, https://doi.org/10.1142/S0218202519410057.
48. J. Yan, B. Augier, A. Korobenko, J. Czarnowski, G. Ketterman, and Y. Bazilevs, "FSI modeling of a propulsion system based on compliant hydrofoils in a tandem configuration", *Computers and Fluids*, **141** (2016) 201–211, https://doi.org/10.1016/j.compfluid.2015.07.013.

49. T.A. Helgedagsrud, I. Akkerman, Y. Bazilevs, K.M. Mathisen, and O.A. Oiseth, "Isogeometric modeling and experimental investigation of moving-domain bridge aerodynamics", *ASCE Journal of Engineering Mechanics*, **145** (2019) 04019026.

50. J. Yan, A. Korobenko, A.E. Tejada-Martinez, R. Golshan, and Y. Bazilevs, "A new variational multiscale formulation for stratified incompressible turbulent flows", *Computers & Fluids*, **158** (2017) 150–156, https://doi.org/10.1016/j.compfluid.2016.12.004.

51. F. Xu, G. Moutsanidis, D. Kamensky, M.-C. Hsu, M. Murugan, A. Ghoshal, and Y. Bazilevs, "Compressible flows on moving domains: Stabilized methods, weakly enforced essential boundary conditions, sliding interfaces, and application to gas-turbine modeling", *Computers & Fluids*, **158** (2017) 201–220, https://doi.org/10.1016/j.compfluid.2017.02.006.

52. T.M. van Opstal, J. Yan, C. Coley, J.A. Evans, T. Kvamsdal, and Y. Bazilevs, "Isogeometric divergence-conforming variational multiscale formulation of incompressible turbulent flows", *Computer Methods in Applied Mechanics and Engineering*, **316** (2017) 859–879, https://doi.org/10.1016/j.cma.2016.10.015.

53. Y. Bazilevs, K. Takizawa, and T.E. Tezduyar, *Computational Fluid–Structure Interaction: Methods and Applications*. Wiley, February 2013, ISBN 978-0470978771.

54. T.E. Tezduyar, "Computation of moving boundaries and interfaces and stabilization parameters", *International Journal for Numerical Methods in Fluids*, **43** (2003) 555–575, https://doi.org/10.1002/fld.505.

55. T.E. Tezduyar and S. Sathe, "Modeling of fluid–structure interactions with the space–time finite elements: Solution techniques", *International Journal for Numerical Methods in Fluids*, **54** (2007) 855–900, https://doi.org/10.1002/fld.1430.

56. M.-C. Hsu, Y. Bazilevs, V.M. Calo, T.E. Tezduyar, and T.J.R. Hughes, "Improving stability of stabilized and multiscale formulations in flow simulations at small time steps", *Computer Methods in Applied Mechanics and Engineering*, **199** (2010) 828–840, https://doi.org/10.1016/j.cma.2009.06.019.

57. K. Takizawa, T.E. Tezduyar, S. McIntyre, N. Kostov, R. Kolesar, and C. Habluetzel, "Space–time VMS computation of wind-turbine rotor and tower aerodynamics", *Computational Mechanics*, **53** (2014) 1–15, https://doi.org/10.1007/s00466-013-0888-x.

58. K. Takizawa, T.E. Tezduyar, and Y. Otoguro, "Stabilization and discontinuity-capturing parameters for space–time flow computations with finite element and isogeometric discretizations", *Computational Mechanics*, **62** (2018) 1169–1186, https://doi.org/10.1007/s00466-018-1557-x.

59. T.J.R. Hughes, J.A. Cottrell, and Y. Bazilevs, "Isogeometric analysis: CAD, finite elements, NURBS, exact geometry, and mesh refinement", *Computer Methods in Applied Mechanics and Engineering*, **194** (2005) 4135–4195.

60. K. Takizawa and T.E. Tezduyar, "Space–time computation techniques with continuous representation in time (ST-C)", *Computational Mechanics*, **53** (2014) 91–99, https://doi.org/10.1007/s00466-013-0895-y.

61. T. Kanai, K. Takizawa, T.E. Tezduyar, T. Tanaka, and A. Hartmann, "Compressible-flow geometric-porosity modeling and spacecraft parachute computation with isogeometric discretization", *Computational Mechanics*, **63** (2019) 301–321, https://doi.org/10.1007/s00466-018-1595-4.

62. T. Sasaki, K. Takizawa, and T.E. Tezduyar, "Medical-image-based aorta modeling with zero-stress-state estimation", *Computational Mechanics*, **64** (2019) 249–271, https://doi.org/10.1007/s00466-019-01669-4.

63. K. Takizawa, T.E. Tezduyar, and T. Sasaki, "Isogeometric hyperelastic shell analysis with out-of-plane deformation mapping", *Computational Mechanics*, **63** (2019) 681–700, https://doi.org/10.1007/s00466-018-1616-3.

64. Y. Otoguro, K. Takizawa, and T.E. Tezduyar, "A general-purpose NURBS mesh generation method for complex geometries", in T.E. Tezduyar, editor, *Frontiers in Computational Fluid–Structure Interaction and Flow Simulation: Research from Lead Investigators under Forty – 2018*, Modeling and Simulation in Science, Engineering and Technology, 399–434, Springer, 2018, ISBN 978-3-319-96468-3, https://doi.org/10.1007/978-3-319-96469-0_10.

65. K. Takizawa, T.E. Tezduyar, S. Asada, and T. Kuraishi, "Space–time method for flow computations with slip interfaces and topology changes (ST-SI-TC)", *Computers & Fluids*, **141** (2016) 124–134, https://doi.org/10.1016/j.compfluid.2016.05.006.
66. Y. Bazilevs, J.R. Gohean, T.J.R. Hughes, R.D. Moser, and Y. Zhang, "Patient-specific isogeometric fluid–structure interaction analysis of thoracic aortic blood flow due to implantation of the Jarvik 2000 left ventricular assist device", *Computer Methods in Applied Mechanics and Engineering*, **198** (2009) 3534–3550.
67. M. Esmaily-Moghadam, Y. Bazilevs, T.-Y. Hsia, I.E. Vignon-Clementel, and A.L. Marsden, "A comparison of outlet boundary treatments for prevention of backflow divergence with relevance to blood flow simulations", *Computational Mechanics*, **48** (2011) 277–291.
68. C.A. Taylor, T.J.R. Hughes, and C.K. Zarins, "Finite element modeling of blood flow in arteries", *Computer Methods in Applied Mechanics and Engineering*, **158** (1998) 155–196.
69. T.E. Tezduyar and M. Senga, "Stabilization and shock-capturing parameters in SUPG formulation of compressible flows", *Computer Methods in Applied Mechanics and Engineering*, **195** (2006) 1621–1632, https://doi.org/10.1016/j.cma.2005.05.032.
70. Y. Bazilevs, V.M. Calo, T.E. Tezduyar, and T.J.R. Hughes, "YZβ discontinuity-capturing for advection-dominated processes with application to arterial drug delivery", *International Journal for Numerical Methods in Fluids*, **54** (2007) 593–608, https://doi.org/10.1002/fld.1484.
71. S.B. Raknes, X. Deng, Y. Bazilevs, D.J. Benson, K.M. Mathisen, and T. Kvamsdal, "Isogeometric rotation-free bending-stabilized cables: Statics, dynamics, bending strips and coupling with shells", *Computer Methods in Applied Mechanics and Engineering*, **263** (2013) 127–143.
72. K. Takizawa, T. Spielman, and T.E. Tezduyar, "Space–time FSI modeling and dynamical analysis of spacecraft parachutes and parachute clusters", *Computational Mechanics*, **48** (2011) 345–364, https://doi.org/10.1007/s00466-011-0590-9.
73. C.C. Long, M. Esmaily-Moghadam, A.L. Marsden, and Y. Bazilevs, "Computation of residence time in the simulation of pulsatile ventricular assist devices", *Computational Mechanics*, **54** (2014) 911–919, https://doi.org/10.1007/s00466-013-0931-y.
74. T.E. Tezduyar, M. Behr, S. Mittal, and A.A. Johnson, "Computation of unsteady incompressible flows with the finite element methods: Space–time formulations, iterative strategies and massively parallel implementations", in *New Methods in Transient Analysis*, PVP-Vol.246/AMD-Vol.143, ASME, New York, (1992) 7–24.
75. T. Tezduyar, S. Aliabadi, M. Behr, A. Johnson, and S. Mittal, "Parallel finite-element computation of 3D flows", *Computer*, **26** (10) (1993) 27–36, https://doi.org/10.1109/2.237441.
76. "Life tower", http://cosmosunfarm.co.jp/lifetower.html.
77. Y. Bazilevs, A. Korobenko, X. Deng, and J. Yan, "Novel structural modeling and mesh moving techniques for advanced FSI simulation of wind turbines", *International Journal for Numerical Methods in Engineering*, **102** (2015) 766–783, https://doi.org/10.1002/nme.4738.
78. M.-C. Hsu, C. Wang, A.J. Herrema, D. Schillinger, A. Ghoshal, and Y. Bazilevs, "An interactive geometry modeling and parametric design platform for isogeometric analysis", *Computers and Mathematics with Applications*, **70** (2015) 1481–1500.
79. B.R. Resor, "Definition of a 5MW/61.5m wind turbine blade reference model", Technical Report SAND2013-2569, Sandia National Laboratories, Albuquerque, NM, 2013.
80. Y. Bazilevs, M.-C. Hsu, J. Kiendl, R. Wüchner, and K.-U. Bletzinger, "3D simulation of wind turbine rotors at full scale. Part II: Fluid–structure interaction modeling with composite blades", *International Journal for Numerical Methods in Fluids*, **65** (2011) 236–253.
81. J.F. Manwell, J.G. McGowan, and A.L. Rogers, *Wind Energy Explained: Theory, Design and Application*. John Wiley & Sons, 2009.

Variational Multiscale Flow Analysis in Aerospace, Energy and Transportation Technologies

Kenji Takizawa, Yuri Bazilevs, Tayfun E. Tezduyar, and Artem Korobenko

1 Introduction

Computational flow analysis is now a valuable engineering tool in aerospace, energy and transportation technologies. It is bringing solution in many classes of challenging problems. Examples are spacecraft parachute analysis for the landing-stage parachutes [1], cover-separation parachutes [2] and the drogue parachutes [3], spacecraft aerodynamics [2], ram-air parachutes [4], compressible-flow spacecraft parachute aerodynamics [5], thermo-fluid analysis of ground vehicles and their tires [6], flow around tires with road contact and deformation [7], thermo-fluid analysis of disk brakes [8], flow analysis of turbocharger turbines [9], wind-turbine aerodynamics and fluid–structure interaction (FSI) [10], more specifically, vertical-axis wind turbines [11], floating wind turbines [12], wind turbines in atmospheric boundary layer (ABL) flow [13], and fatigue damage in wind-turbine blades [14].

K. Takizawa (✉)
Department of Modern Mechanical Engineering, Waseda University, Tokyo, Japan
e-mail: Kenji.Takizawa@tafsm.org

Y. Bazilevs
School of Engineering, Brown University, Providence, RI, USA
e-mail: yuri_bazilevs@brown.edu

T. E. Tezduyar
Mechanical Engineering, Rice University, Houston, TX, USA

Faculty of Science and Engineering, Waseda University, Shinjuku-ku, Tokyo, Japan
e-mail: tezduyar@tafsm.org

A. Korobenko
Department of Mechanical and Manufacturing Engineering, University of Calgary, Calgary, AB, Canada
e-mail: artem.korobenko@ucalgary.ca

© Springer Nature Switzerland AG 2020
A. Grama, A. H. Sameh (eds.), *Parallel Algorithms in Computational Science and Engineering*, Modeling and Simulation in Science, Engineering and Technology, https://doi.org/10.1007/978-3-030-43736-7_8

235

The computational challenges encountered in these classes of problems include complex geometries, moving boundaries and interfaces (MBI), FSI, turbulent flows, rotational flows, and large problem sizes.

Our core methods in addressing the computational challenges are the Residual-Based Variational Multiscale (RBVMS) [15, 16], ALE-VMS [17] and Space–Time VMS (ST-VMS) [18]. methods. We supplement the core methods with a number of special methods targeting specific classes of problems. The special methods used in combination with the ST-VMS include the ST Slip Interface (ST-SI) method [19], ST Isogeometric Analysis (ST-IGA) [20, 21], Multi-Domain Method (MDM) [22], and the "ST-C" data compression method [23]. The special methods used in combination with the ALE-VMS include weak enforcement of no-slip boundary conditions [24] and "sliding interfaces" [25] (the acronym "SI" will also indicate that).

As examples of the challenging computations performed, we present aerodynamic analysis of a ram-air parachute, thermo-fluid analysis of a freight truck and its rear set of tires, and aerodynamic and FSI analysis of two back-to-back wind turbines in ABL flow.

1.1 ST-VMS and ST-SUPS

The ST-VMS and ST-SUPS are versions of the Deforming-Spatial-Domain/Stabilized ST (DSD/SST) method [26], which was introduced for computation of flows with MBI, including FSI. The ST-SUPS is a new name for the original version of the DSD/SST, with "SUPS" reflecting its stabilization components, the Streamline-Upwind/Petrov-Galerkin (SUPG) [27] and Pressure-Stabilizing/Petrov-Galerkin (PSPG) [26] stabilizations. The ST-VMS is the VMS version of the DSD/SST. The VMS components of the ST-VMS are from the RBVMS. The five stabilization terms of the ST-VMS include the three that the ST-SUPS has, and therefore the ST-VMS subsumes the ST-SUPS. In MBI computations the ST-VMS and ST-SUPS function as moving-mesh methods. Moving the fluid mechanics mesh to follow an interface enables mesh-resolution control near the interface and, consequently, high-resolution boundary-layer representation near fluid–solid interfaces. Because of the higher-order accuracy of the ST framework (see [18]), the ST-SUPS and ST-VMS are desirable also in computations without MBI.

The ST-SUPS and ST-VMS have been applied to many classes of challenging FSI, MBI, and fluid mechanics problems (see [28] for a comprehensive summary of the computations prior to July 2018). The classes of problems include spacecraft parachute analysis for the landing-stage parachutes [1], cover-separation parachutes [2] and the drogue parachutes [3], wind-turbine aerodynamics for horizontal-axis wind turbine (HAWT) rotors [29], full HAWTs [30] and vertical-axis wind turbines (VAWTs) [19], flapping-wing aerodynamics for an actual locust [31], bioinspired MAVs [32] and wing-clapping [33], blood flow analysis of cerebral aneurysms [34], stent-treated aneurysms [35], aortas [36] and heart valves [37], spacecraft aerodynamics [2], thermo-fluid analysis of ground vehicles and their tires [6], thermo-fluid analysis of disk brakes [8], flow-driven string dynamics in turbomachinery [38], flow analysis of turbocharger turbines [9], flow around tires with road contact and deformation [7], fluid films [39], ram-air parachutes [4], and compressible-flow spacecraft parachute aerodynamics [5].

In the flow analyses presented here, the ST framework provides higher-order accuracy in a general context. The VMS feature of the ST-VMS addresses the computational challenges associated with the multiscale nature of the unsteady flow. The moving-mesh feature of the ST framework enables high-resolution computation near the truck body as it undergoes heave motion.

1.2   ALE-VMS, RBVMS, and ALE-SUPS

The ALE-VMS [17] is the VMS version of the ALE [40]. It succeeded the ST-SUPS [26] and ALE-SUPS [41] and preceded the ST-VMS. The VMS components are from the RBVMS [15, 16]. It is the moving-mesh extension of the RBVMS formulation of incompressible turbulent flows proposed in [16], and as such, it was first presented in [17] in the FSI context. The ALE-SUPS, RBVMS, and ALE-VMS have also been applied to many classes of challenging FSI, MBI, and fluid mechanics problems. The classes of problems include ram-air parachute FSI [41], wind-turbine aerodynamics and FSI [42, 43], more specifically, VAWTs [43], floating wind turbines [12], wind turbines in atmospheric boundary layers [43], and fatigue damage in wind-turbine blades [14], patient-specific cardiovascular fluid mechanics and FSI [44, 45], biomedical-device FSI [46, 47], ship hydrodynamics with free-surface flow and fluid–object interaction [48], hydrodynamics and FSI of a hydraulic arresting gear [49], hydrodynamics of tidal-stream turbines with free-surface flow [50], passive-morphing FSI in turbomachinery [51], bioinspired FSI for marine propulsion [52], bridge aerodynamics and fluid–object interaction [53], stratified incompressible flows [54], and compressible-flow gas-turbine analysis [55]. Recent advances in stabilized and multiscale methods may be found for stratified incompressible flows in [54], for divergence-conforming discretizations of incompressible flows in [56], and for compressible flows with emphasis on gas-turbine modeling in [55].

In the flow analyses presented here, the VMS feature of the ALE-VMS addresses the computational challenges associated with the multiscale nature of the unsteady flow. The moving-mesh feature of the ALE framework enables high-resolution computation near the wind-turbine blades.

1.3   ALE-SI and ST-SI

The ALE-SI was introduced in [25] to retain the desirable moving-mesh features of the ALE-VMS in computations with spinning solid surfaces, such as a turbine rotor. The mesh covering the spinning surface spins with it, retaining the high-resolution representation of the boundary layers. The method was in the context of incompressible-flow equations. Interface terms added to the ALE-VMS to account for the compatibility conditions for the velocity and stress at the SI accurately connect the two sides of the solution. The ST-SI was introduced in [19], also in the context of incompressible-flow equations, to retain the desirable moving-mesh features of the ST-VMS and ST-SUPS in computations with spinning solid surfaces. The starting point in its development was the ALE-SI. Interface terms similar to those in the ALE-SI are added to the ST-VMS to accurately connect the two sides of the solution. An ST-SI version where the SI is between fluid and solid domains was also presented in [19]. The SI in this case is a "fluid–solid SI" rather than a standard "fluid–fluid SI" and enables weak enforcement of the Dirichlet boundary conditions for the fluid. The ST-SI introduced in [8] for the coupled incompressible-flow and thermal-transport equations retains the high-resolution representation of the thermo-fluid boundary layers near spinning solid surfaces. These ST-SI methods have been applied to aerodynamic analysis of vertical-axis wind turbines [19], thermo-fluid analysis of disk brakes [8], flow-driven string dynamics in turbomachinery [38], flow analysis of turbocharger turbines [9], flow around tires with road contact and deformation [7], fluid films [39], aerodynamic analysis of ram-air parachutes [4], and flow analysis of heart valves [37].

In another ST-SI version presented in [19] the SI is between a thin porous structure and the fluid on its two sides. This enables dealing with the porosity in a fashion consistent with how the standard fluid–fluid SIs are dealt with and how the Dirichlet conditions are enforced weakly with fluid–solid SIs. This version also enables handling thin structures that have T-junctions. This method has been applied to incompressible-flow aerodynamic analysis of ram-air parachutes with fabric porosity [4]. The compressible-flow ST-SI methods were introduced in [5], including the version where the SI is between a thin porous structure and the fluid on its two sides. Compressible-flow porosity models were also introduced in [5]. These, together with the compressible-flow ST SUPG method [57], extended the ST computational analysis range to compressible-flow aerodynamics of parachutes with fabric and geometric porosities. That enabled ST computational flow analysis of the Orion spacecraft drogue parachute in the compressible-flow regime [58].

In the computations here, with the ALE-SI we are able to handle the interaction between the spinning rotor and stationary tower. The ST-SI enables dealing with the fabric porosity of the ram-air parachute.

1.4 Stabilization Parameters

The ST-SUPS, ALE-SUPS, RBVMS, ALE-VMS, ST-VMS, ALE-SI, and ST-SI all have some embedded stabilization parameters that play a significant role (see [19, 59]). There are many ways of defining these stabilization parameters (for examples, see [6, 7, 60–64]). The stabilization-parameter definitions used in the computations reported in this article can be found from the references cited in the sections where those computations are described.

1.5 Discontinuity-Capturing Term

The thermo-fluid analysis methods based on the SUPG/PSPG formulation of the coupled incompressible-flow and thermal-transport equations were presented in [61]. The methods were described in the ALE context since the description followed a section on the ALE formulation with SUPG and PSPG stabilizations and a reader who sees the methods in the ALE context can easily imagine them in the ST context. The methods presented in [61] included discontinuity-capturing (DC) options for both sets of equations as well as stabilization and DC parameters for the thermal-transport equation. The options for the DC parameters were based on those introduced with the "DCDD stabilization" [60] and "YZβ shock-capturing" [65]. These thermo-fluid analysis methods were successfully used in [66] in a number of 2D test computations as well as in a 3D computation with a simplified model of air circulation and cooling in a small data center. A new element length scale option applicable to the stabilization parameters for both the incompressible-flow and thermal-transport equations was introduced in [6]. The new length scale option is applicable also to the DC parameter for the thermal-transport equation.

In the flow analyses presented here, we use the YZβ shock-capturing in the thermo-fluid analysis of a freight truck and its rear set of tires. We use it for the thermal-transport equation. The DC parameter is the one given in [61], which was based on the DC parameter introduced with the YZβ shock-capturing, with the element length scale option introduced in [6].

1.6 ST-IGA

The ST-IGA is the integration of the ST framework with isogeometric discretization, motivated by the success of NURBS meshes in spatial discretization [17, 25, 44, 67]. It was introduced in [18]. Computations with the ST-VMS and ST-IGA were first reported in [18] in a 2D context, with IGA basis functions in space for flow past an airfoil, and in both space and time for the advection equation. Using higher-order basis functions in time enables getting full benefit out of using higher-order basis functions in space (see the stability and accuracy analysis given in [18] for the advection equation).

The ST-IGA with IGA basis functions in time enables, as pointed out and demonstrated in [18, 20], a more accurate representation of the motion of the solid surfaces and a mesh motion consistent with that. It also enables more efficient temporal representation of the motion and deformation of the volume meshes, and more efficient remeshing. These motivated the development of the ST/NURBS Mesh Update Method (STNMUM) [20, 63]. The STNMUM has a wide scope that includes spinning solid surfaces. With the spinning motion represented by quadratic NURBS in time, and with sufficient number of temporal patches for a full rotation, the circular paths are represented exactly. A "secondary mapping" [18] enables also specifying a constant angular velocity for invariant speeds along the circular paths. The ST framework and NURBS in time also enable, with the ST-C, extracting a continuous representation from the computed data and, in large-scale computations, efficient data compression [23]. We will describe that in Sect. 1.9. The STNMUM and the ST-IGA with IGA basis functions in time have been used in many 3D computations. The classes of problems solved are flapping-wing aerodynamics for an actual locust [31], bioinspired MAVs [32] and wing-clapping [33], separation aerodynamics of spacecraft [2], aerodynamics of HAWTs [34] and VAWTs [19], thermo-fluid analysis of ground vehicles and their tires [6], thermo-fluid analysis of disk brakes [8], flow-driven string dynamics in turbomachinery [38], and flow analysis of turbocharger turbines [9].

The ST-IGA with IGA basis functions in space enables more accurate representation of the geometry and increased accuracy in the flow solution. It accomplishes that with fewer control points, and consequently with larger effective element sizes. That in turn enables using larger time-step sizes while keeping the Courant number at a desirable level for good accuracy. It has been used in ST computational flow analysis of turbocharger turbines [9], flow-driven string dynamics in turbomachinery [38], ram-air parachutes [4], spacecraft parachutes [58], aortas [36], heart valves [37], tires with road contact and deformation [7], and fluid films [39]. Using IGA basis functions in space is now also a key part of some of the newest ZSS estimation methods [68] and related shell analysis [69].

For more on the ST-IGA, see [21]. In the computational flow analyses presented here, the ST-IGA enables more accurate representation of the ram-air parachute geometry, increased accuracy in the flow solution, and using larger time-step sizes. Integration of the ST-SI with the ST-IGA enables dealing with the fabric porosity of the ram-air parachute, and we will describe the ST-SI-IGA in Sect. 1.7.

1.7 ST-SI-IGA

The turbocharger turbine analysis [9] and flow-driven string dynamics in turbomachinery [38] were based on the integration of the ST-SI and ST-IGA. The IGA basis functions were used in the spatial discretization of the fluid mechanics equations and also in the temporal representation of the rotor and spinning-mesh motion. That enabled accurate representation of the turbine geometry and rotor motion and increased accuracy in the flow solution. The IGA basis functions were used also in the spatial discretization of the string structural dynamics equations. That enabled increased accuracy in the structural dynamics solution, as well as smoothness in the string shape and fluid dynamics forces computed on the string.

The ram-air parachute analysis [4] and spacecraft parachute compressible-flow analysis [58] were based on the integration of the ST-IGA, the ST-SI version that weakly enforces the Dirichlet conditions, and the ST-SI version that accounts for the porosity of a thin structure. The ST-IGA with IGA basis functions in space enabled, with relatively few number of unknowns, accurate representation of the parafoil and parachute geometries and increased accuracy in the flow solution. The volume mesh needed to be generated both inside and outside the parafoil. Mesh generation inside was challenging near the trailing edge because of the narrowing space. The spacecraft parachute has a very complex geometry, including gores and gaps. Using IGA basis functions addressed those challenges and still kept the element density near the trailing edge of the parafoil and around the spacecraft parachute at a reasonable level. In the heart valve analysis [37], the ST-SI-IGA, beyond enabling a more accurate representation of the geometry and increased accuracy in the flow solution, kept the element density in the narrow spaces near the leaflet contact areas at a reasonable level. In computational analysis of flow around tires with road contact and deformation [7], the ST-SI-IGA enables a more accurate representation of the geometry and motion of the tire surfaces, a mesh motion consistent with that, and increased accuracy in the flow solution. It also keeps the element density in the tire grooves and in the narrow spaces near the contact areas at a reasonable level. In addition, we benefit from the mesh generation flexibility provided by using SIs.

An SI provides mesh generation flexibility by accurately connecting the two sides of the solution computed over nonmatching meshes. This type of mesh generation flexibility is especially valuable in complex-geometry flow computations with isogeometric discretization, removing the matching requirement between the NURBS patches without loss of accuracy. This feature was used in the flow analysis of heart valves [37], turbocharger turbines [9], and spacecraft parachute compressible-flow analysis [58].

For more on the ST-SI-IGA, see [4]. In the computations presented here, the ST-SI-IGA is used for the reasons given and as described in the first paragraph of this section.

1.8   MDM

The MDM [22] was introduced for flow computations where the purpose is to predict the long-wake flow generated by a primary object and, in some cases, also to determine the influence of this wake flow on a secondary object placed far downstream. In the MDM, the problem domain is divided into a sequence of overlapping subdomains. The primary object is placed in the primary subdomain. The subsequent subdomains are used for computing the long-wake flows and flow past secondary objects. The inflow-boundary condition for the primary subdomain is the free-stream velocity. The inflow-boundary condition for each subsequent subdomain is the velocity extracted from the subdomain preceding it. If the outflow boundary of a subsequent subdomain is also within the subdomain preceding it, then the stress condition there is also extracted from the preceding subdomain. Computations over subdomains with no object can be carried out with special, structured meshes or special flow solvers that take into account the special nature of the mesh or with completely different flow solvers.

The 3D applications of the MDM method included flow around a small wing placed in the wake of a larger wing [22], flow in the wake of a circular cylinder up to 300 diameters downstream [70], aerodynamics [71], and FSI [72] of a parachute crossing the far wake of an aircraft. In the case of the cylinder problem, at Reynolds number 140, it was shown that with the MDM the computations can be extended sufficiently downstream, and with sufficient accuracy, to successfully capture the second phase of the Karman vortex street observed in laboratory experiments. In the case of the parachute crossing the aircraft wake, the computations were based on the DSD/SST method, and the subdomain containing the parachute was fully inside the subdomain preceding it.

In the flow analyses presented here, we use a spatially multiscale version of the MDM in the thermo-fluid analysis of a freight truck and its rear set of tires. The full global domain serves as the primary subdomain, and the local domain containing the set of rear tires serves as the secondary subdomain. In this case the secondary subdomain is fully inside the primary subdomain. The thermo-fluid computation over the global domain with a reasonable mesh refinement is followed by a higher-resolution computation over the local domain, with the boundary and initial conditions coming from the data computed over the global domain. The large time-history data from the global computation is stored using the ST-C, which we will explain in Sect. 1.9. The MDM is also used in the aerodynamic and FSI analysis of two back-to-back wind turbines in ABL flow.

1.9 ST-C

The ST-C [23], which serves here as a data compression method, is based on continuous temporal representation of the computed data using NURBS basis functions, with the letter "C" indicating "continuous." As we compute the flow field, we store the computed time-dependent data with the ST-C. With the ST-C, we can represent the data with fewer temporal control points, resulting in reduced computer storage cost. In one of the two ST-C versions introduced in [23], the continuous representation is extracted by projection from a solution already computed. Because we use a successive-projection technique (SPT), with a small number of temporal NURBS basis functions at each projection, the extraction can take place as the original solution is being computed, without the need to first complete the computation and store all that data. This version was named "ST-C-SPT" in [23].

In the flow analyses presented here, the ST-C-SPT is used in the thermo-fluid analysis of a freight truck and its rear set of tires. The large time-history data from the thermo-fluid computation over the global domain of the MDM is stored using the ST-C-SPT. The stored data is used in the thermo-fluid computation over the local domain containing the rear set of tires.

1.10   Examples of the Challenging Computations Performed

1.10.1   Aerodynamic Analysis of a Ram-Air Parachute

This computation is from [4]. A ram-air parachute is a parafoil inflated by the airflow through the inlets at the leading edge. The parafoil behaves like a wing and has better control and gliding capability compared to a round parachute. Its usage is quite common in sports parachuting and special-purpose parachuting that requires good gliding control and landing precision. Their usage is less common at larger sizes, and experience with their design, testing and performance evaluation becomes less and less as the size increases. Wind tunnel testing is not an option for very large ram-air parachutes, and drop tests would be very costly. That generated a demand for computational analysis and motivated the development of methods for reliable analysis (see, for example, [41]).

Reliable analysis of ram-air parachutes, at any practical size, involves a number of computational challenges. They include accurate representation of the parafoil geometry, fabric porosity and the complex, multiscale flow behavior encountered in this class of problems. The FSI between the parachute and the airflow is another computational challenge, with the challenge level increasing with the parachute size. Ram-air parachute computations were the earliest reported 3D, coupled parachute aerodynamics and parachute dynamics computations [73] with the ST-SUPS, and among the earliest reported 3D parachute FSI computations [41] with the ST-SUPS.

Here we use the ST-VMS and ST-SI-IGA. We use a special-purpose NURBS mesh generation techniques for the parachute structure and the flow field inside and outside the parafoil. The special-purpose mesh generation techniques enable NURBS representation of the structure and fluid domains with significant geometric complexity. The test computations we present from [4] are for building a starting parachute shape and a starting flow field associated with that parachute shape, which are the first two key steps in FSI analysis.

1.10.2   Thermo-Fluid Analysis of a Freight Truck and Its Rear Set of Tires

This computation is from [6]. Increasing the accuracy in calculating the heat transfer rates from the tires is the main objective. The multiscale challenges are due to the turbulent nature of the flow and due to the tires being rather small compared to the entire truck. The thermo-fluid-structure analysis of a tire is very complex. Here, we assume that the tire temperature is given. This assumption is justified because the tire temperature depends on the driving history, which represents a much longer time scale compared to the time scale of the surrounding air. To make the point, the truck body is 12 m long, and at a driving speed of 80 km/h, a fluid particle takes only 0.54 s to travel the full length of the truck. With that, we can decouple the problem into thermo-fluid analysis and tire thermo-structure analysis. Here we focus only on the thermo-fluid part.

In our thermo-fluid analysis, the road-surface temperature is higher than the free-stream temperature, and the tire-surface temperature is even higher. The analysis includes the heat from the engine and exhaust system. This is done with a reasonably realistic representation of the rate by which that heat transfer takes place and the surface geometry of the engine and exhaust system over which the heat transfer takes place. The analysis also includes the heave motion of the truck body, prescribed as a periodic motion with a given semi-amplitude and frequency.

1.10.3 Aerodynamic and FSI Analysis of Two Back-to-Back HAWTs in Turbulent ABL Flow

This computation is from [74]. To obtain high-fidelity predictive simulation results for wind turbines, 3D modeling is essential. However, simulation of wind turbines at full scale engenders a number of challenges. The flow is fully turbulent, requiring highly accurate methods and increased grid resolution. The presence of boundary layers, where turbulence is created, complicates the situation further. Wind-turbine blades are long and slender structures, with complex distribution of material properties, for which the numerical approach must have good efficiency and approximation power, and avoid locking. Wind-turbine simulations involve moving and stationary components, and the fluid–structure coupling must be accurate, efficient, and robust to preclude divergence of the computations.

Additional modeling challenges stem from realistic scenarios of wind turbines arranged in arrays, and operating in complex turbulent ABL flows with a wide range of energy-containing scales and in different atmospheric stability regimes. Wind turbines positioned downstream operate in the wakes generated by upstream turbines, and have been observed to generate less power compared to the upstream turbines. In addition, downstream turbines experience higher variations in aerodynamic loads, which tend to shorten their fatigue life, leading to premature blade failure. Depending on the atmospheric stability regime, spacing between turbines, the underlying surface topology, turbulence intensity, and wind direction and speed, the power-generation deficit for the downstream turbines may be as high as 40%.

We adopt the MDM technique to carry out the aerodynamic and FSI simulations of two full-scale, back-to-back HAWTs operating in a stably stratified ABL. The simulations produce novel data for the rotor structural response as it operates in shear flow induced by thermal stratification. The simulations also clearly show the evolution of the upstream-turbine wake leading to a velocity deficit responsible for a 15% drop in the downstream-turbine efficiency.

1.11   Outline of the Remaining Sections

We provide the governing equations in Sect. 2. The thermo-fluid ST-VMS is described in Sect. 3, the ALE-VMS in Sect. 4, and some of the special methods used in the computations in Sect. 5. In Sect. 6 we provide some brief comments on the parallel computations. In Sects. 7 and 8, as examples of ST computations, we present aerodynamic analysis of a ram-air parachute and thermo-fluid analysis of a freight truck and its rear set of tires. In Sect. 9, as an example of ALE computations, we present the aerodynamic and FSI simulations of two full-scale, back-to-back HAWTs operating in a stably stratified ABL. The concluding remarks are given in Sect. 10.

2   Governing Equations

The Navier–Stokes equations of incompressible flows with thermal coupling and Boussinesq approximation and the thermal-transport (energy) equation can be written on the spatial domain Ω_t as

$$\rho \left(\frac{\partial \mathbf{u}}{\partial t} + \mathbf{u} \cdot \nabla \mathbf{u} - \mathbf{f} \right) - \nabla \cdot \boldsymbol{\sigma} = \mathbf{0}, \tag{1}$$

$$\nabla \cdot \mathbf{u} = 0, \tag{2}$$

$$\rho C_{\mathrm{p}} \left(\frac{\partial \theta}{\partial t} + \mathbf{u} \cdot \nabla \theta \right) - \nabla \cdot (\kappa \nabla \theta) = 0, \tag{3}$$

where

$$\rho \mathbf{f} = \rho \left(1 - \beta_\theta \left(\theta - \theta_{\mathrm{ref}} \right) \right) \mathbf{a}_{\mathrm{GRAV}}. \tag{4}$$

In the momentum equation, ρ, $\mathbf{u}$, and $\mathbf{f}$ are the density, velocity, and body force. The stress tensor $\boldsymbol{\sigma}(\mathbf{u}, p) = -p\mathbf{I} + 2\mu\boldsymbol{\varepsilon}(\mathbf{u})$, where p is the pressure, $\mathbf{I}$ is the identity tensor, $\mu = \rho\nu$ is the viscosity, ν is the kinematic viscosity, and the strain rate $\boldsymbol{\varepsilon}(\mathbf{u}) = \left(\nabla \mathbf{u} + (\nabla \mathbf{u})^T \right)/2$. In the energy equation, C_{p}, θ, and κ are the constant-pressure specific heat, temperature, and thermal conductivity. In the expression for the body force, β_θ, θ_{ref}, and $\mathbf{a}_{\mathrm{GRAV}}$ are the thermal-expansion coefficient, reference temperature, and gravitational acceleration. In this mathematical model, ρ and C_{p} are assumed to be constants.

The essential and natural boundary conditions associated with Eq. (1) are represented as $\mathbf{u} = \mathbf{g}$ on $(\Gamma_t)_g$ and $\mathbf{n} \cdot \boldsymbol{\sigma} = \mathbf{h}$ on $(\Gamma_t)_h$, where $(\Gamma_t)_g$ and $(\Gamma_t)_h$ are complementary subsets of the boundary Γ_t, $\mathbf{n}$ is the unit outward normal vector, and $\mathbf{g}$ and $\mathbf{h}$ are given functions. The essential and natural boundary conditions associated with Eq. (3) are represented as $\theta = g_\theta$ on $(\Gamma_t)_{g_\theta}$, and $\kappa\mathbf{n} \cdot \nabla\theta = \mathbf{q}$ on $(\Gamma_t)_{h_\theta}$, where $(\Gamma_t)_{g_\theta}$ and $(\Gamma_t)_{h_\theta}$ are complementary subsets of the boundary Γ_t, and g_θ and $\mathbf{q}$ are given functions.

Remark 1 If the "1" in Eq. (4) is omitted, then p represents the pressure after the static-fluid part at θ_{ref} is subtracted.

In deriving the multiscale ST formulation associated with Eqs. (1)–(3), we find it more convenient to start from the conservation-law form of the momentum and energy equations:

$$\frac{\partial(\rho\mathbf{u})}{\partial t} + \nabla \cdot (\mathbf{u}\rho\mathbf{u}) - \rho\mathbf{f} - \nabla \cdot \boldsymbol{\sigma} = \mathbf{0}, \tag{5}$$

$$\frac{\partial(\rho C_{\text{p}}\theta)}{\partial t} + \nabla \cdot \left(\mathbf{u}\rho C_{\text{p}}\theta\right) - \nabla \cdot (\kappa\nabla\theta) = 0. \tag{6}$$

2.1 Structural Mechanics

In this article we will not provide any of our formulations requiring fluid and structure definitions simultaneously; we will instead give reference to earlier journal articles where the formulations were presented. Therefore, for notation simplicity, we will reuse many of the symbols used in the fluid mechanics equations to represent their counterparts in the structural mechanics equations. To begin with, $\Omega_t \subset \mathbb{R}^{n_{\text{sd}}}$ and Γ_t will represent the structure domain and its boundary. The structural mechanics equations are then written, on Ω_t and $\forall t \in (0, T)$, as

$$\rho\left(\frac{\mathrm{d}^2\mathbf{y}}{\mathrm{d}t^2} - \mathbf{f}\right) - \nabla \cdot \boldsymbol{\sigma} = \mathbf{0}, \tag{7}$$

where $\mathbf{y}$ and $\boldsymbol{\sigma}$ are the displacement and Cauchy stress tensor. The essential and natural boundary conditions for Eq. (7) are represented as $\mathbf{y} = \mathbf{g}$ on $(\Gamma_t)_{\text{g}}$ and $\mathbf{n} \cdot \boldsymbol{\sigma} = \mathbf{h}$ on $(\Gamma_t)_{\text{h}}$. The Cauchy stress tensor can be obtained from

$$\boldsymbol{\sigma} = J^{-1}\mathbf{F}\mathbf{S}\mathbf{F}^T, \tag{8}$$

where $\mathbf{F}$ and J are the deformation gradient tensor and its determinant, and $\mathbf{S}$ is the second Piola–Kirchhoff stress tensor. It is obtained from the strain-energy density function φ as follows:

$$\mathbf{S} \equiv \frac{\partial\varphi}{\partial\mathbf{E}}, \tag{9}$$

where $\mathbf{E}$ is the Green–Lagrange strain tensor:

$$\mathbf{E} = \frac{1}{2}\left(\mathbf{C} - \mathbf{I}\right), \tag{10}$$

and $\mathbf{C}$ is the Cauchy–Green deformation tensor:

$$\mathbf{C} \equiv \mathbf{F}^T \cdot \mathbf{F}. \tag{11}$$

From Eqs. (9) and (10),

$$\mathbf{S} = 2 \frac{\partial \varphi}{\partial \mathbf{C}}. \tag{12}$$

3   Thermo-Fluid ST-VMS

The ST-VMS formulation of Eqs. (1)–(3) was derived in [6] starting from Eqs. (5), (2), and (6). The formulation is written as follows: find $\mathbf{u}^h \in (\mathcal{S}_u^h)_n$, $p^h \in (\mathcal{S}_p^h)_n$ and $\theta^h \in (\mathcal{S}_\theta^h)_n$, such that $\forall \; \mathbf{w}^h \in (\mathcal{V}_u^h)_n$, $q^h \in (\mathcal{V}_p^h)_n$ and $w_\theta^h \in (\mathcal{V}_\theta^h)_n$:

$$\int_{\Omega_n} (\mathbf{w}^h)_n^+ \cdot \rho \left((\mathbf{u}^h)_n^+ - (\mathbf{u}^h)_n^- \right) d\Omega + \int_{Q_n} \mathbf{w}^h \cdot \rho \left(\frac{\partial \mathbf{u}^h}{\partial t} + \mathbf{u}^h \cdot \nabla \mathbf{u}^h - \mathbf{f}^h \right) dQ$$

$$- \int_{(P_n)_h} \mathbf{w}^h \cdot \mathbf{h}^h dP + \int_{Q_n} \nabla \mathbf{w}^h : \sigma(\mathbf{u}^h, p^h) dQ + \int_{Q_n} q^h \nabla \cdot \mathbf{u}^h dQ$$

$$+ \int_{\Omega_n} (w_\theta^h)_n^+ \rho C_{\mathrm{p}} \left((\theta^h)_n^+ - (\theta^h)_n^- \right) d\Omega + \int_{Q_n} w_\theta^h \rho C_{\mathrm{p}} \left(\frac{\partial \theta^h}{\partial t} + \mathbf{u}^h \cdot \nabla \theta^h \right) dQ$$

$$- \int_{(P_n)_{h_\theta}} w_\theta^h \mathbf{q}^h dP + \int_{Q_n} \nabla w_\theta^h \cdot \kappa \nabla \theta^h dQ$$

$$+ \sum_{e=1}^{(n_{\mathrm{el}})_n} \int_{Q_n^e} \frac{\tau_{\mathrm{SUPS}}}{\rho} \left(\rho \left(\frac{\partial \mathbf{w}^h}{\partial t} + \mathbf{u}^h \cdot \nabla \mathbf{w}^h \right) + \nabla q^h \right) \cdot \mathbf{r}_{\mathrm{M}} \left(\mathbf{u}^h, p^h, \theta^h \right) dQ$$

$$+ \sum_{e=1}^{(n_{\mathrm{el}})_n} \int_{Q_n^e} \rho \nu_{\mathrm{LSIC}} \nabla \cdot \mathbf{w}^h r_{\mathrm{C}} \left(\mathbf{u}^h \right) dQ$$

$$- \sum_{e=1}^{(n_{\mathrm{el}})_n} \int_{Q_n^e} \frac{(\tau_{\mathrm{SUPG}})_\theta}{C_{\mathrm{p}}} w^h \cdot \beta_\theta(\theta^h) r_{\mathrm{E}} \left(\mathbf{u}^h, \theta^h \right) \mathbf{a}_{\mathrm{GRAV}} dQ$$

$$- \sum_{e=1}^{(n_{\mathrm{el}})_n} \int_{Q_n^e} \tau_{\mathrm{SUPS}} \mathbf{w}^h \cdot \left(\mathbf{r}_{\mathrm{M}} \left(\mathbf{u}^h, p^h, \theta^h \right) \cdot \nabla \mathbf{u}^h \right) dQ$$

$$- \sum_{e=1}^{(n_{\mathrm{el}})_n} \int_{Q_n^e} \frac{\tau_{\mathrm{SUPS}}^2}{\rho} \nabla \mathbf{w}^h : \mathbf{r}_{\mathrm{M}} \left(\mathbf{u}^h, p^h, \theta^h \right) \mathbf{r}_{\mathrm{M}} \left(\mathbf{u}^h, p^h, \theta^h \right) dQ$$

$$+ \sum_{e=1}^{(n_{\mathrm{el}})_n} \int_{Q_n^e} (\tau_{\mathrm{SUPG}})_\theta \left(\frac{\partial w_\theta^h}{\partial t} + \mathbf{u}^h \cdot \nabla w_\theta^h \right) r_{\mathrm{E}} \left(\mathbf{u}^h, \theta^h \right) \mathrm{d}Q$$

$$- \sum_{e=1}^{(n_{\mathrm{el}})_n} \int_{Q_n^e} \tau_{\mathrm{SUPS}} C_{\mathrm{p}} w_\theta^h \left(\mathbf{r}_{\mathrm{M}} \left(\mathbf{u}^h, p^h, \theta^h \right) \cdot \nabla \theta^h \right) \mathrm{d}Q$$

$$- \sum_{e=1}^{(n_{\mathrm{el}})_n} \int_{Q_n^e} \frac{\tau_{\mathrm{SUPS}} (\tau_{\mathrm{SUPG}})_\theta}{\rho} \mathbf{r}_{\mathrm{M}} \left(\mathbf{u}^h, p^h, \theta^h \right) \cdot \nabla w_\theta^h r_{\mathrm{E}} \left(\mathbf{u}^h, \theta^h \right) \mathrm{d}Q = 0.$$

$$(13)$$

Here Q_n is the slice of the ST domain between the time levels n and $n+1$, P_n is the lateral boundary of Q_n, $(P_n)_g$ and $(P_n)_h$ are the complementary subsets of P_n for the momentum equation, $(P_n)_{g_\theta}$ and $(P_n)_{h_\theta}$ are the complementary subsets of P_n for the energy equation, and $(\mathcal{S}_u^h)_n$, $(\mathcal{S}_p^h)_n$, $(\mathcal{S}_\theta^h)_n$, $(\mathcal{V}_u^h)_n$, $(\mathcal{V}_p^h)_n$, and $(\mathcal{V}_\theta^h)_n$ are the finite-dimensional ST trial and test function spaces. The functions in these spaces are continuous within a ST slab, but discontinuous from one ST slab to another, and the subscript n implies that corresponding to different ST slabs we might have different discretizations. The notation $(\cdot)_n^-$ and $(\cdot)_n^+$ denotes the function values at t_n as approached from below and above. Each Q_n is decomposed into elements Q_n^e, where $e = 1, 2, \ldots, (n_{\mathrm{el}})_n$, and the subscript n used with n_{el} is for the general case where the number of ST elements may change from one ST slab to another.

The residuals are defined as

$$\mathbf{r}_{\mathrm{M}}(\mathbf{u}, p, \theta) = \rho \left(\frac{\partial \mathbf{u}}{\partial t} + \mathbf{u} \cdot \nabla \mathbf{u} - \mathbf{f} \right) - \nabla \cdot \boldsymbol{\sigma}(\mathbf{u}, p), \tag{14}$$

$$r_{\mathrm{C}}(\mathbf{u}) = \nabla \cdot \mathbf{u}, \tag{15}$$

$$r_{\mathrm{E}}(\mathbf{u}, \theta) = \rho C_{\mathrm{p}} \left(\frac{\partial \theta}{\partial t} + \mathbf{u} \cdot \nabla \theta \right) - \nabla \cdot (\kappa \nabla \theta). \tag{16}$$

Remark 2 The 10th integration is the SUPG/PSPG stabilization.

Remark 3 The 11th integration is the LSIC stabilization, with the notation "LSIC," introduced in [75], denoting the stabilization based on least-squares on incompressibility constraint.

Remark 4 Because β_θ is a function of temperature, in general it will have a fine-scale component in the VMS formulation. The simplified form seen in the 12th integration in Eq. (13) was reached by dropping that fine-scale component. In our current computations, we simplify the model even more and just drop that term. For the more general case, see [6].

Remark 5 The 13th and 14th integrations are for the cross and Reynolds stresses.

Remark 6 The 15th integration is the SUPG stabilization for the energy equation.

Remark 7 The 16th and 17th integrations represent the interaction between the energy equation and the fine-scale velocity.

Remark 8 If we exclude the 13th, 14th, 15th, and 17th integrations, the method reduces to the thermo-fluid ST-SUPS, which is the ST version of the thermo-fluid ALE-SUPS method given in [61].

There are various ways of defining the stabilization parameters τ_{SUPS}, ν_{LSIC}, and $(\tau_{SUPG})_\theta$. The stabilization-parameter definitions used in the computations reported in this article are related to those given in [6, 60, 61]. The precise definitions can be found from the references cited in the sections where those computations are described. For more ways of calculating the stabilization parameters in finite element computation of flow problems, see [7, 62, 64]).

4   ALE-VMS

The ALE-VMS formulation of stratified incompressible flows is posed on a spatial domain Ω that is discretized into elements Ω^e. While $\{\Omega^e\}$, Ω, and its boundary Γ are time-dependent, when there is no risk of confusion, we drop the subscript t to simplify notation. The superscript h indicates association with discrete function spaces defined over Ω, which moves with the velocity $\hat{\mathbf{u}}^h$. The semi-discrete formulation is given as

$$
\int_\Omega \mathbf{w}^h \cdot \rho \left(\left. \frac{\partial \mathbf{u}^h}{\partial t} \right|_{\hat{x}} + \left(\mathbf{u}^h - \hat{\mathbf{u}}^h \right) \cdot \nabla \mathbf{u}^h - \mathbf{f}^h \right) d\Omega
$$

$$
- \int_\Omega \mathbf{w}^h \cdot \mathbf{b}^h \, d\Omega + \int_\Omega \boldsymbol{\varepsilon} \left(\mathbf{w}^h \right) : \boldsymbol{\sigma} \left(\mathbf{u}^h, p^h \right) d\Omega
$$

$$
- \int_\Gamma \mathbf{w}^h \cdot \mathbf{h}^h \, d\Gamma + \int_\Omega q^h \nabla \cdot \mathbf{u}^h \, d\Omega
$$

$$
+ \int_\Omega w_\theta^h \left(\left. \frac{\partial \theta^h}{\partial t} \right|_{\hat{x}} + \left(\mathbf{u}^h - \hat{\mathbf{u}}^h \right) \cdot \nabla \theta^h - f^h \right) d\Omega
$$

$$
- \int_\Omega \nabla w_\theta^h \cdot \nu_\theta \nabla \theta^h \, d\Omega - \int_\Gamma w_\theta^h h^h \, d\Gamma
$$

$$
+ \sum_{e=1}^{n_{el}} \int_{\Omega^e} \tau_{SUPS} \left(\left(\mathbf{u}^h - \hat{\mathbf{u}}^h \right) \cdot \nabla \mathbf{w}^h + \frac{\nabla q^h}{\rho} \right) \cdot \mathbf{r}_M \left(\mathbf{u}^h, p^h, \theta^h \right) d\Omega
$$

$$
+ \sum_{e=1}^{n_{el}} \int_{\Omega^e} \rho \nu_{LSIC} \nabla \cdot \mathbf{w}^h r_C(\mathbf{u}^h) \, d\Omega
$$

$$-\sum_{e=1}^{n_{el}} \int_{\Omega^e} \tau_{\text{SUPS}} \mathbf{w}^h \cdot \left(\mathbf{r}_{\text{M}} \left(\mathbf{u}^h, p^h, \theta^h \right) \cdot \nabla \mathbf{u}^h \right) \, d\Omega$$

$$-\sum_{e=1}^{n_{el}} \int_{\Omega^e} \frac{\nabla \mathbf{w}^h}{\rho} : \left(\tau_{\text{SUPS}} \mathbf{r}_{\text{M}} \left(\mathbf{u}^h, p^h, \theta^h \right) \right) \otimes \left(\tau_{\text{SUPS}} \mathbf{r}_{\text{M}} \left(\mathbf{u}^h, p^h, \theta^h \right) \right) \, d\Omega$$

$$+\sum_{e=1}^{n_{el}} \int_{\Omega^e} \tau_{\text{SUPG}} \left(\mathbf{u}^h - \hat{\mathbf{u}}^h \right) \cdot \nabla w_\theta^h r_{\text{E}} \left(\mathbf{u}^h, \theta^h \right) d\Omega = 0. \tag{17}$$

Here $\mathbf{r}_{\text{M}}$, r_{C}, and r_{E} are the residuals of the momentum, continuity, and temperature equations, given as

$$\mathbf{r}_{\text{M}}(\mathbf{u}, p, \theta) = \rho \left(\left. \frac{\partial \mathbf{u}}{\partial t} \right|_{\hat{x}} + (\mathbf{u} - \hat{\mathbf{u}}) \cdot \nabla \mathbf{u} - \mathbf{f} \right) - \mathbf{b} - \nabla \cdot \boldsymbol{\sigma} (\mathbf{u}, p), \tag{18}$$

$$r_{\text{C}}(\mathbf{u}) = \nabla \cdot \mathbf{u}, \tag{19}$$

$$r_{\text{E}}(\mathbf{u}, \theta) = \left. \frac{\partial \theta}{\partial t} \right|_{\hat{x}} + (\mathbf{u} - \hat{\mathbf{u}}) \cdot \nabla \theta - \nabla \cdot (\nu_\theta \nabla \theta) - f, \tag{20}$$

and $\left. \frac{\partial (\cdot)}{\partial t} \right|_{\hat{\mathbf{x}}}$ is the time derivative taken with respect to the fixed reference coordinates $\hat{\mathbf{x}}$ of the spatial configuration.

The Boussinesq forcing term $\mathbf{b}$ in the second line of Eq. (17) takes on the form

$$\mathbf{b} = \rho g \frac{\theta - \bar{\theta}}{\theta_0} \mathbf{e}_3, \tag{21}$$

where $\bar{\theta}$ is a prescribed background temperature field varying only in the x_3-direction (i.e., vertical direction), θ_0 is the reference temperature assumed constant in the Boussinesq approximation, g is the gravitational-acceleration magnitude, and $\mathbf{e}_3$ is the Cartesian basis vector pointing in the vertical direction. The diffusivity ν_θ in the fifth line of Eq. (17) is given by

$$\nu_\theta = \frac{\kappa}{\rho C_{\text{p}}}. \tag{22}$$

Also in Eq. (17), τ_{SUPS}, ν_{LSIC}, and τ_{SUPG} are the stabilization parameters defined in [17] as

$$\tau_{\text{SUPS}} = \left(\frac{4}{\Delta t^2} + \left(\mathbf{u}^h - \hat{\mathbf{u}}^h \right) \cdot \mathbf{G} \left(\mathbf{u}^h - \hat{\mathbf{u}}^h \right) + C_I \nu^2 \mathbf{G} : \mathbf{G} \right)^{-1/2}, \tag{23}$$

$$\nu_{\text{LSIC}} = (\text{tr}\mathbf{G}\, \tau_{\text{SUPS}})^{-1}, \tag{24}$$

$$\tau_{\text{SUPG}} = \left(\frac{4}{\Delta t^2} + \left(\mathbf{u}^h - \hat{\mathbf{u}}^h \right) \cdot \mathbf{G} \left(\mathbf{u}^h - \hat{\mathbf{u}}^h \right) + C_I \nu_\theta^2 \mathbf{G} : \mathbf{G} \right)^{-1/2}, \tag{25}$$

where $\mathbf{G}$ is the element metric tensor and

$$\text{tr}\mathbf{G} = \sum_{i=1}^{d} G_{ii}, \tag{26}$$

Δt is the time-step size, and C_I is a positive constant, independent of the mesh size, derived from an appropriate element-wise inverse estimate (see, e.g., [76]).

Remark 9 In ABL simulations, the Earth rotation effects may be important. For this, the Coriolis force is added to the momentum-balance equation and takes the form

$$\mathbf{f} = f_c \epsilon_{ij3} u_j \mathbf{e}_i, \tag{27}$$

where f_c is the Coriolis parameter, ϵ_{ijk}'s are the Cartesian components of the alternator tensor, and $\mathbf{e}_i$ is the ith Cartesian basis vector.

Remark 10 In the VMS framework, coupling between Navier–Stokes and temperature equations brought about by the Boussinesq approximation gives rise to additional modeling terms. In particular, it can be shown that the x_3-component of the linear-momentum equation and incompressibility constraint are coupled with the residual of the advection–diffusion equation, and the following terms are added to the left-hand side of Eq. (17):

$$+ \sum_{e=1}^{n_{\text{el}}} \int_{\Omega^e} \left(\left(\mathbf{u}^h - \hat{\mathbf{u}}^h \right) \cdot \nabla w_3^h + \frac{1}{\rho} \frac{\partial q^h}{\partial x_3} \right) \bar{\tau} r_{\text{E}}(\mathbf{u}^h, \theta^h) \, d\Omega. \tag{28}$$

The stabilization parameter $\bar{\tau}$ may be obtained following the developments in stabilized methods for advective–diffusive systems presented in [77–79], which gives the following expression for $\bar{\tau}$:

$$\bar{\tau} = -\frac{a_2}{a_1 \sqrt{a_3} + a_3 \sqrt{a_1}}, \tag{29}$$

where a_i's are given as

$$a_1 = \frac{4}{\Delta t^2} + \left(\mathbf{u}^h - \hat{\mathbf{u}}^h\right) \cdot \mathbf{G}\left(\mathbf{u}^h - \hat{\mathbf{u}}^h\right) + C_I \nu^2 \mathbf{G} : \mathbf{G},$$

$$a_2 = \frac{4}{\Delta t}\frac{\rho g}{\theta_0},$$

$$a_3 = \frac{4}{\Delta t^2} + \left(\mathbf{u}^h - \hat{\mathbf{u}}^h\right) \cdot \mathbf{G}\left(\mathbf{u}^h - \hat{\mathbf{u}}^h\right) + C_I \nu_\theta^2 \mathbf{G} : \mathbf{G}. \tag{30}$$

Although the numerical examples presented in this article do not make use of these additional terms, a recent study of stratified turbulent flows [54] showed that these additional VMS modeling terms can appreciably improve the performance of the ALE-VMS for this class of problems.

5 Special Computational Methods

5.1 *YZβ DC*

We add to the formulation given by Eq. (13) the following DC term:

$$\int_{Q_n} \nabla w_\theta^h \cdot \boldsymbol{\kappa}_{\mathrm{DC}} \cdot \nabla \theta^h \mathrm{d}Q, \tag{31}$$

where

$$\boldsymbol{\kappa}_{\mathrm{DC}} = \rho C_{\mathrm{p}}(\nu_{\mathrm{DC}})_\theta \mathbf{I}. \tag{32}$$

The DC parameter is the one given in [61] based on the DC parameter introduced with the YZβ shock-capturing [65]:

$$\rho C_{\mathrm{p}}(\nu_{\mathrm{DC}})_\theta = \frac{|Z|}{Y} \frac{Y^2}{\left\|\nabla \theta^h\right\|^2} \left(\left(\frac{(h_{\mathrm{RGN}})_\theta}{2}\right)^2 \frac{\left\|\nabla \theta^h\right\|^2}{Y^2}\right)^{\frac{\beta}{2}}, \tag{33}$$

where Y is a reference value for θ, and $Z = r_{\mathrm{E}}\left(\mathbf{u}^h, \theta^h\right)$. To avoid singularity when $\left\|\nabla \theta^h\right\| = 0$, we modify the expression as

$$\rho C_{\mathrm{p}}(\nu_{\mathrm{DC}})_\theta = \frac{|Z|}{Y} \frac{Y^2}{\left\|\nabla \theta^h\right\|^2 + \epsilon_\theta \frac{4Y^2}{(h_{\mathrm{RGN}})_\theta^2}} \left(\left(\frac{(h_{\mathrm{RGN}})_\theta}{2}\right)^2 \frac{\left\|\nabla \theta^h\right\|^2}{Y^2}\right)^{\frac{\beta}{2}}, \tag{34}$$

where ϵ_θ is a small nondimensional number. For the definition of the element length $(h_{\mathrm{RGN}})_\theta$, see [6].

5.2 ST-SI Version for Porosity Modeling

In the ST-SI version where the SI is between a thin porous structure and the fluid on its two sides (see Sect. 1.3), the porosity velocity is expressed as a function of the pressure difference between the two sides of the SI:

$$u_{\text{PORO}}^h = -\mathcal{P}(\Delta p^h), \tag{35}$$

$$\Delta p^h \equiv p_{\text{A}}^h - p_{\text{B}}^h. \tag{36}$$

The velocity in $\mathbf{n}_{\text{B}}$ direction is defined as the positive porosity velocity. The expression $\mathcal{P}(\Delta p^h)$ can take different forms depending on the nature of the porosity. Here we use the form

$$\mathcal{P}(\Delta p^h) = k_{\text{PORO}} \Delta p^h, \tag{37}$$

where k_{PORO} is the porosity coefficient (see [1]). The normal component of the velocity is assumed to be continuous, and the tangential component is set to the tangential component of $\mathbf{u}_{\text{S}}^h$, where $\mathbf{u}_{\text{S}}^h$ is the structure velocity, and the volume flux is imposed with the porosity velocity taken into account (see [4] for the complete set of equations).

5.3 Spatially Multiscale MDM

In the spatially multiscale version of the MDM we have here, the full global domain serves as the primary subdomain, and the local domain containing the rear set of tires serves as the secondary subdomain. In this version, the secondary subdomain is fully inside the primary subdomain. First the thermo-fluid computation is carried out over the global domain, with a reasonable mesh refinement. The inflow-boundary conditions are the free-stream velocity and temperature, the outflow-boundary conditions are zero stress and zero normal heat flux, and the conditions at the top and side computational boundaries are zero normal velocity, zero tangential stress, and zero normal heat flux. The large amount of time-history data from the global computation is stored using the ST-C-SPT.

This is followed by a higher-resolution computation over the local domain. This gives us increased accuracy in the thermo-fluid analysis, including increased accuracy in the heat transfer rates from the tires. The boundary conditions at the inflow and top and side computational boundaries at each time step of the computation are the velocity and temperature extracted from the stored global data at the corresponding time. The extraction is based on evaluating the temporal NURBS representation of the velocity at that corresponding time. At the outflow boundary, the stress condition is extracted from the stored global data, and the normal heat flux is set to zero. In general the nodal points of the local-domain boundaries do not coincide with the nodal points of the global domain. Therefore, spatially, the data extraction is based on the least-squares projection. If a local-domain boundary coincides with the global-domain boundary, the boundary condition there is from the values specified for the global domain.

We note that the MDM gives us the option of using different formulations in the global and local computations. That includes the option of using different stabilization and DC parameters, and even different stabilization and DC methods.

5.4 ST-C

As we compute the flow field, the computed data is stored with the ST-C-SPT [23]. This method, which serves as an efficient data compression method in large-scale computations, is one of the versions of the ST-C method introduced in [23]. In the ST-C, the ST framework and NURBS in time are used in extracting a continuous representation from the computed data. With the ST-C, we can represent the data with fewer temporal control points, resulting in reduced computer storage cost. In the ST-C-SPT, the continuous representation is extracted by projection from the solution already computed. Because we use a successive-projection technique, with a small number of temporal NURBS basis functions at each projection, the extraction can take place as the original solution is being computed, without the need to first complete the computation and store all that data.

6 Parallel Computations

Parallel computations with the ST methods go as far back as 1992 [80], with the 3D computations reported as early as 1993 [81]. All computations reported in this chapter were carried out on parallel computing platforms. The number of cores used in a typical computation ranges from 96 to 576. Because the computations were mostly for the purpose of testing a new computational method, parallel efficiency was not a high priority. Still the efficiencies we see are high enough to justify the use of the maximum number of cores available in the computer resources we have.

7 ST Computation: Aerodynamic Analysis of a Ram-Air Parachute

This computation is from [4].

7.1 Structural Mechanics Computation

A ram-air parachute consists of three parts: canopy, suspension lines, and stabilizers. Figure 1 shows the parachute. The canopy size is approximately 8 m × 3 m. The canopy is made of fabric, which is modeled as membrane, with 19 airfoil-shaped ribs, 17 separating the air cells, and 2 at the ends. Figure 2 shows one of the ribs. The suspension lines, used by the parachutist to control the parachute, are modeled as cables. Fabric patches attached to the parachute sides serve as stabilizers.

Figure 3 shows the undeformed configuration, where the membrane parts consist of mostly flat patches. We note that in the configuration we selected, the suspension lines have not yet been reeled in.

Figure 4 shows the control mesh in the NURBS representation of the undeformed configuration. To represent such a complex shape or to add a cable attached to a surface, some control points coalesce (see Fig. 5). Because of the coalescing, at some element boundaries we have only C^0 continuity. The number of control points is 3296, with 2250 elements in the membrane parts and 222 elements in the cables. We use quadratic NURBS.

Fig. 1 Ram-air parachute

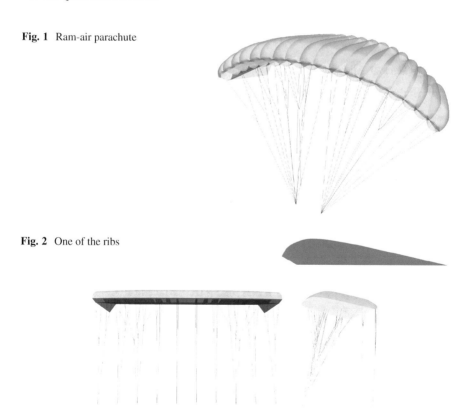

Fig. 2 One of the ribs

Fig. 3 Undeformed geometry. Front (*left*) and side (*right*) views

Fig. 4 Undeformed control mesh. Front (*left*) and side (*right*) views

Fig. 5 Undeformed control
mesh with *red* spheres
representing the control
points that coalesced

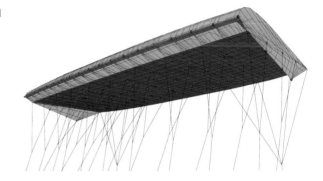

Table 1 Material properties

	Young's modulus (Pa)	Density (kg/m³)	Poisson's ratio	Thickness (in)	Cross-section area (mm²)
Membrane	3.8×10^8	5.0×10^2	0.3	3×10^{-3}	–
Cable	7.6×10^{10}	1.4×10^3	–	–	8.0

The structural mechanics formulation based on the membrane and cable models
(see [1]) is supplemented with wrinkling and slacking models (see [82]). The
material properties are given in Table 1.

In the computation, we specify the pressure difference between the two sides of
the parafoil surfaces and reel the ends of the suspension lines to the center. Figure 6
shows the pressure difference for the control points of the canopy structure mesh.
We have three different values, and they are, in Pa, 0, 94.1, and 117. The stress vector
is formed based on the control variables, using the surface normal and interpolated
value at each surface location.

The structural mechanics solution is symmetrized with respect to the central
vertical plane by averaging. In addition, we apply an upward body force to keep
the parachute in an upright position. The solution is obtained by computing with a
time-marching algorithm until a steady state is reached.

Figure 7 shows, for the steady-state solution, the control mesh and the surface
represented by that mesh.

Fig. 6 Pressure difference for the control points of the canopy structure mesh. Surface membranes have been removed in the *right* half of the picture to make the ribs visible. The values are, in Pa, 0 (*red*), 94.1 (*green*), and 117 (*blue*)

Fig. 7 Deformed configuration at the steady state. Control mesh and surface represented by that mesh

7.2 Fluid Mechanics Computations

The density and kinematic viscosity are 1.237 kg/m^3 and 1.449×10^{-5} m^2/s. The glide speed is 12.5 m/s. The computational-domain size is 100 m × 100 m × 100 m. The parachute is located at 30 m from the inflow boundary.

The surface mesh is the same as the canopy structural mechanics mesh. The volume mesh needs to be generated both inside and outside the parafoil. Mesh generation inside is challenging near the trailing edge because of the narrowing space. Using NURBS meshes for the fluid mechanics computation addresses that challenge. This keeps the element density near the trailing edge at a reasonable level. We create the volume mesh in two steps: first we generate a mesh using the undeformed parafoil shape, which is relatively easier, and then deform that mesh as the parafoil deforms in the structural mechanics computation. Figure 8 shows the mesh obtained in these two steps. The number of control points and elements are 149,568 and 233,378. We use quadratic NURBS.

To represent the pressure jump across a parafoil surface, the control variables on the surfaces have split values. The mesh deformation is computed with the Jacobian-based stiffening method [81]. Volume meshes for different values of the angle of attack (α), ranging from $-2.0°$ to $12°$, are obtained by deforming the mesh for $\alpha = 0°$, and the deformation is driven by the rotation of the parachute canopy from $\alpha = 0°$ to the other values of α.

At the inflow boundary we set the velocity, based on the glide speed, to 12.5 m/s, and at the outflow boundary we set the stress to zero. We use slip conditions at the

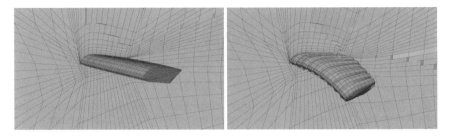

Fig. 8 Fluid mechanics control mesh before (*left*) and after (*right*) the structural deformation

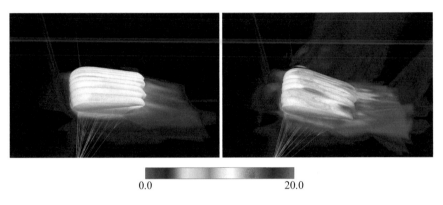

0.0   20.0

Fig. 9 Vorticity magnitude (s^{-1}) at an instant for $\alpha = 0°$ (*left*) and 12° (*right*)

lateral boundaries, and no-slip condition on the parafoil surfaces. We use the ST-VMS method. For the stabilization parameters used, see [4]. The time-step size is 5.34×10^{-3} s, and the number of nonlinear iterations per time step is 3. The number of GMRES [83] iterations per nonlinear iteration is 300.

Figure 9 shows the vorticity at an instant for $\alpha = 0°$ and 12°. Despite the coarseness of the meshes, the solutions are smooth and capture well the attached flow when $\alpha = 0°$ and the separated flow when $\alpha = 12°$. Figures 10 and 11 show the pressure coefficient at an instant for $\alpha = -2°$, 0°, 2°, 4°, 6°, 8°, 10°, and 12°. The picture plane is cutting the 9th cell from the right, roughly bisecting it. The scaling used in computing the pressure coefficient gives a value of 1.0 as the stagnation (i.e., maximum) pressure. Figure 12 shows the moment coefficient around the parachutist and the lift/drag ratio. For each α value, the data displayed was obtained by averaging from the last 2.5 s of the computation.

For $\alpha = 0°$, we also compute the flow field with porosity at all parafoil surfaces except for the top canopy surface. The porosity coefficient is 1.5 CFM. At the top canopy surface, we enforce the no-slip condition weakly. Figure 13 shows the vorticity at an instant. Figure 14 shows the normal component of the velocity at an instant.

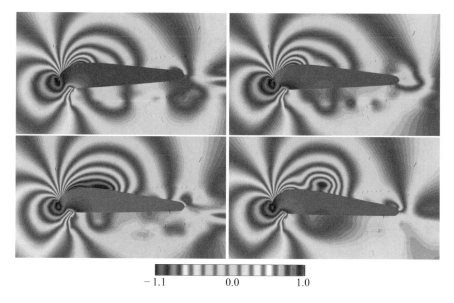

Fig. 10 Pressure coefficient at an instant for $\alpha = -2°, 0°, 2°$, and $4°$

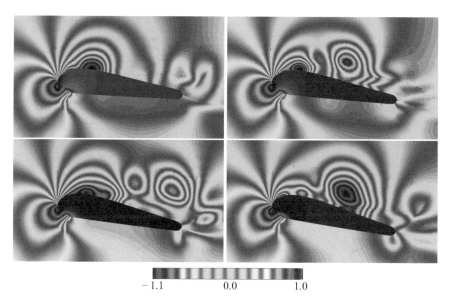

Fig. 11 Pressure coefficient at an instant for $\alpha = 6°, 8°, 10°$, and $12°$

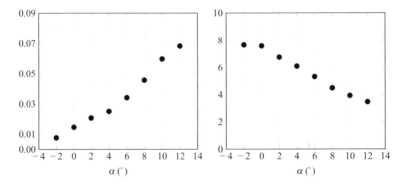

Fig. 12 Moment coefficient (*left*) and lift/drag ratio (*right*)

Fig. 13 Computation with
porosity. Vorticity magnitude
(s^{-1}) at an instant for $\alpha = 0°$

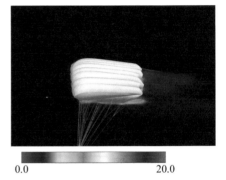

Fig. 14 Computation with
porosity. Normal component
of the velocity (m/s) at an
instant for $\alpha = 0°$

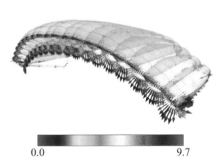

8 ST Computation: Thermo-Fluid Analysis of a Freight Truck and Its Rear Set of Tires

This computation is from [6].

8.1 Problem Setup

The model we use for the freight truck is shown in Fig. 15. The truck is 12.14 m long, 2.41 m wide, and 3.77 m high. The tire diameter, D_{tire}, is 1.07 m, and the tire width is 0.35 m. There are eight rear tires; the inner and outer tires are 0.01 m apart, and the leading and trailing tires are 0.24 m apart. The truck speed is 80 km/h and there is no wind, making the free-stream airflow speed relative to the truck, $\|\mathbf{u}_\infty\|$ = 80 km/h. This is also the speed of the road relative to the truck. In specifying the tire rotation, the angular velocity is calculated as $\omega = 2\|\mathbf{u}_\infty\|/D_{\text{tire}}$. The tires are axisymmetric in the model here, and therefore the tire surface does not need to be represented by a moving computational boundary. We take into account the heave motion of the truck body, prescribed as a periodic motion with a semi-amplitude of 10 mm and a frequency of 1.92 Hz.

The free-stream air temperature, $\theta_\infty = 30\,°C$, and the road-surface temperature, $\theta_{\text{road}} = 50\,°C$. The truck-surface temperature is the same as the free-stream air temperature, and the tire-surface temperature, $\theta_{\text{tire}} = 80\,°C$. The density, based on θ_∞, is 1.205 kg/m³. The viscosity is a function of temperature based on Sutherland's formula:

$$\mu = \mu_0 \frac{\theta_0 + \theta_S}{\theta + \theta_S} \left(\frac{\theta}{\theta_0} \right)^{3/2}, \tag{38}$$

where $\mu_0 = 1.716 \times 10^{-5}$ Pa · s, $\theta_0 = 273$ K, and Sutherland's temperature, $\theta_S = 111$ K. The thermal conductivity is calculated from the constant-pressure specific heat, viscosity, and Prandtl number, Pr:

$$\kappa = \frac{C_p}{\text{Pr}} \mu, \tag{39}$$

where $C_p = 1007$ J/(kg · K) and Pr = 0.71. The thermal-expansion coefficient, $\beta_\theta = 1/\theta$, the gravitational acceleration is 9.8 m/s², and $\theta_{\text{ref}} = \theta_\infty$.

Figure 16 shows the global domain and its dimensions. The inflow boundary is 6.01 m away from the front end of the truck. The conditions at the computational

Fig. 15 Truck model. The truck is 12.14 m long, 2.41 m wide, and 3.77 m high

Fig. 16 Global domain

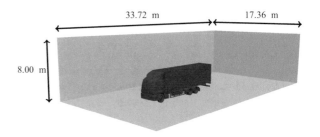

boundaries are specified as described in Sect. 5.3. On the tire, road, and truck surfaces, the velocity and temperature are specified. The exception is that, on the truck surfaces associated with the engine bottom and exhaust pipe, the normal heat flux is specified based on the heat loss rates given below.

The drag loss for the truck is about 50% of the total loss. Consequently, the required engine power, P, can be calculated as

$$P = \frac{\mathbf{F} \cdot \mathbf{u}_\infty}{50\%}, \tag{40}$$

where $\mathbf{F}$ is the drag force. From that, we can write

$$P = \frac{\frac{1}{2}\rho C_D S \|\mathbf{u}_\infty\|^3}{50\%}, \tag{41}$$

where C_D is the drag coefficient and S is the projected area of the truck. For C_D, we use an approximate value of 1.1, obtained from an earlier computation we carried out with a somewhat coarser mesh.

Assuming a thermal efficiency of 35%, the total heat generation is $P/(35\%)$, with 65% of that assumed to be lost from the exhaust system and from underneath the engine. The split between the two is 35% for the exhaust system and 30% for the engine. This split is based on the experimental data (mass flow rate and temperature at the engine exit) from a single-cylinder direct-injection diesel engine at Waseda University. The loss from the exhaust system is split into two: 10% from the exhaust pipe surface and 25% from the exhaust pipe end. This split is based on a simple heat transfer model along the pipe. We note that all percentages are based on the total heat generation. Figure 17 shows the three heat loss locations. These heat losses are used in specifying the boundary conditions on the engine lower surface, on the exhaust pipe surface, and at the exhaust pipe end. The boundary conditions are in terms of the normal heat flux on the engine lower surface and exhaust pipe surface, and the velocity and temperature at the exhaust pipe end, which are 18.9 m/s and 186 °C.

The local domain contains the four rear tires on the left side of the truck, and is positioned as shown in Fig. 18. Figure 19 shows the dimensions of the local domain. The boundary conditions are specified consistent with the procedure described in

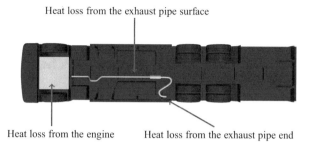

Fig. 17 Heat loss locations used in specifying the boundary conditions

Fig. 18 Local domain containing the four rear tires on the left side of the truck

Fig. 19 Dimensions of the local domain

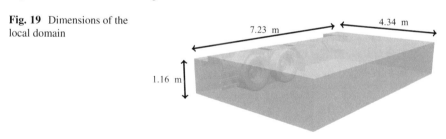

Sect. 5.3. The inflow boundary is 0.80 m away from the front end of the tires. The outflow boundary is 4.05 m away from the back end of the tires. The top boundary is 0.08 m away from the top of the tires. The bottom boundary is the road surface. The inner side boundary starts at the center plane of the truck and is 0.48 m away from the inner face of the tires. The outer side boundary is 3.14 m away from the outer face of the tires, and that leaves a distance of 4.34 m from the global-domain boundary. We note that according to the procedure described in Sect. 5.3, the boundary conditions on the tire, road, and truck surfaces are from the values specified for the global domain.

8.2 Computations and Results

The meshes are made of all tetrahedral elements. The number of nodes and elements
in the global and local meshes are given in Table 2. We have about 58 million
unknowns in the local-domain computation.

We have layers of refined mesh near the tire surface. The number of layers is
about 5 for the global mesh and 10 for the local mesh. The first-element thickness
in the normal direction near the tire is 10 mm for the global domain and 0.2 mm
for the local domain. In both meshes, for each layer the element thickness in the
normal direction is increasing with a progression ratio of 1.5. The mesh near the
tire is shown in Fig. 20. We note that the mesh in the space between the two closest
faces of the inner and outer tires is treated in a special way. In that narrow space,
there are at least 2 elements in the global computation and 12 elements in the local
computation.

Remark 11 To make the mesh generation simpler, we have a 13-mm gap between
the tire and the road surface.

We use the ST-SUPS in the global computation, and the ST-VMS in the local
computation. For the stabilization parameters used, see [6]. In calculation of the DC
parameter, in the global computation, $Y = \theta_{\text{tire}} - \theta_\infty = 50\,^\circ\text{C}$, $\beta = 1$, and $\epsilon_\theta = 10^{-20}$,
In the local computation, $Y = 50\,^\circ\text{C}$, $\beta = 2$, and $\epsilon_\theta = 10^{-20}$. The time-step size is
4.33×10^{-3} s in the global computation, and 1.08×10^{-3} s in the local computation.
The number of nonlinear iterations per time step is 3, and the number of GMRES
iterations per nonlinear iteration is 300. In storing the large amount of time-history
data from the global computation with the ST-C-SPT, we use approximately four
times less points in time, but the data representation is with C^2-continuous temporal
basis functions.

Table 2 Number of nodes
(*nn*) and elements (*ne*) in the
global and local meshes

	nn	*ne*
Global domain	918,753	4,857,973
Local domain	5,809,813	32,986,249

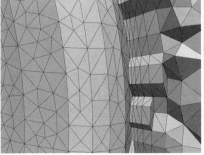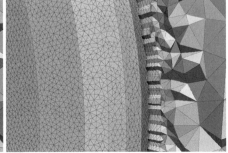

Fig. 20 Mesh resolution in the tangential and normal directions for the global and local domains

In representing the heave motion, we use cubic B-splines in time, with five control points for the cycle. There are only three independent control meshes, corresponding to the up, down, and middle points of the heave motion. The control meshes corresponding to the up and down points are calculated by solving the steady-state structural mechanics equations based on the neo-Hookean model with Jacobian-based stiffening [81]. The mesh corresponding to the middle point of the heave motion is used as the reference configuration in this nonlinear model.

The global computation is for 1.56 s, and the local computation is from 0.52 s to 1.56 s. The first 35 time steps of the global computation is used for ramping the Reynolds number linearly from a value 100 times smaller to its full value. Figures 21 and 22 show the temperature at an instant from the global and local computations.

Fig. 21 Temperature (°C) at $t = 1.04$ s from the global computation. Colors from *blue* to *red* indicate temperature values from low to high

Fig. 22 Temperature (°C) at $t = 1.04$ s from the local computation. Colors from *blue* to *red* indicate temperature values from low to high

In reporting the heat transfer rates from the tires, the heat transfer coefficient is calculated from the expression

$$\alpha = \frac{\hat{h}_\theta}{\theta_{\text{tire}} - \theta_\infty}$$ (42)

and is normalized to the Nusselt number:

$$\text{Nu} = \frac{\alpha D_{\text{tire}}}{\kappa},$$ (43)

where $\hat{h}_\theta$ is the normal heat flux on the tire surface, calculated as $\hat{h}_\theta = \kappa \mathbf{n} \cdot \nabla \theta$. All tire-specific results are for the outer leading tire. Figure 23 shows the Nusselt number at an instant from the local computation and the time history of the spatially averaged Nusselt number from the global and local computations.

Figure 24 shows the time- and circumferentially averaged Nusselt number from the global and local computations. Because the mesh on the tire surface does not have axial symmetry, the circumferential averaging is done with a rotating "averaging mesh." The data is projected from the actual mesh to the averaging mesh as it makes a full rotation with 500 equal rotation increments, and the values collected to each node during the rotation are averaged. In general the averaging mesh does not have the same construction as the actual mesh; in the averaging we do here, it does. The time averaging is done over the last heave period.

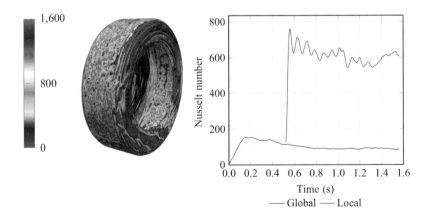

Fig. 23 Nusselt number at $t = 1.04$ s from the local computation and time history of the spatially averaged Nusselt number from the global and local computations

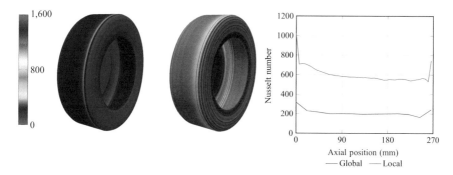

Fig. 24 Time- and circumferentially averaged Nusselt number from the global (*left*) and local (*center*) computations and the profiles (*right*) in the axial direction

9   ALE Computation: Aerodynamic and FSI Analysis of Two Back-to-Back HAWTs in Turbulent ABL Flow

In this section, the techniques described are applied to the simulation of two back-to-back NREL 5 MW wind turbines [84] operating in ABL flow. Each turbine has a rotor with 61 m blades mounted on an 80 m tower and operating at constant, fixed rotor speed of 9 rpm. This rotor speed gives the optimal tip-speed ratio for 8 m/s wind [84], which is also the geostrophic wind speed used in the present computations. The material presented in this section is taken from [74].

9.1   Computational Setup and Boundary Conditions

Two wind turbines are positioned one behind the other at a distance of 480 m, which corresponds to four rotor diameters. The wake generated by the upstream turbine needs to be accurately computed over a long domain before it impacts the downstream turbine, which poses a significant computational challenge due to a very large problem size. To circumvent this difficulty, the MDM is adopted in the present work to efficiently separate the two turbine domains. In the present work the MDM is employed as follows. The problem domain is divided into three subdomains (see Fig. 25 for dimensions and notation). Domains labeled *Turbine 1* and *Turbine 2* contain the upstream and downstream turbines, respectively, and domain labeled *Box* contains the space between the turbines. The three domains are simulated in a sequential manner. Velocity and temperature boundary conditions at the inflow boundary of *Turbine 1*, as well as lateral boundaries of all subdomains, are obtained from a standalone 3D LES computation of a stratified ABL with a uniform grid size of 5 m. This stratified flow computational model [85], which can be run in DNS or LES modes, makes use of a mixed spectral/finite-difference algorithm and a subgrid model based on dynamic eddy viscosity and diffusivity. Nodal values

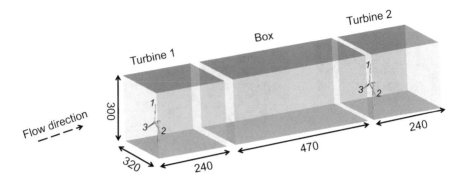

Fig. 25 The three subdomains in the multi-domain wind-turbine simulation. Dimensions are in m

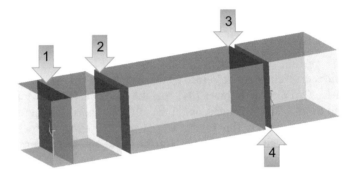

Fig. 26 Data flow between the subdomains. Velocity and temperature collected at location 1 are applied as inlet boundary conditions at location 2. Velocity and temperature collected at location 3 are applied as inlet boundary conditions at location 4. At all lateral boundaries velocity and temperature boundary conditions come from a standalone spectral/fine-difference LES

of the velocity and temperature boundary conditions are obtained by interpolating the finite-difference solution from the structured grid of the LES simulation to the unstructured grids of the wind-turbine simulations. This data transfer strategy, employing the same dataset as in the present work, was successfully tested for the rotor-only ABL simulation in [13]. The background temperature $\bar{\theta}$ is set to 260 K up to 100 m with an overlying inversion of strength 0.01 K/m for all domains. The geostrophic wind speed is set to 8 m/s, and the Coriolis parameter to $f_c = 1.39 \times 10^{-4}$. Velocity and temperature inflow-boundary conditions for *Box* are obtained using a similar data transfer strategy, where, in this case, the data is obtained by interpolating the solution on a plane positioned 10 m behind the turbine during pure aerodynamic simulation on *Turbine 1*. Inflow-boundary conditions for *Turbine 2* are obtained by interpolating the solution on the outflow plane of *Box* (see Fig. 26 for details.)

Fig. 27 Pressure profile used at the outflow boundary of all subdomains

Traction boundary conditions are prescribed at the outlet boundaries of all sub-domains. To generate the traction values, a simulation in *Turbine 1* is performed first with the wind turbine removed, and with zero outlet traction boundary conditions. The inlet tractions produced as a result of this computation, shown in Fig. 27, are then assigned as outlet boundary conditions for all subdomains. A similar strategy was successfully employed in [13], as well as in [6] to perform a detailed thermo-fluid analysis of the rear tires of a ground vehicle.

The subdomains are discretized using triangular prisms in the boundary layer region near the wind-turbine rotors, and tetrahedra elsewhere (see Fig. 28). For *Turbine 1* and *Turbine 2* the boundary-layer mesh design is based on that reported in [10]. For *Turbine 1* a total of 7,824,602 elements are used with a 4 m element length on the outer boundaries. A finer grid resolution with 2 m element length is used on a plane behind the upstream turbine where inlet data is collected for the *Box* simulation. The *Box* domain, which has a refined inner region to more accurately represent the wake turbulence, is discretized using 15,436,631 elements. The *Turbine 2*, with a total of 9,153,426 elements, also contains a refined inner region in front of the turbine for better resolution. The time-step size is set to 10^{-4} s for the *Turbine 1* and *Turbine 2* simulations, and to 10^{-2} s for the *Box* simulation.

9.2 Aerodynamics Simulation

Pure aerodynamics simulation results, which are also referred as "CFD," are reported in this section. During the CFD simulations the wind turbine rotor is considered as a rigid body. Figure 29 shows the velocity and temperature contours on the domain center plane. No discernible discontinuity between the subdomains is observed. A slight growth of the shear layer from the upper edge of the upstream-

Fig. 28 Meshes used in the multi-domain wind-turbine aerodynamics and FSI simulations. *Top-to-bottom: Turbine 1, Box, Turbine 2*

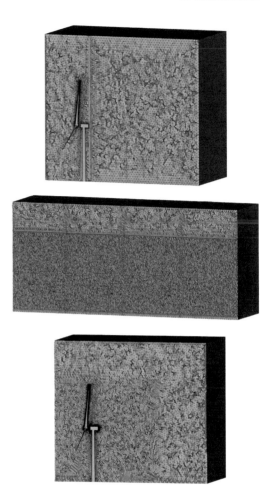

turbine rotor can also be seen in Fig. 29. The bottom shear layer grows much more rapidly, due to higher turbulent mixing and presence of the tower.

Figure 30 shows the vorticity isosurfaces. Rotor-tip vortices of the upstream turbine maintain a helical pattern for a distance of about one rotor diameter. They later break up, and eventually merge with vortices shed from the root and tower to form larger structures at a distance between two and three rotor diameters (see Fig. 30). These larger flow structures impact the downstream-turbine rotor and tower, and break up together with the rotor-tip vortices. The helical pattern of the rotor-tip vortices for the downstream turbine is only maintained for a short distance behind the rotor. This enhanced turbulent mixing gives a faster growth of the shear layer behind the downstream turbine.

Remark 12 When simulating ABL flows, the computational domain should be large enough to account for the wake drift due to side wind and Coriolis force. Figure 31

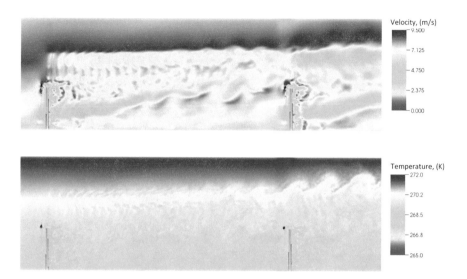

Fig. 29 2D slice of the air speed (*top*) and temperature (*bottom*)

Fig. 30 3D view of the vorticity isosurfaces colored by the air speed

shows the front view of the vorticity isosurfaces, where the wake drift is clearly seen. While in the present simulations wake drift is not as significant, for stronger side winds the computational domain needs to have a larger spanwise dimension.

Figure 32 shows the air speed, averaged over six rotor revolutions, at different locations along the centerline as a function of the vertical coordinate. Air speed

Fig. 31 Front view of the
vorticity isosurfaces colored
by the air speed

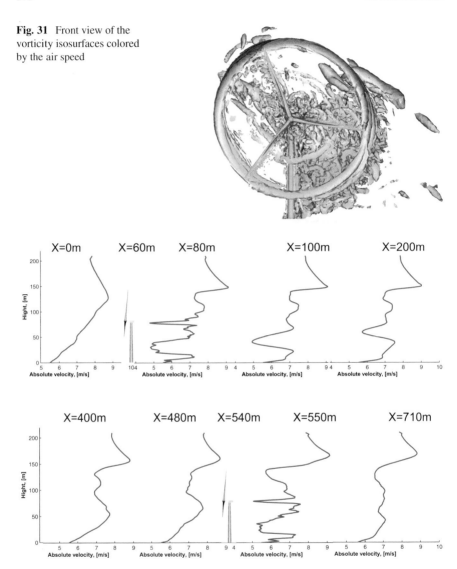

Fig. 32 Air speed averaged over six rotor revolutions and plotted at different locations along the
centerline as a function of the vertical coordinate

profile at the inlet corresponds to that imposed from the LES simulation. A short
distance past *Turbine 1* the profile appears distorted, and slowly begins to recover
with increasing distance from the upstream turbine. By the location of *Turbine 2*
the profile begins to recover up to the hub height and above the upper-blade tip.
However, qualitative differences w.r.t. the inflow profile, e.g., less near-ground shear
and a higher shear above the top of the upper rotor, may be observed. In between
the hub height and upper-blade tip locations, one can clearly see the velocity deficit,

which is on the order of 1–2 m/s. This velocity deficit leads to the power-production drop, as discussed in the next section.

9.3 FSI Simulation

In this section we present FSI simulations of the same multi-domain setup. The wind-turbine geometry, materials, and mesh, which is comprised of 13,273 quadratic NURBS shell elements, are taken from [74]. Figure 33 shows the aerodynamic torque acting on each blade of the upstream-turbine rotor, and compares the pure aerodynamics (labeled "CFD") and FSI results. The FSI simulation curves exhibit low frequency modes coming from the blade flapwise bending motions, as well as high-frequency modes coming from the blade axial torsion motions. These modes are obviously not present in the CFD curves, which underscore the importance of including FSI in the wind-turbine modeling, especially if one is interested in predicting the remaining useful fatigue life of wind-turbine structural components (see, e.g., [14]).

Figure 34 shows a comparison of the aerodynamic torque acting on the upstream and downstream turbines. The results confirm power losses for the downstream turbine of 10–15% relative to the upstream turbine, which are due to the velocity deficit in the upstream-turbine wake. Also note that the amplitude of high-frequency

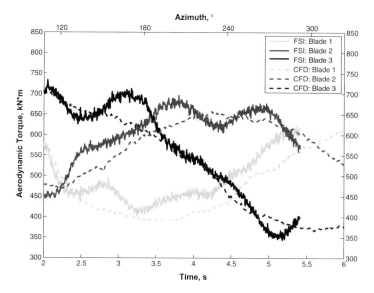

Fig. 33 Time history of the aerodynamic torque for each blade of the upstream turbine. Comparison of pure aerodynamics (labeled "CFD") and FSI simulation results. See Fig. 25 for the blade numbering

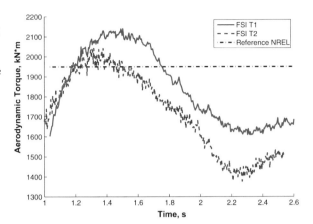

Fig. 34 Time history of the aerodynamic torque from FSI simulations of the upstream (T1) and downstream (T2) turbines. Aerodynamic torque for uniform wind speed of 8 m/s from [84] is shown for comparison

oscillations due to the blade torsional motions is a little higher for the downstream turbine, which is due to higher turbulence intensity in the upstream-turbine wake than in the free stream. The nominal aerodynamic torque from the NREL baseline design for a uniform wind speed of 8 m/s [84] is also plotted for comparison to underscore the importance of including realistic boundary layer flow in the aerodynamics and FSI modeling of wind turbines at full scale.

10 Concluding Remarks

We have described how we are addressing the challenges faced in computational flow analysis in aerospace and transportation technologies, bringing solution in challenging problems such as aerodynamics of parachutes, thermo-fluid analysis of ground vehicles and tires, and wind turbines operating in turbulent ABL flows. The computational challenges include complex geometries, MBI, FSI, turbulent flows, rotational flows, and large problem sizes. Our core computational methods in addressing the computational challenges are the RBVMS, ALE-VMS, and ST-VMS. The special methods used in combination with the core methods include the ALE-SI, ST-SI, YZβ DC, ST-IGA, ST-SI-IGA, MDM, and ST-C data compression. We described the core methods and some of the special methods. We presented, as examples of challenging computations performed with these methods, aerodynamic analysis of a ram-air parachute, thermo-fluid analysis of a freight truck and its rear set of tires, and aerodynamic and FSI analysis of two back-to-back wind turbines operating in thermally stratified ABL flow. The examples show the power and scope of the core and special methods in computational flow analysis in aerospace, energy and transportation technologies.

Acknowledgments This work was supported (first author) in part by Grant-in-Aid for Challenging Exploratory Research 16K13779 from Japan Society for the Promotion of Science;

Grant-in-Aid for Scientific Research (S) 26220002 from the Ministry of Education, Culture, Sports, Science and Technology of Japan (MEXT); Council for Science, Technology and Innovation (CSTI), Cross-Ministerial Strategic Innovation Promotion Program (SIP), "Innovative Combustion Technology" (Funding agency: JST); and Rice–Waseda research agreement. The mathematical model and computational method parts of the work were also supported (third author) in part by ARO Grant W911NF-17-1-0046, ARO DURIP Grant W911NF-18-1-0234, and Top Global University Project of Waseda University. The second author was partially supported by NSF Grant 1854436, and the fourth author was partially supported by NSERC Discovery Grant RGPIN-2017-03781.

References

1. K. Takizawa and T.E. Tezduyar, "Computational methods for parachute fluid–structure interactions", *Archives of Computational Methods in Engineering*, **19** (2012) 125–169. https://doi.org/10.1007/s11831-012-9070-4.
2. K. Takizawa, D. Montes, M. Fritze, S. McIntyre, J. Boben, and T.E. Tezduyar, "Methods for FSI modeling of spacecraft parachute dynamics and cover separation", *Mathematical Models and Methods in Applied Sciences*, **23** (2013) 307–338. https://doi.org/10.1142/S0218202513400058.
3. K. Takizawa, T.E. Tezduyar, and R. Kolesar, "FSI modeling of the Orion spacecraft drogue parachutes", *Computational Mechanics*, **55** (2015) 1167–1179. https://doi.org/10.1007/s00466-014-1108-z.
4. K. Takizawa, T.E. Tezduyar, and T. Terahara, "Ram-air parachute structural and fluid mechanics computations with the space–time isogeometric analysis (ST-IGA)", *Computers & Fluids*, **141** (2016) 191–200. https://doi.org/10.1016/j.compfluid.2016.05.027.
5. K. Takizawa, T.E. Tezduyar, and T. Kanai, "Porosity models and computational methods for compressible-flow aerodynamics of parachutes with geometric porosity", *Mathematical Models and Methods in Applied Sciences*, **27** (2017) 771–806. https://doi.org/10.1142/S0218202517500166.
6. K. Takizawa, T.E. Tezduyar, and T. Kuraishi, "Multiscale ST methods for thermo-fluid analysis of a ground vehicle and its tires", *Mathematical Models and Methods in Applied Sciences*, **25** (2015) 2227–2255. https://doi.org/10.1142/S0218202515400072.
7. T. Kuraishi, K. Takizawa, and T.E. Tezduyar, "Tire aerodynamics with actual tire geometry, road contact and tire deformation", *Computational Mechanics*, **63** (2019) 1165–1185. https://doi.org/10.1007/s00466-018-1642-1.
8. K. Takizawa, T.E. Tezduyar, T. Kuraishi, S. Tabata, and H. Takagi, "Computational thermo-fluid analysis of a disk brake", *Computational Mechanics*, **57** (2016) 965–977. https://doi.org/10.1007/s00466-016-1272-4.
9. Y. Otoguro, K. Takizawa, T.E. Tezduyar, K. Nagaoka, and S. Mei, "Turbocharger turbine and exhaust manifold flow computation with the Space–Time Variational Multiscale Method and Isogeometric Analysis", *Computers & Fluids*, **179** (2019) 764–776. https://doi.org/10.1016/j.compfluid.2018.05.019.
10. Y. Bazilevs, A. Korobenko, X. Deng, and J. Yan, "Novel structural modeling and mesh moving techniques for advanced FSI simulation of wind turbines", *International Journal for Numerical Methods in Engineering*, **102** (2015) 766–783. https://doi.org/10.1002/nme.4738.
11. Y. Bazilevs, A. Korobenko, X. Deng, J. Yan, M. Kinzel, and J.O. Dabiri, "FSI modeling of vertical-axis wind turbines", *Journal of Applied Mechanics*, **81** (2014) 081006. https://doi.org/10.1115/1.4027466.
12. J. Yan, A. Korobenko, X. Deng, and Y. Bazilevs, "Computational free-surface fluid–structure interaction with application to floating offshore wind turbines", *Computers and Fluids*, **141** (2016) 155–174. https://doi.org/10.1016/j.compfluid.2016.03.008.

13. Y. Bazilevs, A. Korobenko, J. Yan, A. Pal, S.M.I. Gohari, and S. Sarkar, "ALE–VMS formulation for stratified turbulent incompressible flows with applications", *Mathematical Models and Methods in Applied Sciences*, **25** (2015) 2349–2375. https://doi.org/10.1142/S0218202515400114.

14. Y. Bazilevs, A. Korobenko, X. Deng, and J. Yan, "FSI modeling for fatigue-damage prediction in full-scale wind-turbine blades", *Journal of Applied Mechanics*, **83** (6) (2016) 061010.

15. T.J.R. Hughes, "Multiscale phenomena: Green's functions, the Dirichlet-to-Neumann formulation, subgrid scale models, bubbles, and the origins of stabilized methods", *Computer Methods in Applied Mechanics and Engineering*, **127** (1995) 387–401.

16. Y. Bazilevs, V.M. Calo, J.A. Cottrell, T.J.R. Hughes, A. Reali, and G. Scovazzi, "Variational multiscale residual-based turbulence modeling for large eddy simulation of incompressible flows", *Computer Methods in Applied Mechanics and Engineering*, **197** (2007) 173–201.

17. Y. Bazilevs, V.M. Calo, T.J.R. Hughes, and Y. Zhang, "Isogeometric fluid–structure interaction: theory, algorithms, and computations", *Computational Mechanics*, **43** (2008) 3–37.

18. K. Takizawa and T.E. Tezduyar, "Multiscale space–time fluid–structure interaction techniques", *Computational Mechanics*, **48** (2011) 247–267. https://doi.org/10.1007/s00466-011-0571-z.

19. K. Takizawa, T.E. Tezduyar, H. Mochizuki, H. Hattori, S. Mei, L. Pan, and K. Montel, "Space–time VMS method for flow computations with slip interfaces (ST-SI)", *Mathematical Models and Methods in Applied Sciences*, **25** (2015) 2377–2406. https://doi.org/10.1142/S0218202515400126.

20. K. Takizawa, B. Henicke, A. Puntel, T. Spielman, and T.E. Tezduyar, "Space–time computational techniques for the aerodynamics of flapping wings", *Journal of Applied Mechanics*, **79** (2012) 010903. https://doi.org/10.1115/1.4005073.

21. K. Takizawa, T.E. Tezduyar, Y. Otoguro, T. Terahara, T. Kuraishi, and H. Hattori, "Turbocharger flow computations with the Space–Time Isogeometric Analysis (ST-IGA)", *Computers & Fluids*, **142** (2017) 15–20. https://doi.org/10.1016/j.compfluid.2016.02.021.

22. Y. Osawa, V. Kalro, and T. Tezduyar, "Multi-domain parallel computation of wake flows", *Computer Methods in Applied Mechanics and Engineering*, **174** (1999) 371–391. https://doi.org/10.1016/S0045-7825(98)00305-3.

23. K. Takizawa and T.E. Tezduyar, "Space–time computation techniques with continuous representation in time (ST-C)", *Computational Mechanics*, **53** (2014) 91–99. https://doi.org/10.1007/s00466-013-0895-y.

24. Y. Bazilevs and T.J.R. Hughes, "Weak imposition of Dirichlet boundary conditions in fluid mechanics", *Computers and Fluids*, **36** (2007) 12–26.

25. Y. Bazilevs and T.J.R. Hughes, "NURBS-based isogeometric analysis for the computation of flows about rotating components", *Computational Mechanics*, **43** (2008) 143–150.

26. T.E. Tezduyar, "Stabilized finite element formulations for incompressible flow computations", *Advances in Applied Mechanics*, **28** (1992) 1–44. https://doi.org/10.1016/S0065-2156(08)70153-4.

27. A.N. Brooks and T.J.R. Hughes, "Streamline upwind/Petrov-Galerkin formulations for convection dominated flows with particular emphasis on the incompressible Navier-Stokes equations", *Computer Methods in Applied Mechanics and Engineering*, **32** (1982) 199–259.

28. T.E. Tezduyar and K. Takizawa, "Space–time computations in practical engineering applications: A summary of the 25-year history", *Computational Mechanics*, **63** (2019) 747–753. https://doi.org/10.1007/s00466-018-1620-7.

29. K. Takizawa, B. Henicke, D. Montes, T.E. Tezduyar, M.-C. Hsu, and Y. Bazilevs, "Numerical-performance studies for the stabilized space–time computation of wind-turbine rotor aerodynamics", *Computational Mechanics*, **48** (2011) 647–657. https://doi.org/10.1007/s00466-011-0614-5.

30. K. Takizawa, "Computational engineering analysis with the new-generation space–time methods", *Computational Mechanics*, **54** (2014) 193–211. https://doi.org/10.1007/s00466-014-0999-z.

31. K. Takizawa, B. Henicke, A. Puntel, N. Kostov, and T.E. Tezduyar, "Space–time techniques for computational aerodynamics modeling of flapping wings of an actual locust", *Computational Mechanics*, **50** (2012) 743–760. https://doi.org/10.1007/s00466-012-0759-x.

32. K. Takizawa, T.E. Tezduyar, and N. Kostov, "Sequentially-coupled space–time FSI analysis of bio-inspired flapping-wing aerodynamics of an MAV", *Computational Mechanics*, **54** (2014) 213–233. https://doi.org/10.1007/s00466-014-0980-x.
33. K. Takizawa, T.E. Tezduyar, and A. Buscher, "Space–time computational analysis of MAV flapping-wing aerodynamics with wing clapping", *Computational Mechanics*, **55** (2015) 1131–1141. https://doi.org/10.1007/s00466-014-1095-0.
34. K. Takizawa, Y. Bazilevs, T.E. Tezduyar, M.-C. Hsu, O. Øiseth, K.M. Mathisen, N. Kostov, and S. McIntyre, "Engineering analysis and design with ALE-VMS and space–time methods", *Archives of Computational Methods in Engineering*, **21** (2014) 481–508. https://doi.org/10.1007/s11831-014-9113-0.
35. K. Takizawa, K. Schjodt, A. Puntel, N. Kostov, and T.E. Tezduyar, "Patient-specific computational analysis of the influence of a stent on the unsteady flow in cerebral aneurysms", *Computational Mechanics*, **51** (2013) 1061–1073. https://doi.org/10.1007/s00466-012-0790-y.
36. K. Takizawa, T.E. Tezduyar, H. Uchikawa, T. Terahara, T. Sasaki, and A. Yoshida, "Mesh refinement influence and cardiac-cycle flow periodicity in aorta flow analysis with isogeometric discretization", *Computers & Fluids*, **179** (2019) 790–798. https://doi.org/10.1016/j.compfluid.2018.05.025.
37. K. Takizawa, T.E. Tezduyar, T. Terahara, and T. Sasaki, "Heart valve flow computation with the integrated Space–Time VMS, Slip Interface, Topology Change and Isogeometric Discretization methods", *Computers & Fluids*, **158** (2017) 176–188. https://doi.org/10.1016/j.compfluid.2016.11.012.
38. T. Kanai, K. Takizawa, T.E. Tezduyar, K. Komiya, M. Kaneko, K. Hirota, M. Nohmi, T. Tsuneda, M. Kawai, and M. Isono, "Methods for computation of flow-driven string dynamics in a pump and residence time", *Mathematical Models and Methods in Applied Sciences*, **29** (2019) 839–870. https://doi.org/10.1142/S021820251941001X.
39. T. Kuraishi, K. Takizawa, and T.E. Tezduyar, "Space–Time Isogeometric flow analysis with built-in Reynolds-equation limit", *Mathematical Models and Methods in Applied Sciences*, **29** (2019) 871–904. https://doi.org/10.1142/S0218202519410021.
40. T.J.R. Hughes, W.K. Liu, and T.K. Zimmermann, "Lagrangian–Eulerian finite element formulation for incompressible viscous flows", *Computer Methods in Applied Mechanics and Engineering*, **29** (1981) 329–349.
41. V. Kalro and T.E. Tezduyar, "A parallel 3D computational method for fluid–structure interactions in parachute systems", *Computer Methods in Applied Mechanics and Engineering*, **190** (2000) 321–332. https://doi.org/10.1016/S0045-7825(00)00204-8.
42. Y. Bazilevs, M.-C. Hsu, I. Akkerman, S. Wright, K. Takizawa, B. Henicke, T. Spielman, and T.E. Tezduyar, "3D simulation of wind turbine rotors at full scale. Part I: Geometry modeling and aerodynamics", *International Journal for Numerical Methods in Fluids*, **65** (2011) 207–235. https://doi.org/10.1002/fld.2400.
43. A. Korobenko, Y. Bazilevs, K. Takizawa, and T.E. Tezduyar, "Computer modeling of wind turbines: 1. ALE-VMS and ST-VMS aerodynamic and FSI analysis", *Archives of Computational Methods in Engineering*, published online, https://doi.org/10.1007/s11831-018-9292-1, September 2018.
44. Y. Bazilevs, V.M. Calo, Y. Zhang, and T.J.R. Hughes, "Isogeometric fluid–structure interaction analysis with applications to arterial blood flow", *Computational Mechanics*, **38** (2006) 310–322.
45. M.-C. Hsu and Y. Bazilevs, "Blood vessel tissue prestress modeling for vascular fluid–structure interaction simulations", *Finite Elements in Analysis and Design*, **47** (2011) 593–599.
46. M.-C. Hsu, D. Kamensky, F. Xu, J. Kiendl, C. Wang, M.C.H. Wu, J. Mineroff, A. Reali, Y. Bazilevs, and M.S. Sacks, "Dynamic and fluid–structure interaction simulations of bioprosthetic heart valves using parametric design with T-splines and Fung-type material models", *Computational Mechanics*, **55** (2015) 1211–1225. https://doi.org/10.1007/s00466-015-1166-x.
47. D. Kamensky, M.-C. Hsu, D. Schillinger, J.A. Evans, A. Aggarwal, Y. Bazilevs, M.S. Sacks, and T.J.R. Hughes, "An immersogeometric variational framework for fluid-structure interac-

tion: Application to bioprosthetic heart valves", *Computer Methods in Applied Mechanics and Engineering*, **284** (2015) 1005–1053.

48. I. Akkerman, J. Dunaway, J. Kvandal, J. Spinks, and Y. Bazilevs, "Toward free-surface modeling of planing vessels: simulation of the Fridsma hull using ALE-VMS", *Computational Mechanics*, **50** (2012) 719–727.

49. M.C.H. Wu, D. Kamensky, C. Wang, A.J. Herrema, F. Xu, M.S. Pigazzini, A. Verma, A.L. Marsden, Y. Bazilevs, and M.-C. Hsu, "Optimizing fluid–structure interaction systems with immersogeometric analysis and surrogate modeling: Application to a hydraulic arresting gear", *Computer Methods in Applied Mechanics and Engineering*, **316** (2017) 668–693.

50. J. Yan, X. Deng, A. Korobenko, and Y. Bazilevs, "Free-surface flow modeling and simulation of horizontal-axis tidal-stream turbines", *Computers and Fluids*, **158** (2017) 157–166. https://doi.org/10.1016/j.compfluid.2016.06.016.

51. A. Castorrini, A. Corsini, F. Rispoli, K. Takizawa, and T.E. Tezduyar, "A stabilized ALE method for computational fluid–structure interaction analysis of passive morphing in turbomachinery", *Mathematical Models and Methods in Applied Sciences*, **29** (2019) 967–994. https://doi.org/10.1142/S0218202519410057.

52. J. Yan, B. Augier, A. Korobenko, J. Czarnowski, G. Ketterman, and Y. Bazilevs, "FSI modeling of a propulsion system based on compliant hydrofoils in a tandem configuration", *Computers and Fluids*, **141** (2016) 201–211. https://doi.org/10.1016/j.compfluid.2015.07.013.

53. T.A. Helgedagsrud, I. Akkerman, Y. Bazilevs, K.M. Mathisen, and O.A. Oiseth, "Isogeometric modeling and experimental investigation of moving-domain bridge aerodynamics", *ASCE Journal of Engineering Mechanics*, Accepted for publication.

54. J. Yan, A. Korobenko, A.E. Tejada-Martinez, R. Golshan, and Y. Bazilevs, "A new variational multiscale formulation for stratified incompressible turbulent flows", *Computers & Fluids*, **158** (2017) 150–156. https://doi.org/10.1016/j.compfluid.2016.12.004.

55. F. Xu, G. Moutsanidis, D. Kamensky, M.-C. Hsu, M. Murugan, A. Ghoshal, and Y. Bazilevs, "Compressible flows on moving domains: Stabilized methods, weakly enforced essential boundary conditions, sliding interfaces, and application to gas-turbine modeling", *Computers & Fluids*, **158** (2017) 201–220. https://doi.org/10.1016/j.compfluid.2017.02.006.

56. T.M. van Opstal, J. Yan, C. Coley, J.A. Evans, T. Kvamsdal, and Y. Bazilevs, "Isogeometric divergence-conforming variational multiscale formulation of incompressible turbulent flows", *Computer Methods in Applied Mechanics and Engineering*, **316** (2017) 859–879. https://doi.org/10.1016/j.cma.2016.10.015.

57. T.E. Tezduyar, S.K. Aliabadi, M. Behr, and S. Mittal, "Massively parallel finite element simulation of compressible and incompressible flows", *Computer Methods in Applied Mechanics and Engineering*, **119** (1994) 157–177. https://doi.org/10.1016/0045-7825(94)00082-4.

58. T. Kanai, K. Takizawa, T.E. Tezduyar, T. Tanaka, and A. Hartmann, "Compressible-flow geometric-porosity modeling and spacecraft parachute computation with isogeometric discretization", *Computational Mechanics*, **63** (2019) 301–321. https://doi.org/10.1007/s00466-018-1595-4.

59. Y. Bazilevs, K. Takizawa, and T.E. Tezduyar, *Computational Fluid–Structure Interaction: Methods and Applications*. Wiley, February 2013, ISBN 978-0470978771.

60. T.E. Tezduyar, "Computation of moving boundaries and interfaces and stabilization parameters", *International Journal for Numerical Methods in Fluids*, **43** (2003) 555–575. https://doi.org/10.1002/fld.505.

61. T.E. Tezduyar and S. Sathe, "Modeling of fluid–structure interactions with the space–time finite elements: Solution techniques", *International Journal for Numerical Methods in Fluids*, **54** (2007) 855–900. https://doi.org/10.1002/fld.1430.

62. M.-C. Hsu, Y. Bazilevs, V.M. Calo, T.E. Tezduyar, and T.J.R. Hughes, "Improving stability of stabilized and multiscale formulations in flow simulations at small time steps", *Computer Methods in Applied Mechanics and Engineering*, **199** (2010) 828–840. https://doi.org/10.1016/j.cma.2009.06.019.

63. K. Takizawa, T.E. Tezduyar, S. McIntyre, N. Kostov, R. Kolesar, and C. Habluetzel, "Space–time VMS computation of wind-turbine rotor and tower aerodynamics", *Computational Mechanics*, **53** (2014) 1–15. https://doi.org/10.1007/s00466-013-0888-x.

64. K. Takizawa, T.E. Tezduyar, and Y. Otoguro, "Stabilization and discontinuity-capturing parameters for space–time flow computations with finite element and isogeometric discretizations", *Computational Mechanics*, **62** (2018) 1169–1186. https://doi.org/10.1007/s00466-018-1557-x.
65. T.E. Tezduyar and M. Senga, "Stabilization and shock-capturing parameters in SUPG formulation of compressible flows", *Computer Methods in Applied Mechanics and Engineering*, **195** (2006) 1621–1632. https://doi.org/10.1016/j.cma.2005.05.032.
66. T.E. Tezduyar, S. Ramakrishnan, and S. Sathe, "Stabilized formulations for incompressible flows with thermal coupling", *International Journal for Numerical Methods in Fluids*, **57** (2008) 1189–1209. https://doi.org/10.1002/fld.1743.
67. T.J.R. Hughes, J.A. Cottrell, and Y. Bazilevs, "Isogeometric analysis: CAD, finite elements, NURBS, exact geometry, and mesh refinement", *Computer Methods in Applied Mechanics and Engineering*, **194** (2005) 4135–4195.
68. T. Sasaki, K. Takizawa, and T.E. Tezduyar, "Medical-image-based aorta modeling with zero-stress-state estimation", *Computational Mechanics*, **64** (2019) 249–271. https://doi.org/10.1007/s00466-019-01669-4.
69. K. Takizawa, T.E. Tezduyar, and T. Sasaki, "Isogeometric hyperelastic shell analysis with out-of-plane deformation mapping", *Computational Mechanics*, **63** (2019) 681–700. https://doi.org/10.1007/s00466-018-1616-3.
70. T. Tezduyar and Y. Osawa, "Methods for parallel computation of complex flow problems", *Parallel Computing*, **25** (1999) 2039–2066. https://doi.org/10.1016/S0167-8191(99)00080-0.
71. T. Tezduyar and Y. Osawa, "The Multi-Domain Method for computation of the aerodynamics of a parachute crossing the far wake of an aircraft", *Computer Methods in Applied Mechanics and Engineering*, **191** (2001) 705–716. https://doi.org/10.1016/S0045-7825(01)00310-3.
72. T. Tezduyar and Y. Osawa, "Fluid–structure interactions of a parachute crossing the far wake of an aircraft", *Computer Methods in Applied Mechanics and Engineering*, **191** (2001) 717–726. https://doi.org/10.1016/S0045-7825(01)00311-5.
73. V. Kalro, S. Aliabadi, W. Garrard, T. Tezduyar, S. Mittal, and K. Stein, "Parallel finite element simulation of large ram-air parachutes", *International Journal for Numerical Methods in Fluids*, **24** (1997) 1353–1369. https://doi.org/10.1002/(SICI)1097-0363(199706)24:12<1353::AID-FLD564>3.0.CO;2-6.
74. A. Korobenko, J. Yan, S.M.I. Gohari, S. Sarkar, and Y. Bazilevs, "FSI simulation of two back-to-back wind turbines in atmospheric boundary layer flow", *Computers & Fluids*, **158** (2017) 167–175. https://doi.org/10.1016/j.compfluid.2017.05.010.
75. T.E. Tezduyar and Y. Osawa, "Finite element stabilization parameters computed from element matrices and vectors", *Computer Methods in Applied Mechanics and Engineering*, **190** (2000) 411–430. https://doi.org/10.1016/S0045-7825(00)00211-5.
76. C. Johnson, *Numerical solution of partial differential equations by the finite element method*. Cambridge University Press, Sweden, 1987.
77. F. Shakib, T.J.R. Hughes, and Z. Johan, "A multi-element group preconditionined GMRES algorithm for nonsymmetric systems arising in finite element analysis", *Computer Methods in Applied Mechanics and Engineering*, **75** (1989) 415–456.
78. T.J.R. Hughes and M. Mallet, "A new finite element formulation for computational fluid dynamics: III. The generalized streamline operator for multidimensional advective-diffusive systems", *Computer Methods in Applied Mechanics and Engineering*, **58** (1986) 305–328.
79. T.J.R. Hughes, G. Scovazzi, and T.E. Tezduyar, "Stabilized methods for compressible flows", *Journal of Scientific Computing*, **43** (2010) 343–368. https://doi.org/10.1007/s10915-008-9233-5.
80. T.E. Tezduyar, M. Behr, S. Mittal, and A.A. Johnson, "Computation of unsteady incompressible flows with the finite element methods: Space–time formulations, iterative strategies and massively parallel implementations", in *New Methods in Transient Analysis*, PVP-Vol.246/AMD-Vol.143, ASME, New York, (1992) 7–24.

81. T. Tezduyar, S. Aliabadi, M. Behr, A. Johnson, and S. Mittal, "Parallel finite-element computation of 3D flows", *Computer*, **26** (10) (1993) 27–36. https://doi.org/10.1109/2.237441.
82. M.L. Accorsi, J.W. Leonard, R. Benney, and K. Stein, "Structural modeling of parachute dynamics", *AIAA Journal*, **38** (2000) 139–146.
83. Y. Saad and M. Schultz, "GMRES: A generalized minimal residual algorithm for solving nonsymmetric linear systems", *SIAM Journal of Scientific and Statistical Computing*, **7** (1986) 856–869.
84. J. Jonkman, S. Butterfield, W. Musial, and G. Scott, "Definition of a 5-MW reference wind turbine for offshore system development", Technical Report NREL/TP-500-38060, National Renewable Energy Laboratory, 2009.
85. B. Gayen and S. Sarkar, "Direct and large-eddy simulations of internal tide generation at a near-critical slope.", *Journal of Fluid Mechanics*, **681** (2011) 48–79.

Multiscale Crowd Dynamics Modeling and Safety Problems Towards Parallel Computing

Bouchra Aylaj and Nicola Bellomo

1 Plan of the Chapter

The modeling and simulations of the dynamics of human crowds are challenging research areas which have stimulated a rapidly growing interest of scientists active not only in mathematics and numerical analysis, but also in different fields of applied and natural sciences, for instance, technology, psychology, and safety sciences. This interest is induced not only by scientific motivations, but also by a collective search of well-being in our society, specifically of the search for safety strategies in crisis situations which might appear in high density flow of crowds. Examples of even tragic crisis situations are well known in very recent events in Europe.

As it is known [3], the modeling approach to crowd dynamics can be developed at the microscopic, mesoscopic, and macroscopic scales which correspond, respectively, to individual based, kinetic, and hydrodynamical models.

In more detail, the micro-scale corresponds to *individual based models* which describe the dynamics of walkers represented by their individual position and velocity; the meso-scale corresponds to *kinetic models* which define the dynamics of a probability distribution function over time, position, and velocity, namely over the individual *microscopic state* of each walker; and the macroscopic scale corresponds to *hydrodynamic models*, where the dependent variables are the local density and

B. Aylaj
Laboratoire de Modélisation, Analyse, Contrôle et Statistiques, Faculté des Sciences Ain-Chock, Hassan II University of Casablanca, Maarif, Casablanca, Maroc
e-mail: bouchra.aylaj@univh2c.ma

N. Bellomo (✉)
Politecnico of Torino and Collegio Carlo Alberto, Torino, Italy
e-mail: nicola.bellomo@polito.it

© Springer Nature Switzerland AG 2020
A. Grama, A. H. Sameh (eds.), *Parallel Algorithms in Computational Science and Engineering*, Modeling and Simulation in Science, Engineering and Technology,
https://doi.org/10.1007/978-3-030-43736-7_9

281

the local mean velocity, while the independent variables are the dimensionless time and position.

This multiscale aspect requires, in most cases, computational tools related to parallel computing. Important motivations are induced by strategies towards safety management which generally need a reduction of the computational time to support crisis managers by visualization of simulations delivered by models in time intervals possibly shorter than the real ones. This achievement allows crisis managers to take correct decisions to improve safety conditions.

This brief introduction, given above, is sufficient to define the aims and the plan of this chapter which presents a tutorial to crowd modeling with the aim of working out some mathematical tools related to parallel computing. As we shall see, this objective might appear, at a first naive glance, simply related to technical issues. On the other hand, it generates some requirements for the structure of models that are posed to applied mathematicians as challenging research objectives.

Bearing in mind this brief introduction, our chapter is proposed as a contribution to the following topics:

1. *Multiscale vision* *of crowd dynamics by a unified approach to modeling individual based, kinetic, and hydrodynamic models corresponding, respectively, to the microscopic, mesoscopic, and macroscopic scales.*
2. *Development of a* *systems approach* *to crowd dynamics in complex venues, where walkers move across different areas which present different quality and geometry.*
3. *Speculations on the use of modeling and simulations to support crisis managers in evacuation dynamics as well as on related* *parallel computing to support safety problems* *and develop* *devices of artificial intelligence.*

The aforementioned key problems are treated in the next three sections, respectively, within a general framework, where the study and modeling of complexity features of all systems where human behavior plays an important role in the overall dynamics of the systems under considerations [6, 8]. Section 5 closes the chapter by a critical analysis focused on the research perspectives presented in this chapter.

Bearing all the above in mind, let us now add a few remarks to enlighten the aims and style of this chapter.

Interacting living entities are called *active particles* [11]. Propagation of stress conditions in crowds, which is occasionally called panic, has been recently studied by various authors [13, 15], while a broad literature has been developed on safety problems [35, 36, 44] which include elements of artificial intelligence to produce algorithms to guide the decision process of crisis managers.

The aim of this chapter consists of opening a dialogue involving applied mathematicians devoted to modeling complex systems constituted by a large number of interacting living entities; crisis managers devoted to care about safety conditions of people who might find themselves in critical situations such as evacuation induced by incidents, and experts in scientific computing deemed to develop computational codes to provide simulations in the aforementioned crisis situations.

The presentation of the chapter takes advantage of concepts rather than of mathematical formalization so that the interdisciplinary dialogue can follow the rationale proposed in this paper without being lost in a framework of a complex mathematical model. However, detailed bibliographic indications are given with the aim of directing the interested reader to all mathematical details.

2   On a Multiscale Vision of Crowd Dynamics (Key Problem 1)

Let us consider the first key problem, namely the development of a multiscale vision of human crowds. The presentation is simply qualitative on the main features of the approach. Some bibliographic indications, which do not claim to be exhaustive, are given for each scale with the aim of addressing the interested reader to the pertinent literature.

Let us now present the specific features of each scale and the mathematical structures which provide the conceptual frameworks for the derivation of models.

- **Individual based models** describe the dynamics of each individual based walker under the action of the other walkers and under the conditioning imposed by obstacles and walls of the venue, where the dynamics occur. Mathematical models consist of large systems of ordinary differential equations which describe the dynamics of position and velocity of each walkers.

The key problem in the derivation of models consists of describing the acceleration term, occasionally called force, acting on each individual entity [26]. An example of the acceleration term is given by the social force model [27]. Additional literature is reported and critically analyzed in the survey [3] which includes also a review on vehicular traffic and swarms.

Mathematical Structures at the Microscopic Scale
The description of the system is given, for each i-th walker with $i \in \{1, \ldots, N\}$, by position $\mathbf{x}_i = \mathbf{x}_i(t) = (x_i(t), y_i(t))$ and velocity $\mathbf{v}_i = \mathbf{v}_i(t) = (v_i^x(t), v_i^y(t))$. The specific mathematical structure underlying models is as follows:

$$\begin{cases} \dfrac{d\mathbf{x}_i}{dt} = \mathbf{v}_i \, , \\[2mm] \dfrac{d\mathbf{v}_i}{dt} = \mathbf{F}_i(\mathbf{x}_1, \ldots, \mathbf{x}_N, \mathbf{v}_1, \ldots, \mathbf{v}_N; \Sigma), \end{cases} \tag{1}$$

(continued)

where $\mathbf{F}_i$ is a psycho-mechanical acceleration acting on the i-th walker based on the action of other walkers zone as well as on walker's interaction with the geometrical properties of the venue which are formally expressed by Σ.

- **Kinetic models** describe the dynamics in time and space of a probability distribution function over the individual *microscopic state* of each walker. These models have been introduced in [9] for the dynamics in unbounded domain and in [12] for a walls, exists, and obstacles. The microscopic state is delivered by position and velocity. Recent studies [13, 15, 42] include an additional variable modeling the said emotional states. It has been shown that the strategy developed by walkers in stress conditions is subject to important modifications that might even induce unsafe situations [13]. The conceptual framework towards modeling social dynamics is delivered in [2].

The derivation of models, namely of the dynamics of the aforementioned probability distribution function, is obtained by equating the transport term to the interaction term suitable to describe the inlet and outlet flows of particles in the elementary volume of the space of microscopic state. The said flows are determined by interactions modeled by a stochastic game theory approach. The mathematical structure of models is delivered by integro-differential equations which present some analogy with classical kinetic theory [20].

Mathematical Structures at the Mesoscopic Scale
The description of the system is given by the probability distribution function $f = f(t, \mathbf{x}, \mathbf{v})$, while the mathematical structure is as follows:

$$(\partial_t + \mathbf{v} \cdot \partial_{\mathbf{x}}) \, f(t, \mathbf{x}, \mathbf{v}) = J[\mathbf{f}](t, \mathbf{x}, \mathbf{v})$$

$$= \int \eta[f](\mathbf{x}, \mathbf{v}_*, \mathbf{v}^*; \Sigma) \mathscr{P}[f](\mathbf{v}_* \to \mathbf{v} | \mathbf{v}_*, \mathbf{v}^*; \Sigma) \, f(t, \mathbf{x}, \mathbf{v}_*) f(t, \mathbf{x}, \mathbf{v}^*) \, d\mathbf{v}_* \, d\mathbf{v}^*$$

$$- f(t, \mathbf{x}, \mathbf{v}) \int \eta[f](\mathbf{x}, \mathbf{v}, \mathbf{v}^*) \, f(t, \mathbf{x}, \mathbf{v}^*) \, d\mathbf{v}^*. \tag{2}$$

Macroscopic quantities are obtained by velocity weighted moments.

- **Hydrodynamical models** describe the dynamics in time and space of the dependent variables, namely the local density and the local mean velocity, while the independent variables are the dimensionless time and position. The mathematical framework is defined by two partial differential equations corresponding to conservation of mass and linear momentum equilibrium, where the acceleration term can be modeled by accounting for the interaction of walkers in the elementary space volume with the surrounding crowd.

A simplified structure is defined by the equation corresponding to mass conservation only linked to a phenomenological model expressing how the local mean velocity depends on the local density conditions and on the geometry of the venue. The literature on the derivation of models at the macroscopic scale is reviewed in [22].

Mathematical Structures at the Macroscopic Scale
The description of the system is given by the *local density* $\rho = \rho(t, \mathbf{x})$ and the *mean velocity* $\xi = \xi(t, \mathbf{x})$. The general mathematical structure is as follows:

$$\begin{cases} \partial_t \rho + \nabla_{\mathbf{x}} \cdot (\rho\,\xi) = 0, \\ \\ \partial_t \xi + (\xi \cdot \nabla_{\mathbf{x}})\xi = \mathscr{A}[\rho, \xi;\, \Sigma], \end{cases} \tag{3}$$

where $\mathscr{A}[\rho, \xi;\, \Sigma]$ is a psycho-mechanical acceleration acting on walkers.

The derivation of models requires, at each scale, a deep understanding to be cast into a mathematical framework of the strategy developed by walkers to move in the venues, namely trajectories and speed along them. The literature that has been cited above provides useful approaches for handling this problem which requires mixing the study of social and emotional behaviors with mechanics.

In more detail, the modeling approach should include some relevant features, for instance, the ability of individuals to develop a walking strategy and role of emotional states over such strategy. Models should also show how the specific features of the venues can modify the said strategy in view of the development of safety problems. Last, but not least, models should be validated to be effectively reliable.

Issues that are focused on the modeling approach, at each scale, are listed below:

1. *Strategy:* Walkers develop a strategy, which is heterogeneously distributed in the crowd, by which they continuously modify their direction of motion and speed due to nonlinearly additive and nonlocal interactions. The quality, the geometry of the environment as well as the emotional state of the walkers modify quantitatively and qualitatively the said geometry. The model, proposed in [12], indicates that the choice of the direction is induced by a continuous selection between four stimuli, namely trend towards a target, search of less crowded region, attraction towards the mainstream, and search of trajectories that avoid the contact with walls. The selection depends on local density conditions, on local emotional states, and on geometrical parameters such as the distance from

targets and walls. Subsequently, walkers adapt their speed to the new density conditions perceived in the new direction of the motion.

2. *Emotional state:* The level of stress in critical situations modifies the strategy described in Item 1. Referring to the model proposed in [12], high value of stress induces walkers to be attracted by the mainstream rather than by the search of less crowded regions. This behavior is basically irrational as it creates overcrowded areas which in some specific conditions affect the safety. In some cases, irrational behavior of a few entities can generate large deviations from the standard dynamics observed in rational situations. This behavioral issue can be taken into account by introducing a parameter $\beta \in [0, 1]$ to model the intensity of emotional state, for instance, stress, where $\beta = 0$ denotes the lowest level which indicates not only search for an uncongested area but also low speed, while $\beta = 1$ denotes the highest stress conditions which indicate congestion as well as high speed. Joining these features leads to highly unsafe conditions [12].

3. *Role of the venue:* The strategy described in Item 1 is quantitatively modified by the geometry and quality of the venue. In fact, different geometries imply different trajectories and speed as walkers attempt to avoid high density flows near walls, while the quality of the venue has a direct influence on the speed which decreases/increases with decreasing/increasing quality of the area where walkers move.

   The use of the parameter $\alpha \in [0, 1]$ has been introduced in [12] to account for this specific feature, where $\alpha = 0$ denotes worse conditions which prevent motion, while $\alpha = 1$ denotes the best conditions which allow high speed. Simulations should be developed to understand the interplay between quality and stress. For instance, stress conditions are not induced only by incidents, but also by overcrowding which, in turn, might be induced by poor geometries. Indeed, it is useful to study the dynamics of a crowd in low stress conditions to investigate the role of the geometry of venues to understand fluidity (opposite to congestion) of venues.

4. *Safety problems:* As mentioned, an important aspect of crowd modeling consists of providing models and simulations to support crisis situations, for example, forced evacuations due to incidents. In these specific cases simulations should run at least at the same speed of real flow so that crisis managers can make rapid decisions to create possibly safer conditions. Actually, these specific problems have recently received great attention due to studying evacuation dynamics related to social riots where safety and security problems interact.

   The requirement of real time simulations imposes the need for a careful development of computational problems. The approach deals with deterministic methods in the case of individual based, hydrodynamical models [14, 25], and stochastic particle methods in the case of kinetic models [5, 7, 16, 34, 42]. The multiscale vision suggests that both methodological approaches should coexist, while a parallel computing vision suggests that different models should be simultaneously used.

5. *Validation of models:* The validation of models is a necessary step to lead to their reliable use in safety problems. A general rationale towards validation has been

proposed in [12] based on the requirements that models reproduce quantitatively empirical data for solutions corresponding to steady uniform flow and that reproduce qualitatively emerging behaviors which are repetitively observed. In addition, models should include parameters that account for the specific role of the venue over the dynamics of the crowd. The main difficulty consists of the fact that quantitative results are available in steady flow (it might be equilibrium) conditions; however, models are required to provide quantitative results in real conditions which are generally far from equilibrium. In addition, the behavior of the crowd is strongly venue dependent and that different results are observed when the geometry and the quality of the venue change.

It is worth stressing that crisis situations can appear suddenly and unsuspectedly as they are generated by localized stress induced by incidents or even unusual behavior of a few individuals. Stress propagates rapidly in the crowd and can become the main cause of serious incidents.

This very concise overview of the literature allows to enlighten some conceptual differences of the three model classes. A motivation to develop the proposals reported in the following is induced by a recurrent dispute which involves scientists regarding the choice of the most appropriate scale to be used. It is often argued that only the microscopic scale is the most appropriate for systems with finite number of degrees, while the macroscopic scale requires unrealistic assumptions on the continuity of the matter and kills some heterogeneity features of the individual behaviors. The intermediate kinetic theory approach needs the assumption of continuity of the said probability over position and velocity of the particles which is reasonable only when their number is sufficiently large.

A **multiscale vision**, *which has been idealized in Figure 1, consists of developing a bottom-up derivation of models, where a detailed modeling of individual based interactions, namely at the microscopic scale, is used to model the acceleration term $\mathbf{F}_i$ at the microscopic scale and subsequently to implement the interaction terms η and $\mathscr{P}$ to derive kinetic type models, namely at the mesoscopic scale. Subsequently asymptotic methods can be developed to obtain hydrodynamical models from the underlying description at the lower scale.*

The micro-macro derivation looks ahead to a unified approach to physical sciences as inspired by the sixth Hilbert problem [28]. In fluid dynamics, this problem has been interpreted as the derivation of hydrodynamical models from the description delivered by the Boltzmann equation Saint-Raymond [37]. The recent literature on the application of this approach to large systems of active particles can be found in [17, 18]. The rationale to understand how local interaction dynamics can be transferred to collective, self-organized, motion is proposed in [29]. Experimental studies of this challenging objective have been developed in [33]. An overview of the merits of the kinetic theory approach with respect to individual based and hydrodynamic models has been proposed in [24].

In addition, let us stress that multiscale vision requires using dimensionless quantities both for dependent, independent/and microscopic variables. The approach can be developed by referring the said variables to:

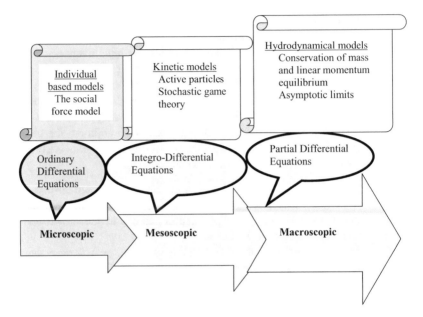

Fig. 1 From individual based to kinetics and hydrodynamics

1. ℓ is a characteristic length to be taken as the diameter of the circle containing the domain Ω containing the overall venue;
2. v_M which is the highest individual speed which can be reached by a very fast walker in a free flow in high quality venues;
3. ξ_M is the highest mean speed which can be reached by walkers by a free flow in a high quality venues;
4. $T = \ell/v_M$ is the characteristic time corresponding to the time by which a fast walker can cover the distance ℓ;
5. ρ_M denotes the maximal number density (occupancy) of walkers packed in a square meter.

Accordingly, linear space variables are referred to ℓ, real time to T, speed is referred to v_M. In addition, macroscopic quantities, specifically density and mean velocity, are divided, respectively, by ρ_M and v_M, while the probability distribution function f is referred to ρ_M. This normalization implies that the order of magnitude of all quantities is of the order of one, but also poses an important constraint to the solutions of mathematical problems. For instance, the physical meaning of density and velocity is below one.

An additional feature of the multiscale vision is that models at all scale should use the same parameters corresponding to interactions at the same scales.

3   On a Systems Approach to Crowds Modeling (Key Problem 2)

Let us now consider the second key problem, namely a systems approach to crowd modeling of dynamics of people in complex venues constituted by K interconnected areas each of them characterized by different geometrical and qualitative features. If V denotes the overall venue and V_k each specific walking area with $k = 1, \ldots, K$, then V is constituted by all venues V_k and the interconnections among them.

A possible strategy towards a systems approach and some technical remarks are proposed in the following:

1. *Modeling the quality of the venue:* A minimal description of the quality of each walking area V_k can be obtained by introducing, for each area, a parameter $\alpha_k \in [0, 1]$, where $\alpha_k = 0$ corresponds to very low quality which prevents motion, while $\alpha_k = 1$ to very high quality allowing fast motion. The speed of walkers is heterogeneously distributed within a minimal and a maximal value in each V_k, where the following simple model $v_\ell = \alpha_k \, v_M$ has been proposed [13] to denote the highest, venue dependent, speed. Namely, the limit velocity of the fastest walkers is reduced by α_k with respect to the maximal speed observed in a high quality walking area.

2. *Emotional state:* As mentioned, the development of the walking strategy depends on the level of stress shared by the crowd which can be taken into account by a parameter $\beta_k \in [0, 1]$, where $\beta_k = 0$ corresponds to very low stress, while $\beta_k = 1$ to very high stress [12]. The level of stress can differ across the walking areas V_k due not only to localized incidents, but also to the specific features of each V_k.

3. *Further comments on the parameters α and β:* For instance, a detailed analysis of the role of the two aforementioned parameters has been developed by kinetic type models, but not exhaustively studied in the case of individual based and hydrodynamical models. The use of a scalar parameter in each venue is a simplification of physical reality. In fact, different variables can play a role over the dynamics of crowds, as an example the quality of the venues can be modified by specific incidents which reduce the visibility of each walker. An additional example referring to the emotional states appears in crowds where political contrasts are present so that stress conditions are mixed with the intensity of political expression.

4. *Selection of the modeling scale for each V_k and subsequent modeling and simulation of the overall dynamics:* The selection of the specific model to be used in each V_k depends on the specific features of the venue and of the flow. In addition, the selection of the scale out of the three possible ones should also account for the computational time required by simulations.

5. *Safety and related computational problems:* The systems approach can be developed according to the rationale presented in the preceding items. However, additional requirements are generally imposed by the specific features of the

application-investigation his need is generally motivated by the specific investigation object of the modeling and computational approach. According to our choice and regressed experience, the study is focused on safety problems in evacuation. Focusing on evacuation dynamics, the term "safety" is here used to identify local density below a threshold and evacuation time as much reduced as possible consistently with the former requirement. Then, this application requires the development of the multiscale vision summarized in Item 4. In addition, real time simulations are necessary to tackle safety problems. Parallel computing can contribute to this objective as it will be indicated in the next section.

This strategy can be regarded as a research perspective rather than an achievement known in the literature as only specific problems have been treated generally at one scale only, as examples selected among various ones, sparse control problems have been treated in [4, 30], dynamics with reduced visibility in [21], space propagation of emotional states in [13, 42], and various others.

However, a systematic study of the systems approach described in the above issues is still missing, while it would be an interesting research perspective in view of dealing with support crisis managements by artificial intelligence methods.

4 Parallel Computing and Artificial Intelligence (Key Problem 3)

This final section presents, out of the survey and critical analysis proposed in the preceding sections, some practical indications focused on crisis managements and parallel computing. Hence this section tackles the third key problem proposed in Sect. 1. Firstly, some perspective ideas to support crisis situations are presented, subsequently we examine how the implication of the approach on some specific aspect of parallel computing.

Essentially we can distinguish between two possible supports which can be delivered by modeling and computing of crowd dynamics, namely the support to training crisis managers and a direct support to crisis management. In the former case, the aim of computer modeling consists of providing simulations, where priority is given to precise description of the flow patterns at variable parameters of the geometry and quality of the venue, while in the latter case, which includes evacuations due to incidents, simulations should run at least at the same speed of real flow so that crisis managers can take rapid decisions to create possibly safe conditions for the individuals under critical situations.

As mentioned, the dynamics of evacuation can generate a crisis situation whenever safe conditions appear to be lost, generally when the onset of stress conditions generates overcrowding due to the trend of walkers to imitate the behavior of the others rather than looking for less overcrowded paths. Bearing all above in mind, let us provide two immediate remarks concerning each of the specific uses of modeling and simulations that have been mentioned above.

Remark 1 Crisis managers can train themselves by developing simulations of well-defined real cases and subsequently by inserting in the dynamics possible control action by visual or vocal indications. Therefore, simulations allow a comparison among a variety of possible actions and identify the most appropriate selection. In addition, experienced managers can address this training to support the learning action of less experienced ones. Real time computing is not a strict requirement of simulation, while a systems approach appears to be necessary.

Remark 2 Crisis managers can use visualization of real flows stored in databases and refer them to simulations related to the case study under consideration. Subsequently an artificial intelligence process can lead managers to select the most appropriate safety action.

Several technical difficulties need to be tackled. For instance, in most cases, decision-making has to be developed in a very short time, while generally the information delivered by experiment is not complete. Some aspects of the general problem can be enlightened accounting for the two aforementioned use of computational modeling:

- *Training crisis managers* by big-data *database* repository of simulations corresponding to different venues, crowd features, and specific actions addressed to safety. Simulations should refer to evacuation dynamics [1, 38], should be specifically related to support crisis [39, 44], and should include, for each case study, a variety of actions to support crisis management. In addition, simulations need to be validated also referring to the specific venue, where the dynamics occur [31, 35, 41].
- *Selection of the most appropriate actions* during an evacuation process should be achieved by the design of a *predictive engine* to support the aforementioned selection of safety actions by optimality criteria suitable to minimize the evacuation time and local overcrowding.
- The *design of the predictive engine* should account for a large variety of simulations to be compared with the real case under consideration to achieve the selection of the most appropriate action.
- Comparisons can take advantage of a *distance (metrics)* between the main properties of different dynamical systems. Such a distance can provide the correct information to decide how far a simulation is close to the real dynamics observed by the crisis manager.

The various methods to treat these large amounts of data still need to be properly developed to define an emerging data science [23, 40] which aims at improving the decision-making process towards cost reductions and reduced risk. As a matter of fact, crisis management is not yet sufficiently developed up to a commonly shared theory. Still it can be stated that the development of crowd simulations should take into account human behaviors within the general framework of behavioral sciences and specifically social sciences [2] and collective learning [19, 32]. Let us summarize the sequential steps for the use of crowd modeling in the decision process to support safety according to [10, 30].

1. Assessment of the main features of the evacuation venues and of the crowd;
2. Implementation of a number of possible simulations corresponding to both aforementioned classifications;
3. Implementation of a number of possible actions to make safe the evacuation process;
4. Scoring for each simulation the output of the dynamics corresponding to different safety actions;
5. Define a metric to compare a real situation with those stored in the database;
6. Select a number of simulations close to the real situation and choose from among them the most appropriate action based on a weighted combination of the score and the said metric distance.

This process succeeds in providing a technical response which goes beyond the heuristic approach based on personal bias. The reliability of the learning machine depends on the validity of the simulations stored; hence, it depends on the validity of models which generate the simulations. This concept imposes a filter on the stored data not only in the case study under consideration, but as a general rule for the use of stored data to be used to support human well-being and safety [43].

5 Closure

This chapter has been devoted to a survey and critical analysis of the modeling and simulation of human crowds in crisis situations induced by evacuation dynamics somehow forced by a sudden not predictable incident. The contents have been proposed focusing on three key problems concerning the following topics: a multiscale vision of the modeling of human crowds; a systems approach to crowd modeling; support that modeling and simulations can provide to crisis managers.

This chapter has shown that the literature in the field is a valuable resource for the aforementioned topics; however, several problems are still open and need of further research activity. Possible perspectives have already been given in the preceding three sections.

The final closure of the chapter provides some additions that may contribute to the development of any of the aforementioned perspectives.

1. *Understanding human behavior in crowds* is a key step in the derivation of models. This important hint indicates that understanding social and dynamical behavior of a crowd is the absolutely necessary basis for any decision process related to safety. We have put in evidence that any approach should consider the crowd as a living complex system. Hence, understanding the complexity features of human crowd is very important also in designing computational models. Knowledge of the emotional behavior of individuals in a crowd should be reflected in the modeling of the walking strategy of these individuals.
2. *A multiscale vision* should be developed to derive models at the three different scales based on the same rationale by which individuals organize their walking

strategy. Accordingly models at each scale should include analogous parameters to account for the specific features of the aforementioned strategy.

3. *A systems approach to crowd modeling* appears necessary to model the dynamics in complex venues, where walkers move across different areas of the network of the overall venue. The systems approach should also consider using different scales in different interconnected areas of venues.

4. A *new vision of parallel computing* has been introduced where parallelization corresponds to models derived according to a scale selection which can differ in each area of the overall venue.

References

1. J.P. Agnelli, F. Colasuonno and D. Knopoff, A kinetic theory approach to the dynamics of crowd evacuation from bounded domains, *Mathematical Models and Methods in Applied Sciences*, **25**(1), 109–129, (2015).
2. G. Ajmone Marsan, N. Bellomo and L. Gibelli, Stochastic evolutionary differential games toward a systems theory of behavioral social dynamics, *Mathematical Models and Methods in Applied Sciences*, **26**(6), 1051–1093, (2016).
3. G. Albi, N. Bellomo, L. Fermo, S.-Y. Ha, J. Kim, L. Pareschi, D. Poyato, and J. Soler, Traffic, crowds, and swarms. From kinetic theory and multiscale methods to applications and research perspectives *Mathematical Models and Methods in Applied Sciences*, **29**, 1901–2005, (2019).
4. G. Albi, M. Bongini, E. Cristiano and D. Kalise, Invisible control of self-organizing agents leaving unknown environments, *SIAM J. Applied Mathematics*, **76**(4), 1683–1710, (2016).
5. V.V. Aristov, **Direct Methods for Solving the Boltzmann Equation and Study of Nonequilibrium Flows**, Springer-Verlag, New York, (2001).
6. P. Ball **Why Society is a Complex Matter**, Springer-Verlag, Heidelberg, (2012).
7. P. Barbante, A. Frezzotti and L. Gibelli, A kinetic theory description of liquid menisci at the microscale *Kinetic and Related Models*, **8**(2), 235–254, (2015).
8. N. Bellomo, A. Bellouquid, L. Gibelli and N. Outada, **A Quest Towards a Mathematical Theory of Living Systems**, Birkhäuser, New York, (2017).
9. N. Bellomo, A. Bellouquid and D. Knopoff, From the micro-scale to collective crowd dynamics, *Multiscale Modelling Simulation*, **11**(3), 943–963, (2013).
10. N. Bellomo, D. Clarke, L. Gibelli, P. Townsend and B.J. Vreugdenhil, Human behaviours in evacuation crowd dynamics: From modelling to "big data" toward crisis management, *Physics of Life Review*, **18**, 1–21, (2016).
11. N. Bellomo, P. Degond and E. Tadmor, Eds., **Active Particles, Volume 1 - Advances in Theory, Models, and Applications**, *Modeling and Simulation in Science Engineering and Technology*, Birkhäuser, New York, (2017).
12. N. Bellomo and L. Gibelli, Toward a mathematical theory of behavioral-social dynamics for pedestrian crowds, *Mathematical Models and Methods in Applied Sciences*, **25**(13), 2417–2437, (2015).
13. N.Bellomo, L. Gibelli and N. Outada, On the interplay between behavioral dynamics and social interactions in human crowds, *Kinetic and Related Models*, **12**, 397–409, (2019).
14. N. Bellomo, B. Lods, R. Revelli and L. Ridolfi, **Generalized Collocation Methods - Solution to Nonlinear Problems**, Birkhauser-Springer, Boston, (2008).
15. A.L. Bertozzi, J. Rosado, M.B. Short and L. Wang, Contagion shocks in one dimension, *J. Statistical Physics*, **158**(3), 647–664, (2015).
16. G.A. Bird, **Molecular Gas Dynamics and the Direct Simulation of Gas Flows**, Oxford University Press, (1994).

17. D. Burini and N. Chouhad, Hilbert method toward a multiscale analysis from kinetic to macroscopic models for active particles, *Mathematical Models and Methods in Applied Sciences*, **27**(7), 1327–1353, (2017).

18. D. Burini and N. Chouhad, A multiscale view of nonlinear diffusion in biology: From cells to tissues, *Mathematical Models and Methods in Applied Sciences*, **29**, 791–823, (2019).

19. D. Burini and S. De Lillo, On the complex interaction between collective learning and social dynamics, *Symmetry*, **11**, 967; https://doi.org/10.3390/sym11080967, (2019).

20. C. Cercignani, R. Illner and M. Pulvirenti, **The Mathematical Theory of Dilute Gases**, Springer-Verlag, Heidelberg, New York, (1994).

21. M. Colangeli, A. Muntean, O. Richardson and T. Thieu, Modelling interactions between active and passive agents moving through heterogeneous environments, Chapter 2 in **Crowd Dynamics, Volume 1 - Theory, Models, and Safety Problems**, *Modeling and Simulation in Science, Engineering, and Technology*, Birkhäuser, New York, (2018).

22. E. Cristiani, B. Piccoli and A. Tosin, **Multiscale Modeling of Pedestrian Dynamics**, Springer, (2014).

23. H. De Sterck and C. Johnson, Data science: What is it and how is it thought?. *SIAM News*, **48**, 1–6, (2015).

24. A. Elaiw, Y. Al-Turki, and M. Alghamdi, A critical analysis of behavioural crowd dynamics: From a modelling strategy to kinetic theory methods, *Symmetry*, **11**, 851; https://doi.org/10.3390/sym11070851, (2019).

25. J.H. Ferziger and M. Peric, **Computational Methods for Fluid Dynamics**, Springer Science & Business Media, (2002).

26. D. Helbing, Traffic and related self-driven many-particle systems, *Review Modern Physics*, **73**, 1067–1141, (2001).

27. D. Helbing and P. Molnár, Social force model for pedestrian dynamics *Physics Rev. E, Stat. Phys. Plasmas Fluids Relat. Interdiscip. Topics*, **51**(5), 4282–4286, (1995).

28. D. Hilbert, Mathematical problems, *Bulletin American Mathematical Society*, **8**(10), 437–479, (1902).

29. S.P. Hoogendoorn., F. L.M. van Wageningen-Kessels, W. Daamen and D.C. Duives, Continuum modelling of pedestrian flows: From microscopic principles to self-organised macroscopic phenomena *Physica A*, **416**, 684–694, (2014).

30. C. Kecai and C. Yangquan **Fractional Order Crowd Dynamics: Cyber-Human System Modeling and Control**, Walter de Gruyter GmbH & Co KG, **4**, (2018).

31. J. Lin and T.A. Lucas, A particle swarm optimization model of emergency airplane evacuation with emotion, *Networks and Heterogenous Media*, **10**(3), 631–646, (2015).

32. V. Mayer-Schönberg and K. Cukier, **Learning with Big Data**, Kindle-Single-ebook.

33. M. Moussaid, D. Helbing, S. Garnier, A. Johanson, M. Combe, and G. Theraulaz, Experimental study of the behavioral mechanism underlying self-organization in human crowd, *Proceedings Royal Society B: Biological Sciences*, **276**, 2755–2762, (2009).

34. L. Pareschi and G. Toscani, **Interacting Multiagent Systems. Kinetic Equations and Monte Carlo Methods**, Oxford University Press, (2013).

35. F. Ronchi, F. Nieto Uriz, X. Criel and P. Reilly, Modelling large-scale evacuation of music festivals. *Case Studies in Fire Safety*, **5**, 11–19, (2016).

36. E. Ronchi, P.A. Reneke and R.D. Peacock, A conceptual fatigue-motivation model to represent pedestrian movement during stair evacuation, *Applied Mathematical Modelling*, **40**(7-8), 4380–4396, (2016).

37. L. Saint-Raymond, **Hydrodynamic Limits of the Boltzmann Equation**, Lecture Notes in Mathematics n.1971, Springer, Berlin, (2009).

38. Twarogowska M, Goatin P., Duvigneau R. Macroscopic modeling and simulations of room evacuation. *Applied Mathematical Modelling*, **38**(24), 5781–5795, (2014).

39. H. Vermuyten, J. Belien, L. De Boeck, G. Reniers, and T. Wauters, A review of optimisation models for pedestrian evacuation and design problems, *Safety Science*, **87**, 167–178, (2016).

40. J. Zhou, H. Pei and H. Wu, Early warning of human crowds based on query data from Baidu maps: Analysis based on Shanghai stampede, Chapter 2 in **Big Data Support of Urban Planning and Management**, *Advances in Geography Information Science*, Springer, New York, (2018).
41. A.U.K. Wagoum, M. Chraibi, J. Mehlich, A. Seyfried and A. Scadschneider, Efficient and validated simulation of crowds for an evacuation assistant. *Computer Animation & Virtual Worlds*, **23**(1), 3–15, (2012).
42. L. Wang, M.B. Short and A.L. Bertozzi, Efficient numerical methods for multiscale crowd dynamics with emotional contagion, *Mathematical Models and Methods in Applied Sciences*, **27**(1), 205–230, (2017).
43. "Web Source", OECD, Organization for Economic Co-Operation and Development, Paris, France. **Data-Driven Innovation, Big Data for Growth and Well-Being**, OECD Publishing-www.oecd.org/sti/ieconomy/data-driven-innovation, 2015.
44. N. Wijermans, C. Conrado, M. van Steen, C. Martella and J.-L. Li, A landscape of crowd management support: An integrative approach, *Safety Science*, **86** 142–164, (2016).

HPC for Weather Forecasting

John Michalakes

1 Introduction: Weather and HPC

Numerical weather prediction (NWP) and high-performance computing have grown up together. Even before the computational means existed, L. F. Richardson of the UK Met Office had published a numerical foundation for forecasting the weather [33]. The first computer-generated forecast of the atmosphere had to wait until 1950 and was conducted by Jule Charney's meteorology group within John von Neumann's ENIAC project at Princeton's Institute for Advanced Study. By the 1970s, advances in models and computing capability allowed the skill of numerically generated forecasts to outpace forecasting that relied solely on expert meteorologists interpreting weather observations [17, 38]. Today the list of major weather services that develop and run operational weather forecasting systems includes the European Center for Medium Range Weather Forecasts (ECMWF) and its member national services, the U.S. National Weather Service within NOAA, the U.S. Navy's Fleet Numerical Meteorology and Oceanography Center (FNMOC) and Naval Research Laboratory (NRL), the U.K. Met Office (UKMO), Meteo France, the German National Weather Service (DWD), Environment Canada, the Japan Meteorological Agency, the Korea Meteorological Administration, and the China Meteorological Administration.

Historically, an *exponential* rate of increase in supercomputing power has fueled a *linear* pace of forecast skill improvement (Fig. 1). Each decade's 1000-fold increase in computing power has enabled larger numbers of higher resolution forecasts, better representations of the physics of the atmosphere, and more sophisticated assimilation of greater volumes of observational data to provide better

J. Michalakes (✉)
University Corporation for Atmospheric Research, Boulder, CO, USA
e-mail: michalak@ucar.edu

© Springer Nature Switzerland AG 2020
A. Grama, A. H. Sameh (eds.), *Parallel Algorithms in Computational Science and Engineering*, Modeling and Simulation in Science, Engineering and Technology, https://doi.org/10.1007/978-3-030-43736-7_10

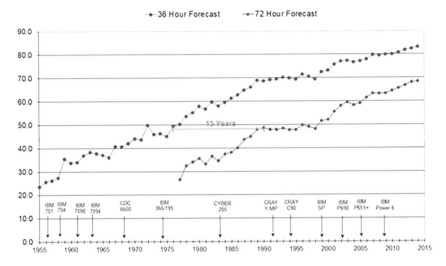

Fig. 1 Anomaly correlation, a measure of forecast skill (100 = perfect), increases linearly as computing increases exponentially over successive generations of supercomputer at the U.S. National Weather Service [20]

initial conditions. The result has been to add 1 day of forecast skill every decade for the last 40 years [3]. Five-day forecasts today are as accurate as 3-day forecasts 20 years ago. Today, twenty of the fastest 500 supercomputers in the world are dedicated to weather forecasting, consuming 7% (60 PFLOPs) of the total compute capacity of the Top500 list in November, 2017[1]. Continuing this trend into the era of exascale supercomputers is the ongoing challenge for weather forecast centers.

Operational weather forecasting involves running a large suite of applications: preprocessors, post-processors, and the model itself (Fig. 2). Preprocessors combine data streaming in from weather stations, aircraft, and satellites with archives of climatological data and previously generated forecasts to produce initial conditions for the new forecast. The forecast model takes this initial state of the atmosphere and computes an approximation of the future state over a succession of many small time intervals until the desired end time of the forecast is reached—as little as 12 h or as long as 16 days, depending on the needs of the center (climate predictions run longer still from seasonal to decadal scales). Periodic output over the course of the forecast is fed into a myriad of post-processors and downstream models that produce specialized products with analysis and visualization for use by forecasters

---

[1] https://www.top500.org/lists/2017/11/.

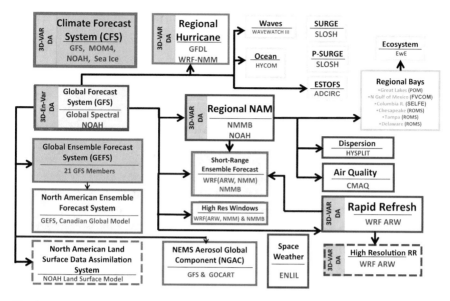

Fig. 2 Production suite at the National Centers for Environmental Prediction (NCEP) in 2014, presented as part of a NOAA annual review. The model itself, the Global Forecast System (GFS), appears as the red box in the first column. Illustration by William Lapenta, NOAA/NWS. Used with permission [44]. The diagram is drawn this way to illustrate the system was becoming too complex. In fact, the system is actually more complex than shown, since data assimilation and other model preprocessors are subsumed within the blue area of the small GFS box

and end users. End-to-end forecasting involves both large amounts of data handling and large amounts of computational horsepower. The focus of this chapter is on the computational requirements and challenges of the forecast model at the heart of the operational weather forecast system.

Models vary according to their use. Climate models simulate characteristics of the atmosphere from seasonal to century time scales at relatively low resolutions. Models designed for real-time weather forecasting run at higher resolution over time scales short enough to fit within the limits of predictability for weather forecasting, from several hours to usually no longer than 2 weeks [43]. The domains for weather and climate models also vary. Models may forecast the entire global atmosphere or a specific region. Global models are constrained by available computing to relatively modest resolutions (currently grid cells no smaller than 9–13 km to a side). Finer than this and the model will not run fast enough for the forecast to be timely[2]. Regional model domains are smaller and can run at higher resolutions but typically require data generated by global models to provide lateral boundary conditions. As computers become more powerful a convergence has begun such that global models

---

[2]The U.S. National Weather Service requires a forecast rate of 8.5 min per forecast day.

will soon capture finer-scale turbulent and convective processes important for local
weather, especially severe storm forecasting.

2 History

The history of NWP is tied closely to the steady but occasionally disruptive
evolution of supercomputing over the last half-century. As today's computational
scientists scratch their heads wondering how to design efficient codes for exascale
systems on the horizon, it is comforting to realize that every previous generation
of supercomputer forced scientific programmers to devise codes and data structures
tortured in some way to run efficiently on the HPC architecture at hand.

Until the 1990s, supercomputers used for NWP were expensive room-size
devices with a single processor. Very fast for their day, speed came from high clock
rates (hundreds of megahertz!) and special vector processing units, hardware that
could perform many floating point operations over successive data elements during
each clock cycle. Today's architectural analogs are vector or SIMD[3] instructions
on conventional CPU cores (e.g. Intel's AVX instruction set) and fine-grained
parallelism over warps of threads on GPUs. Multi-port memories and high capacity
buses were needed to provide the bandwidth necessary to keep up with the
processors. These high-performance memory systems contributed further to the
already high cost (millions of dollars) of vector supercomputers in the 1970s,
1980s, and 1990s. The impact on software design was also considerable. Weather
calculations most naturally expressed in one dimension of the domain had to be
rewritten to operate over whatever dimension happened to be vectorizable. For
example, subroutines that computed a vertical process such as convection up and
down a single column of grid cells had to be rewritten to run horizontally over
multiple columns because data dependencies in the vertical inhibited vectorization.

Later, faster but similarly architected systems were developed by connecting
several vector processors to the same memory for parallelism over different tasks
(task parallelism) or different sections of the domain (data parallelism). This more
coarse-grained mechanism, called "microtasking" at the time, is analogous to
medium-grain thread parallelism (e.g. OpenMP, pthreads) today. The move to thread
parallelism was not overly disruptive since the codes had already been restructured
for vectors. Then as now, however, contention for memory bandwidth meant that
only a few processors could be added to provide more speed. In other words, the
systems could not scale. Ultimately, the cost of building and operating successively
faster supercomputers using vector/shared-memory designs became prohibitive.

In the 1990s, a new design for constructing supercomputers from many more
less powerful processors pushed past the shared-memory scaling barrier. Processors
were organized as nodes on a network, each accessing data exclusively from its own

---

[3] Single-instruction, multiple-data stream.

memory to avoid the scaling bottleneck. The nodes worked in parallel by exchanging messages over a network, the carrying capacity of which (bandwidth) increased with the number of nodes. Distributed memory message passing was the third and coarsest level of parallelism and could scale to arbitrarily large configurations.

The move to distributed memory supercomputers was unavoidable but deeply disruptive for the NWP community. Significant effort was expended during the 1990s to update models that had been developed for vector supercomputers. Each of the major weather services undertook programs to rapidly convert their large investments in modeling software but struggled with their earlier legacy software designs. The U.S. Department of Energy founded an entire program to convert atmospheric and ocean models used for climate prediction to these new systems [8, 22]. Global address spaces had to be decomposed—that is, broken up— into distributed memory subdomains to be run as separate processes (tasks) over many nodes. Data dependencies needed to be analyzed and explicit mechanisms implemented to buffer and exchange data as messages between separate processes. Entirely new problems of debugging and profiling parallel programs at scale remain areas of active computer science research and development today.

Today's supercomputers are still built as networks of coarse-grain parallel nodes exchanging messages, but also incorporate the other two earlier forms of parallelism: fine-grained within each processor core (vectors or GPU threads) and medium-grained between processors on a node (OpenMP and pthreads).[4] There is no longer any limit to scaling other than money, electrical power, and, more fundamentally, the fraction of parallelism available in the application itself (Amdahl's law). And herein lie both the practical and fundamental disruptions for weather prediction going forward into the exascale era.

The practical disruption is simply that current and next generation supercomputing architectures will require so much parallelism (estimates go to millions of threads) that there is not any level of parallelism that can be ignored: fine-grain vector parallelism at the loop level all the way out to hitherto underexploited coarse-grain distributed memory parallelism over the vertical grid dimension, between different physics subroutines, and between the components (atmosphere, ocean, land, sea-ice, the ionosphere, and other physical systems) in coupled earth system models. In many cases this will mean rediscovering and implementing fine-grained parallelism (discussed in a later section) that was disregarded when microprocessor-based clusters replaced vector supercomputers.

Mining all available parallelism also means rediscovering and implementing shared-memory thread programming, which was largely discarded because the first generations of distributed memory supercomputers had only a single-core processor on each node. Even as nodes with multiple multicore processors have become prevalent, hybrid MPI/OpenMP has only recently begun to show better performance

---

[4]There is also at least a fourth: instruction level parallelism at the processor core level that exists to some extent even in otherwise outwardly sequential programs. ILP is limited and generally hidden from and outside the control of the programmer.

than parallelizing entirely over single-threaded MPI parallel tasks. This is partly because of improvements to memory systems on new generations of multicore processors and nodes. ECMWF reported optimal performance using eight OpenMP threads per MPI task running the current IFS model on 48 nodes of their Phase-1 Cray XC30 system [46]. But hybrid MPI/OpenMP programming is also taking hold because it will be unavoidable: scaling to larger problem sizes, models run out of pure-MPI parallelism. ECMWF also reported that running the IFS at high resolution was not possible with only MPI parallelism because too much memory was needed to replicate data over many MPI tasks on each node.

The more fundamental disruption from moving to more powerful generations of HPC systems is that future increases in supercomputing speed must come solely from increased parallelism, and that a real-time weather forecast is not weakly scalable. Weak scaling is the ability of an application to run at the same speed using more processors as problem size is increased. For a weather model, increasing problem size means adding grid points and, for a global weather model, the only way to add grid points is to increase resolution. But increasing spatial resolution requires increased temporal resolution: many smaller time steps are needed to produce the same length of forecast. Since time steps must be executed sequentially, complexity increases with resolution in one more dimension, the temporal, than the available parallelism. The cost for higher resolution balloons in terms of number of processors and electricity needed (Fig. 3).

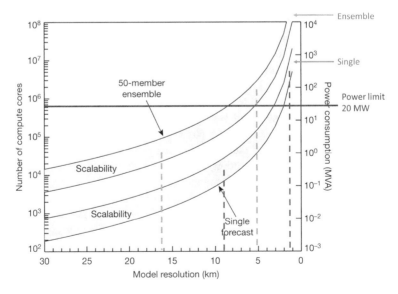

Fig. 3 Projected resources required to scale operational forecasting to higher resolution [3]. Global weather models are not weakly scalable with resolution because the temporal dimension must also be refined and is inherently sequential. Additional notations are from [23]

Note, this lack of weak scalability is the result of running the model *deterministically*: one run from a single set of initial conditions to predict one possible future state of the atmosphere. There is considerable value to consumers from *probabilistic* weather information. For example output from an *ensemble* of many runs of a hurricane forecast, each with a perturbed set of initial conditions, is used to generate a Cone of Uncertainty[5] for where the storm will make landfall. Adding members to the ensemble increases parallelism without constraining the time step, so ensemble forecasts are weakly scalable, at least computationally. The volume of model output generated by the ensemble increases with more members too, so the scaling problem does not disappear but instead shifts to I/O. Moreover, if one also increases the resolution of each ensemble member, as shown in Fig. 3, weak scaling is again problematic.

Parallel-in-time algorithms that can exploit scale separation in partial differential equations to provide parallelism over the time dimension are possible [19] but application to operational weather forecasting is likely distant.

3   Models, Grids, and Parallelization

The specific approach to parallelizing a weather model depends on the choice of numerical scheme and how mapping the mesh onto a spherical geometry is addressed. Various grid geometries and numerical fixes have been developed to adapt "numerical methods to the spherical geometry of the earth, which presents unique problems, usually and vaguely referred to collectively as the pole problem" [48]. For example, in a latitude/longitude grid (Fig. 4) the narrowing of grid cells approaching the two poles requires a smaller time step or unwanted filtering for stability. Icosahedral (soccer ball) meshes have 12 pentagons. Cubed-sphere meshes have corners. The main types of model are grid point, spectral, and finite element (which includes spectral element).

Grid-point models using finite-difference and finite-volume methods evolve the model state (wind velocities, temperature, pressure, moisture, and other tracers) in physical space, directly at each cell of the grid.

Spectral models avoid distortions and singularities by first transforming the gridded representation of the global state to a series of spherical harmonics. Additionally, spectrally computed derivatives are higher-order and non-local, providing more accuracy than finite-difference methods for a given cost. Three dimensional Helmholtz solvers are expensive in grid-point models but essentially free in spectral models [40].

A third type, *spectral element* models are a hybrid formulation: finite-volume methods between the elements and spectral methods local within each element [29].

---

[5]https://www.nhc.noaa.gov/aboutcone.shtml.

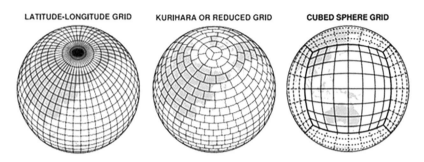

Fig. 4 Sampling of grids used in weather and climate models [48]

The first weather models prior to the 1970s were grid-point formulations, but numeric and computational advantages of globally spectral methods ushered in a heyday that is only now beginning to wane as grid-point methods resurge. Scientific and numerical factors have been central to the progression, but as noted above and in the discussion that follows, a key driver has been the disruptive evolution of HPC architectures. This section derives heavily from [48], an authoritative and to a significant extent eyewitness account of the evolution and types of models used for weather and climate modeling.

3.1   Spectral Dynamics

The spectral transform method was first developed in 1970 [9, 30] and became the dominant dynamical core for global weather and climate modeling for the two decades that followed. Today, ECMWF's world-leading IFS model is the premier example of a spectral model. Other examples include the U.S. National Weather Service's Global Forecast System[6], the U.S. Navy's Global Environmental Model (NAVGEM), and the Japan Meteorological Agency's Global Spectral Model.

Whereas grid-point models represent fields as values at discrete points, spectral models use expansions on a series of spherical harmonics. Each vertical layer of a horizontal field is represented as an M by N array of spectral coefficients. The M dimension corresponds to increasing wave numbers in the zonal (west-east) dimension of the domain; N corresponds to increasing wave numbers in the meridional (equator to pole) dimension. The M and N dimensions of spectral space extend to infinity but are truncated for computational purposes above a certain wave number. If the truncation is the same in both M and N dimensions, it is said to be triangular and has the favorable property of being isotropic and not subject to the

---

[6]The spectral dynamics in the U.S. weather service's Global Forecast System has reached end of life and has been replaced by FvGFS, a grid point model.

pole problem: discontinuities and time steps constrained by narrowing grid lines near the poles in grid-point models. Higher truncation limits correspond to higher spatial resolution. The spectral dynamics in use in the GFS at the U.S. National Weather Service truncated M and N at wave number 1534, equivalent to a grid of slightly over three million grid points covering the earth's surface at a resolution of 13 km. ECMWF's IFS model achieves finer (9 km) spatial resolution for physics and advection using fewer (1279) waves in M and N through the use of a cubic rather than linear mapping between grid points and the highest frequency wave in spectral space, and by using an octahedral adaptation to IFS's reduced physical-space grid.

Only a portion of a spectral model time step is computed in the spectral domain. Non-linear terms of the Eulerian dynamics, semi-Lagrangian transport, and physics—subgrid-scale radiative heating and cooling, convection, turbulence, surface drag, and other physical processes—are computed on grid points. To move between spectral and grid-point representations, a forward and inverse spectral transform is computed every model time step. The grid point to spectral transform first applies an FFT to each west-east circle of grid values along latitude lines of the domain, producing vectors of Fourier coefficients that correspond to wave numbers in the M spectral dimension. Next, a Legendre transform is applied in the equator-to-pole dimension to construct the N dimension of spectral space. Each resulting m,n spectral coefficient is the sum of the products of the m element of each Fourier vector times the Legendre coefficient for the Gaussian latitude from which the Fourier vector was computed. The Legendre transform is algorithmically equivalent to a matrix multiply, and can be implemented using calls to DGEMM in LAPACK.

The computational complexity of the combined N Fourier transforms is $O(N^2 \log N)$, where N is the truncation number. Overall, the spectral transform is dominated by the $O(N^3)$ complexity of the Legendre transform. ECMWF has been able to devise a "Fast" Legendre Transform (FLT) that is $O(N^2 \log N^3)$. The FLT exploits similarities of associated Legendre polynomials at all the Gaussian latitudes but with different wave number and then precomputes and reuses an approximate representation of the matrices. The FLT is less efficient than DGEMM at lower resolutions but breaks even and continues to improve at T2047 and higher [45].

Sensitivity to rounding error requires that Gaussian weights used in the Legendre transform be computed using double (64 bit) floating point precision, even when other parts of the model are computed at lower precision [7]. Other operations in the spectral transform and other parts of the model may be computed using single (32-bit) floating point precision but weather centers are only now beginning to explore reducing precision for better computational efficiency [31].

The spectral transform method has advantages for parallel computing and software engineering because virtually all computations in spectral dynamics are dependency-free and perfectly parallel, both over wave components in spectral space and in the two horizontal dimensions of physical grid space. From a software point of view, the parallelism in a spectral model is highly encapsulated. Code to implement message passing is compact and isolated to within the subroutines that transform back and forth between grid and spectral space each model time step.

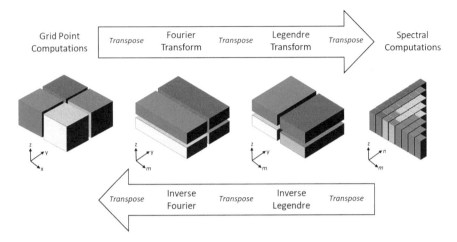

Fig. 5 Schematic of the forward and reverse spectral transforms in a time step of a spectral model, showing the decompositions of the physical, Fourier, and spectral domains over four tasks with transposes between the decompositions

Parallelizing the spectral transforms involves interprocessor communication that can be implemented in either of two ways: distribute the FFTs and DGEMMs themselves or transpose the data between decompositions to allow serial Fourier and Legendre transforms to be used. The advantage of using distributed FFT and DGEMM packages is that parallelism is built-in, the routines are portable and are likely to have been optimized for the computational platform. On the other hand, transpose implementations of spectral transforms are more flexible and general, permitting the use of non-power of two serial FFT packages. Transposes also ensure identical order of operations in the transforms giving results that are bit-for-bit reproducible on different numbers of MPI tasks. Generally, transpose implementations are favored because they send less data than distributed implementations, an advantage on systems where transfer (bandwidth) costs are high relative to message startup costs [11]. Figure 5 shows parallel transposes and data layouts for one time step of a typical spectral transform model.

Modern spectral models incorporate semi-implicit semi-Lagrangian transport (SLT) for advection because SLT is unconditionally stable and allows longer time steps. As with fully Lagrangian methods, SLT involves calculating the trajectories of parcels of a fluid over time; the difference being that SLT interpolates forward or backward trajectories relative to a fixed grid at each time step.

The issue for parallelizing SLT occurs when parcels flow to an area of the domain on another processor. As with finite-difference and finite-volume methods, these dependencies and dependencies associated with interpolation stencils are addressed by communicating with neighboring tasks to update halo- or ghost-regions around a task's subdomain. Anisotropy of the domain closer to the poles will require more data to be sent to update increasingly wide halos. Fortunately, the reduced grid used for computational efficiency elsewhere in the model (Fig. 4) also helps address the communication costs near the poles for SLT.

Spectral models have been extremely successful since the 1970s and are still deployed in major forecast centers. Nevertheless, the spectral method is approaching obsolescence on new generations of supercomputers that will require applications that can exploit 10^5 to 10^6 way parallelism without losing efficiency to parallel overheads such as interprocessor communication. Of course, any model of fluid flow requires communication between processors, but communication cost for local methods such as explicit finite-difference and finite-volume remains constant with increasing domain sizes and numbers of processors. Cost for non-local communication in spectral models increases as a function of domain size. Communication cost for high-resolution 5 km and 2.5 km experimental runs of ECMWF's IFS on up to a quarter-million processor cores of the TITAN supercomputer at Oak Ridge National Laboratory reached 75% of the total cost of the spectral transforms [46]. ECMWF estimates that stopgap improvements such as the Fast Legendre transform, the cubic grid-to-spectral mapping, and an octahedral grid-reduction geometry can extend the spectral IFS model's life for a time, but are exploring grid-point formulations for scaling to higher resolution [42].

3.2   Grid-Point Dynamics

The first computer models of the atmosphere were grid-point models, which flourished during a period of active development beginning with the U.S. Joint Numerical Weather Prediction Unit in the 1950s and lasting until a two-decade hiatus around the advent of spectral transform models in the 1970s. A key focus was to address the pole problem inherent in Cartesian latitude-longitude grids, leading to development of novel and promising quasi-uniform mesh geometries such as those shown in Fig. 4. These included composite and overset grids, icosahedral and geodesic grids, reduced latitude-longitude grids, Fibonacci grids (these were later), and regular polyhedra circumscribed to the sphere, most commonly the cubed sphere. The new approaches were generally successful at addressing the pole problem but presented other issues for solution quality: noise and interpolation error at boundaries of overset meshes, the edges and corners of faces on the cubed sphere or at the 12 pentagons in hexagonal meshes. Numerous schemes to reduce or eliminate these issues were developed and the topic remains an active focus of research and development today.[7]

Generating grids that are composed of Cartesian grids involves projecting the component grids onto curvilinear coordinates of the sphere. The cubed-sphere grid in the U.S. National Weather Service's next model, FvGFS, is composed of six

---

[7]The PDEs on the Sphere workshop series (https://pdes2017.sciencesconf.org) have focused on the problems of grids and numerical methods for weather, climate, and ocean circulation since 1990.

Cartesian grid faces of a cube inflated out to the surface of an enclosing sphere. The global version of the NCAR Weather Research and Forecast (WRF) model [41] is an overset mesh scheme comprised of two Cartesian meshes, one covering each hemisphere and then projected onto polar stereographic coordinates. In both cases, and aside from the extra work involved to handle the corners and edges on the cubed sphere and the overlap regions of the overset mesh, the component grids themselves are Cartesian and straightforward from a coding point of view. Traversing the domain and accessing values for neighboring grid cells is done using array index arithmetic inside multiply nested loops that are easily recognized and optimized by modern compilers.

Approaches for non-Cartesian grids are more complex and interesting from numerical, geometric, and computational points of view. The Non-hydrostatic Icosa-hedral Model (NIM) developed at NOAA uses an icosahedral mesh constructed mostly of hexagons with 12 pentagons. The Model for Prediction Across Scales (MPAS) developed by NCAR and Los Alamos National Laboratory uses a cen-troidal Voronoi tessellation (CVT) of arbitrary polygons that aligns to an icosahedral hexagonal mesh but that allows further in-place refinement (Fig. 6) to focus higher resolution over an area of interest: for example, the Gulf of Mexico and western Atlantic during hurricane season. The CVT in MPAS obeys the additional constraint that lines connecting neighboring cell centers bisect the neighbor edges and intersect at right angles. This supports an unstructured generalization of Arakawa C-grid staggering used to overcome problems with the representation of gravity waves in collocated grids while addressing problems reproducing geostrophic balance that stem from the discretization of the Coriolis force [36].

The unstructured horizontal dimensions in the MPAS grid are represented as arrays of vertical columns of the domain. The relationships between adjacent cell centers, edges, and vertices are computed when a mesh is generated and stored as integer arrays for each column. Traversing the grid and accessing neighbor values

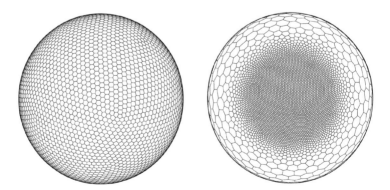

Fig. 6 Centroidal Voronoi Tessellation (CVT) of a quasi-uniform (left) and variable resolution MPAS mesh. The meshes shown contain the same number of grid cells [32]. © American Meteorological Society. Used with permission

Algorithm 1 (Lloyd method) *Given a domain Ω, a density function $\rho(\mathbf{x})$ defined on Ω, and a positive integer n.*

0. *Select an initial set of n points $\{\mathbf{x}_i\}_{i=1}^n$ on Ω ;*

1. *Construct the Voronoi regions $\{V_i\}_{i=1}^n$ of Ω associated with $\{\mathbf{x}_i\}_{i=1}^n$;*

2. *Deterimine the (constrained) mass centroids of the Voronoi regions $\{V_i\}_{i=1}^n$; these centroids form the new set of points $\{\mathbf{x}_i\}_{i=1}^n$;*

3. *If the new points meet some convergence criterion, return $\{(\mathbf{x}_i, V_i)\}_{i=1}^n$ and terminate; otherwise, go to step 1.*

Fig. 7 Lloyd algorithm for constructing centroidal Voronoi tessellations used in the MPAS model [35]

for computation require indirect indexing, a computational penalty compared to iterating over Cartesian meshes. This impact can be offset by ordering grid points in memory to be stored successively in vertical vectors that can be vector-parallel on CPUs [25] and thread parallel on GPUs.

Relative to Cartesian meshes, generating unstructured meshes is complicated and expensive and is typically done offline. The MPAS grid generation program is based on a method originally developed at Bell Laboratories for signal processing (Lloyd 1982) and applied to generating CVTs on the sphere (Fig. 7). The Lloyd method is sequential and essentially trial and error, so that creating a global mesh at a new quasi-uniform resolution or generating a global mesh with new areas of refinement may require days of computer time to converge[8]. Fortunately, once generated, the meshes can be reused, rotating to a different orientation if necessary to expose a different part of the domain to the area of mesh refinement. Improvements in mesh-generation speed have been obtained using GPUs and work to improve the quality and speed of generated CVTs is ongoing [21] (Engwirda 2017 JIGSAW-GEO).

3.2.1 Domain Decomposition

Once generated, decomposing unstructured grids involves finding a partitioning that assigns approximately equal numbers of grid columns to MPI tasks for load balance while minimizing the surface area to adjacent partitions to minimize the volume of data communicated. The MPAS model uses the METIS package from the University of Minnesota to decompose its domain over tasks (Fig. 8).

---

[8]Skamarock, W., personal communication.

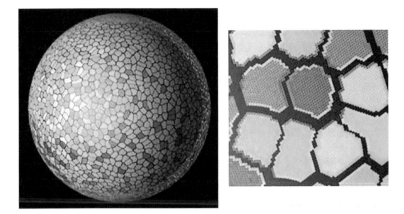

Fig. 8 MPAS unstructured grid decomposition that minimizes the amount of overlap between subdomain edges, thus minimizing the amount of data that must be communicated. Reproduced with permission [34]. Used with permission. Partitioning was generated using the METIS package from U. Minnesota [24]

METIS uses recursive bisection and K-way partitioning to find a decomposition that minimizes computational imbalance which results from uneven distribution of work and communication imbalance which results from edge-cuts between vertices of the mesh.

The icosahedral mesh of the NIM model consists of the ten rhombus-shaped faces and is decomposed over tasks in two steps. First, each face is assigned to a separate set of MPI tasks. Then each face is decomposed over its set of tasks in checkerboard Hex[9]-board? fashion. The decomposition originally required NIM to run on multiples of 10 MPI tasks. Subsequent refinements using 20 rhombuses allowed multiples of 1, 2, or 5 MPI tasks. Ordering the grid columns on each task in spirals eliminated the need copy cells into buffers to send and receive data to neighboring tasks through MPI (Fig. 9).

Space filling curves have been used to order and decompose elements in the HOMME and NEPTUNE spectral element models on cubed-sphere grids [4, 6].

3.2.2 Load Imbalance

Partitioning may also need to account for varying amounts of work for a given column depending on location in the domain and the time of the simulation. Sources of load imbalance can be static or can vary over the course of a simulation. An example of a static imbalance occurs when the number of processors does not divide the number of grid columns without a remainder. Or with limited-area

---

[9]https://www.hexwiki.net.

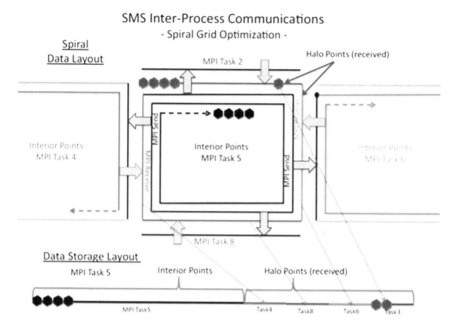

Fig. 9 Spiral traversal of icosahedral grid cells for MPI tasks in the NOAA NIM model, allowing interior data and data stored for interprocessor halo exchanges stored contiguously in memory. The ordering allows exchange sections of the computational storage to be passed directly to MPI without additional copies. © American Meteorological Society. Used with permission [18]

(not global) domains, the computations on the lateral boundaries may involve less work. Static imbalances can be addressed by assigning different numbers of grid columns to processors. Dynamic imbalances may be associated with processes such as cloud physics that require more computation around convective systems (storms).

Arriving at a perfectly balanced load is usually not possible. A weather model is multi-phasic, performing different physical or dynamical processes in the course of a time step and also different mixes of these processes from one time step to another. Since each phase of computation may have a different load profile, using an optimal decomposition to balance each phase is impractical. Inefficiency from load imbalance in the 5–20% range is usually not enough to justify the cost of redistributing work and data between each phase. An exception is the imbalance associated with the diurnal cycle in radiation physics that is computed only in the sunlit half of the domain. Here the imbalance is large and regular enough for load balancing to provide a benefit, even with the cost for relocating the data [10, 37].

3.3   Element-Based Dynamics

Finite element and spectral element methods, used widely in aerospace and other applications of computational fluid dynamics, are being applied to weather modeling because they are high order local methods that scale well computationally. Each element computes a local solution to the desired level of accuracy using an expansion on a set of orthogonal basis functions, not unlike the calculations done globally in spectral models described above. The element-local solutions are combined to form a global solution using either a continuous or discontinuous Galerkin method: the local solution at points along the faces of each element is summed with the edge solutions of the element's neighbors in a process called direct stiffness summation (DSS). DSS requires only nearest neighbor communication and the amount of data communicated is constant with respect to numerical order. Thus, element-based methods provide the accuracy and high computational intensity of globally spectral methods but without domain-wide interprocessor communication that inhibits scalability. Element-based methods are also well suited to complex geometries and lend themselves to adaptive mesh refinement [2, 12, 14, 15, 29]. Examples of models using element-based dynamical cores include the NUMA[10] dynamical core in the U.S. Navy's NEPTUNE and the HOMME dynamical core used in the Community Earth System Model[11] and the Department of Energy's Energy Exascale Earth System Model (E3SM)[12]. The UK Met Office is developing the finite element Gung-Ho[13] dynamical core for its new LFRic[14] modeling system.

Benchmarking NOAA's Next Forecast Model
In 2015 the National Weather Service needed to replace its aging Global Spectral Model. Six dynamical cores from development teams in the USA were evaluated: NOAA/GFDL's FV3, NOAA/NCEP's NMM-UJ, NOAA/ESRL's NIM, NCAR's MPAS, and Naval Research Laboratory's NEPTUNE based on the Naval Postgraduate School's NUMA model. FV3, NMM-UJ, and NEPTUNE used a cubed-sphere grid; NIM and MPAS used icosahedral/unstructured. Numerically, NMM-UJ used finite-difference; FV3, NIM, and MPAS were finite-volume; and NEPTUNE/NUMA used spectral elements. ECMWF's spectral/semi-Lagrangian IFS was included for comparison.

Computational performance and scaling were benchmarked on Edison, a large Cray supercomputer at the Department of Energy's NERSC facility. The first chart shows performance results for the models running a 13 km resolution workload (up to 3.5 million cells). The horizontal dotted at 1.0 is the speed threshold for forecasting. The second chart shows strong scaling efficiency for a higher resolution

---

[10]http://faculty.nps.edu/fxgirald/projects/NUMA/Introduction_to_NUMA.html.

[11]http://www.cesm.ucar.edu.

[12]https://e3sm.org/.

[13]https://www.metoffice.gov.uk/research/foundation/dynamics/next-generation.

[14]https://www.metoffice.gov.uk/research/modelling-systems/lfric.

3 km resolution workload (up to 65-million cells) expected to be commonplace within the next decade.

The fastest models scaled the least well, a not unexpected result. Computationally heavy models like MPAS and NEPTUNE perform more work per processor making the overhead from communication proportionately less costly. Non-local communication in IFS's spectral transforms hindered its scaling. FV3 ran 1.36 times faster (and scaled less well) at single precision than double precision and gave acceptable results [27].

The evaluation concluded in 2016 with selection of GFDL's FV3. The reports from all phases of testing are available online from the National Weather Service.

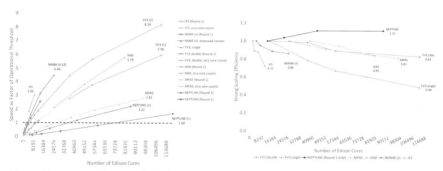

(https://www.weather.gov/sti/stimodeling_nggps).

For a given forecast configuration, element-based methods are more costly in terms of floating point operations than finite-difference and finite-volume based approaches but provide greater accuracy and scalability to large numbers of parallel threads on current and next generation HPC architectures. The NUMA spectral element dynamical core was the first ever to achieve operational forecast speed at a uniform global resolution of 3 km (1.8 billion cells), scaling with 99% efficiency to the full 786-thousand cores of the IBM Blue Gene/Q Mira system at Argonne National Laboratory [28]. In NOAA's 2015 intercomparison to choose the next dynamical core for the U.S. National Weather Service, the NUMA/NEPTUNE dynamical core was the most costly but also the most efficient running up to the full number of processors available (see "Benchmarking NOAA's Next Forecast Model").

3.4 Physics

The parts of a weather model that provide forcing terms that drive atmospheric dynamics—radiative heating, evaporation, condensation, convection, chemistry, turbulence, surface drag, and other physical processes—are collectively known as physics (the usage may be singular or plural). Physics differentiate weather and climate models from more general computational fluid dynamics applications.

Physics packages in a model are parameterizations because they are simplified representations of processes that occur at subgrid scales, too fine to be resolved by the dynamics. Physics is where much of the predictive skill of a model resides. Adapting a physics package to a particular forecast application involves tuning—adjusting parameters within the physics package—to remove forecast error and biases at a given forecast scale with respect to observations.

Physics usually represents processes that act only in the vertical dimension, and is perfectly parallel between adjacent columns in the horizontal domain dimensions; however, physics work-per-column depends on the state of the atmosphere and is a major source of load imbalance. There are opportunities for parallelism over different physics packages—for example, running radiative transfer concurrently with convention and other physics. Parallelism in the vertical dimension is typically limited or non-existent.

The computational cost of physics is a significant fraction of the overall cost of a model run, anywhere from 20% to half of a typical forecast depending on the configuration. Radiative transfer (Fig. 10) and cloud microphysics (Fig. 11) are typically the most expensive physics components unless chemistry is also employed. In that case, for air quality and pollution predictions, the cost of simulating chemical reactions in the atmosphere and for advecting large numbers of chemical tracers

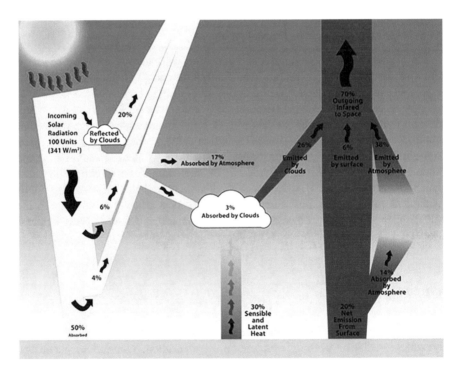

Fig. 10 Heating and cooling from incoming shortwave and outgoing longwave solar radiation as modeled by the Rapid Radiative Transfer Model. Illustration by AER Corp. Used with permission

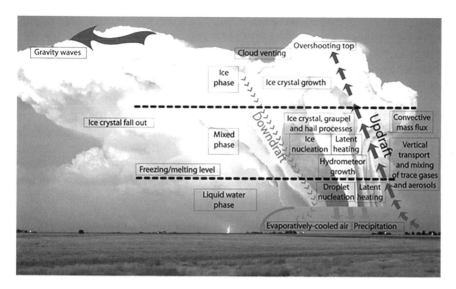

Fig. 11 Cloud microphysics models subgrid-scale moisture processes governing production of precipitation in multiple forms and thermodynamic feedbacks from evaporation and condensation in active convection. Illustration by Rob Seigel, Colorado State University. Used with permission

may be several times greater than the cost of the entire rest of the model. Physics is also state-heavy. While dynamics requires no more than a half-dozen or so prognostic variables per grid cell, the combined working set for a full-physics meteorological application is at least an order of magnitude larger (two orders larger with chemistry). In spite of large working sets, physics makes greater use than dynamics of exponent, log, square root, power, and other intrinsic operations and is therefore more computationally intense than the model overall.

From a software point of view, physics code must be updated more frequently than dynamics and is a source of inconsistency in a model's software repository. The physics packages within a given model may have been developed and contributed by groups of experts outside a model development team using different vertical coordinates, representations of physical fields, and coding practices.

4 Challenges for Next-Generation HPC

HPC systems are increasingly out of balance. Floating point capability is increasing but the usable percentage is decreasing because only the number of floating point units that can be constructed and powered for a given area of silicon and watt of electricity continues to increase exponentially (and that may end soon). Rates of increase for memory system, network, and I/O performance have slowed. The 8 billion transistor Knights Landing (KNL) processor, Intel's most recent (and last) generation of Intel's Many-Integrated Core (MIC) architecture, was rated at up to three TFLOPs peak performance (2.2 TFLOPs measured). To achieve that, however, an application would need to perform seven floating point operations for every byte accessed from KNL's high-bandwidth (700 GB/s) MCDRAM memory. The number one ranked system on the Top500[15] list at this writing was the 200 PFLOPs Summit system at Oak Ridge National Laboratory, which comprises 28,000 NVidia Volta (V100) GPUs. To reach the rated 7 TFLOPs peak performance on the V100 GPU, an application must perform nine operations for every byte accessed. By contrast, the highest computational intensity (CI) measured for a full NWP model (non-spectral transform) is 0.7 operations per byte [28], an order of magnitude gap between application intensity and realizable floating point performance that is widening with time.

One may argue that realized percentage of peak performance is an artificial metric and that time-to-solution is what matters. If a model is scalable, why not use larger numbers of processors to reach the required simulation speed? In the first place, as discussed above, real-time deterministic weather forecasting does not weakly scale with resolution because the sequential temporal dimension must also be refined. But even within this fundamental scaling limit, there are also sound practical reasons to worry about efficiency. Although it may be possible to run a forecast using 15–20 MW of electricity, wasting all but a few percent is difficult to justify. And application parallelism itself is a limited resource. Scaling to more tasks and threads without using the resources available to each thread efficiently leaves performance on the table and limits additional speedup unnecessarily.

Roofline analysis [47] characterizes realizable performance in terms of how an application maps to the memory system and computational capabilities of a processor. The idea, illustrated in the roofline plot for a Knights Landing (KNL) processor (Fig. 13), is that performance (vertical axis) is bound by memory system performance in the sloping part of the roofline. In that region, the memory system cannot provide operands fast enough to keep up with the floating point units of the processor. When CI is high enough and the roofline is flat, the application is bound only by its ability to saturate the speed of the floating point units. The several sloping parts of the roofline in the figure correspond to levels of the KNL memory system from the fastest and smallest level one cache out to the DRAM main-memory on the KNL device. The memory image of a weather model is too small to fit entirely within the L1 and L2 caches but does fit within the 16 GB MCDRAM, the high-bandwidth (nominally 400 GB/s, 387 GB/s measured) on-chip memory of the KNL. Thus, MCDRAM bandwidth is the limiting factor on KNL for applications with CI of <7 FLOP/byte. The shaded area shows how little of the KNL's peak performance is used by weather models with an overall CI of <1. Optimization involves restructuring loops and data structures to increase memory locality, moving CI to the right; and then increasing vector utilization to use as much of the increased headroom under the roofline as possible.

---

[15]June 2018 Top500 list, https://www.top500.org/lists/2018/06/.

The example in the figure is the roofline plot of an expensive subroutine from the NEPTUNE model's profile that diffuses energy cascading to wave numbers too high to be resolved. The routine already has better than average CI, but additional improvement was obtained using an AoS to SoA (array of structures to structure of arrays) transformation. The original version of the code copied from a model array into local spectral element arrays and back again. Fields in the state array were stored together for each point (AoS), so that traversing a field required non-stride-one accesses. In the optimized code, fields for each element were stored in an element structure (SoA). The element structures were stored as an array (AoSoA) that replaced the original state array. The diffusion routine was modified to be called over each element structure in the state array and compute using the field data in place without copying in and out of the routine. This optimized memory restructuring moved CI to the right from 1.1 to 1.5 FLOP/byte.

Remaining optimization involved increasing vector and FMA (fused multiply-add) utilization by reorganizing loops to make it easier for the compiler to generate vector and FMA instructions, nearly doubling performance. A key benefit of roofline analysis is signaling when the programmer has more work to do. In this case, Fig. 12 shows considerable unexploited headroom remained beneath the 700 GFLOPs roofline for MCDRAM, suggesting other sources of inefficiency, for example, incomplete vector utilization, instruction latency, or load imbalance between threads.

Figure 13 shows the end result of a several month cycle of AoS to SoA and other optimizations to improve CI and vectorization in the NEPTUNE model. The solid bars are performance of the unoptimized code for a workload small enough for a single node running on generations of Intel Xeon processors and Cavium Corp's ARM-based ThunderX2 processor. The hatched bars show the increase in simulation rate (simulated time over wall clock) after optimization.

4.1 Next Generation HPC and the Programming Challenge

In the absence of increasing clock rates, effort now focuses on-processor fine-grained parallelism: threads on GPUs and vector instructions on CPU cores. On GPUs, approaches have ranged from inserting OpenACC or OpenMP directives to offload computation to the GPU to complete recoding into NVidia's CUDA programming language. Because of the difficulty of generating and maintaining a separate GPU version, the only instance of an entire weather model converted to CUDA by hand was the Japan Meteorological Agency's ASUCA model [39]. The authors showed their code running 80 times faster on the GPU, but with caveats. The comparison was relative to original Fortran code running on a single CPU core. Moreover, single-precision GPU performance was compared to performance of the original code at double precision. Taking this into account, one estimates that a node-for-node GPU to CPU comparison with equivalent configurations would have

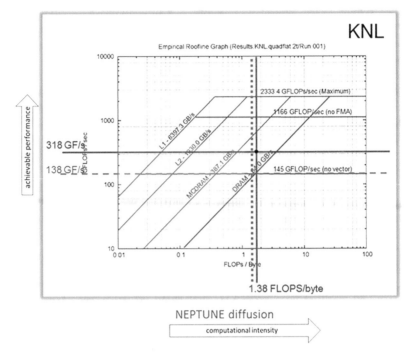

Fig. 12 Roofline plot for a costly diffusion routine in the NEPTUNE spectral element model. The plot was generated by running the UC Berkeley's Empirical Roofline Toolkit on an Intel Xeon Knights Landing processor, then annotated with computational intensity (CI) and performance that was measured for the diffusion kernel. Dotted lines show original measurements, solid show after optimization. The shaded box shows the portion of the KNL's theoretical peak performance that can be utilized by weather models having overall CI of <1.0

yielded two- to four-times speedup, a ratio that has remained consistent with other NWP codes on successive generations of hardware.

Directive-based approaches using OpenACC and OpenMP allow code to be implemented, maintained, and optimized on both CPU and GPU architectures. NOAA's Earth System Research Laboratory developed a single-source implementation of the NIM model (one of the models described in the box) using OpenACC directives [18]. The authors showed a two to three times performance benefit for GPUs compared to conventional multicore CPUs and 1.3 times compared to MIC on a device-to-device basis. This was without accounting for the additional cost of moving data between the GPU device and its host processor. The authors showed up to a $2\times$ GPU to CPU benefit in terms of hardware cost in dollars, after internode communication and overhead for transferring data between the host and GPU device were addressed.

Meteo Suisse has deployed a GPU version of the COSMO model that was implemented using Gridtools (formerly STELLA), a domain specific framework of C++ templates and libraries developed by the Swiss National Supercomputing

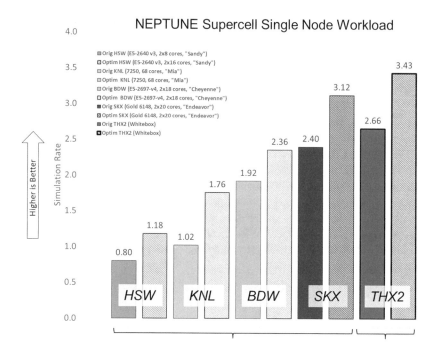

Fig. 13 Original and optimized performance (as simulation rate) of the NEPTUNE model on single-node workload over a successive generations of multicore CPUs

Center. Gridtools uses template metaprogramming to embed the DSL within the C++ host language. At compile time the DSL is translated into an executable with OpenMP threading on CPU architectures and CUDA for GPUs [13]. Physics in the COSMO model was adapted to GPU outside of Gridtools using OpenACC. The authors reported between two and three times faster performance on the GPU compared to multicore CPU, depending on the amount of work (number of grid points) per node.

On the MIC architecture, application speedups relative to conventional multicore CPUs are similar to speedups seen on GPU, but with considerably less programming effort. This is because vector and parallel programming on the Knights Landing is fundamentally the same as for conventional multicore Xeon processors. WRF and other models able to use both MPI for message passing and OpenMP for threading ported easily to MIC. Programmers can focus attention on exploiting fine-grain parallelism, usually by helping the compiler recognize and generate vector instructions and by restructuring code and data to make more efficient use of cache and memory. The Knights Landing version of the MIC ran a standard WRF benchmark 1.7 times faster than one node (two sockets) of an Intel Xeon (Broadwell) processor [16].

Porting and optimizing NWP codes for next generation architectures remain areas of active effort and research, and are the focus of numerous conference and workshop series.[16, 17, 18] The European Union's Energy-efficient Scalable Algorithms for Weather Prediction at Exascale (ESCAPE) is a 3-year project to address the problem for weather and climate services in the EC, stated as follows:

> Existing extreme-scale application software of weather and climate services is ill-equipped to adapt to the rapidly evolving hardware. This is exacerbated by other drivers for hardware development, with processor arrangements not necessarily optimal for weather and climate simulations.[19]

A key activity within ESCAPE has been to identify and package kernel benchmarks called Weather and Climate Dwarfs, after the original Berkeley Dwarfs [1], to focus co-design efforts between the applications and HPC research and manufacturing communities. In the USA, the HPC working group of the multi-agency Earth System Prediction Capability (ESPC) program comprises model developers and users from NOAA, NASA, DOE, the Dept. of Defense, and the National Science Foundation. Less far along than the European efforts, at this writing the ESPC group had defined initial requirements on which to undertake effort along the lines of the EC program [5].

5 Summary

> "Modern weather prediction is perhaps the most cooperative activity of our species"
> – Prof. Clifford Mass, Dept. of Atmos. Sciences. U. Washington [26]

Today, the numerically generated weather information available through print, radio, television, and the internet is public's most direct experience with high-performance computing. The half-century history of numerical weather prediction is a story of massive scientific and technical investment on an international scale; of steady progress fraught with technological disruption harnessing a billion-fold increase in computer power; and of the challenges for continuing to add value from numerically generated forecasts into the exascale era.

Acknowledgements Thank you to Tom Henderson, Kevin Viner, Jim Doyle, John Dennis, Michael Duda, Jacques Middlecoff, and Jordon Powers for reading the manuscript and providing their valuable advice.

---

[16]NCAR Multicore Workshop series: www2.cisl.ucar.edu/events.

[17]ECMWF Workshop on HPC in Meteorology: events.ecmwf.int.

[18]AMS Symposium on HPC for Weather, Water and Climate: ams.confex.com.

[19]http://www.hpc-escape.eu/.

References

1. Asonovic, K., Bodik, R., Catanzaro, B., Gebis, J., Husbands, P., Keutzer, K., . . . Yellick, K. (2006). *The landscape of parallel computing research: a view from Berkeley.* Electrical Engineering and Computer Sciences. University of California at Berkeley. Retrieved from http://www.eecs.berkeley.edu/Pubs/TechRpts/2006/EECS-2006-183.html

2. Bao, L., Nair, R., & Tufo, H. (2014). A mass and momentum flux-form high-order discontinuous Galerkin shallow water model on the cubed-sphere. *Journal of Computational Physics, 271,* 224-243.

3. Bauer, P., Thorpe, A., & Brunet, G. (2015, September 3). A quiet revolution of numerical weather prediction. *Nature, 525,* 47-55. Retrieved from www.nature.com/doifinder/10.1038/nature14956

4. Burstedde, C., Wilcox, L., & Ghattas, O. (2011, May). p4est: Scalable Algorithms for Parallel Adaptive Mesh Refinement on Forests of Octrees. *SIAM J. Sci. Comput., 33*(3), 1103-1133. Retrieved from https://doi.org/10.1137/100791634

5. Carman, J., Clune, T., Giraldo, F., Govette, M., Gross, B., Kamrath, A., . . . Whitcomb, T. (2017). *Position paper on high performance computing needs in earth system prediction.* ESPC position paper, National Earth System Prediction Capability, Silver Spring, MD. doi:10.7289/V5862DH3

6. Dennis, J. (2003). Inverse space-filling curve partitioning of a global ocean model. *Proceedings fo IEEE International Parallel and Distributed Processing Symposium,* (p. 7). doi:10.1109/IPDPS.2003.1213486

7. Drake, J., Flanery, R., Semararo, D., Worley, P., Foster, I., Michalakes, J., . . . Williamson, D. (1995). *Parallel Community Climate Model: Description and User's Guide.* Oak Ridge National Laboratory.

8. Drake, J., Semeraro, B., Worley, P., Foster, I., Michalakes, J., Toonen, B., . . . & Williamson, D. (1994). *PCCM2: A GCM adapted for scalable parallel computers.* Chicago, IL: Argonne National Laboratory. Retrieved from https://www.osti.gov/biblio/10114472

9. Eliasen, E., Machenhauer, B., & Rasmussen, E. (1970). *On a numerical method for integration of the hydrodynamical equations with a spectral representation of the horizontal fields.* Report No. 2, University of Copenhagen, Institute for Teoretisk Meteorologi.

10. Foster, I., & Toonen, B. (1995). *Load Balancing Algorithms for the Parallel Community Climate Model.* Technial memorandum ANL/MCS-TM-190, Argonne National Laboratory.

11. Foster, I., & Worley, P. (1997). Parallel algorithms for the spectral transform method. *SIAM Journal on Scientific Computing, 18,* 806-837. doi:10.2172/10168301

12. Fournier, A., Taylor, M., & Tribbia, J. (2004). A spectral element atomsopheric model (SEAM). *Monthly Weather Review, 132,* 726-748.

13. Fuhrer, O., Osuna, C., Lapillonne, X., Gysi, T., Cumming, B., Bianco, M., . . . Schulthess, T. (2014). Towards a performance portable, architecture agnostic implementation strategy for weather and climate models. *Supercomputing Frontiers and Innovations, 1*(1), 45-62. doi:10.14529/jsfi140103

14. Gaberšek, S., Giraldo, F., & Doyle, J. (2012). Dry and moist experiments with a two-dimensional spectral element model. *Mon. Wea. Rev., 140,* 3163-3182.

15. Giraldo, F., & Rosmond, T. (2004, January). A scalable spectral element Eulerian atmospheric model (SEE-AM) for NWP. *Monthly Weather Review, 132,* 133-153.

16. Gokhale, I., & Michalakes, J. (2016). Weather Research and Forecasting (WRF). In J. Jeffers, J. Reinders, & A. Sodani, *Intel Xeon Phi Processor High Performance Programming: Knights Landing Edition* (pp. 499-509). Morgan Kaufman.

17. Golding, B., Mylne, K., & Clark, P. (2004). The history and future of numerical weather prediction in the Met Office. *Weather, 59*(11), 299-3-6. doi:10.1256/wea.113.04

18. Govett, M., Rosinski, J., Middlecoff, T., Lee, J., MacDonald, A., Wang, N., . . . Duarte, A. (2017). Parallelization and performance of the NIM weather model on CPU, GPU and MIC

processors. *Bulletin of the American Meteorology Society* © *American Meteorological Society. Used with permission*, 2201-2213.

19. Haut, T., & Wingate, B. (2014). An Asymptotic Parallel-in-Time Method for Highly Oscillatory PDEs. *SIAM Journal on Scientific Computing, 32*(2), 693-713. Retrieved from https://doi.org/10.1137/130914577

20. Henderson, T., Michalakes, J., Gokhale, I., & Jha, A. (2015). Numerical weather prediction optimization. In J. Reinders, & J. Jeffers, *High Performance Parallelism Pearls, Volume 2* (pp. 7-23). Morgan Kaufman.

21. Jacobsen, D., Gunzburger, M., Ringler, T., Burkardt, J., & Peterson, J. (2013). Parallel algorithms for planar and spherical Delaunay construction. *Geosci. Model Dev., 6*, 1353-1365. doi:10.5194/gmd-6-1353-2013

22. Jones, P. (1996). The Los Alamos Parallel Ocean Program (POP) and Coupled Model on MPP and Clustered SMP Architectures. In G.-R. Hoffman, & N. Krietz (Ed.), *Seventh ECMWF Workshop on the Use of Parallel Processors in Meteorology* (pp. 226-238). Reading, UK: World Scientific.

23. Källén, E. (2016). Weather Prediction and the Scalability Challenge. *EASC 2016: Exascale Applications & Software Conference*. Stockholm, Sweden. Retrieved from https://youtu.be/WAqR4aUzpgo

24. Karypis, G., & Kumar, V. (1998). *METIS 4.0: Unstructured graph partitioning and sparse matrix ordering system*. University of Minnesota, Dept. of Computer Science and Engineering. University of Minnesota. Retrieved from http://glaros.dtc.umn.edu/gkhome/metis/metis/overview

25. MacDonald, A., Middlecoff, J., Henderson, T., & Lee, J. (2010). A general method for modeling on irregular grids. *International Journal of High Performance Computing Applications*. Retrieved from http://journals.sagepub.com/doi/abs/10.1177/1094342010385019

26. Mass, C. (2014, December 25). Is Numerical Weather Prediction One of Mankind's Greatest Achievements? Retrieved from http://cliffmass.blogspot.com/2014/12/

27. Michalakes, J., Govett, M., Benson, R., Black, T., Juang, H., Reinecke, A., & Skamarock, W. (2015). *AVEC Report: NGGPS Level-1 Benchmarks and Software Evaluation*. Technical report, NOAA, Office of Science and Technology Integration. Retrieved from https://www.weather.gov/sti/stimodeling_nggps_implementation_atmdynamics

28. Müller, A., Koera, M., Marras, S., Wilcox, L., Isaac, T., & Giraldo, F. (2018, April 5). Strong scaling for numerical weather prediction at petascale with the atmospheric model NUMA. *International Journal of High Performance Computing Applications*, 31. Retrieved from https://doi.org/10.1177/1094342018763966

29. Nair, R., Thomas, S., & Loft, R. (2005, April). A discontinuous Galerkin global shallow water model. *Monthy Weather Review*, 876-888. Retrieved from https://doi.org/10.1175/MWR2903.1

30. Orszag, S. A. (1970). Transform method for calculation of vector coupled sums: Application to the spectral form of the vorticity equation. *Journal of Atmospheric Science, 27*, 890-895.

31. Palmer, T., & Düben, P. (2014, August). The use of imprecise processing to improve accuracy in weather & climate prediction. *Journal of Computational Physics, 271*, 2-18. Retrieved from https://doi.org/10.1016/j.jcp.2013.10.042

32. Raucher, S., Ringler, T., Skamarock, W., & Mirin, A. (2012). Exploring a Global Multi-Resolution Modeling Approach Using Aquaplanet Simulations. *Journal of Climate* © *American Meteorological Society. Used with permission*, 2432-2452. doi:10.1175/JCLI-D-12-00154.1

33. Richardson, L. (1922). *Weather Predication by Numerical Process*. Cambridge University Press.

34. Ringler, T. (2018). Personal communication.

35. Ringler, T., Ju, L., & Gunzburger, M. (2008). A multiresolution method for climate system modeling: application of spherical centroidal Voronoi tessellations. *Ocean Dynamics, 58*(5-6), 475-498. doi:10.1007/s10236-008-0157-2

36. Ringler, T., Thuburn, J., Klemp, J., & Skamarock, W. (2010, May). A unified approach to energy conservation and potential vorticity dynamics for arbitrarily-structured C-grids. *Journal of Computational Physics, 229*(9), 3065-3090. Retrieved from https://doi.org/10.1016/j.jcp.2009.12.007

37. Rodrigues, E., Navaux, P., Panetta, J., Fazenda, A., Mendes, C., & Kale, L. (2010). A comparitive analysis of load balancing algorithms applied to a weather forecast model. *22nd International Symposium on Computer Architecture and High Performance Computing*. Petropolis, Brazil. doi:10.1109/SBAC-PAD.2010.18

38. Schuman, F. G. (1989). History of Numerical Weather Prediction at the National Meteorlogical Center. *AMS Weather and Forecasting, 4*, 286-296. Retrieved from https://journals.ametsoc.org/doi/pdf/10.1175/1520-0434%281989%29004%3C0286%3AHONWPA%3E2.0.CO%3B2

39. Shimokawabe, T., Aoki, T., Muroi, C., Ishida, J., Kawano, K., Endo, T., ... Matsuoka, S. (2010). An 80-fold speeup 15.0 TFlops, full GPU acceleration of non-hydrostatic weather model ASUCA production code. *Proceeddings of the 2010 ACM/IEEE conference on Super-computing (SC'10)*. New Orleans, LA.

40. Simmons, A., Burridge, D., Jarraud, M., Girard, C., & Wergen, W. (1989). The ECMWF medium range prediction models, development of the numerical formulations and the impact of increased resolution. *Meteorol. Atmos. Phys., 40*, 28-60.

41. Skamarock, W., Klemp, J., Dudhia, J., Gill, D., Barker, D., Wang, W., & Powers, J. (2005). *A description of the advanced research WRF version 2*. Technical report. Retrieved from http://www.dtic.mil/docs/citations/ADA487419

42. Smolarkiewicz, P., Deconinck, W., Hamrud, M., Kühnlein, C., Mizdzynski, G., Szmelter, J., & Wedi, N. (2015, Autumn). An all-scale, finite-volume module for the IFS. *ECMWF Newsletter*, pp. 24-29. Retrieved from https://www.ecmwf.int/en/elibrary/14589-newsletter-no-145-autumn-2015

43. Stern, H., & Davidson, N. (2015, October). Trends in the skill of weather prediction at lead times of 1-14 days. *Q. J. R. Meteorol. Soc.*, 2726-2736. doi:10.1002/qj.2559

44. Tolman, H. (2017). *The production suite: looking forward*. National Oceanic and Atmospheric Administration, OSTI, College Park, MD. Retrieved from http://www.emc.ncep.noaa.gov/annualreviews/day-2/01a-Tolman_NPSR_2017_townhalls.pdf

45. Wedi, N. P., Hamrud, M., & Mozdzynski, G. (2013). A fast spherical harmoics transform for global NWP and climate models. *Monthly Weather Review, 141*, 3450-3461.

46. Wedi, N., Bauer, P., Deconinck, W., Diamantakis, M., Hamrud, M., Kühnlein, C., ... Smolarkiewicz, P. (2015). *The modeling infrastructure of the Integrated Forecast System: Recent advances and future challenges*. Technical Memorandum 760, European Centre for Medium-Range Weather Forecasts, Reading, UK. Retrieved from https://www.ecmwf.int/sites/default/files/elibrary/2015/15259-modelling-infrastructure-integrated-forecasting-system-recent-advances-and-future-challenges.pdf

47. Williams, S., Waterman, A., & Patterson, D. (2009). Roofline: an insightful visual performance model for multicore architectures. *Communications of the ACM, 52*, 65-76. Retrieved from https://www.osti.gov/servlets/purl/963540

48. Williamson, D. (2007). The Evolution of Dynamical Cores for Global Atmospheric Models. *Journal of the Meteorological Society of Japan. Used with permission, 85B*, 241-269. Retrieved from https://pdfs.semanticscholar.org/5a42/471e95ec434e0eb08bf03380da4a578c420d.pdf

A Simple Study of Pleasing Parallelism on Multicore Computers

Yanfei Ren and David F. Gleich

1 Introduction

Current single-machine computing environments are a mixture of high-power CPUs and GPUs mixed to large quantities of memory of various speeds. Often these are subsequently networked together into large distributed computational platforms. Cloud computing further complicates the scenario as advanced resources can be purchased for the time needed. These environments present a wide-range of opportunities to schedule what are often called *pleasingly parallel* computations, namely, those that have a large amount of independent computation that can be scheduled simultaneously.

We wish to investigate how to leverage and utilize such resources in the context of a large graph computation. While the focus on large graph computation is often in terms of solving problems on massive graphs with distributed computation, the downside to such computations is that they often involve *nearly linear-time* algorithms [16]. These have runtimes such as $O(n \log n)$ and typically involve a small number of passes over all the edges of the graph, for instance, running a connected components analysis or computing a PageRank vector. Consequently, the algorithm performance is largely dominated by how well the computation maps to the IO and memory system strategies of the platform.

Instead, the computation we investigate is the all-to-all personalized PageRank computation. Given an n-node graph, this involves computing the personalized PageRank vector associated with each node. We state the problem formally in Sect. 2. Consequently, there are n such computations that are all independent and decoupled. In terms of the scale, we are targeting graphs with up to a 100 million

Y. Ren · D. F. Gleich (✉)
Department of Computer Science, Purdue University, West Lafayette, IN, USA
e-mail: ren105@purdue.edu; dgleich@purdue.edu

© Springer Nature Switzerland AG 2020
A. Grama, A. H. Sameh (eds.), *Parallel Algorithms in Computational Science and Engineering*, Modeling and Simulation in Science, Engineering and Technology,
https://doi.org/10.1007/978-3-030-43736-7_11

edges and with up to 10 million nodes. Real-world instances of such graphs are the LiveJournal social network crawl with around 4 million vertices and 67 million edges and the Orkut social network crawl with 3 million vertices and around 220 million edges.

Because the output from the all-to-all problem would be $O(n^2)$ data, we seek to output only summary statistics of the personalized PageRank vectors including inverse participation ratios for the solutions that serve as a soft-measure of the number of non-zeros, as well as the largest 1000 entries of the vectors. The large values are commonly used as latent measures of node similarity [14, 41, 46]. Hence, a simple strategy for this computation is to load the graph into memory on all computers available, take the *fastest* single-core algorithm for personalized PageRank, and run it as many places as possible.

This picture becomes more interesting in light of the heterogeneous nature of computers. For instance, we can use vector or SIMD instructions to potentially compute multiple PageRank vectors at once, if the algorithm used is amenable to it. Second, large shared memory machines may have a large number of computing cores (over 200 is possible with commercially available systems that cost less than $250k). However, many of these cores share memory bandwidth resources that can impede some algorithms. This suggests that sharing access to a single graph may not scale. Furthermore, GPUs are constantly changing their underlying compute resources. Fourth, the algorithm performance itself is likely to be sensitive to choices of data distribution within the graph due to memory locality. Hence, even for this simple setting there is a rich set of complications to simplistically expecting a pleasingly parallel algorithm to scale.

Our goal is to investigate these performance differences in the context of the simple personalized PageRank computation. We chose that computation as it is representative of a wide swath of related computations on graphs including scalable methods for all-pairs shortest paths. Moreover, the algorithms to compute it are simple. They are specializations of well-known matrix computation algorithms including the power method and Gauss–Seidel method [19]. We can easily investigate a diverse collection of possible implementations that have different memory characteristics. Our focus was to keep the investigation simple and reflective of what might be expected from an informed, but non-expert, user of the algorithms. This is someone who understand how the algorithms works, where the relevant bottlenecks might be, but does not want to attempt to re-engineer the algorithm for the absolute maximum level of parallelism or performance. This individual is optimistic that the pleasingly parallel nature of the computation will be sufficient to drive performance. Towards that end, we discounted using GPUs at the moment as the toolkit for graph computations on GPUs is still evolving.

We have done all of these experiments in the Julia computing environment to make it easy for others to further investigate our ideas. It is also a high-level programming language that makes it simple to implement a variety of algorithms in a consistent fashion. Regarding the idea that high-level languages may be slow, we initially benchmarked the Julia implementation against a C++ implementation of a similar algorithm [25] and found the runtimes of the methods to be within 10% of each other.

2 The PageRank and AllPageRank Problem

The PageRank problem begins with a graph G, which could be both weighted and directed. However, in the interest of simplicity, we take G to be unweighted, directed, and strongly connected. This greatly simplifies the setting and puts the focus on the relevant pieces of the computation. Let us note that we lose no generality by doing so: a PageRank computation on a graph with multiple strongly connected components can be reduced to a sequence of PageRank computations on the individual strong components, and usually, these additional computations are much smaller because most real-world networks only have a single large strongly connected component (see, among others who make this observation, [35]).

Fix an ordering for G's vertices from 1 to n and identify each vertex with its index in this order. Let A be the resulting adjacency matrix of G

$$A_{ij} = \begin{cases} 1 & (i, j) \text{ is a directed } i \to j \text{ edge} \\ 0 & (i, j) \text{ is not an edge.} \end{cases}$$

We use the following additional notation:

$\mathbf{d} = $ vector of degrees $\quad d_i = \sum_j A_{ij}$

$\mathbf{p} = $ vector of inverse degrees $\quad p_i = 1/d_i$

$P = $ the stochastic transition matrix $\quad P = A^T \text{Diag}(\mathbf{p})$

$\mathbf{e}_i = $ the ith column of the identity, $\mathbf{e}_i$ has a 1 in the ith row and 0 elsewhere,

where we use the $\text{Diag}(\cdot)$ operator to put the argument along the matrix diagonal. The PageRank problem [14, 28, 39, 40] is to compute the stationary distribution of a random walk that with probability α follows a standard random walk model on G and with probability $1 - \alpha$ jumps according a teleportation distribution vector $\mathbf{v}$, where $\mathbf{v}$ encodes the probability of jumping to each node. Typically α is between 0.5 and 0.99. Throughout this paper, we use what became the *standard value* of 0.85 [28]. The stationary distribution corresponds to a solution of the following nonsingular linear system:

$$(I - \alpha P)\mathbf{x} = (1 - \alpha)\mathbf{v}.$$

Personalized, or seeded, PageRank problems set $\mathbf{v}$ to be a single node, or in this case, a column of the identity matrix $\mathbf{e}_i$ and the linear system

$$(I - \alpha P)\mathbf{x}_i = (1 - \alpha)\mathbf{e}_i. \tag{1}$$

As an aside, we note that a standard feature of most PageRank constructions [14] is the dangling correction vector $\mathbf{c}$. In this case, we do not have this correction vector because we assume that we are given a strongly connected graph.

Our goal is to compute $\mathbf{x}_i$ for all i from 1 to n, or more simply, the matrix

$$X = (1 - \alpha)(I - \alpha P)^{-1}.$$

We wish the entries of X to be of high accuracy, and intend to compute each column of X such that the 1-norm error is provably less than $(1 - \alpha)/n$. Because the graph is strongly connected, the matrix X is dense when computed exactly. For a graph with one million vertices this graph is too large to store even on a large shared memory machine. We thus define the AllPageRank problem.

Problem 1 (AllPageRank) Fix a graph G, let A be a binary adjacency matrix indicating the presence of an edge, let P be a column stochastic matrix giving transition probabilities on the same graph. The AllPageRank problem is to compute the following entries of $(1 - \alpha)(I - \alpha P)^{-1}$:

- the participation ratio for each column $\mathbf{x}_i$, which is a soft measure of the number of non-zeros in the column
- the non-zero values of $X \odot A$ and $X^T \odot A$, and
- the k largest entries in each column, for $k = 100$ or $k = 1000$.

Here $\odot$ is the elementwise, or Hadamard, product. Note that the transition matrix P need not come from the transition described above and could come from anywhere, such as the common stochastic transformations in PageRank [14]. Nonetheless, we will always use $P = A^T \text{Diag}(\mathbf{p})$ in this manuscript. The results of AllPageRank could be used to form a nearest neighbor approximation, to form PageRank affinities [46], or simply as a diffusion approximation of the underlying graph. Additional applications of such an output involve similar methods that solve protein function inference problems [23, 31]. Finally, this can be related to some idea of a "PageRank effective resistance" on an edge.

We stress that there *are* applications of the output for PageRank, but that our general goal is to use PageRank as a model computation that is representative of the challenges faced by more general numerical computing problems on graphs. This is akin to PageRank's widespread use to evaluate the performance of distributed graph computation engines [1, 9, 26, 33, 43, 44]. See additional examples of related computations in Sect. 5.

3   PageRank Algorithms

There are a few classic algorithms for PageRank computations: the power method, the Gauss–Seidel method, and the *push method* in two variations. We briefly explain these algorithms, give a small pseudocode for the computation, as well as an easy-to-compute error bound.

For the following set of algorithms, we will describe how to use them to compute a single vector, although we note that all of them are amenable to computing multiple vectors simultaneously as discussed in Sect. 3.6. We will use the notation $\mathbf{x}$ to refer to the solution vector to $(\mathbf{I} - \alpha \mathbf{P})\mathbf{x} = (1 - \alpha)\mathbf{v}$ where $\mathbf{v} = \mathbf{e}_i$ for some fixed i. Each iterate in a high-level description of the method will be written $\mathbf{x}^{(k)}$; what exactly constitutes an iteration may vary among the discussions. For instance, for Gauss–Seidel and the Push Methods, it is often helpful to analyze a single update step within an iteration. We have endeavored to keep the discussion consistent and try to point to the pseudocode to clarify any ambiguities. Note that, in the pseudocode and discussions about it, however, we will be more clear about memory and use $\mathbf{x}$, $\mathbf{y}$, and $\mathbf{r}$ to denote *vectors* of memory associated with an iteration rather than their interpretations about the solution.

3.1 The Power Method

What is usually called the power method for PageRank is probably better called the Richardson method for the linear system formulation of PageRank [14] because the two iterations are exactly identical in the scenario that each iterate is a probability distribution. The idea underlying both is to unwrap the linear system (1) into the fixed point iteration

$$\mathbf{x}_i^{(0)} = \mathbf{e}_i \qquad \mathbf{x}_i^{(k+1)} = \alpha \mathbf{P} \mathbf{x}_i^{(k)} + (1 - \alpha)\mathbf{e}_i.$$

The main work at each iteration is the matrix vector product $\mathbf{P}\mathbf{x}^{(k)}$. This can be done either by computing a sparse matrix $\mathbf{P}$ where the non-zero value is the probability $\mathbf{A}^T \text{Diag}(\mathbf{p})$ or instead, by storing just the graph structure of $\mathbf{A}^T$ alone without any values for the non-zero entries along with the vector $\mathbf{p}$. To find a point where $\|\mathbf{x}^{(k)} - \mathbf{x}\|_1 \leq \tau$, this method requires at most $2 \log(\tau)/\log(\alpha)$ iterations [7]. As noted in [14], we can terminate this earlier when $\|\mathbf{x}^{(k+1)} - \mathbf{x}^{(k)}\|_1 \leq \tau(1 - \alpha)$ because that guarantees the same error condition. This helpful circumstance arises due to the relationship between $\|\mathbf{x}^{(k+1)} - \mathbf{x}^{(k)}\|_1$ and the residual of the linear system.

In our implementation, this iteration is implemented using two vectors of memory for a compressed sparse column representation of the adjacency matrix $\mathbf{A}$. The pseudocode is in Fig. 1. In this algorithm, we store an iteration in $\mathbf{x}$ and use the memory in $\mathbf{y}$ to compute the next iterate $\mathbf{x}^{(k+1)}$. After the entire update is done, we compare the vectors and swap.

3.2 Gauss–Seidel

Gauss–Seidel is a simple variant of the power method where we update the solution vector immediately after computing the value `update` in Fig. 1. This requires only

```
 1  Inputs:
 2   - adjacency matrix A (in compressed column storage),
 3   - p as the inverse degree vector of A,
 4   - 0 < α < 1,
 5   - v is an integer giving the column of X to compute
 6   - error tolerance τ = (1 − α)/n
 7   - two vectors of memory x and y
 8  initialize x as 0
 9  set x[v] = 1
10  set maxiter = 2 log(τ)/ log(α)
11  for iter=1:maxiter
12    for i=1:n
13      update = 0
14      for j in nonzeros(A[:,i])
15        update += x[j] * p[j]
16      end
17      y[i] ← α * update
18    end
19    y[v] ← y[v] + 1 − α
20    set δ = ‖x − y‖₁
21    x, y ← y, x  # swap x, y
22    if δ/(1 − α) ≤ τ
23      return x and converged
24    end
25  end
```

Fig. 1 Pseudocode for the power method to compute the vth column of $X = (1 − α)(I − αP)^{-1}$. This algorithm takes two vectors of memory and performs random reads from the memory in **x** and **p**, but then linearly ordered writes to the memory **y**

one vector of memory. Writing this update formally is often annoyingly intricate—it involves an idea called a regular splitting [45]—but is an extremely simple change in terms of the code. Thus, we start with the pseudocode in Fig. 2.

There are a few subtle differences from the pseudocode of the power method. First, we initialize the vector **x** from zero. This choice will turn out to make tracking the error in the Gauss–Seidel iteration much easier [8]. Second, the algorithm actually stores $\mathbf{x}^{(k)} \odot \mathbf{p}$ in the memory **x** where $\odot$ is an elementwise product. This choice is made so that we can compute the quantities in update on lines 14–17 without looking up the values in **p**. Note that we could have done the same transformation for the power method, but we found it slightly decreased performance. Here, the value of δ tracks the total sum of $\mathbf{x}^{(k)} = \mathbf{x} \odot \mathbf{d}$ after the loop 13–25. This corresponds to the sum of an iterate $\mathbf{x}^{(k)}$ of the Gauss–Seidel method. As we will see next, for Gauss–Seidel starting from 0, we have that $\sum_{i=1}^{n}[\mathbf{x}^{(k)}]_i$ gives the 1-norm of the error.

The error analysis of the method is fairly straightforward. The iterations we analyze are the *unscaled iterations* that would correspond to multiplying **x** in the code by **d** (elementwise) at each step. We call these $\mathbf{x}^{(k)}$ as discussed in the previous

```
1   Inputs:
2   -  adjacency matrix A (in compressed column storage)
3   -  p as the inverse degree vector of A,
4   -  d as the degree vector of A,
5   -  0 < α < 1,
6   -  v is an integer giving the column of X to compute
7   -  error tolerance τ = (1 - α)/n
8   -  one vector of memory x
9   initialize x as 0
10  set maxiter = 2 log(τ)/ log(α)
11  for iter=1:maxiter
12     set δ = 0
13     for i=1:n
14        update = 0
15        for j in nonzeros(A[:,i])
16           update += x[j]
17        end
18        update *= α
19        if i == v  # handle  the  right  hand  side
20           update += 1 - α
21        end
22        δ ← δ + update
23        x[i] ← update * p[i]
24     end
25     if δ ≥ 1 - τ
26        x ← x ⊙ d  # scale x by d element-wise
27        return x and converged
28     end
29  end
```

Fig. 2 Pseudocode for the Gauss–Seidel method to compute the vth column of $X = (1 - \alpha)(I - \alpha P)^{-1}$. This algorithm takes one vector of memory. It maintains $\mathbf{x}$ as the Gauss–Seidel iterate elementwise scaled by $\mathbf{p}$. This performs random reads from the memory in $\mathbf{x}$, and then linearly ordered writes to the same memory $\mathbf{x}$. This works like the power-method from Fig. 1 where the updates are immediately applied in the vector $\mathbf{x}$

paragraph. In what follows $\mathbf{x}$ is the solution vector. However, let us note that what constitutes an iteration is not the loop on line 11, but the loop on line 13. This is because this method is easiest to analyze if we only consider what happens when a single element of $\mathbf{x}^{(k)}$ is changed on Line 23. In our analysis, we will show that each iterate $\mathbf{x}^{(k)}$ is bounded above by the true solution $\mathbf{x}$. Formally, this can be stated as $\mathbf{x}^{(k)} \leq \mathbf{x}$. We will establish this by showing that iterates only increase the value of $\mathbf{x}^{(k)}$ and they never get too large. If $\mathbf{x}^{(k)} \leq \mathbf{x}$ is the case, then the error

$$\|\mathbf{x} - \mathbf{x}^{(k)}\|_1 = \sum_i [\mathbf{x} - \mathbf{x}^{(k)}]_i = \sum_i [\mathbf{x}]_i - \sum_i [\mathbf{x}^{(k)}]_i = 1 - \sum_i [\mathbf{x}^{(k)}]_i.$$

Here, we only used that the sum of the entire PageRank vector $\sum_i [\mathbf{x}]_i = 1$ for the true solution on a strongly connected graph. Note that in line 22, we update δ which is tracking the sum of the unscaled vector $\mathbf{x}^{(k)}$ and after the full loop on 13–24, we have computed $\delta = \sum_i [\mathbf{x}^{(k)}]_i$. Consequently, this termination criteria maps to what we use in the algorithm.

Now, it remains to show that we indeed have the solution upper bounding each unscaled iterate. Note that, because $\mathbf{x}^{(0)} = 0$ we immediately have $\mathbf{x}^{(0)} \leq \mathbf{x}$. We will also strengthen our setup and note that the residual of the linear system (1)

$$\mathbf{r}^{(0)} = (1 - \alpha)\mathbf{v} - (I - \alpha P)\mathbf{x}^{(0)}$$

is also non-negative. The importance of the relationship with the residual is that the residual and error satisfy the following system of equations:

$$(I - \alpha P)(\mathbf{x} - \mathbf{x}^{(k)}) = \mathbf{r}^{(k)}.$$

The matrix $(I - \alpha P)$ is an M-matrix [27] with a non-negative inverse, so the error vector $\mathbf{x} - \mathbf{x}^{(k)} \geq 0$ when $\mathbf{r}^{(k)} \geq 0$. Thus, it suffices to show that $\mathbf{r}^{(k+1)} \geq 0$ given $\mathbf{r}^{(k)} \geq 0$. In this case, we know that $\mathbf{x}^{(k)}$ and $\mathbf{x}^{(k+1)}$ are the same in all but one coordinate. Let u correspond to the index i that is changed in iteration k. We compute

$$\mathbf{x}^{(k+1)} = \mathbf{x}^{(k)} + \mu_k \mathbf{e}_u,$$

where μ_k is the value of $\texttt{update} - \mathbf{e}_u^T \mathbf{x}^{(k)}$. Expanding out the code to get μ_k gives

$$\mu_k = \begin{cases} \alpha \sum_{j \to u} x_j^{(k)} / d_j - x_u^{(k)} & u \neq v \\ \alpha \sum_{j \to u} x_j^{(k)} / d_j + (1 - \alpha) - x_u^{(k)} & u = v \end{cases}.$$

Note that μ_k is exactly the uth element of the residual of the linear system (1)

$$\mathbf{r}^{(k)} = (1 - \alpha)\mathbf{v} - (I - \alpha P)\mathbf{x}^{(k)} \Rightarrow \mu_k = \mathbf{e}_u^T \mathbf{r}^{(k)}. \tag{2}$$

We have that $\mu_k \geq 0$ because $\mathbf{r}^{(k)} \geq 0$ by assumption. At this point, we still need to show that $\mathbf{r}^{(k+1)} \geq 0$, and we have

$$\mathbf{r}^{(k+1)} = (1 - \alpha)\mathbf{v} - (I - \alpha P)(\mathbf{x}^{(k)} + \mu_k \mathbf{e}_u) = \mathbf{r}^{(k)} - \mu_k (I - \alpha P)\mathbf{e}_u$$

$$= (\mathbf{r}^{(k)} - \mu_k \mathbf{e}_u) + \underbrace{\mu_k \alpha P \mathbf{e}_u}_{\text{non-negative}}$$

Now, we also have that

$$[\mathbf{r}^{(k)} - \mu_k \mathbf{e}_u]_i = \begin{cases} 0 & i = u \\ [\mathbf{r}^{(k)}]_i & i \neq u \end{cases} \geq 0$$

because μ_k is the uth component of $\mathbf{r}^{(k)}$ (see (2)). Thus we have $\mathbf{r}^{(k+1)} \geq 0$.

This justifies that the algorithm in Fig. 2, if it terminates, will have the correct error. To see that it will terminate, note that this same analysis shows that we reduce the sum of the residual at each step of the algorithm. We can also get convergence through classical results about the convergence of Gauss–Seidel on M-matrices [45].

Although there is no sub-asymptotic theory about Gauss–Seidel compared with the power method, ample empirical evidence suggests that, for most graphs, Gauss–Seidel runs in about half the iterations of PageRank. The asymptotic theory in Varga [45] shows that Gauss–Seidel is asymptotically faster than the power method. However, this is in terms of the spectral radius alone. This asymptotic theory, however, can be misleading for PageRank as an example with a random graph from [13] shows. To foreshadow our results, Gauss–Seidel will be the method to beat for computing PageRank with a single thread. This mirrors results found in other scenarios as well [15, 36].

3.3 The Cyclic Push Method

One challenge with Gauss–Seidel is that it requires in-neighbor access to the edges of the graph. These are still accessed consecutively, which makes streaming solutions a possibility. There are nevertheless many graph systems that provide the most efficient access to the out-neighbors of a directed graph. It turns out that there is a way to implement the Gauss–Seidel for these systems using something called the *push method* for PageRank, the big difference, however, is that we maintain two vectors of memory. The first variant of the *push* method we will describe will exactly map to the Gauss–Seidel computation above. The key difference is that it explicitly maintains a residual vector.

Suppose we kept a solution vector $\mathbf{x}^{(k)}$ along with a residual vector $\mathbf{r}^{(k)}$. Then the single-entry update in Gauss–Seidel corresponds to

$$\mathbf{x}^{(k+1)} = \mathbf{x}^{(k)} + \mathbf{e}_u \mathbf{e}_u^T \mathbf{r}^{(k)}.$$

(This expression arises from (2) combined with the μ_k variation on the Gauss–Seidel update.) This is easy to compute, but then we have to update $\mathbf{r}^{(k)}$ to get the new residual $\mathbf{r}^{(k+1)}$. In the push method, this second update dominates the work.

Recall the expression for the residual update that arose in our theory on Gauss–Seidel

$$\mathbf{r}^{(k+1)} = (\mathbf{r}^{(k)} - \mu_k \mathbf{e}_u) + \mu_k \alpha \mathbf{P} \mathbf{e}_u.$$

```
 1   Inputs:
 2     - adjacency matrix A (in compressed row storage),
 3     - p as the inverse degree vector of A,
 4     - 0 < α < 1,
 5     - v is an integer giving the column of X to compute
 6     - error tolerance τ = (1 − α)/n
 7     - two vector of memory x, r
 8   initialize x, r as 0
 9   set r[v] ← (1 − α)
10   set maxiter = 2 log(τ)/ log(α)
11   for iter=1:maxiter
12     set δ = 0
13     for i=1:n
14       μ = r[i]
15       x[i] += μ
16       δ ← δ + x[i]
17       # now handle the residual update
18       r[i] = 0
19       ρ = α ∗ μ ∗ p[i]
20       for j in nonzeros(A[i,:])
21         r[j] += ρ
22       end
23     end
24     if δ ≥ 1 − τ
25       return x and converged
26     end
27   end
```

Fig. 3 Pseudocode for the cyclic push method to compute the vth column of $X = (1 - \alpha)(I - \alpha P)^{-1}$. This algorithm takes two vectors of memory. It maintains $\mathbf{x}$ as the Gauss–Seidel iterate and $\mathbf{r}$ as the residual $(1 - \alpha)\mathbf{e}_v - (I - \alpha P)$. This performs random writes to the memory in $\mathbf{r}$. Note that this iteration is mathematically identical to Fig. 2, but it uses compressed row storage for A instead of compressed column storage

To perform this update, all we need to do is set the uth element of $\mathbf{r}^{(k)}$ to 0, and then lookup the values of the uth column of P. Note that the matrix $P = A^T \text{Diag}(\mathbf{p})$ and so the uth column is just the uth row of A, which encodes the out-neighbors, scaled by $p[u]$. The resulting algorithm is given by the pseudocode in Fig. 3.

This iteration is mathematically identical to Gauss–Seidel. The iteration in this form was described by McSherry [36] as an alternative way of computing PageRank that was more amenable to optimization because we can use properties of the residual to choose when to revisit or skip updating a node. The term "push" comes from the idea that when you update $x[i]$ you "push" an update out to the neighbors of i in the residual vector.

3.4 The Push Method With a Work Queue

The name "push method" actually comes from [2]. That paper utilized the push method to compute a personalized PageRank vector of an undirected graph in *constant* time (where the constant depends on α and the accuracy τ) for a weaker notion of error. This weak notion corresponds to finding an iterate with error that satisfies $0 \leq \mathbf{x} - \mathbf{x}^{(k)} \leq \tau \mathbf{d}$. So the error on a node with a large degree could be large. This enabled a number of clever ideas to show that this can be done in work that does not depend on the size of the graph. One of the key ideas is that this algorithm maintains a *queue* of vertices to process, and hence, avoids storing or working with vectors that are the size of the graph.

In this case, we adopt similar ideas and add a work queue of vertices that have not yet satisfied their tolerance. In comparison with the cyclic push method, this maintains the same amount of memory, in addition, when the residual associated with a vertex goes above a threshold, we add it to a queue to process in the figure. Namely, if the residual on a node is ω, then we can show that the maximum change to the solution vector due to that element is $\omega(1 - \alpha)$. There might be as many as n items in the residual, so if we want a solution that is accurate to 1-norm error τ, then we can check if the residual is smaller than $(1 - \alpha)\tau/n$. If it is smaller than this, we can show it will not impact the solution.

The pseudocode with the queue is in Fig. 4. The algorithm is identical to Fig. 3, except that we visit vertices in the order that they have been added to the queue. The only small subtlety is that we can check if a vertex is in the queue in order to avoid adding it multiple times based on the current value of the residual. In Line 25, we check if this is the first time that the element increased beyond the threshold ω. The other small detail is that we keep a running sum of the vector $\mathbf{x}$ in δ, which is incremented based on the value of μ at each step. In a low-precision implementation, this sum would need to be accumulated at a higher precision as it involves an extremely large running sum. As such, we can use the previous error analysis which justifies that when the total sum of the vector $\mathbf{x}$ exceeds τ, then we have converged.

3.5 Related Algorithmic Advances

It was [21] and [36] that realized that the *push* formulation offered a number of additional opportunities to accelerate PageRank computation by skipping and optimizing potential updates in a Gauss–Seidel-like fashion. These were later improved upon by [6] and [2] with the idea of the workqueue. The connection to Gauss–Seidel only arose later [8, 11]. The algorithms in our paper do not use the full flexibility of these methods as they are often specialized techniques that arise for web-graphs.

```
 1   Inputs:
 2     - adjacency matrix A (in compressed row storage),
 3     - p as the inverse degree vector of A,
 4     - 0 < α < 1,
 5     - v is an integer giving the column of X to compute
 6     - error tolerance τ = (1 − α)/n
 7     - two vector of memory x, r
 8   initialize x, r as 0
 9   set r[v] ← (1 − α)
10   set maxiter = 2n log(τ)/ log(α)
11   set δ = 0
12   let Q be a queue initialized with vertex v
13   ω = (1 − α)τ/n
14   while Q is not empty
15     i ← next(Q)
16     μ = r[i]
17     x[i] += μ
18     δ ← δ + μ
19     # now handle the residual update
20     r[i] = 0
21     ρ = α * μ * p[i]
22     for j in nonzeros(A[i,:])
23       rj = r[j]
24       r[j] = rj + ρ
25       if rj < ω and rj + ρ ≥ ω
26         add j to the end of the Q
27       end
28     if δ ≥ 1 − τ
29       return x and converged
30     end
31   end
```

Fig. 4 Pseudocode for the push method with a work queue to compute the vth column of $X = (1 − \alpha)(I − \alpha P)^{-1}$. This algorithm takes two vectors of memory. It maintains $\mathbf{x}$ as the Gauss–Seidel iterate and $\mathbf{r}$ as the residual $(1 − \alpha)\mathbf{e}_v − (I − \alpha P)$. This performs random writes to the memory in $\mathbf{r}$ and picks what amounts to a randomly scattered entry of i to process next

We have ignored here a wide class of methods for PageRank that work via Monte Carlo approaches [3–5, 32]. These methods all have trouble getting high accuracy entries, although they tend to get the top-k lists correct and should be considered for applications that only desire that type of information. Krylov methods are only competitive for PageRank when α is extremely large [18]. There are have been numerous attempts to parallelize the computation of a single PageRank vector [17]—especially on graph processing systems [1, 9, 26, 33, 43, 44]. In particular, these methods often utilize ideas closely related to the workqueue notion of the push method. Analysis of these results show that they often fail to be useful parallelizations of the underlying problem and have significant overhead compared to simple implementations [37].

3.6 Multivector Transformations

The algorithms described so far here—and most of the discussions of PageRank that we are aware of—deal with computing a single PageRank or personalized PageRank vector. (The biggest exception are a number of techniques to attempt to approximate all PageRank vectors [4, 21].) With the idea in mind that we are considering an educated, but non-expert, user of these algorithms we note the following idea. Modern processors feature vector execution units often called SIMD (single instruction, multiple data) or simply vector instructions. Because the data access pattern for the power method, Gauss–Seidel, and the cyclic push method are entirely independent of both the choice of the right hand side $\mathbf{e}_i$ and any elements of the vectors, then we can conceptually execute the same iteration on multiple vectors simultaneously. This involves few changes to the code assuming that the language supports some notion of treating a *vector of entries* like a scalar. Thus, for each of the methods above, we create a variation that processes multiple vectors simultaneously. Our technique to do this in the Julia programming language is to replace a one dimensional array of data with a one dimension array of statically sized vectors. This enables the compiler to unroll and auto vectorize code that involves multiple entries at once in a way that is consistent with our informed user persona. The code is essentially unchanged from the previous cases and we refer interested readers to our online codes to reproduce these ideas. (See Sect. 6.)

4 Results

We now conduct a set of experiments using these four PageRank algorithms in the setting of the *AllPageRank* problem. That is, we run them to compute *multiple columns* of the matrix X. The primary performance measure we are considering is the *number of columns computed* per unit time. We run the algorithms for one of two time intervals: 14.4 min and 5 min. Note that 14.4 min is exactly 1/100th of a day, and so the number of vectors computable in 24 h is exactly 100 times greater. For 5 min, the factor is roughly 300 times larger. Note that the *AllPageRank* problem involves a great deal of computation, and so it is natural to, perhaps, think of running this for a few days. Months or weeks are less reasonable, though.

We consider two parallelization strategies: threads and processes. In the threaded implementation, we load the graph information into memory once and use the high-level language's threading library to launch a given number of computation threads. These threads continue to compute single columns, or multiple columns simultaneously, of the solution until the time limit is exhausted. They all access the same shared memory copy. The process scenario is largely the same, except we launch independent processes that all have their own copy of the graph information. Note that we do not consider any parallel setup or IO time; but let us state that this was negligible for our experiments—it might take 1–5 min to set up an execution which we expect to run for hours. Our code for these experiments is all available online (see Sect. 6 for the reference).

Also in keeping with our informed user persona, we did not perform any heroic measures to eliminate all simultaneous usage of the machine. We asked other users not to use the machines during our tests, which, we believe was respected. There were a few processes from other users that would appear to be doing intermittent work. (As an example, we may see someone running the unix "top" command to see if the machine was being used).

4.1 Data and Machines

We report on two datasets, each of which is a strongly connected component of a larger graph. These data come from [29, 38].

- **Orkut** has 2, 997, 355 nodes and 220, 259, 736 directed edges.
- **LiveJournal** has 3, 828, 682 nodes and 65, 825, 429 directed edges.

The two machines we use are:

- A 64-core (4 × 16-core) shared memory server with Xeon E7-8867 v3 (2.50 GHz) CPUs and 2 TB of RAM; this is configured in a fully connected topology. Each processor has four memory channels, 45 MB of L3 cache, and 256 KB × 16 of L2 cache.
- A 192-core (8 × 24-core) shared memory server with Xeon Platinum 8168 (2.70 GHz) CPUs and 6 TB of RAM; this is configured in a hypercube topology with three connections per CPU. Each processor has six memory channels, 33 MB L3 Cache, and 24 × 1 MB of L2 cache.

4.2 Performance on a 64-Core System

We begin our discussion by looking at the results of all the algorithms on the 64-core server as these are the simplest to understand. These are summarized in Table 1, which shows how performance varies on 1, 32, and 64 threads and processes when we compute 1, 8, or 16 vectors simultaneously. In principle, using multiple vectors simultaneously will result in the Julia compiler generating AVX and SIMD instructions on the platform, which can greatly increase the computational power. We see that this increases performance by around a factor of 4 or 5. We see only a

Table 1 Vectors computed on the LiveJournal graph within 14.4 min

(a) *Threads*

	Vectors								
	1			8			16		
	Threads								
Method	1	32	64	1	32	64	1	32	64
Power	25	399	731	88	1984	2752	96	1584	2784
Gauss–Seidel	46	675	1324	152	3128	5320	176	3248	5680
Cyclic push	41	626	849	104	2240	3192	96	1840	3040
Queue push	12	213	323	56	944	1504	64	1232	2048

(b) *Processes*

	Vectors								
	1			8			16		
	Processes								
Method	1	32	64	1	32	64	1	32	64
Power	24	438	697	88	1800	2920	96	1696	2912
Gauss–Seidel	28	690	965	144	3064	5704	176	3616	5632
Cyclic push	27	544	723	96	1992	2848	96	1776	2480
Queue push	9	189	302	40	880	1536	48	1296	1872

These results are from a threads (a) and processes (b) implementation on the 64-core server on the 64-core server. For each method, we vary the number of vectors computed simultaneously among 1, 8, and 16 along with varying the number of threads from 1, 32, to 64

small change going from 8 to 16 vectors computed simultaneously, and sometimes this will decrease performance (see the threaded results on the power method and processes results for Gauss–Seidel). The results with processes are generally, but not always, faster than the results with the same number of threads.

Note that the power method uses more iterations than either the Gauss–Seidel and cyclic push methods, and so we expect it to be slower from an algorithm perspective (although the memory access patterns are more amenable to parallelization). Gauss–Seidel and the cyclic push methods are mathematically identical and so execute the same number of iterations. The difference in performance is entirely due to the memory access patterns. These results show that it is better to have random reads than random writes as the power method is faster than the cyclic push method. Although the Queue Push method should do the least work of all, it seems that the additional cost of maintaining the queue causes the method to run the slowest.

In summary, these results point to challenges in linearly scaling the work involved in this pleasingly parallel computation. They also highlight the need to compute multiple vectors simultaneously. Note that running Gauss–Seidel with one process or thread produces about half the output of the power method with 32 threads computing only one vector at a time.

4.3   Sparse Matrix Ordering

The next experiment we consider is using a sparse matrix ordering scheme to improve the locality of reference among the operations. This is a standard technique in sparse matrix computations that is commonly taught in graduate curricula. We use the METIS algorithm [24] and generate 50 and 100 partitions. We then re-order the matrix so that each partition is a consecutive block. Since the computations with multiple vectors all had uniformly higher performance, we only report the results for the methods that compute eight vectors simultaneously.

Again, these results show a considerable increase in performance for most methods. The performance of Gauss–Seidel increases by 30%, for instance. Notable exceptions include the power method and Queue Push methods on Orkut. The partitions took less than an hour to compute. Since we envision running these computations for over 10 h, the permuted method would overtake the non-permuted one after about 4 h. Consequently, it seems this technique is still worth doing even for these pleasingly parallel computations. In particular, note that the cyclic push method shows a very large change in performance and largely runs faster than the power method in all cases. Given the random write nature of this work, this is perhaps unsurprising, but it is useful to know that this type of algorithm is especially sensitive to ordering (Table 2).

4.4   Performance on a 192-Core System

Next, we investigate how performance changes on a 192-core system for the algorithms that run eight vectors simultaneously. Table 3 shows the results for the threaded and independent process scenarios. This table highlights the problem with scaling threaded computation on this particular system. As the number of threads

Table 2 The change in the number of vectors computed on LiveJournal and Orkut as we vary the sparse matrix order shows that a small bit of careful ordering dramatically improves performance

	(a) LiveJournal			(b) Orkut		
	Ordering			Ordering		
Method	Native	50	100	Native	50	100
Power	2752	3312	3544	904	896	944
Gauss–Seidel	5320	7128	6864	1744	1920	2120
Cyclic push	3192	4272	4928	696	1104	1176
Queue push	1504	1808	1936	512	424	512

These results are from the 64-core server in a threading environment with the 14.4 min interval. The ordering varies from the native order of the file as it emerged from the strongly connected component computation to one computed using 50 and 100 partitions with METIS. The algorithms all compute eight vectors simultaneously. Performance improves for all algorithms except the queue push and power methods on Orkut. Note the dramatic increase in the performance of cyclic push on Orkut

Table 3 Vectors computed by the 192-core server show problems with scaling the threaded computation

(a) *LiveJournal*

Method	Threads or processes					
	1T	1P	96T	96P	192T	192P
Power	104	120	3232	5960	272	7768
Gauss–Seidel	264	272	10,248	13,920	3592	19,160
Cyclic push	160	168	6432	7728	4752	9920
Queue push	64	64	2752	3256	1808	4496

(b) *Orkut*

Method	Threads or processes					
	1T	1P	96T	96P	192T	192P
Power	40	40	1048	1768	0	888
Gauss–Seidel	88	88	2384	4184	0	5800
Cyclic Push	48	48	1584	1864	1040	2016
Queue Push	24	24	680	872	0	920

These results show both the threaded (T) and process (P) environment with the 14.4 min interval. These results all used the ordering computed with 50 partitions and the algorithms all compute eight vectors simultaneously. We repeated the experiment with 192 threads and verified a similar result from another trial; we were unable to determine a cause for why these results showed *no* vectors computed

increases, the performance decreases. We investigate this finding in the next section (Sect. 4.5) as well. In fact, on the Orkut networks, there is no work done when using 192 threads within 14.4 min for most of the trials. We repeated this trial to verify that the result was consistent—it was.

Overall, these results show challenges when using threads on a machine with a more complex memory topology, even when using pleasingly parallel computations.

4.5 Performance Scaling

The final experiment we conduct is a performance scaling study for the best algorithm we found: the SIMD Gauss–Seidel algorithm. We use eight-vectors as there was only a minor performance difference (if any) for the 16-vector variant. Here we also use the data that has been reordered with METIS in order to get a sense of scaling when the computation is performing well. We vary the number of threads or processes in each system and report the scaling results in Fig. 5. These show that the threading performance quickly degrades on the machine with 192 cores and the per-process implementation is needed to get good scaling results. Note also that neither setup *scales* particularly well for a pleasingly parallel computation.

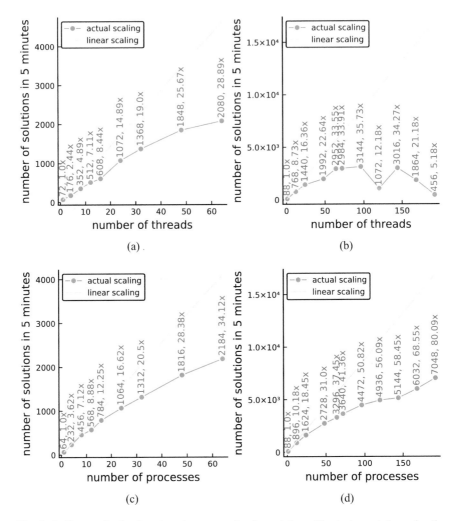

Fig. 5 Scaling results for threads and processes implementations. The text annotations give the raw number of vectors computed by that method within 5 min as well as the speedup ratio over the 1 thread or process result. (**a**) 64-core, threaded. (**b**) 192-core, threaded. (**c**) 64-core, processes. (**d**) 192-core, processes

5   Related Problems

AllPageRank is just a simple instance of a more general need for this type of computation. A closely related methodology underlies the *network community profile* calculation [12, 22, 30, 34]. This setting involves running a local clustering algorithm for hundreds or thousands of times—independently—on a shared graph. These computations often take hours to run on graphs of similar size.

A related computation is the GHOST technique used for network alignment [42]. This calculation extracts a subset of vertices from a large graph and then computes an eigenvalue histogram on the induced subgraph. These histograms are used as a invariant and characteristic feature for network alignment methods. (As an aside, we note that there are better ways to get a related concept called the *network density of states* [10].)

In summary, the style of computation used for the AllPageRank problem occurs repeatedly and is worth understanding given that the computations often consume considerable time and informed users.

6   Conclusion

The focus of this manuscript is on a *pleasingly parallel* computation: the AllPageRank problem we introduce. When we investigated computing the vectors involved in this problem on two shared memory parallel systems, it showed that expecting *linear* speedup on these problems is unrealistic. Even in this simple case, our results show that two ideas are crucial to get reasonable performance:

- computing multiple vectors simultaneously
- using matrix ordering techniques.

Both of these are easy to incorporate into parallel execution libraries that could be designed for this class of tasks, which is distinct from the current focus of distributed graph computation libraries. Our code is available online for others to reproduce our findings on new and emerging systems: https://github.com/YanfeiRen/pagerank

Back to the problem at hand, we are able to compute around 20,000 columns of X for the LiveJournal graph in 14.4 min. This shows that it would take around two days with 192 cores to generate all the information for the AllPageRank problem. For the Orkut graph, it would take around 5 days. We note that both are reasonable and acceptable runtimes to generate an interesting derived dataset. Waiting a week for an experiment is a fairly standard scenario in the physical sciences.

That said, this is still an expensive computation. Making these techniques commonplace on graphs of this scale would likely require another factor of 10 increase in performance so that the results come in 5 h on 192 cores, or say, 15 h on 64 cores. Monte Carlo techniques may be one possibly, along with reduced precision computation. Our experiments all used 64-bit floating point values. The computation may be possible in 32-bit floating point values although it will require some care as values such as $1/4,000,000$ are within a factor of 10 of the unit roundoff value for 32-bit floats. Finally, we note that there are methods that should further accelerate Gauss–Seidel, such as successive over-relaxation. While there is a negative finding about SOR on general PageRank systems [20], there are many PageRank systems and near relative PageRank systems that would use symmetric positive definite matrices [34] where SOR, with the optimal choice of ω, might be productive. Preliminary tests show this yields another 2–3 fold improvement for undirected graphs.

We realize that there are additional strategies that an expert could take to improve performance such as developing custom routines to control memory placement and thread locality. We note, however, that these tools are difficult to access from high-level libraries where our hypothetical informed user resides.

Acknowledgments This work was supported in part by NSF IIS-1546488, CCF-1909528, the NSF Center for Science of Information STC, CCF-0939370, DOE award DE-SC0014543, NASA, and the Sloan Foundation.

References

1. Aberger, C.R., Tu, S., Olukotun, K., Ré, C.: Emptyheaded: A relational engine for graph processing. arXiv **cs.DB**, 1503.02368 (2015). URL http://arxiv.org/abs/1503.02368
2. Andersen, R., Chung, F., Lang, K.: Local graph partitioning using PageRank vectors. In: Proceedings of the 47th Annual IEEE Symposium on Foundations of Computer Science (2006). URL http://www.math.ucsd.edu/~fan/wp/localpartition.pdf
3. Avrachenkov, K., Litvak, N., Nemirovsky, D., Osipova, N.: Monte Carlo methods in PageRank computation: When one iteration is sufficient. SIAM J. Numer. Anal. **45**(2), 890–904 (2007). DOI 10.1137/050643799. URL http://dx.doi.org/10.1137/050643799
4. Avrachenkov, K., Litvak, N., Nemirovsky, D., Smirnova, E., Sokol, M.: Quick detection of top-k personalized PageRank lists. In: A. Frieze, P. Horn, P. Prałat (eds.) Algorithms and Models for the Web Graph, pp. 50–61. Springer Berlin Heidelberg, Berlin, Heidelberg (2011)
5. Bahmani, B., Chakrabarti, K., Xin, D.: Fast personalized PageRank on MapReduce. In: Proceedings of the 2011 international conference on Management of data, SIGMOD '11, pp. 973–984. ACM, New York, NY, USA (2011). DOI 10.1145/1989323.1989425
6. Berkhin, P.: Bookmark-coloring algorithm for personalized PageRank computing. Internet Mathematics **3**(1), 41–62 (2007). URL http://www.internetmathematics.org/volumes/3/1/Berkhin.pdf
7. Bianchini, M., Gori, M., Scarselli, F.: Inside PageRank. ACM Transactions on Internet Technologies **5**(1), 92–128 (2005). DOI 10.1145/1052934.1052938
8. Boldi, P., Vigna, S.: The push algorithm for spectral ranking. arXiv **cs.SI**, 1109.4680 (2011). URL https://arxiv.org/abs/1109.4680
9. Ching, A., Kunz, C.: Giraph: Large-scale graph processing infrastructure on Hadoop. In: Proceedings of the Hadoop Summit (2011)
10. Dong, K., Benson, A.R., Bindel, D.: Network density of states. In: Proceedings of the 25th ACM SIGKDD International Conference on Knowledge Discovery & Data Mining - KDD '19. ACM Press (2019). DOI 10.1145/3292500.3330891. URL https://doi.org/10.1145/3292500.3330891
11. Esfandiar, P., Bonchi, F., Gleich, D.F., Greif, C., Lakshmanan, L.V.S., On, B.W.: Fast Katz and commuters: Efficient approximation of social relatedness over large networks. In: Algorithms and Models for the Web Graph (2010). DOI 10.1007/978-3-642-18009-5_13
12. Fountoulakis, K., Gleich, D.F., Mahoney, M.W.: A short introduction to local graph clustering methods and software. In: Book of Abstracts for 7th International Conference on Complex Networks and Their Applications, pp. 56–59 (2018)
13. Gleich, D.F.: Models and algorithms for PageRank sensitivity. Ph.D. thesis, Stanford University (2009). URL http://www.stanford.edu/group/SOL/dissertations/pagerank-sensitivity-thesis-online.pdf

14. Gleich, D.F.: PageRank beyond the web. SIAM Review **57**(3), 321–363 (2015). DOI 10.1137/140976649
15. Gleich, D.F., Gray, A.P., Greif, C., Lau, T.: An inner-outer iteration for PageRank. SIAM Journal of Scientific Computing **32**(1), 349–371 (2010). DOI 10.1137/080727397
16. Gleich, D.F., Mahoney, M.W.: Mining large graphs. In: P. Bühlmann, P. Drineas, M. Kane, M. van de Laan (eds.) Handbook of Big Data, Handbooks of modern statistical methods, pp. 191–220. CRC Press (2016). DOI 10.1201/b19567-17
17. Gleich, D.F., Zhukov, L.: Scalable computing with power-law graphs: Experience with parallel PageRank. In: SuperComputing 2005 (2005). URL http://www.cs.purdue.edu/homes/dgleich/publications/gleich2005-parallelpagerank.pdf. Poster.
18. Golub, G., Greif, C.: An Arnoldi-type algorithm for computing PageRank. BIT Numerical Mathematics **46**(4), 759–771 (2006). DOI 10.1007/s10543-006-0091-y
19. Golub, G.H., van Loan, C.: Matrix Computations. Johns Hopkins University Press (2013)
20. Greif, C., Kurokawa, D.: A note on the convergence of SOR for the PageRank problem. SIAM Journal on Scientific Computing **33**(6), 3201–3209 (2011). DOI 10.1137/110823523. URL https://doi.org/10.1137/110823523
21. Jeh, G., Widom, J.: Scaling personalized web search. In: Proceedings of the 12th international conference on the World Wide Web, pp. 271–279. ACM, Budapest, Hungary (2003). DOI 10.1145/775152.775191
22. Jeub, L.G.S., Balachandran, P., Porter, M.A., Mucha, P.J., Mahoney, M.W.: Think locally, act locally: Detection of small, medium-sized, and large communities in large networks. Phys. Rev. E **91**, 012821 (2015). DOI 10.1103/PhysRevE.91.012821
23. Jiang, B., Kloster, K., Gleich, D.F., Gribskov, M.: AptRank: an adaptive PageRank model for protein function prediction on bi-relational graphs. Bioinformatics **33**(12), 1829–1836 (2017). DOI 10.1093/bioinformatics/btx029
24. Karypis, G., Kumar, V.: A fast and high quality multilevel scheme for partitioning irregular graphs. SIAM J. Sci. Comput. **20**(1), 359–392 (1998). DOI 10.1137/S1064827595287997
25. Kurokawa, D., Gleich, D.F., Greif, C.: Prpack. Github repository, https://github.com/dgleich/prpack (2013). URL https://github.com/dgleich/prpack
26. Kyrola, A., Bllelloch, G., Guestrin, C.: GraphChi: Large-scale graph computation on just a PC. In: Proceedings of the 10th USENIX Symposium on Operating Systems Design and Implementation (2012)
27. Langville, A.N., Meyer, C.D.: Deeper inside PageRank. Internet Mathematics **1**(3), 335–380 (2004). URL http://www.ams.org/msnmain?fn=130&form=fullsearch&pg4=MR&s4=2111012
28. Langville, A.N., Meyer, C.D.: Google's PageRank and Beyond: The Science of Search Engine Rankings. Princeton University Press (2006)
29. Leskovec, J., Krevl, A.: SNAP Datasets: Stanford large network dataset collection. http://snap.stanford.edu/data (2014)
30. Leskovec, J., Lang, K.J., Dasgupta, A., Mahoney, M.W.: Statistical properties of community structure in large social and information networks. In: WWW '08: Proceeding of the 17th international conference on World Wide Web, pp. 695–704. ACM, New York, NY, USA (2008). DOI 10.1145/1367497.1367591
31. Lin, C.H., Konecki, D.M., Liu, M., Wilson, S.J., Nassar, H., Wilkins, A.D., Gleich, D.F., Lichtarge, O.: Multimodal network diffusion predicts future disease–gene–chemical associations. Bioinformatics p. bty858 (2018). DOI 10.1093/bioinformatics/bty858
32. Lofgren, P.A., Banerjee, S., Goel, A., Seshadhri, C.: FAST-PPR: Scaling personalized PageRank estimation for large graphs. In: Proceedings of the 20th ACM SIGKDD International Conference on Knowledge Discovery and Data Mining, KDD '14, pp. 1436–1445. ACM, New York, NY, USA (2014). DOI 10.1145/2623330.2623745
33. Low, Y., Bickson, D., Gonzalez, J., Guestrin, C., Kyrola, A., Hellerstein, J.M.: Distributed GraphLab: A framework for machine learning and data mining in the cloud. In: Proceedings of the VLDB Endowment, vol. 5, pp. 716–727 (2012)

34. Mahoney, M.W., Orecchia, L., Vishnoi, N.K.: A local spectral method for graphs: With applications to improving graph partitions and exploring data graphs locally. Journal of Machine Learning Research **13**, 2339–2365 (2012). URL http://www.jmlr.org/papers/volume13/mahoney12a/mahoney12a.pdf
35. McGlohon, M., Akoglu, L., Faloutsos, C.: Weighted graphs and disconnected components: patterns and a generator. In: Proceedings of the 14th ACM SIGKDD international conference on Knowledge discovery and data mining, KDD '08, pp. 524–532. ACM, New York, NY, USA (2008). DOI 10.1145/1401890.1401955
36. McSherry, F.: A uniform approach to accelerated PageRank computation. In: Proceedings of the 14th international conference on the World Wide Web, pp. 575–582. ACM Press, New York, NY, USA (2005). DOI 10.1145/1060745.1060829
37. McSherry, F., Isard, M., Murray, D.G.: Scalability! but at what cost? In: 15th Workshop on Hot Topics in Operating Systems (HotOS XV). USENIX Association, Kartause Ittingen, Switzerland (2015). URL http://blogs.usenix.org/conference/hotos15/workshop-program/presentation/mcsherry
38. Mislove, A., Marcon, M., Gummadi, K.P., Druschel, P., Bhattacharjee, B.: Measurement and analysis of online social networks. In: Proceedings of the 7th ACM SIGCOMM Conference on Internet Measurement, IMC '07, pp. 29–42. ACM, New York, NY, USA (2007). DOI 10.1145/1298306.1298311
39. Page, L.: Method for node ranking in a linked database (2001). URL http://www.freepatentsonline.com/6285999.pdf
40. Page, L., Brin, S., Motwani, R., Winograd, T.: The PageRank citation ranking: Bringing order to the web. Tech. Rep. 1999-66, Stanford University (1999). URL http://dbpubs.stanford.edu:8090/pub/1999-66
41. Pan, J.Y., Yang, H.J., Faloutsos, C., Duygulu, P.: Automatic multimedia cross-modal correlation discovery. In: KDD '04: Proceedings of the tenth ACM SIGKDD international conference on Knowledge discovery and data mining, pp. 653–658. ACM, New York, NY, USA (2004). DOI 10.1145/1014052.1014135
42. Patro, R., Kingsford, C.: Global network alignment using multiscale spectral signatures. Bioinformatics **28**(23), 3105–3114 (2012). DOI 10.1093/bioinformatics/bts592
43. Perez, Y., Sosic, R., Banerjee, A., Puttagunta, R., Raison, M., Shah, P., Leskovec, J.: Ringo: Iinteractive graph analytics on big-memory machines. In: Proceedings of the ACM SIGMOD Conference (2015)
44. Pingali, K., Nguyen, D., Kulkarni, M., Burtscher, M., Hassaan, M.A., Kaleem, R., Lee, T.H., Lenharth, A., Manevich, R., Mendez-Lojo, M., Prountzos, D., Sui, X.: The tao of parallelism in algorithms. In: Proceedings of the 32nd Conference on Programming Language Design and Implementation (2011)
45. Varga, R.S.: Matrix Iterative Analysis. Prentice Hall (1962)
46. Voevodski, K., Teng, S.H., Xia, Y.: Spectral affinity in protein networks. BMC Systems Biology **3**(1), 112 (2009). DOI 10.1186/1752-0509-3-112

Parallel Fast Time-Domain Integral-Equation Methods for Transient Electromagnetic Analysis

Yang Liu and Eric Michielssen

1 Introduction

Marching-on-in-time (MOT)-based time-domain (TD) integral-equation (IE) methods provide an appealing avenue for tackling a broad range of transient electromagnetic problems including radiation and scattering analysis [1, 2], electromagnetic interference/compatibility (EMI/EMC) characterization [3–5], electromagnetic material discovery [6, 7], etc. TDIE methods relate transient incident and scattered electromagnetic fields through convolution of the equivalent surface or volume sources with TD Green's functions. Upon discretizing the equivalent sources with space-time localized basis functions and spatially Galerkin testing the pertinent IEs at discrete times, the spatial-temporal unknowns are solved by time marching. Compared to TD differential equation (DE) methods, TDIE methods enjoy several advantages. First, IE methods implicitly impose radiation boundary conditions, whereas DE methods require that the computational domain be truncated using an artificial boundary condition. Second, when applied to surface scatterers, IE methods only require surface unknowns, whereas DE methods use unknowns throughout a volume enclosing scatterers. Unfortunately, the applicability of these TDIE solvers to large-scale real-world problems was oftentimes hindered by their late-time instability and computational inefficiency. Indeed, the computed source and field densities were oftentimes polluted by spurious and non-decaying solutions as the number of time steps became large due to the buildup of numerical errors and the use of unstable IE operators. In addition, the computational and memory costs of the classical MOT scheme scale as $O\left(N_t N_s^2\right)$ and $O\left(N_s^2\right)$ due to the

Y. Liu · E. Michielssen (✉)
Department of Electrical Engineering and Computer Science, University of Michigan, Ann Arbor, MI, USA
e-mail: liuyangz@umich.edu; emichiel@umich.edu

© Springer Nature Switzerland AG 2020
A. Grama, A. H. Sameh (eds.), *Parallel Algorithms in Computational Science and Engineering*, Modeling and Simulation in Science, Engineering and Technology, https://doi.org/10.1007/978-3-030-43736-7_12

need to compute Galerkin-tested scattered fields at each time step generated by the equivalent sources via the use of the TD Green's functions. Here, N_s is the number of spatial unknowns and N_t is the total number of time steps. Not surprisingly, issues pertaining to the accuracy and computational efficiency of TDIE solvers have received significant attention from the research community. In fact, the last decade has experienced unprecedented developments in high-order accurate, late-time stable, rapidly converging, low-complexity, and highly parallel TDIE solvers capable of solving very large and complex electromagnetic problems. This chapter summarizes these recent advances with an emphasis on low-complexity and parallel TDIE solvers.

Many fast algorithms have been developed to accelerate classical MOT-based TDIE solvers, including the plane-wave time-domain (PWTD) algorithm [8], time-domain adaptive integral method (TD-AIM) [9], nonuniform-grid time-domain algorithm (NGTD) [10], accelerated Cartesian expansion method (ACE) [11, 12], Taylor expansion-based algorithm [13], wavelet-based algorithm [14], envelope tracking techniques [15], and hybridized algorithms [16–20]. These fast algorithms can reduce the quadratic computational and memory costs of the classical MOT scheme from $O\left(N_t N_s^2\right)$ and $O\left(N_s^2\right)$ to $O(N_t N_s)$ and $O(N_s \log N_s)$, respectively. Among the family of fast algorithms, the PWTD algorithm, just like its frequency domain (FD) counterpart known as the multilevel fast multipole algorithm (MLFMA), permits fast evaluation of the fields produced by bandlimited source constellations through their expansion into homogeneous plane waves. Compared to other fast algorithms such as the TD-AIM and NGTD, the PWTD requires the least computational resources. Furthermore, PWTD and its various extensions have been applied to the analysis of transient electromagnetic phenomena that involve various types of object materials, surrounding media, and frequency regimes. Not surprisingly, PWTD-accelerated TDIE solvers have been successfully applied to a broad class of complex and large-scale transient electromagnetic problems. Sequential implementations of the PWTD (and other fast algorithms)-accelerated TDIE solvers have been applied to analyze scattering problems involving $N_s \approx 10^5$ spatial unknowns.

The capabilities of the TDIE solvers can be significantly increased through CPU and GPU parallelization. Parallel versions of both classical and fast-algorithm-accelerated TDIE solvers have been developed [21–26]; in this chapter, we focus on CPU-parallel PWTD algorithms. Parallelizing the multilevel PWTD algorithm on distributed memory clusters is a challenging task due to the algorithm's complex nature and heterogeneous structure. While schemes that use spatial [27] and hybrid spatial/angular partitioning [28] strategies to parallelize the algorithm were developed, neither approach scales well on distributed memory clusters with large processor count since their computational, communication, and memory requirements per processor are not inversely proportional to the total processor count. Similar difficulties have been observed in parallelization of the MLFMA-accelerated frequency domain solvers [29]. Recently, however, provably scalable parallelization techniques were developed for MLFMA using a hierarchical partitioning strategy that simultaneously leverages spatial and angular partitioning at each level of the MLFMA tree [30–34]. Their extensions to the PWTD scheme that account for the need to also partition the temporal dimension while using global spherical interpolation/filtering schemes (not present in MLFMA) have been developed [25]. In addition, novel asynchronous communication techniques for reducing the cost and memory requirements of inter-processor communication in PWTD and MLFMA have been developed [25, 34]. By combining the hierarchical partitioning strategy and asynchronous communication technique to achieve CPU and memory load balancing among processors, these newly developed parallel PWTD-accelerated TDIE solvers are capable of analyzing transient scattering from electrically large perfect electrically conducting (PEC) and dielectric objects that are orders of magnitudes larger than before (e.g., well beyond $N_s \approx 10^7$). Not surprisingly, these parallel and fast TDIE solvers have been successfully applied to the analysis of large transient scattering problems, the design of broadband antennas, the characterization of EMI/EMC phenomena, as well as several biomedical applications.

2   TDIE Solvers for Different Scatterers and Media

This section provides a review of TDIE solvers applicable to different types of scatterers and media. Section 2.1 describes the formulation and space-time discretization of standard TDIEs for solving transient problems involving PEC objects. Sections 2.2, 2.3, and 2.4 briefly review TDIE solvers for homogeneous dielectrics residing in lossless and dissipative media and inhomogeneous dielectrics, respectively.

2.1   PEC Scatterers

Consider a closed PEC surface S that resides in a lossless and unbounded background medium with permittivity ε_0 and permeability μ_0 (see Fig. 1). The surface is illuminated by an incident electromagnetic field $\{\mathbf{E}^i(\mathbf{r}, t), \mathbf{H}^i(\mathbf{r}, t)\}$ that is assumed temporally bandlimited to maximum frequency f_{max} and vanishingly small for $t < 0$. The incident field induces an electric current density $\mathbf{J}(\mathbf{r}, t)$ on S that in turn generates the scattered field $\{\mathbf{E}^s(\mathbf{r}, t), \mathbf{H}^s(\mathbf{r}, t)\}$. Enforcing appropriate boundary conditions on total electric and magnetic field tangential to S produces time-domain electric field and magnetic field integral equations (TD-EFIE and TD-MFIE); linearly combining them yields a time-domain combined field integral equation (TD-CFIE):

$$\hat{\mathbf{n}} \times \hat{\mathbf{n}} \times \partial_t \mathbf{E}^i(\mathbf{r}, t) = -\hat{\mathbf{n}} \times \hat{\mathbf{n}} \times \partial_t \mathbf{E}^s(\mathbf{r}, t) = \mathcal{L}_e[\mathbf{J}](\mathbf{r}, t) \quad \forall \mathbf{r} \in S, S^+, S^-$$

$$(1)$$

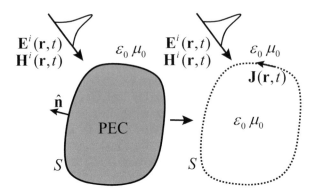

Fig. 1 Transient scattering from a PEC object in a lossless unbounded medium

$$\hat{\mathbf{n}} \times \partial_t \mathbf{H}^i (\mathbf{r}, t) = -\hat{\mathbf{n}} \times \partial_t \mathbf{H}^s (\mathbf{r}, t) = \mathcal{L}_h [\mathbf{J}] (\mathbf{r}, t) \quad \forall \mathbf{r} \in S^-. \tag{2}$$

$$\hat{\mathbf{n}} \times \partial_t \mathbf{H}^i (\mathbf{r}, t) - \alpha/\eta_0 \hat{\mathbf{n}} \times \hat{\mathbf{n}} \times \partial_t \mathbf{E}^i (\mathbf{r}, t) = \mathcal{L}_h [\mathbf{J}] (\mathbf{r}, t) - \alpha/\eta_0 \mathcal{L}_e [\mathbf{J}] (\mathbf{r}, t)$$
$$= \mathcal{L}_c [\mathbf{J}] (\mathbf{r}, t) \quad \forall \mathbf{r} \in S^-. \tag{3}$$

Here, $\hat{\mathbf{n}}$ is the outward unit normal to S, η_0 is the characteristic impedance of the background medium, and S^- and S^+ denote surfaces conformal to but just inside and outside S, respectively. Note that the TD-EFIE in (1) is also valid for open PEC surfaces. The TD-EFIE and TD-MFIE operators are

$$\mathcal{L}_e [\mathbf{J}] (\mathbf{r}, t) = \hat{\mathbf{n}} \times \hat{\mathbf{n}} \times \frac{\mu_0}{4\pi} \int_S dS' \left(\partial_t^2 \mathcal{I} - c_0^2 \nabla \nabla \right) \cdot \frac{\mathbf{J} (\mathbf{r}', \tau)}{R} \tag{4}$$

$$\mathcal{L}_h [\mathbf{J}] (\mathbf{r}, t) = \frac{1}{4\pi} \hat{\mathbf{n}} \times \int_S dS' (\mathbf{r} - \mathbf{r}') \times \left[\frac{1}{c_0 R^2} \partial_t^2 \mathbf{J} (\mathbf{r}', \tau) + \frac{1}{R^3} \partial_t \mathbf{J} (\mathbf{r}', \tau) \right]. \tag{5}$$

Here $R = |\mathbf{r} - \mathbf{r}'|$, c_0 is the speed of light in the background medium, and $\tau = t - R/c_0$ represents retarded time. The two operators inside the bracketed factor in $\mathcal{L}_e$ in (4) constitute the operator's singular (vector potential) and hypersingular (scalar potential) component, respectively.

Since both $\mathcal{L}_e$ and $\mathcal{L}_h$ have a null space that allows for the presence of nonphysical oscillating currents, solutions of the TD-EFIE and TD-MFIE are oftentimes corrupted by spurious resonance modes. The TD-CFIE with operator $\mathcal{L}_c$ in (3) that linearly combines the TD-EFIE and TD-MFIE using a constant α

is free of these spurious resonance modes. Note that the TD-CFIE reduces to the TD-EFIE and TD-MFIE when $\alpha = \infty$ and $\alpha = 0$, respectively.

To numerically solve (3) S is discretized by a planar triangle mesh with minimum edge length Δs chosen to properly resolve both the geometrical details and the wavelength λ corresponding to the incident field's maximum frequency f_{max}, i.e., $\lambda = c_0/f_{max}$. The current density $\mathbf{J}(\mathbf{r}, t)$ is expanded using N_s spatial basis functions and N_t temporal basis functions as

$$\mathbf{J}(\mathbf{r}, t) = \sum_{n=1}^{N_s} f_n(t)\mathbf{S}_n(\mathbf{r}) = \sum_{j=1}^{N_t}\sum_{n=1}^{N_s} I_{j,n} T_j(t)\mathbf{S}_n(\mathbf{r}). \tag{6}$$

Here $f_n(t)$ is the temporal signature associated with spatial basis function $\mathbf{S}_n(\mathbf{r})$ and $I_{j,n}$ is the current expansion coefficient associated with space-time basis function $T_j(t)\mathbf{S}_n(\mathbf{r})$. $\mathbf{S}_n(\mathbf{r})$ is often chosen as the Rao–Wilton–Glisson (RWG) basis function defined on the nth internal edge of the mesh [35]. $T_j(t) = T(t - j\Delta t)$ is the time-shifted local Lagrange interpolant with time step size $\Delta t = 1/(2\chi_t f_{max})$ and $\chi_t > 1$ is the temporal oversampling factor [36]. The Lagrange functions $T_j(t)$ are piecewise smooth for $(k - 1)\Delta t \leq t \leq k\Delta t$, $k = 0, \ldots, d$ and are nonzero for $-\Delta t < t < d\Delta t$ with d denoting the polynomial order.

Upon substituting (6) into (3), spatially testing (3) with spatial basis functions $\mathbf{S}_m(\mathbf{r})$, $m = 1, \ldots, N_s$, and enforcing (3) at discrete times $i\Delta t$, $i = 1, \ldots, N_t$, the following set of linear equations is obtained:

$$\overline{\overline{\mathbf{Z}}}_0 \overline{\mathbf{I}}_i = \overline{\mathbf{V}}_i - \sum_{k=1}^{\min\{i-1, k_{max}\}} \overline{\overline{\mathbf{Z}}}_k \overline{\mathbf{I}}_{i-k}. \tag{7}$$

Here, the entries of the matrices $\overline{\overline{\mathbf{Z}}}_k$, excitation vectors $\overline{\mathbf{V}}_i$, and current coefficient vectors $\overline{\mathbf{I}}_i$ are

$$\left\{\overline{\overline{\mathbf{Z}}}_k\right\}_{mn} = \left. <\mathbf{S}_m(\mathbf{r}), \mathcal{L}_c[\mathbf{S}_n T_{-k}]\left(\mathbf{r}, t\right)> \right|_{t=0} \tag{8}$$

$$\left\{\overline{\mathbf{V}}_i\right\}_m = \left. <\mathbf{S}_m(\mathbf{r}), \hat{\mathbf{n}} \times \partial_t \mathbf{H}^i\left(\mathbf{r}, t\right) - \frac{\alpha}{\eta_0}\hat{\mathbf{n}} \times \hat{\mathbf{n}} \times \partial_t \mathbf{E}^i\left(\mathbf{r}, t\right)> \right|_{t=i\Delta t} \tag{9}$$

and $\left\{\overline{\mathbf{I}}_i\right\}_n = I_{i,n}$. The maximum number of nonzero impedance matrices $\overline{\overline{\mathbf{Z}}}_k$ is approximately $k_{max} = \lceil D_{max}/c_0\Delta t \rceil$, where D_{max} denotes the maximum linear dimension of the scatterer. The above set of linear Eq. (7) can be solved by MOT. First, $\overline{\mathbf{I}}_1$ is computed by solving (7) for $i = 1$ using iterative methods such as the generalized minimal residual (GMRES) algorithm. Then, for $i = 2$, the summation on the right-hand side (RHS) of (7) is computed and the resulting system is solved for $\bar{\mathbf{I}}_2$. This process is repeated to compute $\bar{\mathbf{I}}_3$ and onward.

2.2 Homogeneous Dielectrics in Lossless Media

Transient scattering from homogeneous or piecewise homogeneous dielectrics residing in lossless unbounded media can be analyzed by solving a coupled pair of surface TDIEs involving both equivalent electric and magnetic surface currents radiating in the interior and exterior regions. Among the many choices of how the equations are coupled, the most popular ones are the time domain Müller [37, 38] and Poggio–Miller–Chang–Harrington–Wu–Tsai (PMCHWT) formulations [39–41]. Upon discretizing the electric/magnetic currents using space-time basis functions and spatially testing the equations at discrete times, MOT systems similar to (7) can be obtained [38].

2.3 Homogeneous Dielectrics in Dissipative and Dispersive Media

For dissipative and dispersive media, the computation of scattered fields requires the convolution of surface currents and a Green's function (either available in closed form or numerically constructed) with an infinite temporal tail. TDIE solvers leveraging convolutional quadrature techniques and Laplace transforms [42, 43] to facilitate this temporal convolution have been proposed. Similar methods for modeling electromagnetic interactions with plasmonic structures and graphene sheets using time-domain PMCHWT solvers and their variants [44–46] have been developed as well.

2.4 Inhomogeneous Dielectrics

Transient scattering from inhomogeneous dielectrics in lossless unbounded media can be analyzed using TD volume integral equations (VIE) [47] that are cast in terms of equivalent volume electric polarization current $\mathbf{J}^P(\mathbf{r}, t)$

$$\mathbf{J}^P(\mathbf{r}, t) = \partial_t \mathbf{D}(\mathbf{r}, t) - \varepsilon_0 \partial_t \mathbf{E}(\mathbf{r}, t). \tag{10}$$

where $\mathbf{D}(\mathbf{r}, t)$ is the electric flux density, $\mathbf{E}(\mathbf{r}, t)$ is the total electric field, and ε_0 is the permittivity of the background medium. The flux density and total field are related by the following formulas:

- In lossless dielectrics: $\mathbf{D}(\mathbf{r}, t) = \varepsilon(\mathbf{r})\mathbf{E}(\mathbf{r}, t)$. Here $\varepsilon(\mathbf{r})$ denotes the frequency independent permittivity of the scatterer.
- In lossy dielectrics: $\partial_t \mathbf{D}(\mathbf{r}, t) = \sigma(\mathbf{r})\mathbf{E}(\mathbf{r}, t) + \varepsilon(\mathbf{r})\partial_t\mathbf{E}(\mathbf{r}, t)$. Here the flux density has a conduction current contribution [48] and $\sigma(\mathbf{r})$ denotes the frequency independent conductivity of the scatterer.
- In dispersive dielectrics: $\mathbf{D}(\mathbf{r}, t) = \varepsilon(\mathbf{r}, t) * \mathbf{E}(\mathbf{r}, t)$, where $\varepsilon(\mathbf{r}, t)$ denotes the permittivity of the dispersive scatterer, and * denotes temporal convolution.

The unknown electric flux density can be solved using MOT schemes similar to (7). Note that for lossy and dispersive dielectrics, the total field can be updated from the electric flux density via recursive computation [48, 49].

In addition to the aforementioned generic TDIE solvers, specialized versions applicable to problems involving nonlinear ferromagnetic materials [50], surface scatterers embedded in a half-space [51, 52] or layered media [53], resonant cavities [54], periodic objects [55], 2D scatterers [56], and wire structures [57] have been developed.

3 Accurate, Stable, and Well-Conditioned TDIE Solvers

Classical TDIE solvers, though widely applicable, oftentimes suffer from accuracy, stability, and/or convergence issues when applied to real-world problems. In what follows, we briefly summarize advances in high-order accurate, late-time stable, and well-conditioned TDIE solvers.

3.1 Accurate Space-Time Discretization

The accuracy of a TDIE solver is affected by several error sources rooted in the spatial and temporal schemes used, as well as other aspects related to the computation of the MOT matrix elements in (8).

Accurate Temporal Discretization Local Lagrange temporal basis functions, though commonly used, oftentimes introduce numerical errors due to their discontinuous derivatives at integer multiples of Δt. Other temporal basis functions, including first-order continuous cosine square function [58], continuous exponential [59] and smooth B-spline functions [60–62], have been proposed. In addition, approximate prolate spheroidal wave functions (APS) [63] that are time limited and quasi-bandlimited have been shown [64] to yield spectral accuracy but require carefully designed extrapolation techniques to retrieve the form of MOT equations [64]. These methods not only have been applied to surface scatterers, but also to penetrable volumetric bodies. For example, Sayed et al. [65] report on a TD-VIE using APS functions to model high-contrast scatterers.

High-Order Spatial Discretization High-order spatial basis function and geometry modeling are efficient techniques to avoid excessive mesh refinement to achieve a prescribed solution accuracy. Among the many available choices, divergence conforming Graglia–Wilton–Peterson (GWP) functions [66], constructed as products of a scalar polynomial of the given order and RWG basis functions, are the most popular. The authors in [67] developed high-order TD-EFIE/MFIE/CFIE solvers leveraging GWP functions in conjunction with loop-tree decompositions of the pertinent function space to enhance not only the method's accuracy but also low-frequency stability [68]. Other methods to improve the spatial discretization accuracy in TDIE solvers include a high-order Nyström method [69] and a Generalized Method of Moments scheme that uses mixed spatial basis functions [70]. A high-order space-time discretization scheme has also been developed [65].

Classical TD-MFIE solvers often suffer from inaccuracies due to incorrect nonsolenoidal current scaling in the low-frequency limit. The inaccuracy can be corrected by using mixed discretizations [71] that use RWG functions and Buffa–Christiansen (BC) functions as source and testing bases, respectively.

Accurate Evaluation of MOT Matrix Numerical errors in evaluation of MOT matrix element $\left\{ \overline{\overline{\mathbf{Z}}}_k \right\}_{mn}$ representing nearby source and testing functions can affect the overall accuracy and (more importantly) the stability of the TDIE solvers. The evaluation of the MOT elements for scatterers residing in nondissipative backgrounds calls for the computation of two-dimensional source and test spatial integrals. Accurate evaluation of these four spatial integrals can be very challenging using standard quadrature rules that do not account for the nonsmooth character of the temporal basis functions. Recently, semi-analytical methods that analytically evaluate two [41, 72] or even three [73] out of the four spatial integrals have been developed. These methods assume the usage of RWG spatial basis functions and Lagrange temporal basis functions. Methods that allow more flexible choices of temporal basis functions yielding closed-form expressions of the electric [74], magnetic [75, 76], and combined fields [77] due to impulse excited RWG spatial basis functions have been developed as well. Other methods to accurately evaluate the MOT matrix include polar integration [78], fully numerical integration based on a separable approximation of the convolution kernel [79], and radial source integration/smoothed test integration [80].

3.2 Stabilized TDIE Solvers

TDIE solvers oftentimes are plagued by instabilities, viz., the presence of non-decaying solutions $\overline{\mathbf{I}}_i$ for decaying excitation vectors $\overline{\mathbf{V}}_i$. TDIE instabilities broadly speaking fall into three categories: (1) high-frequency instabilities, i.e., wildly oscillating and exponentially growing solutions, (2) DC instabilities, i.e., constant or slowly growing solutions, and (3) resonant instabilities, i.e., harmonic solutions with oscillating frequencies corresponding to those of interior resonant modes. Among these three types of instabilities, the high-frequency instabilities are mainly due to the numerical discretization errors; they can be remedied using the methods referenced in the previous section in conjunction with techniques discussed below. In contrast, DC and resonance instabilities are rooted in the spectral properties of the pertinent TDIE operators and their removal requires more intrusive changes to the solver.

High-Frequency Instability Traditionally, high-frequency instabilities in TDIE solvers have been remedied via temporal/spatial averaging [81–86], special collocation-in-time schemes [87], or implicit time stepping methods [88, 89]. More recently, these methods have been replaced by techniques that leverage space-time Galerkin testing [90, 91], Laplace/Z-transforms [42, 43, 92], temporal extrapolation [64], and the abovementioned accurate MOT element evaluation schemes. The use of these methods has all but eliminated high-frequency instabilities in MOT solvers. It is also worth mentioning that recently proposed explicit time stepping methods that use predictor–corrector schemes exhibit stability properties on par with implicit schemes [93, 94].

DC Instability DC instabilities occur mostly in TD-EFIE solvers and are caused by the presence of static (or linear-in-time) solenoidal currents that reside in the null space of TD-EFIE operator $\mathcal{L}_e$ (or its differentiated form). Methods that leveraging loop-tree decomposition [64, 67] or augmented TD-EFIE formulations [95, 96] can efficiently suppress the nonphysical static or linear-in-time solenoidal current. More recently, a "Dottrick" scheme that totally eliminates such instability based on Calderón preconditioned TD-EFIE has been reported [97].

Resonant Instability As mentioned in Sect. 2.1, solutions to both the TD-EFIE and MFIE can be corrupted by nonphysical resonant currents that reside in the null space of their respective operators. The TD-CFIE turns out very effective in suppressing such instabilities [1]. In addition, highly accurate iterative solvers and MOT element computation can also reduce numerical errors from building up to these spurious solutions [98].

3.3   Well-Conditioned TDIE Solvers

When solving the matrix Eq. (8) using iterative methods, the number of iterations for the solution to converge is typically proportional to the condition number (the ratio between the largest and smallest singular value) of $\overline{\overline{\mathbf{Z}}}_0$. Unfortunately, standard TD-EFIE/CFIE solvers suffer from two types of breakdowns, viz., the condition number of $\overline{\overline{\mathbf{Z}}}_0$ grows without bound when the time step size $\Delta t \to 0$ (i.e., when the excitation is a low-frequency pulse) or the mesh size $\Delta s \to 0$ (i.e., when intricate geometry features call for a dense/multi-scale mesh).

Low-Frequency Breakdown Low-frequency breakdown occurs in TD-EFIE solvers for reasons relating to the inconsistent asymptotic behavior of the (RWG basis) discretized vector potential component and scalar potential component of the TD-EFIE operator in the low-frequency regime. As a result, the MOT matrix $\overline{\overline{\mathbf{Z}}}_0$ becomes ill-conditioned and the MOT system has a null space for the solenoidal current. This low-frequency breakdown can be avoided by discretizing the current using scaled solenoidal/nonsolenoidal sub-domain spatial basis functions. Methods that leverage loop-star/tree basis transformation techniques [67, 95, 99] and hierarchical nonsolenoidal bases [100–102] have been proposed to cap the condition number of $\overline{\overline{\mathbf{Z}}}_0$.

Dense Mesh Breakdown Dense mesh breakdown occurs in TD-EFIE solvers for reasons relating to the inconsistent asymptotic behaviors of the EFIE operator's singular values associated with solenoidal and nonsolenoidal singular functions that can be supported by a dense mesh. As the mesh size decreases, the singular values associated with solenoidal currents go to zero, while those associated with non-solenoidal currents approach infinity. As a result, the condition number of $\overline{\overline{\mathbf{Z}}}_0$ in the discretized TD-EFIE system grows without bound. This type of breakdown is cured by leveraging the self-regularization property of the time-domain Calderón identity along with BC functions for spatial discretization [97, 103–105]. More recently, schemes that combine Calderón preconditioners, quasi-Helmholtz projectors, and accurate discretization schemes have been developed [106]. High-order Calderón preconditioned TD-EFIEs leveraging GWP functions and high-order BC functions [107] also have been reported [108]. It is worth noting that the TD-CFIE also suffers from dense mesh breakdown due to the presence of the EFIE operator. A Calderón preconditioned TD-CFIE that gives rise to bounded condition number irrespective of mesh density is proposed in [97]. Finally, a Calderón preconditioned single source surface TDIE solver for homogeneous dielectrics also has been developed [109].

4   Fast TDIE Solvers

The computationally most demanding operation in the abovementioned TDIE solvers is the evaluation of the sum on the RHS of (7) during time marching, which requires the computation of tested fields at N_s observers due to N_s sources for N_t time steps. The computational cost of this operation, if performed directly, is prohibitively high and hinders TDIE solvers from simulating transient phenomena involving electrically large objects. Indeed, when applied to problems that involve surface or volumetric objects residing in unbounded lossless media, the computational and memory costs of this operation scale as $O\left(N_t N_s^2\right)$ and $O\left(N_s^2\right)$. When applied to problems involving surface objects embedded in dissipative or structured environments, these costs increase to $O\left(N_t^2 N_s^2\right)$ and $O\left(N_t N_s^2\right)$, due to the infinite temporal tail of the Green's function in these media.

Luckily, the computational efficiency of TDIE solvers can be significantly improved by various fast algorithms. The most popular ones among them, viz., the TD-AIM, NGTD, and PWTD, are discussed in this subsection. Other methods to reduce the computational and/or memory costs of the MOT scheme include the accelerated Cartesian expansion (ACE)-based algorithm that is well-suited for accelerating low-frequency integral kernels [11, 12], the wavelet-based adaptive MOT scheme [14], an envelope tracking technique that permits large time step sizes in the high-frequency regime [110, 111], and hybridization of TDIE methods with physical optics (PO) [16–20] and DE methods [112–116]. Finally, Taylor expansion-based TDIE methods [13, 117] have also been used to reduce the computational cost associated with classical solvers.

4.1 TD-AIM

Just like its frequency domain counterpart known as the FD-AIM [118], the TD-AIM permits the fast evaluation of the radiated fields by projecting them onto auxiliary uniform spatial grids and propagating them using fast Fourier transforms (FFTs). However, unlike the FD-AIM that leverages space-only FFTs, the TD-AIM utilizes multilevel/blocked space-time FFTs as the sparse structure of the MOT matrices needs to be accounted for [9, 119]. When applied to transient analyses involving quasi-planar surface scatterers residing in unbounded lossless media under high-frequency excitations, the computational and memory costs of the TD-AIM scale as $O(N_t N_s \log^2 N_s)$ and $O\left(N_s^{1.5}\right)$, respectively [9]; for more general surfaces, these costs increase to $O\left(N_t N_s^{1.5} \log^2 N_s\right)$ and $O\left(N_s^2\right)$ [9]. Moreover, the TD-AIM can be applied, with minimal modifications, to transient analysis involving surfaces embedded in lossy or half-space media [51, 52]. In [120], the TD-AIM is extended to the low-frequency regime through accelerating the computation of both the RHS and left-hand side (LHS) of (7) by the space-time and space-only FFTs, respectively. More recently, an envelope tracking-based TD-AIM was proposed to efficiently handle band-pass transient scattering problems [15]. The computational costs of these TD-AIM-accelerated TDIE solvers are listed in Table 1. Note that although TD-AIM is asymptotically inferior to other fast algorithms such as the PWTD and NGTD schemes, it is very competitive for many practical problems.

4.2 NGTD

The NGTD algorithm accelerates the computation of fields produced by temporally bandlimited and space-confined sources by representing their delay- and amplitude-compensated fields on a sparse grid surrounding the observers and evaluating the true fields through interpolation and delay/amplitude restoration [121]. The

Table 1 Best available estimates (multiplicative constants omitted) of the computational costs for computing the RHS sum of (7) using the direct scheme and fast algorithms. "General" refers to arbitrarily shaped surface scatterers residing in lossless unbounded media in the high-frequency regime

	General	Low-freq	Quasi-planar	Dissipative	Volumetric
Direct	$N_t N_s^2$	$N_t N_s^2$	$N_t N_s^2$	$N_t^2 N_s^2$	$N_t N_s^2$
TD-AIM	$N_t N_s^{1.5}\log^2 N_s$	$N_t N_s \log N_s$	$N_t N_s \log^2 N_s$	$N_t N_s \log(N_t N_s) \log N_t$	$N_t N_s \log^2 N_s$
NGTD	$N_t N_s \log^2 N_s$	$N_t N_s \log N_s$	$N_t N_s \log^2 N_s$	–	$N_t N_s \log^2 N_s$
PWTD	$N_t N_s \log^2 N_s$	$N_t N_s \log N_s$	$N_t N_s$	$N_t N_s \log N_t \log N_s$	$N_t N_s$

two-level NGTD algorithm was first developed using the spherical nonuniform grid [121]. Later, multilevel NGTD schemes that use a Cartesian nonuniform grid were developed as well [10]. When applied to the computation of fields due to either surface-bound or volumetrically distributed sources that reside in unbounded lossless media, the computational costs of multilevel NGTD scale as $O(N_t N_s \log^\mu N_s)$. Here, $\mu = 2$ in the high-frequency regime, $\mu = 1$ in the low-frequency regime, and $1 < \mu < 2$ for mixed-scale mesh (Table 1). Moreover, the NGTD scheme is remarkably simple to implement compared with other fast algorithms.

4.3 PWTD

This section reviews advances in the PWTD algorithms and their variants for different types of scatterers, background media, and frequency regimes. The PWTD algorithm for general PEC scatterers in lossless media will be explained in detail while its variants are only briefly reviewed.

PEC Object in Lossless Media Consider the transient scattering problem described in Sect. 2.1. In order to accelerate the computation of the RHS of (7), the multilevel PWTD algorithm first recursively subdivides a rectangular box enclosing S into eight boxes until the radius of the smallest boxes is a prescribed fraction of the wavelength at the maximum frequency $\lambda = 2\pi c_0/\omega_{max}$. The resulting N_L-level geometrical octree has N_g^v nonempty boxes with radius R^v at level $v = 1, \ldots, N_L$. For nonfractal surface scatterers, $N_g^1 = O(N_s)$, $R^1 = O(1)$, $N_g^{v+1} \approx N_g^v/4$, and $R^v = 2^{(v-1)}R^1$. Within this PWTD tree, two same-level boxes are labeled a "level-v far-field pair" if the distance between the box centers exceeds γR^v ($3 < \gamma < 6$) and their respective parent boxes do not constitute a far-field pair. Two finest-level (i.e., $v = 1$) boxes that do not constitute a far-field pair are labeled a "near-field pair." The partial sums in the RHS of (7) that correspond to interactions between near-field pairs are directly evaluated using (8); those corresponding to interactions between far-field pairs are handled by PWTD.

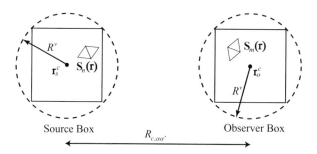

Fig. 2 One far-field box pair in the multilevel PWTD algorithm

Consider a level-v far-field pair (α, α') shown in Fig. 2. Let $R_{c,\alpha\alpha'} = \left|\mathbf{R}_{c,\alpha\alpha'}\right| = \left|\mathbf{r}_o^c - \mathbf{r}_s^c\right|$ denote the distance between the source and observer box centers $\mathbf{r}_s^c$ and $\mathbf{r}_o^c$. The far-field pair contains spatial basis functions $\mathbf{S}_n(\mathbf{r})$, $n \in \alpha$, and $\mathbf{S}_m(\mathbf{r})$, $m \in \alpha'$, respectively. For $\forall n \in \alpha$, the temporal signature $f_n(t)$ is split into N_l^v consecutive subsignals using the APS function $T^{APS}(t)$, which is bandlimited to $\omega_s = \chi_t \omega_{max}$ and quasi-time-limited to $-p_f \Delta t < t < p_f \Delta t$, $5 \le p_f \le 10$, as

$$f_n(t) = \sum_{l=1}^{N_l^v} f_n^l(t) = \sum_{l=1}^{N_l^v} \sum_{j=(l-1)M^v+1}^{lM^v} I_{n,j} T_j^{APS}(t), \tag{11}$$

where $T_j^{APS}(t) = T^{APS}(t - j\Delta t)$ and $N_l^v M^v = N_t$; M^v is chosen such that the duration of each subsignal, $T^v = (M^v + 2p_f)\Delta t$, is less than $(R_{c,\alpha\alpha'} - 2R^v)/c_0$. Let $\mathbf{J}_\alpha^l(\mathbf{r}, t) = \sum_{n \in \alpha} \mathbf{S}_n(\mathbf{r}) f_n^l(t)$ denote the current due to the lth subsignal associated with all source basis functions in box α. The fields tested by any $\mathbf{S}_m(\mathbf{r})$ produced by $\mathbf{J}_\alpha^l(\mathbf{r}, t)$, denoted by $\left\langle \mathbf{S}_m(\mathbf{r}), \mathcal{L}_c\left[\mathbf{J}_\alpha^l\right](\mathbf{r}, t) \right\rangle$, can be computed in three steps as

$$\mathbf{G}_{l,\alpha}^+\left(\hat{\mathbf{k}}_{qp}^v, t\right) = \frac{\partial_t^2}{16\pi^2 c_0^2} \sum_{n \in \alpha} \mathbf{P}_n^+\left(\hat{\mathbf{k}}_{qp}^v, t, \hat{\mathbf{k}}_{qp}^v\right) * f_n^l(t) \quad \text{Construction of outgoing rays} \tag{12}$$

$$\mathbf{G}_{l,\alpha'}^-\left(\hat{\mathbf{k}}_{qp}^v, t\right) = \mathcal{T}\left(\hat{\mathbf{k}}_{qp}^v, t\right) * \mathbf{G}_{l,\alpha}^+\left(\hat{\mathbf{k}}_{qp}^v, t\right) \quad \text{Translation} \tag{13}$$

$$\left\langle \mathbf{S}_m(\mathbf{r}), \mathcal{L}_c\left[\mathbf{J}_\alpha^l\right](\mathbf{r}, t) \right\rangle = \sum_{q=0}^{K^v} \sum_{p=-K^v}^{K^v} \omega_{qp}^v \quad \text{Projection of incoming rays}$$
$$\left[-\alpha \mathbf{P}_m^-\left(\hat{\mathbf{k}}_{qp}^v, t, \hat{\mathbf{k}}_{qp}^v\right) + \mathbf{P}_m^-\left(\hat{\mathbf{k}}_{qp}^v, t, \hat{\mathbf{n}}\right)\right]^T * \mathbf{G}_{l,\alpha'}^-\left(\hat{\mathbf{k}}_{qp}^v, t\right) \tag{14}$$

First, a set of outgoing rays $\mathbf{G}_{l,\alpha}^{+}\left(\hat{\mathbf{k}}_{qp}^{v}, t\right)$ in directions $\hat{\mathbf{k}}_{qp}^{v}$, $q = 0, \ldots, K^{v}$, $p = -K^{v}, \ldots, K^{v}$, are constructed from $f_n^l(t)$ via (12); the total number of directions is $N_k^v = (K^v + 1)(2K^v + 1)$. Next, outgoing rays $\mathbf{G}_{l,\alpha}^{+}\left(\hat{\mathbf{k}}_{qp}^{v}, t\right)$ of box α are translated into incoming rays $\mathbf{G}_{l,\alpha'}^{-}\left(\hat{\mathbf{k}}_{qp}^{v}, t\right)$ of box α' via (13). Finally, the tested fields are computed by projecting incoming rays onto $\mathbf{S}_m(\mathbf{r})$ and summing over all ray directions via (14). In (12)–(14), superscript T denotes transpose, $K^v = \lfloor 2\chi_s \omega_s R^v / c_0 \rfloor + 1$, and χ_s is a spherical oversampling factor. Directions $\hat{\mathbf{k}}_{qp}^{v}$ and weights ω_{qp}^{v} are determined by quadrature rules on the unit sphere. The projection function $\mathbf{P}_{\{m,n\}}^{\pm}\left(\hat{\mathbf{k}}_{qp}^{v}, t, \hat{\mathbf{v}}\right)$ and translation function $\mathcal{T}\left(\hat{\mathbf{k}}_{qp}^{v}, t\right)$ are

$$
\mathbf{P}_{\{m,n\}}^{\pm}\left(\hat{\mathbf{k}}_{qp}^{v}, t, \hat{\mathbf{v}}\right) = \int_{S_{\{m,n\}}} dS' \hat{\mathbf{v}} \times \mathbf{S}_{\{m,n\}}\left(\mathbf{r}'\right) \delta\left(t \pm \hat{\mathbf{k}}_{qp}^{v} \cdot \left(\mathbf{r}' - \mathbf{r}_{\{o,s\}}^{c}\right)/c_0\right).
$$

(15)

$$
\mathcal{T}\left(\hat{\mathbf{k}}_{qp}^{v}, t\right) = \frac{c_0 \partial_t}{2 R_{c,\alpha\alpha'}} \sum_{k=0}^{K^v} (2k+1) \, \Phi_k\left(\frac{c_0 t}{R_{c,\alpha\alpha'}}\right) \Phi_k\left(\frac{\hat{\mathbf{k}}_{qp}^{v} \cdot \mathbf{R}_{c,\alpha\alpha'}}{R_{c,\alpha\alpha'}}\right),
$$

(16)

where $\Phi_k(\cdot)$ is the Legendre polynomial of degree k and $|t| \leq R_{c,\alpha\alpha'}/c_0$.

Note that only outgoing/incoming rays of the finest-level boxes are constructed/projected directly via (12)/(14); those of higher-level boxes are constructed/projected by an exact global vector spherical interpolation/filtering technique described in [122]. The computational cost and memory requirements of a PWTD-accelerated surface TDIE solver generally scale as $O(N_t N_s \log^2 N_s)$ and $O\left(N_s^{1.5}\right)$, respectively.

Homogeneous Dielectrics in Lossless Media For (piecewise) homogeneous dielectrics, the PWTD algorithm accelerates computation of the fields in each dielectric region due to surface electric and magnetic currents [38]. Furthermore, due to different wave speeds in each region, multiple PWTD trees need to be constructed. The computational and memory costs of these PWTD-accelerated TDIE solvers scale as $O(N_t N_s \log^2 N_s)$ and $O\left(N_s^{1.5}\right)$. These cost estimates, though seemingly similar to those of the abovementioned solvers for analysis scattering from PEC scatterers, have larger leading constants due to the presence of extra surface current unknowns and multiple PWTD trees. In [123], a PWTD-accelerated TDIE solver applicable to composite scatterers that involve piecewise homogeneous dielectrics and PEC structures was developed.

Inhomogeneous Dielectrics in Lossless Media In the PWTD-accelerated TD-VIE solvers for analyzing transient scattering from inhomogeneous dielectrics residing in unbounded lossless media, the PWTD algorithm permits fast computation of the fields due to both the electric flux density and the total field contributions in the polarization current $\mathbf{J}^P(\mathbf{r}, t)$ [47–49]. Unlike the above-described PWTD algorithms, here the dielectric volume (instead of the surface) of the scatterer is subdivided using the PWTD tree. The computational and memory costs of the PWTD-accelerated volume TD-VIE solvers scale as $O(N_t N_s)$ and $O(N_s \log N_s)$, respectively (Table 1). A scalar-field PWTD-accelerated TD-VIE with explicit time stepping is reported in [26]. A PWTD-accelerated hybrid surface-volume TDIE solver is also reported in [124].

Surface Scatterers in Lossy Media As discussed at the beginning of Sect. 4, the computational cost of the direct TDIE solvers for analyzing scattering from surfaces embedded in unbounded lossy media scales as $O\left(N_t^2 N_s^2\right)$ due to the infinite temporal tail of the lossy-medium Green's function. Fast PWTD-accelerated TD-EFIE [125] and TD-CFIE [126] solvers were reported leveraging (a) a scalar-field lossy-medium PWTD algorithm that permits rapid computation of the far-field interactions [127] and (b) a Prony series-based scheme that permits fast temporal convolution of the lossy-medium Green's function with space-time basis functions during the evaluation of near-field interactions [128]. The computational costs of these solvers scale as $O(N_t N_s \log N_t \log N_s)$ (Table 1).

Mixed-Scale Scatterers The abovementioned PWTD algorithms will lose their computational efficiency when directly applied to mixed-scale scatterers (residing in unbounded lossless media) due to their inefficiency of field computation in the dense mesh region. An adaptive TDIE solver that leverages the standard PWTD algorithm in the electrically large region and a low-frequency PWTD algorithm in the dense mesh region was developed [129, 130]. As the computational cost of the low-frequency PWTD algorithm scales no worse than $O(N_t N_s \log N_s)$, that of the overall solver can be efficiently capped by $O(N_t N_s \log^2 N_s)$.

Quasi-Planar Scatterers For smooth quasi-planar surface scatterers that reside in unbounded lossless media, the computational and memory costs of the standard PWTD algorithm can be further improved. When illuminated by short-duration and high-frequency temporal pulses, the induced surface current and PWTD ray data often are spatially and temporally localized. Recent work exploits this locality by leveraging local cosine wavelet bases (LCB) to compress PWTD ray data and operations [131]. The computational and memory costs of the LCB-enhanced PWTD algorithm can be reduced to $O(N_t N_s)$ and $O(N_s \log N_s)$, respectively (Table 1).

In addition to the abovementioned PWTD-accelerated TDIE solvers, those applicable to transient scattering problems that involve surface scatterers embedded in half-space or layered media [132], periodic structures [133], 2D objects [56, 134, 135] have been developed. As can be summarized from Table 1, PWTD algorithms attain the best computational complexities among all existing fast algorithms for various scatterer shapes, medium types, and frequency regimes.

5   Parallelization of PWTD-Accelerated TDIE Solvers

The above developments in accurate, stable, well-conditioned, and fast TDIE solvers significantly improve their reliability and efficacy when applied to the analysis of many transient and wideband electromagnetic problems. That said, these sequential implementations remain severely limited in the size of problem they can handle and cannot be applied to problems involving scatterers discretized using millions of spatial basis functions that need to be tracked for tens of thousands of time steps. To address problems of this nature, solvers leveraging distributed-memory parallelization and GPU acceleration of classical TDIE solvers, TD-AIM-accelerated solvers, and PWTD-accelerated solvers have been developed [21–26]. In what follows, we focus on recent progress in distributed-memory parallel PWTD algorithms.

For sake of simplicity, we describe a highly scalable scheme for parallelizing the standard PWTD algorithm for surface scatterers that reside in a lossless unbounded medium. In principle, such parallelization strategy can be easily extended to any other PWTD variant such as scalar-field algorithms for volumetric scatterers [26] and LCB-enhanced schemes for quasi-planar surfaces [131]. Specifically, the parallelization strategy leverages hierarchical partitioning of the multilevel PWTD tree among processors and an asynchronous scheme for memory and cost efficient communications between processors.

5.1   Overview of Parallel PWTD

The optimal distribution of the workload of PWTD operations among processors depends quite heavily on the stage of the algorithm. For example, for near-field computations, one processor may simply take charge of a chunk of finest-level boxes and store the corresponding near-field MOT matrix elements. Unfortunately, when constructing/projecting/translating PWTD ray data, a similarly straightforward dispatch scheme does not achieve a uniform load distribution among processors due to PWTD algorithm's heterogeneous tree structure. Specifically, the PWTD ray data $G_{l,\alpha}^{\pm}\left(\hat{\mathbf{k}}_{qp}^{v}, t\right)$ has $N_g^v = O(N_s/4^v)$ spatial samples (each representing rays in one box), $N_k^v = O(4^v)$ angular samples, and $T^v = O(2^v)$ temporal samples at each PWTD tree level v. Partitioning among any single dimension results in poor load balance at certain levels. A more suitable approach is to adaptively partition ray data among more than one dimension. In other words, more processors are assigned to the angular/temporal and spatial dimension at higher and lower tree levels, respectively.

Assuming a total of N_p processors, let v_b denote the highest possible level at which $N_g^v \geq N_p$. At level $v \leq v_b$, each processor stores the ray data for all angular and temporal samples in approximately N_g^v/N_p boxes; at level $v > v_b$, each processor stores $N_k^v N_g^v/N_p$ angular samples and all temporal samples for one box.

Fig. 3 Partitioning of boxes and their ray data in a five-level PWTD tree among six processors

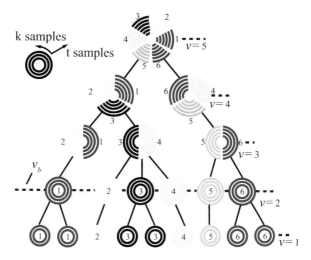

In other words, the ray data of one box at level $v > v_b$ is stored on $N_r^v = \left\lceil N_p / N_g^v \right\rceil$ processors. As an example, consider the five-level PWTD tree that is partitioned among six processors shown in Fig. 3; each set of concentric circles represents one box and its associated ray data. The angular and radial dimensions of the circles correspond to the angular and temporal samples of the ray data, respectively. The number shown near the concentric circles and arcs indicates the ID of the processor in charge of the data marked with a certain color. For this example, $N_g^v = 9, 6, 3, 2, 1$ for $v = 1, \ldots, 5$, and $N_p = 6$, hence $v_b = 2$. Each processor stores the complete ray data in one or two boxes at level $v = 1, 2$, and one-half, one-third, and one-sixth of the angular samples of the ray data in one box at levels $v = 3, 4, 5$, respectively. This parallelization strategy yields excellent computation and memory load balance and produces scalable communication patterns among processors at all levels of the PWTD tree.

5.2 Parallelization of PWTD Stages

With the above-described parallelization strategy, the implementation and analysis of each PWTD stage proceed as follows.

Construction/Projection of Outgoing/Incoming Rays Here only the construction of outgoing rays is described as the projection of incoming rays can be performed similarly. At level $v = 1$, each process constructs outgoing rays $\mathbf{G}_{l,\alpha}^+ \left(\hat{\mathbf{k}}_{qp}^v, t \right)$ directly from basis functions via (12) for N_g^v / N_p boxes and requires no communication. At level $v > 1$, the outgoing rays are constructed using spherical interpolation from rays belonging to child boxes and may require inter-processor communication.

Depending on how the ray data is stored (Fig. 4a), the required computation and communication belongs to one of the three distinct cases: (1) At level $v \leq v_b$, ray data is directly interpolated (into more angular samples) and shifted (in the temporal dimension) from those of the child boxes all by the same processor; clearly this operation requires no communication. (2) At level $v = v_b + 1$, ray data in child boxes are stored completely on one processor different from that one that stores ray data for the parent box; this case requires communication after local interpolation. (3) At level $v > v_b + 1$, the ray data for the child boxes are stored on more than one processor. The construction of the outgoing rays (Fig. 4b) is performed in five steps. Step 1: Ray data of the child boxes stored in N_r^{v-1} processors is exchanged in such a way that each processor handles $O\left(T^v/N_r^{v-1}\right)$ temporal samples of outgoing rays along all N_k^{v-1} directions (Fig. 4b). Step 2: Each processor performs its own spherical interpolation, requiring $O\left(T^v N_k^v \log K^v/N_r^{v-1}\right)$ operations. Step 3: The interpolated ray data of each child box is split along the angular dimension and the resulting data is exchanged between N_r^{v-1} processors. Step 4: The interpolated ray data is sent to the processors in charge of the parent box via non-blocked MPI communication. Step 5: The transferred ray data is locally shifted to the center of the parent box. Note that in Steps 1 and 3, redistribution of ray data from the angular dimension to temporal dimension (and vice versa) requires all-to-all communications among N_r^{v-1} processors. However, each processor only sends partial ray data of size $O\left(T^v N_k^v/N_r^{v-1}\right)$ to $N_r^{v-1} - 1$ other processors and receives data of size $O\left(T^v N_k^v/\left(N_r^{v-1}\right)^2\right)$ from each of the other $N_r^{v-1} - 1$ processors. Therefore the communication volume per processor scales inversely proportional

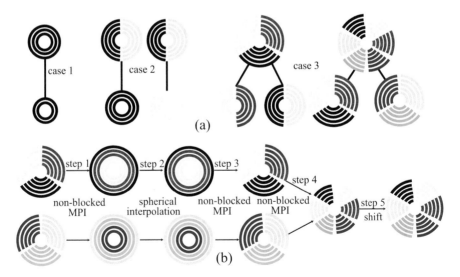

Fig. 4 (**a**) Three possible cases encountered during the construction of outgoing rays of boxes in PWTD tree. Each case requires different communication patterns. (**b**) The steps to construct the outgoing rays in case 3

to N_r^{v-1}. It can be shown [25] that the computational costs CC_1 and communication volume CM_1 per processor for construction and projection of ray data scale as

$$CC_1 = O\left(\frac{N_t N_s \log^2 N_s}{N_p}\right), \quad CM_1 = O\left(\frac{N_s N_t \log N_p}{N_p}\right). \tag{17}$$

Translation Translating one outgoing ray $\mathbf{G}_{l,\alpha}^+\left(\hat{\mathbf{k}}_{qp}^v, t\right)$ from a source box onto the incoming ray $\mathbf{G}_{l,\alpha'}^-\left(\hat{\mathbf{k}}_{qp}^v, t\right)$ of an observer box requires the temporal convolution of the outgoing ray with the translation function in (13). At levels $v \leq v_b$, each processor performs translations along N_k^v directions for its N_g^v/N_p observer boxes; at levels $v > v_b$, each processor performs translations along N_k^v/N_r^v directions for one single observer box. The translation (via convolution) is carried out in the Fourier domain, i.e., by inverse Fourier transforming the product of the ray's and translation function's Fourier transforms [8], and this operation is not further parallelized along the temporal (frequency) dimension. Note that the outgoing ray data of a box residing in a processor is sent to at most $O(1)$ other processors. For each processor, the amount of ray data sent and received during the translation stage for one box scales as $O\left(N_k^v T^v\right)$ for levels $v \leq v_b$ and $O\left(N_k^v T^v/N_r^v\right)$ for levels $v > v_b$. It is easily shown that the computational costs CC_2 and communication volume CM_2 per processor for translation scale as

$$CC_2 = O\left(\frac{N_t N_s \log^2 N_s}{N_p}\right), \quad CM_2 = O\left(\frac{N_t N_s \log N_s}{N_p}\right). \tag{18}$$

Near-Field Calculation Near-field calculations include (a) matrix–vector multiplications on the LHS of (7) at each iteration and time step and (b) partial matrix–vector multiplications on the RHS of (7) at each time step. Each processor is in charge of approximately N_g^1/N_p source boxes at the finest level, each contributing to $O(1)$ near-field pairs. The computational cost and communication volume of the near-field calculations, CC_3 and CM_3, scale as

$$CC_3 = O\left(\frac{N_s N_t}{N_p}\right), CM_3 = O\left(\frac{N_s N_t}{N_p}\right). \tag{19}$$

The abovementioned computational and communication estimates are listed in Table 2. Detailed analysis and their extensions to other PWTD variants can be found in [25, 26, 131].

Table 2 Computational costs and communication volumes for different stages in the highly scalable parallel PWTD algorithms for general surface scatterers in lossless medium

	Ray construction and projection	Translation	Near-field	Overall
Compute	$N_t N_s \log^2 N_s / N_p$	$N_t N_s \log^2 N_s / N_p$	$N_t N_s / N_p$	$N_t N_s \log^2 N_s / N_p$
Volume	$N_t N_s \log N_p / N_p$	$N_t N_s \log N_s / N_p$	$N_t N_s / N_p$	$N_t N_s \log N_s / N_p$

5.3   Asynchronous Task Queue-Based Communication Scheme

Despite the provable scalability of the abovementioned parallelization strategy, achieving good parallel efficiencies on many processors requires an implementation that leverages asynchronous communication, exploits computation/communication overlap, optimally manages memory allocations, etc. The need for carefully tuning the computations is easily recognized during the translation stage, where the number of source boxes that is far-field paired with one observer box is large—in practice it often exceeds 100—and the processor in charge needs to allocate excessively large temporary memory for receiving all outgoing rays of source boxes at different levels. A novel, memory-efficient, and asynchronous communication scheme is developed to overcome this parallelization bottleneck [25]. The scheme is described below for the translation stage but can also be generalized to other stages.

The workflow of this scheme for one processor can be summarized as follows (Fig. 5). First, the processor allocates a "receiving" memory pool containing memory grains of size $O\left(T^v N_k^v\right)$ or $O\left(T^v N_k^v / N_r^v\right)$ to receive outgoing rays of one source box at level $v \leq v_b$ or $v > v_b$, respectively. The memory grains can be marked by one of the four possible labels: "local," "empty," "busy," and "ready." Initially, memory grains containing outgoing rays of one local source box are marked "local," all other grains are marked "empty." In the beginning of the translation stage, the processor sends out outgoing ray data needed by other processors and enqueues all "local" memory grains into a working task queue. The processor then iterates over the following four steps until the translation stage is complete. Step 1: If any outgoing ray data for a source box arrives and there is a suitable sized "empty" memory grain, the processor posts non-blocked receiving, enqueues the memory grain into a receiving task queue, and marks it as "busy." Step 2: The processor dequeues any completed task in the receiving queue, marks it as "ready," and

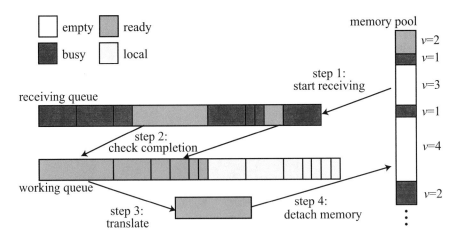

Fig. 5 Task queue-based asynchronous communication for the translation stage

enqueues it into the working queue. The working queue is a priority queue such that the "ready" grains (as opposed to "local" grains) and grains corresponding to higher-level (as opposed to finer-level) source boxes have a higher priority to dequeue. Step 3: The processor dequeues one grain in the working queue and performs the translation operation for the corresponding source box. Step 4: The grain in Step 3 is marked "empty" and returned to the memory pool, becoming available again for Step 1. This queue-based communication scheme ensures that the computation and communication are performed asynchronously, and the maximum amount of temporary memory to be allocated is pre-determined before the stage.

5.4 Numerical Validation

The parallelization performance of the PWTD algorithm is demonstrated via the computation of scattered fields from source points distributed on a square plate with $N_s = 731{,}247$, on a sphere with $N_s = 992{,}766$, and in a cube with $N_s = 804, 357$. All tests were performed on a cluster of Quad-Core 850 MHz PowerPC CPUs with 4 GB/CPU memory located at the King Abdullah University of Science and Technology (KAUST) Supercomputing Laboratory. The parallel efficiencies are computed using $\kappa = N_{ref} T_{N_{ref}} / N_p T_{N_p}$, where $T_{N_{ref}}$ and T_{N_p} are the measured execution times on N_{ref} and N_p processors, respectively. κ is computed for different PWTD stages with $N_p = 128 - 2048$, $N_{ref} = 128$ for the plate (Fig. 6a) and $N_p = 64 - 1024$, $N_{ref} = 64$ for the cube (Fig. 6b). For the plate and the cube, efficiencies of over 90% and 80% are observed at $N_p = 2048$ and $N_p = 1024$ for all PWTD stages, respectively. Figure 7 plots the execution time on each processor for the plate and sphere when $N_p = 2048$; good computation load balance is observed.

6 Applications

This section highlights applications of parallel fast TDIE solvers to large-scale transient electromagnetic problems including scattering analysis, broadband antenna design, EMI/EMC characterization, and biomedical applications.

- *Transient Scattering Analysis*: Fast TDIE solvers have been widely applied to transient scattering problems involving large and complex targets. Sequential and space/hybrid partitioning-based parallel implementations of PWTD-accelerated TDIE solvers are capable of analyzing scattering from real-world targets (e.g., aircraft and vessels) that involve about 5×10^5 spatial unknowns [122, 136]. The above-described scalable parallel PWTD-accelerated TDIE solvers in contrast can be used to solve both canonical and real-world scattering problems involving 10^7 spatial unknowns [25, 131]. In addition, parallel TD-AIM and envelope-tracking-based variants can simulate canonical scattering problems involving

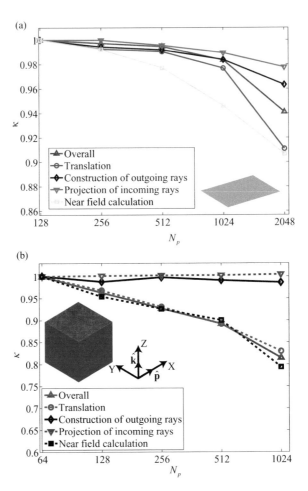

Fig. 6 Parallel efficiencies (($r = 1, \theta = 180°, \phi = 0$)) of PWTD stages for the canonical problem involving (**a**) a square PEC plate and (**b**) a dielectric cube

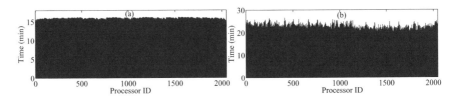

Fig. 7 Execution time on each processor for the canonical problem involving (**a**) a square PEC plate and (**b**) a PEC sphere when $N_p = 2048$

3×10^6 spatial unknowns [9, 15]. As an example, a scalable parallel PWTD-accelerated TD-CFIE solver is applied to the analysis of scattering from an Airbus-A320 model. The Airbus is illuminated by a $\hat{y}$-propagating, $\hat{z}$-polarized electromagnetic plane wave with central frequency of 740 MHz and essential bandwidth of 245 MHz. The current density is discretized using 4, 086, 129 RWG spatial basis functions and fourth-order temporal basis functions. The simulation is performed for 1140 time steps with $\Delta t = 50$ ps. The broadband RCS along the $+z$ direction is computed using the PWTD-accelerated TD-CFIE solver and a FD-CFIE solver; good agreement is observed (Fig. 8a). The snapshot of the current induced on the Airbus at $t = 480 \, \Delta t$ is plotted in Fig. 8b.

- *Broadband Antenna Design*: TDIE methods are well-suited to analyze electromagnetic scattering and radiation from broadband antennas and arrays. In the past, classical TDIE solvers were applied to thin wire antennas [57, 137, 138] and small 3D antennas [139, 140]. In addition, radiation from antennas mounted on electrically large platforms have been studied using hybrid TDIE-PO solvers [19, 20]. The fast and reliable TDIE solvers can be used to analyze large and complex antenna radiation problems. As an example, the PWTD-accelerated TD-EFIE solver is applied to the analysis of radiation from an ultra-wideband (UWB) phased antenna array that consists of 8×8 antenna elements. The geometric configuration of each element is similar to that described in [141] (see Fig. 9a). The antenna array is fed by delta-gap excitations with phase-shifted temporal signatures $G\left(t - \mathbf{r}_n \cdot \hat{\mathbf{k}}/c_0\right)$, where $G(t)$ is a modulated Gaussian pulse with

Fig. 8 (**a**) Broadband RCS and (**b**) the snapshot of the current density (in dB) at $t = 480 \, \Delta t$ for an Aribus-A320 model involving $N_s = 4,086,129$ spatial unknowns. The results are computed by a parallel PWTD-accelerated TD-CFIE solver

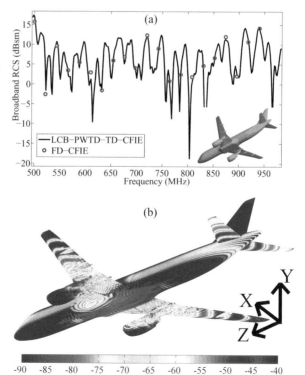

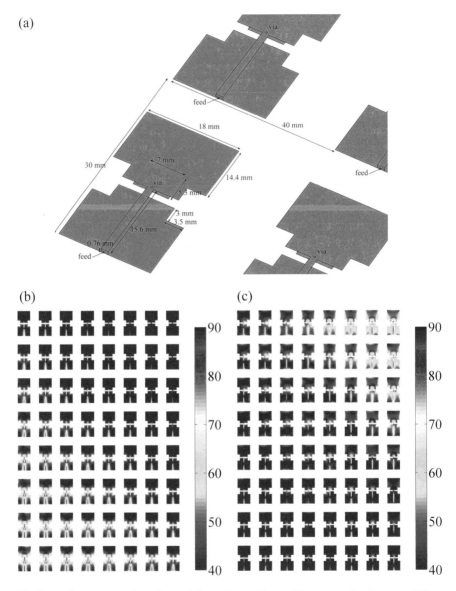

Fig. 9 (**a**) Geometric configuration and (**b, c**) the snapshots of the current density at $t = 440 \, \Delta t$ and $t = 740 \, \Delta t$ of a UWB phased antenna array involving $N_s = 178{,}787$ spatial unknowns. The results are computed by a parallel PWTD-accelerated TD-EFIE solver

central frequency of 9 GHz and essential bandwidth of 4 GHz, $\mathbf{r}_n$ is the location of feed point of the nth element, and $\hat{\mathbf{k}} = \sin\left(45^\circ\right) \hat{\mathbf{y}} + \cos\left(45^\circ\right) \hat{\mathbf{z}}$. The current density induced on the antenna array is discretized using 178,787 RWG spatial basis functions and fourth-order Lagrange basis functions. The simulation is performed for 1000 time steps with $\Delta t = 4.5$ ps. The snapshots of the current density (in dB) induced on the antenna array at times 440 Δt and 740 Δt are plotted in Fig. 9b, c.

- *EMI/EMC*: Real-life EMC/EMI problems often involve complex and multi-scale structures, e.g., radiation components, cables, and microwave circuits that reside in (electrically large) shielding enclosures. TDIE solvers capable of accurately and efficiently modeling PEC surfaces/wires/junctions, homogeneous/inhomogeneous dielectrics, and linear/nonlinear lumped elements are required. In [142], a PWTD-accelerated TDIE solver was applied to the EMC/EMI analysis involving PEC surfaces and wires. A parallel implementation of this solver was developed in [27]. In addition, a PWTD-accelerated hybrid surface-volume TDIE solver capable of modeling composite structures was developed [143]. More recently, EMC/EMI problems involving microwave circuits have been analyzed using a field-circuit simulator that couples PWTD-accelerated TDIE solvers capable of effectively modeling volumes/surfaces/wires/junctions with a modified nodal analysis (MNA)-based circuit solver [144]. Moreover, the TD-AIM scheme has been applied to accelerate a hybrid field-circuit simulator that couples a surface-volume TDIE solver with the modified nodal analysis (MNA)-based circuit solver [145]. In addition, a TD-AIM-accelerated multiconductor transmission line (MTL) simulator that further combines the TDIE solver, a TDIE-based multiconductor transmission line (MTL) solver, and an MNA-based circuit solver was developed for electromagnetic characterization of complex structures [4, 5, 146].

- *Biomedical Applications*: Electromagnetic analysis in support of optical imaging and biomedical applications as well as nano-device modeling traditionally has been performed by FDTD and FDIE methods. A distributed-memory TD-VIE solver was used to model transient scattering from a single red blood cell (RBC) involving 3×10^6 spatial unknowns [21]. More recently, a parallel PWTD-accelerated TD-VIE solver became capable of modeling transient scattering from canonical and real-life objects involving 2.5×10^7 spatial unknowns [26] and was applied to the analysis of scattering from a collection of RBCs excited by a plane wave with central frequency of 400 THz and essential bandwidth of 600 THz. The total electric fields in the cells were discretized with $N_s = 11,746,563$ nodal spatial basis functions and the simulation was carried out for $N_t = 700$ time steps with step size $\Delta t = 0.134$ fs and first-order temporal basis function. The magnitude of the scattered field's $\hat{x}$-component at two observer points is plotted in Fig. 10a. The snapshot of the magnitude of the total electric field inside the cells at $t = 500 \Delta t$ is plotted in Fig. 10b.

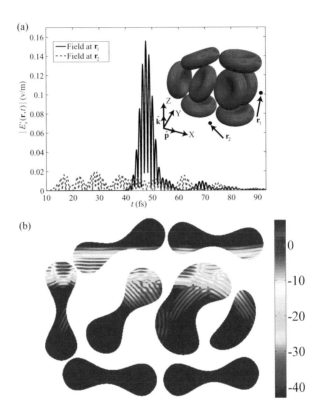

Fig. 10 (**a**) Scattered electric fields near and (**b**) the snapshot of the total fields at $t = 500 \ \Delta t$ inside a RBC aggregation involving $N_s = 11{,}746{,}563$ spatial unknowns. The results are computed by a parallel PWTD-accelerated TD-VIE solver

7 Conclusion

The TDIE solvers have been successfully applied to transient electromagnetic analyses involving conducting objects, homogeneous/inhomogeneous dielectrics, wires, and lumped elements residing in lossless, lossy, dispersive, layered media, cavities, and half-spaces. The recent advances in parallel fast, accurate, stable, and rapidly converging TDIE solvers have significantly improved their reliability and capability in modeling real-life electromagnetic phenomena for scattering, radiation, EMI/EMC, and biomedical applications.

References

1. Shanker, B., et al. (2000). Analysis of transient electromagnetic scattering from closed surfaces using a combined field integral equation. IEEE Trans. Antennas Propag. **48**(7): 1064-1074.
2. Chew, W.C., et al. (2001). Fast and efficient algorithms in computational electromagnetics, Artech House, Inc.
3. Michielssen, E., et al. (2006). Fast time domain integral equation based electromagnetic analysis: A maturing technology. Proc. Eur. Conf. Comput. Fluid Dyn.
4. Bagci, H., et al. (2007). Fast and rigorous analysis of EMC/EMI phenomena on electrically large and complex cable-loaded structures. IEEE Trans. Electromagn. Compat. **49**(2): 361-381.
5. Bagci, H., et al. (2009). A fast Stroud-based collocation method for statistically characterizing EMI/EMC phenomena on complex platforms. IEEE Trans. Electromagn. Compat. **51**(2): 301-311.
6. Chen, N.-W., B. Shanker, and E. Michielssen (2003). Integral-equation-based analysis of transient scattering from periodic perfectly conducting structures. IEE P-Microw. Anten. P. **150**(2): 120-124.
7. Pantoja, M., et al. (2010). TDIE modeling of carbon nanotube dipoles at microwave and terahertz bands. IEEE Antennas Wireless Propag. Lett. **9**: 32-35.
8. Ergin, A.A., B. Shanker, and E. Michielssen (1999). The plane-wave time-domain algorithm for the fast analysis of transient wave phenomena. IEEE Antennas Propag. Mag. **41**(4): 39-52.
9. Yilmaz, A.E., J.-M. Jin, and E. Michielssen (2004). Time domain adaptive integral method for surface integral equations. IEEE Trans. Antennas Propag. **52**(10): 2692-2708.
10. Meng, J., et al. (2010). A multilevel Cartesian non-uniform grid time domain algorithm. J. Comput. Phys. **229**(22): 8430-8444.
11. Vikram, M. and B. Shanker (2007). Fast evaluation of time domain fields in sub-wavelength source/observer distributions using accelerated Cartesian expansions (ACE). J. Comput. Phys. **227**(2): 1007-1023.
12. Shanker, B. and H. Huang (2007). Accelerated Cartesian expansions–A fast method for computing of potentials of the form $R - \nu$ for all real ν. J. Comput. Phys. **226**(1): 732-753.
13. Cheng, G.S. and R.S. Chen (2016). Fast analysis of transient electromagnetic scattering using the Taylor series expansion-enhanced time-domain integral equation solver. IEEE Trans. Antennas Propag. **64**(9): 3943-3952.
14. Zhou, Z. and J.S. Tyo (2005). An adaptive time-domain integral equation method for transient analysis of wire scatterer. IEEE Antennas Wireless Propag. Lett. **4**: 147-150.
15. Kaur, G. and A.E. Ylmaz (2015). Envelope-tracking adaptive integral method for band-pass transient scattering analysis. IEEE Trans. Antennas Propag. **63**(5): 2215-2227.
16. Walker, S.P. and M.J. Vartiainen (1998). Hybridization of curvilinear time-domain integral equation and time-domain optical methods for electromagnetic scattering analysis. IEEE Trans. Antennas Propag. **46**(3): 318-324.
17. Kobidze, G., B. Shanker, and E. Michielssen (2003). Hybrid PO-PWTD scheme for analyzing of scattering from electrically large PEC objects. Proc. IEEE Int. Symp. AP-S/URSI. **3**: 547-550.
18. Meng, R., et al. (2008). Coupled TDIE-PO method for transient scattering from electrically large conducting objects. Electron. Lett. **44**(4): 258-260.
19. Luo, W., et al. (2011). Hybrid TDIE-TDPO method for studying on transient responses of some wire and surface structures illuminated by an electromagnetic pulse. Prog. Electromagn. Res. **116**: 203-219.
20. Luo, W., et al. (2012). Investigation on time-and frequency-domain responses of some complex composite structures in the presence of high-power electromagnetic pulses. IEEE Trans. Electromagn. Compat. **54**(5): 1006-1016.

21. Al-Jarro, A., M. Cheeseman, and H. Bağcı (2012). Distributed-memory parallelization of an explicit time-domain volume integral equation solver on Blue Gene/P. Appl. Comp. Electromag. Soc. J. **27**(2): 132-144.

22. Liu, Y., V. Lomakin, and E. Michielssen (2012). Graphics processing unit-accelerated implementation of the plane wave time domain algorithm. 28th Ann. Rev. Prog. Appl. Computat. Electromagn.

23. Feki, S., et al. (2014). Porting an explicit time-domain volume-integral-equation solver on GPUs with OpenACC. IEEE Antennas Propag. Mag. **56**(2): 265-277.

24. Liu, Y., et al. (2014). Graphics processing unit implementation of multilevel plane-wave time-domain algorithm. IEEE Antennas Wireless Propag. Lett. **13**: 1671-1675.

25. Liu, Y., et al. (2016). A scalable parallel PWTD-accelerated SIE solver for analyzing transient scattering from electrically large objects. IEEE Trans. Antennas Propag. **64**(2): 663-674.

26. Liu, Y., et al. (2016). Parallel PWTD-accelerated explicit solution of the time-domain electric field volume integral equation. IEEE Trans. Antennas Propag. **64**(6): 2378-2388.

27. Aygun, K., et al. (2003). A parallel PWTD accelerated time marching scheme for analysis of EMC/EMI problems. Proc. IEEE Int. Symp. Electromagn. Compat.

28. Liu, N., et al. (2004). The parallel plane wave time domain algorithm-accelerated marching on in time solvers for large-scale electromagnetic scattering problems. Proc. IEEE Int. Symp. AP-S/URSI, IEEE.

29. Song, J., C.-C. Lu, and W.C. Chew (1997). Multilevel fast multipole algorithm for electromagnetic scattering by large complex objects. IEEE Trans. Antennas Propag. **45**(10): 1488-1493.

30. Fostier, J. and F. Olyslager (2008). Provably scalable parallel multilevel fast multipole algorithm. Electron. Lett. **44**(19): 1111-1113.

31. Ergul, O. and L. Gurel (2009). A hierarchical partitioning strategy for an efficient parallelization of the multilevel fast multipole algorithm. IEEE Trans. Antennas Propag. **57**(6): 1740-1750.

32. Fostier, J. and F. Olyslager (2010). An open-source implementation for full-wave 2D scattering by million-wavelength-size objects. IEEE Antennas Propag. Mag. **52**(5): 23-34.

33. Ergul, O. and L. Gurel (2011). Rigorous solutions of electromagnetic problems involving hundreds of millions of unknowns. IEEE Antennas Propag. Mag. **53**(1): 18-27.

34. Melapudi, V., et al. (2011). A scalable parallel wideband MLFMA for efficient electromagnetic simulations on large scale clusters. IEEE Trans. Antennas Propag. **59**(7): 2565-2577.

35. Rao, S., D. Wilton, and A. Glisson (1982). Electromagnetic scattering by surfaces of arbitrary shape. IEEE Trans. Antennas Propag. **30**(3): 409-418.

36. Manara, G., A. Monorchio, and R. Reggiannini (1997). A space-time discretization criterion for a stable time-marching solution of the electric field integral equation. IEEE Trans. Antennas Propag. **45**(3): 527-532.

37. Mieras, H. and C. Bennett (1982). Space-time integral equation approach to dielectric targets. IEEE Trans. Antennas Propag. **30**(1): 2-9.

38. Shanker, B., A.A. Ergin, and E. Michielssen (2002). Plane-wave-time-domain-enhanced marching-on-in-time scheme for analyzing scattering from homogeneous dielectric structures. J. Opt. Soc. Am. A **19**(4): 716-726.

39. Vechinski, D.A., S.M. Rao, and T.K. Sarkar (1994). Transient scattering from three-dimensional arbitrarily shaped dielectric bodies. J. Opt. Soc. Am. A **11**(4): 1458-1470.

40. Pisharody, G. and D.S. Weile (2006). Electromagnetic scattering from homogeneous dielectric bodies using time-domain integral equations. IEEE Trans. Antennas Propag. **54**(2): 687-697.

41. Shanker, B., et al. (2009). Time domain integral equation analysis of scattering from composite bodies via exact evaluation of radiation fields. IEEE Trans. Antennas Propag. **57**(5): 1506-1520.

42. Wang, X., et al. (2008). A finite difference delay modeling approach to the discretization of the time domain integral equations of electromagnetics. IEEE Trans. Antennas Propag. **56**(8): 2442-2452.

43. Wang, X. and D.S. Weile (2010). Electromagnetic scattering from dispersive dielectric scatterers using the finite difference delay modeling method. IEEE Trans. Antennas Propag. **58**(5): 1720-1730.

44. Uysal, I.E., H. Arda Ülkü, and H. Bağci (2016). Transient analysis of electromagnetic wave interactions on plasmonic nanostructures using a surface integral equation solver. J. Acoust. Soc. Am. **33**(9): 1747-1759.

45. Uysal, I.E., H.A. Ülkü, and H. Bağcı (2015). MOT solution of the PMCHWT equation for analyzing transient scattering from conductive dielectrics. IEEE Antennas Wireless Propag. Lett. **14**: 507-510.

46. Zhao, Y., et al. (2018). A time-domain thin dielectric sheet (TD-TDS) integral equation method for scattering characteristics of tunable graphene. IEEE Trans. Antennas Propag. **66**(3): 1366-1373.

47. Gres, N.T., et al. (2001). Volume-integral-equation-based analysis of transient electromagnetic scattering from three-dimensional inhomogeneous dielectric objects. Radio Sci. **36**(3): 379-386.

48. Shanker, B., K. Aygun, and E. Michielssen (2004). Fast analysis of transient scattering from lossy inhomogeneous dielectric bodies. Radio Sci. **39**(2).

49. Kobidze, G., et al. (2005). A fast time domain integral equation based scheme for analyzing scattering from dispersive objects. IEEE Trans. Antennas Propag. **53**(3): 1215-1226.

50. Sayed, S.B., H.A. Ulku, and H. Bagci (2016). Transient analysis of scattering from ferromagnetic objects using Landau-Lifshitz-Gilbert and volume integral equations. 2016 IEEE International Symposium on Antennas and Propagation (APSURSI).

51. Yilmaz, A.E., et al. (2002). Fast analysis of transient scattering in lossy media. IEEE Antennas Wireless Propag. Lett. **1**(1): 14-17.

52. Bagci, H., et al. (2005). Fast solution of mixed-potential time-domain integral equations for half-space environments. IEEE Trans. Geosci. Remote Sens. **43**(2): 269-279.

53. Ghaffari-Miab, M., et al. (2014). Time-domain integral equation solver for planar circuits over layered media using finite difference generated Green's functions. IEEE Trans. Antennas Propag. **62**(6): 3076-3090.

54. Karami, H., et al. (2014). Efficient analysis of shielding effectiveness of metallic rectangular enclosures using unconditionally stable time-domain integral equations. IEEE Trans. Electromagn. Compat. **56**(6): 1412-1419.

55. Gao, J. and B. Shanker (2007). Time domain Weyl's identity and the causality trick based formulation of the time domain periodic Green's function. IEEE Trans. Antennas Propag. **55**(6): 1656-1666.

56. Lu, M., et al. (2004). Fast time domain integral equation solvers for analyzing two-dimensional scattering phenomena; Part I: temporal acceleration. Electromagnetics **24**(6): 425-449.

57. Ghaffari-Miab, M., S.M.H. Haddad, and R. Faraji-Dana (2009). A new fast and accurate time domain formulation of the method of moment (TD-MoM) for thin-wire antennas. Asia Pac. Microw. Conf.: 72-75.

58. Hu, J.-L. and C.H. Chan (1999). Improved temporal basis function for time domain electric field integral equation method. Electron. Lett. **35**(11): 883-885.

59. Hu, J.-L., C.H. Chan, and Y. Xu (2001). A new temporal basis function for the time-domain integral equation method. IEEE Microw. Wireless Compon. Lett. **11**(11): 465-466.

60. van't Wout, E., et al. (2013). Design of temporal basis functions for time domain integral equation methods with predefined accuracy and smoothness. IEEE Trans. Antennas Propag. **61**(1): 271-280.

61. Geranmayeh, A., W. Ackermann, and T. Weiland (2009). Temporal discretization choices for stable boundary element methods in electromagnetic scattering problems. Appl. Numer. Math. **59**(11): 2751-2773.

62. Wang, P., et al. (2007). Time-domain integral equation solvers using quadratic B-spline temporal basis functions. Microw. Opt. Techn. Lett. **49**(5): 1154-1159.

63. Knab, J. (1979). Interpolation of band-limited functions using the approximate prolate series. IEEE Trans. Inf. Theory **25**(6): 717-720.
64. Weile, D.S., et al. (2004). A novel scheme for the solution of the time-domain integral equations of electromagnetics. IEEE Trans. Antennas Propag. **52**(1): 283-295.
65. Sayed, S.B., H.A. Ülkü, and H. Bağcı (2015). A stable marching on-in-time scheme for solving the time-domain electric field volume integral equation on high-contrast scatterers. IEEE Trans. Antennas Propag. **63**(7): 3098-3110.
66. Graglia, R.D., D.R. Wilton, and A.F. Peterson (1997). Higher order interpolatory vector bases for computational electromagnetics. IEEE Trans. Antennas Propag. **45**(3): 329-342.
67. Wildman, R.A., et al. (2004). An accurate scheme for the solution of the time-domain integral equations of electromagnetics using higher order vector bases and bandlimited extrapolation. IEEE Trans. Antennas Propag. **52**(11): 2973-2984.
68. Wildman, R.A. and D.S. Weile (2004). An accurate broad-band method of moments using higher order basis functions and tree-loop decomposition. IEEE Trans. Antennas Propag. **52**(11): 3005-3011.
69. Wildman, R.A. and D.S. Weile (2005). Two-dimensional transverse-magnetic time-domain scattering using the Nyström method and bandlimited extrapolation. IEEE Trans. Antennas Propag. **53**(7): 2259-2266.
70. Nair, N.V., A.J. Pray, and B. Shanker (2010). Analysis of transient scattering from PEC objects using the generalized method of moments. Proc. IEEE Int. Symp. AP-S/URSI: 1-4.
71. Ulku, H.A., et al. (2017). Mixed discretization of the time-domain MFIE at low frequencies. IEEE Antennas Wireless Propag. Lett. **16**: 1565-1568.
72. Lu, M. and E. Michielssen (2002). Closed form evaluation of time domain fields due to Rao-Wilton-Glisson sources for use in marching-on-in-time based EFIE solvers. Proc. IEEE Int. Symp. AP-S/URSI. **1**: 74-77.
73. Shi, Y., et al. (2011). Stable electric field TDIE solvers via quasi-exact evaluation of MOT matrix elements. IEEE Trans. Antennas Propag. **59**(2): 574-585.
74. Yucel, A.C. and A.A. Ergin (2006). Exact evaluation of retarded-time potential integrals for the RWG bases. IEEE Trans. Antennas Propag. **54**(5): 1496-1502.
75. Ulku, H.A. and A.A. Ergin (2011). On the singularity of the closed-form expression of the magnetic field in time domain. IEEE Trans. Antennas Propag. **59**(2): 691-694.
76. Ulku, H.A. and A.A. Ergin (2007). Analytical evaluation of transient magnetic fields due to RWG current bases. IEEE Trans. Antennas Propag. **55**(12): 3565-3575.
77. Ulku, H.A. and A.A. Ergin (2011). Application of analytical retarded-time potential expressions to the solution of time domain integral equations. IEEE Trans. Antennas Propag. **59**(11): 4123-4131.
78. Pingenot, J., S. Chakraborty, and V. Jandhyala (2006). Polar integration for exact space-time quadrature in time-domain integral equations. IEEE Trans. Antennas Propag. **54**(10): 3037-3042.
79. Pray, A.J., N.V. Nair, and B. Shanker (2012). Stability properties of the time domain electric field integral equation using a separable approximation for the convolution with the retarded potential. IEEE Trans. Antennas Propag. **60**(8): 3772-3781.
80. Zhu, M.-D., X.-L. Zhou, and W.-Y. Yin (2013). Efficient evaluation of double surface integrals in time-domain integral equation formulations. IEEE Trans. Antennas Propag. **61**(9): 4653.
81. Rynne, B.P. (1985). Stability and convergence of time marching methods in scattering problems. IMA J. Appl. Math. **35**(3): 297-310.
82. Rynne, B.P. and P.D. Smith (1990). Stability of time marching algorithms for the electric field integral equation. J. Electromagnet. Wave **4**(12): 1181-1205.
83. Smith, P.D. (1990). Instabilities in time marching methods for scattering: Cause and rectification. Electromagnetics **10**(4): 439-451.
84. Vechinski, D.A. and S.M. Rao (1992). A stable procedure to calculate the transient scattering by conducting surfaces of arbitrary shape. IEEE Trans. Antennas Propag. **40**(6): 661-665.
85. Sadigh, A. and E. Arvas (1993). Treating the instabilities in marching-on-in-time method from a different perspective. IEEE Trans. Antennas Propag. **41**(12): 1695-1702.

86. Davies, P.J. (1998). A stability analysis of a time marching scheme for the general surface electric field integral equation. Appl. Numer. Math. **27**(1): 33-57.
87. Davies, P.J. and D.B. Duncan (2004). Stability and convergence of collocation schemes for retarded potential integral equations. SIAM J. Numer. Anal. **42**(3): 1167-1188.
88. Bluck, M.J. and S.P. Walker (1997). Time-domain BIE analysis of large three-dimensional electromagnetic scattering problems. IEEE Trans. Antennas Propag. **45**(5): 894-901.
89. Dodson, S.J., S.P. Walker, and M.J. Bluck (1998). Implicitness and stability of time domain integral equation scattering analysis. Appl. Comp. Electromag. Soc. J. **13**: 291-301.
90. Beghein, Y., et al. (2013). A space-time mixed Galerkin marching-on-in-time scheme for the time-domain combined field integral equation. IEEE Trans. Antennas Propag. **61**(3): 1228-1238.
91. Abboud, T., J. Nedelec, and J. Volakis (2001). Stable solution of the retarded potential equations. 17th Ann. Rev. Prog. Appl. Computat. Electromagn.
92. Wang, X. and D.S. Weile (2011). Implicit Runge-Kutta methods for the discretization of time domain integral equations. IEEE Trans. Antennas Propag. **59**(12): 4651-4663.
93. Ülkü, H.A., H.B.ğ.c. ı, and E. Michielssen (2013). Marching on-in-time solution of the time domain magnetic field integral equation using a predictor-corrector scheme. IEEE Trans. Antennas Propag. **61**(8): 4120-4131.
94. Al-Jarro, A., et al. (2012). Explicit solution of the time domain volume integral equation using a stable predictor-corrector scheme. IEEE Trans. Antennas Propag. **60**(11): 5203-5214.
95. Pisharody, G. and D.S. Weile (2005). Robust solution of time-domain integral equations using loop-tree decomposition and bandlimited extrapolation. IEEE Trans. Antennas Propag. **53**(6): 2089-2098.
96. Pisharody, G. and D.S. Weile (2005). Electromagnetic scattering from perfect electric conductors using an augmented time-domain integral-equation technique. Microw. Opt. Techn. Let. **45**(1): 26-31.
97. Andriulli, F.P., et al. (2009). Time domain Calderón identities and their application to the integral equation analysis of scattering by PEC objects part II: Stability. IEEE Trans. Antennas Propag. **57**(8): 2365-2375.
98. Shi, Y., H. Bagci, and M. Lu (2013). On the internal resonant modes in marching-on-in-time solution of the time domain electric field integral equation. IEEE Trans. Antennas Propag. **61**(8): 4389-4392.
99. Chen, N.-W., K. Aygun, and E. Michielssen (2001). Integral-equation-based analysis of transient scattering and radiation from conducting bodies at very low frequencies. IEE P-Microw. Anten. P. **148**(6): 381-387.
100. Andriulli, F.P., et al. (2007). A marching-on-in-time hierarchical scheme for the solution of the time domain electric field integral equation. IEEE Trans. Antennas Propag. **55**(12): 3734-3738.
101. Andriulli, F.P., et al. (2009). Analysis and regularization of the TD-EFIE low-frequency breakdown. IEEE Trans. Antennas Propag. **57**(7): 2034-2046.
102. Bagci, H., et al. (2010). A well-conditioned integral-equation formulation for efficient transient analysis of electrically small microelectronic devices. IEEE Trans. Adv. Packag. **33**(2): 468-480.
103. Cools, K., et al. (2009). Time domain Calderón identities and their application to the integral equation analysis of scattering by PEC objects Part I: Preconditioning. IEEE Trans. Antennas Propag. **57**(8): 2352-2364.
104. Cools, K., et al. (2007). Calderón preconditioned time-domain integral equation solvers. Proc. IEEE Int. Symp. AP-S/URSI: 4565-4568.
105. Andriulli, F.P., et al. (2007). Calderón stabilized time domain integral equation solvers. Proc. IEEE Int. Symp. AP-S/URSI: 4573-4576.
106. Beghein, Y., K. Cools, and F.P. Andriulli (2015). A DC-stable, well-balanced, Calderón preconditioned time domain electric field integral equation. IEEE Trans. Antennas Propag. **63**(12): 5650-5660.

107. Valdes, F., et al. (2011). High-order div-and quasi curl-conforming basis functions for Calderón multiplicative preconditioning of the EFIE. IEEE Trans. Antennas Propag. **59**(4): 1321-1337.
108. Valdes, F., et al. (2013). High-order Calderón preconditioned time domain integral equation solvers. IEEE Trans. Antennas Propag. **61**(5): 2570-2588.
109. Valdes, F., et al. (2013). Time-domain single-source integral equations for analyzing scattering from homogeneous penetrable objects. IEEE Trans. Antennas Propag. **61**(3): 1239-1254.
110. Kaur, G. and A.E. Yilmaz (2011). On the performance of envelope-tracking surface-integral equation solvers. Proc. IEEE Int. Symp. AP-S/URSI: 2716-2719.
111. Mohan, A. and D.S. Weile (2005). A hybrid method of moments-marching on in time method for the solution of electromagnetic scattering problems. IEEE Trans. Antennas Propag. **53**(3): 1237-1242.
112. Jiao, D., et al. (2001). A fast time-domain finite element-boundary integral method for electromagnetic analysis. IEEE Trans. Antennas Propag. **49**(10): 1453-1461.
113. Jiao, D., et al. (2002). A fast higher-order time-domain finite element-boundary integral method for 3-D electromagnetic scattering analysis. IEEE Trans. Antennas Propag. **50**(9): 1192-1202.
114. McCowen, A., A.J. Radcliffe, and M.S. Towers (2003). Time-domain modeling of scattering from arbitrary cylinders in two dimensions using a hybrid finite-element and integral equation method. IEEE Trans. Magn. **39**(3): 1227-1229.
115. Monorchio, A., et al. (2004). A hybrid time-domain technique that combines the finite element, finite difference and method of moment techniques to solve complex electromagnetic problems. IEEE Trans. Antennas Propag. **52**(10): 2666-2674.
116. Yilmaz, A.E., et al. (2007). A single-boundary implicit and FFT-accelerated time-domain finite element-boundary integral solver. IEEE Trans. Antennas Propag. **55**(5): 1382-1397.
117. Cheng, G.S., D.Z. Ding, and R.S. Chen (2017). An efficient fast algorithm for accelerating the time-domain integral equation discontinuous Galerkin method. IEEE Trans. Antennas Propag. **65**(9): 4919-4924.
118. Bleszynski, E., M. Bleszynski, and T. Jaroszewicz (1996). AIM: Adaptive integral method for solving large-scale electromagnetic scattering and radiation problems. Radio Sci. **31**(5): 1225-1251.
119. Yilmaz, A.E., et al. (2002). A hierarchical FFT algorithm (HIL-FFT) for the fast analysis of transient electromagnetic scattering phenomena. IEEE Trans. Antennas Propag. **50**(7): 971-982.
120. Yilmaz, A.E., J.-M. Jin, and E. Michielssen (2007). Analysis of low-frequency electromagnetic transients by an extended time-domain adaptive integral method. IEEE Trans. Adv. Packag. **30**(2): 301-312.
121. Boag, A., V. Lomakin, and E. Michielssen (2006). Nonuniform grid time domain (NGTD) algorithm for fast evaluation of transient wave fields. IEEE Trans. Antennas Propag. **54**(7): 1943-1951.
122. Shanker, B., et al. (2003). Fast analysis of transient electromagnetic scattering phenomena using the multilevel plane wave time domain algorithm. IEEE Trans. Antennas Propag. **51**(3): 628-641.
123. Gao, J., et al. (2003). Analysis of transient scattering from multiregion bodies using the plane wave time domain algorithm. Proc. IEEE Int. Symp. AP-S/URSI.
124. Hu, Y.L. and R.S. Chen (2016). Analysis of scattering from composite conducting dispersive dielectric objects by time-domain volume-surface integral equation. IEEE Trans. Antennas Propag. **64**(5): 1984-1989.
125. Jiang, P.L. and E. Michielssen (2005). Multilevel plane wave time domain-enhanced MOT solver for analyzing electromagnetic scattering from objects residing in lossy media. Proc. IEEE Int. Symp. AP-S/URSI.
126. Jiang, P.-L. and E. Michielssen (2006). Multilevel PWTD-enhanced CFIE solver for analyzing EM scattering from PEC objects residing in lossy media. Proc. IEEE Int. Symp. AP-S/URSI: 2967-2970.

127. Jiang, P., et al. (2003). An improved plane wave time domain algorithm for dissipative media. Proc. IEEE Int. Symp. AP-S/URSI. **3**: 563-566.

128. Jiang, P.-L. and E. Michielssen (2005). Temporal acceleration of time-domain integral-equation solvers for electromagnetic scattering from objects residing in lossy media. Microw. Opt. Techn. Let. **44**(3): 223-230.

129. Aygun, K., et al. (2000). Analysis of PCB level EMI phenomena using an adaptive low-frequency plane wave time domain algorithm. Proc. IEEE Int. Symp. Electromagn. Compat. **1**: 295-300.

130. Aygun, K., B. Shanker, and E. Michielssen (2001). Low frequency plane wave time domain kernels. Proc. Int. Conf. Electromagn. Adv. Appl.: 782-869.

131. Liu, Y., et al. (2018). A wavelet-enhanced PWTD-accelerated time-domain integral equation solver for analysis of transient scattering from electrically large conducting objects. IEEE Trans. Antennas Propag. **66**(5): 2458-2470.

132. Lu, M., A.A. Ergin, and E. Michielssen (2000). Fast evaluation of transient fields in the presence of two half spaces using a plane wave time domain algorithm. Proc. IEEE Int. Symp. AP-S/URSI.

133. Chen, N., et al. (2003). Fast integral-equation-based analysis of transient scattering from doubly periodic perfectly conducting structures. Proc. IEEE Int. Symp. AP-S/URSI.

134. Lu, M., E. Michielssen, and B. Shanker (2004). Fast time domain integral equation solvers for analyzing two-dimensional scattering phenomena; Part II: full PWTD acceleration. Electromagnetics **24**(6): 451-470.

135. Lu, M., et al. (2000). Fast evaluation of two-dimensional transient wave fields. J. Comput. Phys. **158**(2): 161-185.

136. Shanker, B., et al. (2000). Analysis of transient electromagnetic scattering phenomena using a two-level plane wave time-domain algorithm. IEEE Trans. Antennas Propag. **48**(4): 510-523.

137. Bost, F., L. Nicolas, and G. Rojat (2000). A time-domain integral formulation for the scattering by thin wires. IEEE Trans. Magn. **36**(4): 868-871.

138. Ji, Z., et al. (2004). A stable solution of time domain electric field integral equation for thin-wire antennas using the Laguerre polynomials. IEEE Trans. Antennas Propag. **52**(10): 2641-2649.

139. Excell, P.S., A.D. Tinniswood, and R.W. Clarke (1999). An independently fed log-periodic antenna for directed pulsed radiation. IEEE Trans. Electromagn. Compat. **41**(4): 344-349.

140. Boryssenko, A.O. and D.H. Schaubert (2004). Time-domain integral-equation-based solver for transient and broadband problems in electromagnetics. Proc. Int. Conf. UWB/SP Electro-magn., Springer: 239-249.

141. Thomas, K.G. and M. Sreenivasan (2010). A simple ultrawideband planar rectangular printed antenna with band dispensation. IEEE Trans. Antennas Propag. **58**(1): 27-34.

142. Aygun, K., et al. (2002). A two-level plane wave time-domain algorithm for fast analysis of EMC/EMI problems. IEEE Trans. Electromagn. Compat. **44**(1): 152-164.

143. Aygun, K., B. Shanker, and E. Michielssen (2002). Fast time-domain characterization of finite size microstrip structures. Int. J. Numer. Model. El. **15**(5-6): 439-457.

144. Aygun, K., et al. (2004). A fast hybrid field-circuit simulator for transient analysis of microwave circuits. IEEE Trans. Microw. Theory Tech. **52**(2): 573-583.

145. Yilmaz, A.E., J.-M. Jin, and E. Michielssen (2005). A parallel FFT accelerated transient field-circuit simulator. IEEE Trans. Microw. Theory Tech. **53**(9): 2851-2865.

146. Bagci, H., A.E. Yilmaz, and E. Michielssen (2010). An FFT-accelerated time-domain multiconductor transmission line simulator. IEEE Trans. Electromagn. Compat. **52**(1): 199-214.

Parallel Optimization Techniques
for Machine Learning

Sudhir Kylasa, Chih-Hao Fang, Fred Roosta, and Ananth Grama

Increasing processing power of the bare metal hardware has motivated significant interest in the development of optimization techniques for machine learning problems on massively large datasets. Applications such as autonomous vehicles, artificial intelligence (Google's GO system [38], IBM's Deep Blue [34], IBM's Project Debator [35]), image classification, and cybersecurity have been enabled by developments in optimization techniques and their parallel implementations. Many current applications mentioned above are modeled as either convex or non-convex optimization problems. These problems have rich theoretical foundations, as well as algorithmic (both serial and parallel) contributions. In this chapter, we focus on *finite-sum minimization* problems in the context of convex and non-convex formulations. We discuss state-of-the-art optimization methods for these problems under real-world assumptions on parallel platforms. We also highlight the need for hardware accelerators, such as GPUs, in significantly accelerating solutions.

S. Kylasa
Department of Electrical and Computer Engineering, Purdue University, West Lafayette, IN, USA
e-mail: skylasa@purdue.edu

C.-H. Fang · A. Grama (✉)
Department of Computer Science, Purdue University, West Lafayette, IN, USA
e-mail: fang150@cs.purdue.edu; ayg@cs.purdue.edu

F. Roosta
School of Mathematics and Physics, University of Queensland, Brisbane, QLD, Australia
e-mail: fred.roosta@uq.edu.au

© Springer Nature Switzerland AG 2020
A. Grama, A. H. Sameh (eds.), *Parallel Algorithms in Computational Science and Engineering*, Modeling and Simulation in Science, Engineering and Technology,
https://doi.org/10.1007/978-3-030-43736-7_13

1 Introduction and Motivation

Finite-sum optimization problems can be written in the form:

$$\min_{\mathbf{x}\in\mathbb{R}^d} F(\mathbf{x}) \triangleq \sum_{i=1}^{n} f_i(\mathbf{x}). \tag{1}$$

Here, each $f_i(\mathbf{x})$ is a smooth convex function, representing a loss (or misfit) corresponding to the ith observation (or measurement). These problems are well studied in the machine learning community [7, 27, 66]. In such applications, F in Eq. (1) corresponds to the *empirical risk* [65], and the goal of solving Eq. (1) is to obtain a solution with small generalization error, i.e., high predictive accuracy on "unseen" data. We consider Eq. (1) at scale, where the values of n and d are large. In such settings, the mere computation of the first- and second-order statistics (gradient and the Hessian, respectively) of F increases linearly in n. In large-scale settings, operations involving these statistics constitute the main computational bottleneck. In such cases, randomized sub-sampling has been shown to be highly successful in reducing computational and memory costs of the state-of-the-art optimizers to be effectively independent of n.

The most commonly used optimization technique in machine learning is gradient descent and its stochastic version, stochastic gradient descent (SGD). Gradient descent is a simple iterative procedure that takes steps in the direction of the negative gradient of the function, evaluated at the current point, using a step-size that is chosen to satisfy appropriate descent conditions. The stochastic variant of gradient descent estimates the gradient using mini-batches, as opposed to the entire training dataset. Algorithms such as gradient descent, that solely rely on gradient information, are often referred to as first-order methods. In typical problem settings, gradient descent does not offer good convergence results owing to a number of limitations: (1) approximating proper learning rate (a.k.a. step-size), (2) same learning rate schedule is applied to all components of the parameters, when in most cases some components of parameters change frequently compared to other components that change slowly, and (3) minimizing highly non-convex error functions associated with deep learning problems, like neural networks, are known to be dominated by saddle points surrounded by high error plateaus, which make it very hard to escape from these regions for methods like SGD [22]. To address these challenges, several first-order alternatives have been proposed in recent literature such as SGD with Momentum (henceforth referred to as Momentum) [60], Adam [42], Adagrad [24], Adadelta [81], and RMSProp [32, 68]. However the hyper-parameter space for these methods becomes large and the methods become difficult to tune.

Compared with first-order alternatives, second-order methods use additional curvature information in the form of the Hessian matrix. As a result of incorporating such information, in addition to faster convergence rates, second-order methods offer a variety of, rather more subtle, benefits. For example, unlike first-order methods, Newton-type methods have been shown to be highly resilient to increasing problem ill-conditioning [63, 64, 72]. Furthermore, second-order methods typically require fewer parameters (e.g., inexactness tolerance for the subproblem solver and line-search parameters) and are less sensitive to their specific settings [4, 71]. By using curvature information at each iteration, these methods scale the gradient so that it is a more suitable direction to follow. Consequently, they typically require much fewer iterations, as compared to first-order counterparts.

A key challenge in optimization for machine learning problems is the large, often, distributed nature of the training dataset. It may be infeasible to collect the entire training set at a single node and process it serially because of resource constraints (the training set may be too large for a single mode), privacy (data may be constrained to specific locations), or the need for reducing optimization time. In each of these cases, there is a need for optimization methods that are suitably adapted to the parallel and distributed computing environments.

Distributed optimization solvers adopt one of the two strategies: (1) executing each operation in conventional solvers (e.g., SGD or (quasi) Newton) in a distributed environment, e.g., [15, 18, 20, 23, 29, 40, 61, 69, 83]; or (2) executing an ensemble of local optimization procedures that operate on their own data, with a coordinating procedure that harmonizes the models over iterations, e.g., [74, 75]. The trade-offs between these two methods are relatively well understood in the context of existing solvers—namely that the communication overhead of methods in the first class is higher, whereas the convergence rate of the second class of methods is compromised. For this reason, methods in the first class are generally preferred in tightly coupled data-center type environments, whereas methods in the latter class are preferred for wide area deployments.

A method that occupies the middle ground between first- and second-order methods relies on the natural gradient [36, 37, 76], proposed by Shun-chin Amari. This work posits that in fitting probabilistic models, the underlying parametric distributions can be thought of as belonging to a manifold, whose geometry is governed by the Fisher information matrix. Under this hypothesis, scaling the gradient using the Fisher information matrix can result in more effective directions for navigating the manifold of the parametric probability densities. However, in high-dimensional settings, using the exact Fisher matrix can be intractable. To remedy this, Martens et al. [30, 52] proposed a method, called Kronecker Factored Approximated Curvature (KFAC), to approximate the Fisher information matrix and its *inverse*-vector product, and applied it to applications in neural networks and reinforcement learning. It was shown that KFAC can significantly outperform many of the first-order alternatives.

Deep learning models such as convolution neural networks, residual neural networks, and LSTM [33] have millions of parameters for state-of-the-art network architectures and training such networks is a time-consuming proposition partic-ularly when massively large datasets are used for training. Higher-order solvers that use higher-order statistics of the underlying networks are often prohibitively expensive at scale. In this context more effective solvers which would yield better, if not similar, results in same number of epochs, as well as speedup in processing the mini-batches are critical to the performance of the optimizer. Higher-order solvers like Newton-type methods and KFAC methods have been shown to achieve significantly better results compared to first-order solvers for convex and non-convex optimization problems. GP-GPUs provide powerful platforms for realizing these results in practice. With thousands of compute cores, associated high performance memory architecture, single-instruction-multiple-thread (SIMT), and programming semantics, GPUs are capable of handling large compute intensive tasks with significant performance gains over traditional CPU cores. In fact without hardware accelerators, like GPUs, training state-of-the-art deep learning networks is often not possible in practice.

The rest of this chapter is organized as follows: Section 2 provides an overview of the existing methods for convex and non-convex problems in machine learning. Section 4 provides a discussion of higher-order methods for convex optimization; Sect. 6 extends these results to distributed settings, dealing with massively large datasets. Finally, in Sect. 5 we provide an in-depth analysis of a hybrid method, which uses Fisher information matrix of the objective function, in the context of deep convolution neural networks. These developments have motivated the development of distributed optimizers for non-convex applications, which involve deep networks with millions of model parameters and trained on massively large datasets.

2 Related Research

SGD [6] is the most commonly used first-order method in machine learning, owing to its simplicity and inexpensive per-iteration cost. Iterations in SGD require computation of the gradient on a mini-batch scaled by a predetermined learning schedule and possibly Nesterov-accelerated momentum [55]. It has been argued that high-dimensional non-convex functions such as those arising in deep learning are riddled with undesirable saddle points [2, 21, 22, 39]. For instance, convolutional neural networks, CNNs, display structural symmetry in their parameter space, which leads to an abundance of saddle points [3, 30, 51]. First-order methods, such as SGD, are known to "zig-zag" in high curvature areas and "stagnate" in low curvature regions [3, 22]. In these regions step-size (or learning rate) plays a critical role. Perturbed gradient based methods [28, 39, 45], where random noise is injected in the gradient computation have been proposed and shown to converge to second-order stationary points. However, their computational cost is often worse compared to second-order methods.

One of the primary reasons for the susceptibility of first-order methods to getting trapped in saddle points or nearly flat regions is their reliance on gradient information. Indeed, navigating around saddle points and plateau-like regions can become a challenge for these methods because the gradient is close to zero in most directions [22]. To this end, a number of alternate methods have been proposed in recent times, which using history of gradients aim to approximate curvature information, and hence maintaining the simplicity of SGD. Such methods include Adam [42] and Adagrad [24]. However, such approximations of the Hessian do not always scale the gradient according to the entire curvature information. Hence, these methods suffer from similar deficiencies near saddle points and flat regions. More effective variants of these curvature approximations are those in quasi-Newton methods such as SR1 [58], DFP [58], and BFGS [48, 58], which use rank-1 and rank-2 updates to iteratively approximate the Hessian. Aided by line-search methods, typically satisfying Strong-Wolfe [58] conditions, these methods yield good results compared to first-order methods for convex problems [43]. However, these methods remain as topics of active investigation in the non-convex regime.

Newton-type optimizers have been developed as alternatives to first-order methods. These optimizers can effectively navigate the steep and flat regions of the optimization landscape. By incorporating curvature information in the form of the Hessian matrix, e.g., negative curvature directions, these methods can escape saddle points [2, 21, 70, 73, 77, 79, 80]. Nocedal and Wright [58] propose the use of absolute Hessian matrix, **H**, to update parameters. Dauphin et al. [21] propose a *saddle-free Newton* method that optimizes first-order Taylor series approximation of the objective function in a trust-region framework constrained by the distance between successive updates measured by the curvature, $|H|$ of the objective function. In order to make this computationally tractable, the *Lanczos*-method is used to compute the eigenvectors corresponding to few highest eigenvalues as an approximation to **H**. Negative curvature descent methods, where the eigenvector corresponding to the least eigenvalue is used to traverse past the parameter manifold around saddle points, have been proposed by Yaodong Yu et al. [80]. Negative curvature can be embedded in gradient descent based methods, which upon encountering saddle points injects random perturbations in the gradient to navigate past the saddle points (perturbed gradient). Neon [2, 73] and Flash [79] methods also use negative curvature direction in a novel form in stochastic methods to navigate past the saddle points. However, such methods need to compute the least eigen-pair for each iteration, which is computationally expensive. To avoid explicitly forming the Hessian matrices, Hessian-free methods [14, 50, 54, 82] have been proposed, which only require Hessian-vector products. Arguably, a highly effective among these methods is the trust-region based method that comes with attractive theoretical guarantees and is relatively easy to implement [16, 70, 71, 77].

Several distributed optimization solvers have been developed recently [15, 18, 20, 23, 29, 40, 61, 69, 83]. Among these, [15, 23, 29, 40] are classified as first-order methods. Although they incur low computational costs, they have higher communication costs due to a large number of messages exchanged per mini-batch and high total iteration counts. Second-order variants [18, 20, 61, 69, 83] are designed to improve convergence rate, as well as to reduce communication costs (because of more accurate descent direction leading to fewer epochs to reach convergence). DANE [20] and the accelerated scheme AIDE [61] use SVRG [41] as the subproblem solver to approximate the Newton direction. These methods are often sensitive to the fine-tuning of SVRG. DiSCO [83] uses distributed preconditioned conjugate gradient (PCG) to approximate the Newton direction. The number of communications across nodes per PCG call is proportional to the number of PCG iterations. In contrast to DiSCO, GIANT [69] executes CG at each node and approximates the Newton direction by averaging the solution from each CG call. Empirical results have shown that GIANT outperforms DANE, AIDE, and DiSCO. The solver of Dunner et al. [25] is shown to outperform GIANT; however, it is restricted to sparse datasets. More recently, DINGO [18] has been developed, which unlike GIANT can be applied to a class of non-convex functions, namely invex [19], which includes convexity as a special sub-class. However, in the absence of invexity, the method can converge to undesirable stationary points.

A popular choice in distributed settings is ADMM [9], which combines dual ascent method and the method of multipliers. ADMM only requires one round of communication per iteration. However, ADMM's performance is affected by the selection of the penalty parameter [74, 75], as well as the choice of local subproblem solvers.

Lying on the spectrum between first- and second-order methods is Amari's natural gradient method [36, 37]. This method provided a new direction in the context of high-dimensional optimization of probabilistic models. In this work, Amari showed that natural gradient descent yields Fisher efficient estimate of the parameters; he subsequently applied the method to multi-layer perceptrons for solving blind source detection problems. However, computing Fisher matrix and its inverse in high-dimensional settings is computationally expensive both in terms of memory and computational resources. RMSProp [32, 68] methods use a diagonal approximation of Fisher matrix of the objective function to compute the descent direction. These methods incur little overhead with regard to diagonal approximation but nevertheless fail to make progress relative to SGD in some cases. Martens et al. [30, 51, 52] proposed the KFAC method, which approximates the natural gradient using Kronecker products of smaller matrices formed during backpropagation. KFAC method and its distributed counterpart [3] have been shown to outperform well-tuned SGD in many applications.

For non-convex optimization, we discuss an optimizer that couples the advantages of trust-region and KFAC methods and propose a stochastic optimization framework involving trust region objective computed on a mini-batch, constrained to directions that are aligned with those obtained from KFAC. Major computational tasks in updating the parameters in our method are Hessian-vector products involving the solution of the trust region subproblem, as well as finding the KFAC direction. Our Hessian-vector products can be computed at a similar cost as that of gradient computation using back-propagation. Furthermore, the Fisher matrix approximation and its inverse are only needed once every few mini-batches, thus reducing average iteration cost significantly. Invariance to re-parameterization, as well as immunity to large batch sizes, makes this method a suitable alternative to first-order methods for practitioners.

3 Notation and Assumptions

In the rest of this chapter, vectors are denoted by bold lowercase letters, e.g., $\mathbf{v}$, and matrices or random variables are denoted by bold uppercase letters, e.g., $\mathbf{V}$. For a vector $\mathbf{v}$ and a matrix $\mathbf{V}$, $\|\mathbf{v}\|$ and $\|\mathbf{V}\|$ denote the vector ℓ_2 norm and matrix spectral norm, respectively, while $\|\mathbf{V}\|_F$ is the matrix Frobenius norm. $\nabla f(\mathbf{x})$ and $\nabla^2 f(\mathbf{x})$ are the gradient and the Hessian of f evaluated at $\mathbf{x}$, respectively, and $\mathbb{I}$ denotes the identity matrix. For two symmetric matrices $\mathbf{A}$ and $\mathbf{B}$, $\mathbf{A} \succeq \mathbf{B}$ indicates that $\mathbf{A} - \mathbf{B}$ is symmetric positive semi-definite. The superscript, e.g., $\mathbf{x}^{(k)}$, denotes iteration counter and $ln(x)$ is the natural logarithm of x. $\mathcal{S}$ denotes a collection of indices from $\{1, 2, \cdots, n\}$, with potentially repeated items and its cardinality is denoted by $|\mathcal{S}|$.

We assume that each f_i is twice-differentiable, smooth, and convex, i.e., for some $0 < K_i < \infty$ and $\forall \mathbf{x} \in \mathbb{R}^p$

$$0 \preceq \nabla^2 f_i(\mathbf{x}) \preceq K_i \mathbb{I}. \tag{2a}$$

We also assume that F is smooth and strongly convex, i.e., $0 < \gamma \geq K < \infty$ and $\forall \mathbf{x} \in \mathbb{R}^p$

$$\gamma \mathbb{I} \preceq \nabla^2 F(\mathbf{x}) \preceq K \mathbb{I}. \tag{2b}$$

Note that assumption (2b) implies uniqueness of the minimizer, $\mathbf{x}^*$, which is assumed to be attained. The quantity

$$\kappa = \frac{K}{\gamma} \tag{3}$$

is known as the condition number of the problem.

For an integer $1 \leq q \leq n$, let Q be the set of indices corresponding to q largest K_i's and define the "sub-sampling" condition number as

$$\kappa_q = \frac{\hat{K}_q}{\gamma}, \tag{4}$$

where

$$\hat{K}_q = \frac{1}{q} \sum_{j \in Q} K_j, \tag{5}$$

It is easy to see that for any two integers q and r such that $1 \leq q \leq r \leq n$, we have $\kappa \leq \kappa_r \leq \kappa_q$. Finally, define

$$
\tilde{\kappa} = \begin{cases} \kappa_1 & : \text{If sample } S \text{ is drawn with replacement} \\ \kappa_{|S|} & : \text{If sample } S \text{ is drawn without replacement .} \end{cases}
$$

4   Convex Optimization Problems

The standard deterministic or full gradient method, which dates back to
Cauchy [13], for minimizing (1) uses iterates of the form:

$$
\mathbf{x}^{(k+1)} = \mathbf{x}^{(k)} - \alpha_k \nabla F(\mathbf{x}^{(k)}).
$$

Here, α_k is the step-size at iteration k. However, when $n \gg 1$, the full gradient
method can be inefficient because its iteration cost scales linearly in n. In addition,
when $p \gg 1$ or when individual functions f_i are complicated (e.g., evaluating each
f_i may require the solution of a partial differential equation), the mere evaluation of
the gradient can be computationally prohibitive. Consequently, a stochastic variant
of full gradient descent, stochastic gradient descent (SGD) was developed [5, 6, 8,
17, 46, 62]. In such methods a subset $S \subset \{1, 2, \cdots, n\}$ is chosen at random and
the update is obtained by

$$
\mathbf{x}^{(k+1)} = \mathbf{x}^{(k)} - \alpha_k \sum_{j \in S} \nabla f_j(\mathbf{x}^{(k)}).
$$

When $|S| \ll n$ (e.g., $|S| = 1$ for simple SGD), the iteration cost of stochastic
gradient methods is independent of n and can be much cheaper than the full gradient
methods, making them suitable for modern problems with large n. This class of
methods is referred to as *first-order* methods, since only the gradient information is
used at each iteration. By incorporating curvature information (e.g., Hessian) as a
form of scaling the gradient, i.e.,

$$
\mathbf{x}^{(k+1)} = \mathbf{x}^{(k)} - \alpha_k D_k \nabla F(\mathbf{x}^{(k)}),
$$

we can significantly improve convergence rate. This class of methods, which take
curvature information into account, are known as *second-order* methods. Compared
to first-order methods, they enjoy superior convergence rate in theory, as well as in
real application scenarios. This is because of implicit local scaling of components
at a given $\mathbf{x}$, which is determined by the local curvature of F. This local curvature
determines the condition number of a F at $\mathbf{x}$. Consequently, second-order methods
can rescale the gradient direction so that it is a better direction to traverse. Second-
order methods have long been used in many machine learning applications [6, 11,
12, 47, 50, 78].

   The canonical example of second-order methods, Newton's method [10, 55, 57],
uses a step-size of one and scales the gradient by the inverse of the Hessian, i.e.,

$$\mathbf{x}^{(k+1)} = \mathbf{x}^{(k)} - \left[\nabla^2 F(\mathbf{x}^{(k)}) \right]^{-1} \nabla F(\mathbf{x}^{(k)}).$$

It is well known that for a smooth and strongly convex function F, the Newton direction is always a descent direction and with a suitable step-size, α_k, global convergence is guaranteed. In addition, for cases when F is not strongly convex, Levenberg–Marquardt type regularization [44, 49] of the Hessian can be used to obtain a globally convergent algorithm. Newton's method exhibits *scale invariance*, i.e., for some new parameterization $\tilde{\mathbf{x}} = \mathbf{A}\mathbf{x}$ for invertible matrix $\mathbf{A}$, optimal search direction in the new coordinate system is $\tilde{\mathbf{p}} = \mathbf{A}\mathbf{p}$, where $\mathbf{p}$ is the original optimal search direction. In contrast the search direction produced by gradient descent methods behaves in an opposite fashion $\tilde{\mathbf{p}} = \mathbf{A}^{-\mathsf{T}}\mathbf{p}$. This scale invariance property is important for effectively optimizing poorly scaled parameters; see [50] for an intuitive explanation of this phenomenon. However, when $n, p \gg 1$, the per-iteration cost of this algorithm is significantly higher than that of first-order methods.

We now discuss a *sub-sampling* based method that approximates the gradient and Hessian of the objective function and present analyses of bounds on sample sizes. We then present results for an accelerated sub-sampled Newton's method over a range of real-world datasets and show that such methods can be highly competitive for machine learning applications.

For the optimization problem Eq. (1), in each iteration, consider selecting two sample sets of indices from $\{1, 2, \ldots, n\}$, uniformly at random *with* or *without* replacement. Let $\mathcal{S}_{\mathbf{g}}$ and $\mathcal{S}_{\mathbf{H}}$ denote the sample collections, and define $\mathbf{g}$ and $\mathbf{H}$ as

$$\mathbf{g}(\mathbf{x}) \triangleq \frac{n}{|\mathcal{S}_{\mathbf{g}}|} \sum_{j \in \mathcal{S}_{\mathbf{g}}} \nabla f_j(\mathbf{x}), \tag{6a}$$

$$\mathbf{H}(\mathbf{x}) \triangleq \frac{n}{|\mathcal{S}_{\mathbf{H}}|} \sum_{j \in \mathcal{S}_{\mathbf{H}}} \nabla^2 f_j(\mathbf{x}) \tag{6b}$$

to be the sub-sampled gradient and Hessian, respectively.

Lemma 1 (Uniform Hessian Sub-sampling) *Given any* $0 < \epsilon_{\mathbf{H}} < 1, 0 < \delta < 1$, $\mathbf{x} \in \mathbb{R}^p$, *and assumption* (2a) *holds, if*

$$|\mathcal{S}_{\mathbf{H}}| \geq \frac{2\kappa_1 log(p/\delta)}{\epsilon_{\mathbf{H}}^2},$$

then for $\mathbf{H}(\mathbf{x})$ *defined in* (6b), *we have*

$$Pr\left((1 - \epsilon_{\mathbf{H}})\gamma \leq \lambda_{\min}(\mathbf{H}(\mathbf{x}))\right) \geq 1 - \delta,$$

where γ *and* κ_1 *are defined in* (2b) *and* (4), *respectively.*

Using random matrix concentration inequalities, Roosta et al., [63, 64] derive lower bounds on the sample sizes for gradient and Hessian computation to probabilistically guarantee their utility in sub-sampled Newton-type methods. Depending on κ_1, the sample size $|S_H|$ can be smaller than n. In addition, we can always guarantee that the sub-sampled Hessian is uniformly positive definite and, consequently, the direction given by it, indeed, yields a direction of descent. Note that the sample size $|S_H|$ here grows only linearly in κ_1 compared to quadratically as in [63, 64].

Lemma 2 (Uniform Gradient Sub-Sampling) *For a given* $\mathbf{x} \in \mathbb{R}^p$, *let:*

$$\|\nabla f_i(\mathbf{x})\| \leq G(\mathbf{x}), \ i = 1, 2, \cdots, n .\tag{7}$$

For any $0 < \epsilon_\mathbf{g} < 1$ *and* $0 < \delta < 1$, *if*

$$|S_\mathbf{g}| \geq \frac{G(\mathbf{x})^2}{\epsilon_\mathbf{g}^2}\left(1 + \sqrt{8 ln \frac{1}{\delta}}\right)^2,\tag{8}$$

then for $\mathbf{g}(\mathbf{x})$ *defined in* (6a), *we have*

$$Pr\left(\|\nabla F(\mathbf{x}) - \mathbf{g}(\mathbf{x})\| \leq \epsilon_\mathbf{g}\right) \geq 1 - \delta.$$

Lemma 2 assumes that sampling preserves as much first-order information from the full gradient as possible. Note that in each iteration, $G(\mathbf{x})$ is required to guarantee the theoretical bounds on the gradient sample size, $|S_\mathbf{g}|$. Fortunately this can be estimated for most of the well-known objective functions [63].

With the bounds in Lemmas 1 and 2 on the size of the samples, $|S_\mathbf{g}|$ and $|S_H|$, one can, with high probability, ensure that $\mathbf{g}$ and $\mathbf{H}$ are "suitable" approximations to the full gradient and Hessian, in an algorithmic sense [63, 64]. For each iterate $\mathbf{x}^{(k)}$, using the corresponding sub-sampled approximations of the full gradient, $\mathbf{g}(\mathbf{x}^{(k)})$, and the full Hessian, $\mathbf{H}(\mathbf{x}^{(k)})$, we consider *inexact* Newton-type iterations of the form

$$\mathbf{x}^{(k+1)} = \mathbf{x}^{(k)} + \alpha_k \mathbf{p}_k,\tag{9a}$$

where $\mathbf{p}_k$ is a search direction satisfying

$$\|\mathbf{H}(\mathbf{x}^{(k)})\mathbf{p}_k + \mathbf{g}(\mathbf{x}^{(k)})\| \leq \theta \|\mathbf{g}(\mathbf{x}^{(k)})\|,\tag{9b}$$

for some inexactness tolerance $0 < \theta < 1$ and α_k is the largest $\alpha \leq 1$ such that:

$$F(\mathbf{x}^{(k)} + \alpha \mathbf{p}_k) \leq F(\mathbf{x}^{(k)}) + \alpha\beta \mathbf{p}_k^T \mathbf{g}(\mathbf{x}^{(k)}),\tag{9c}$$

for some $\beta \in (0, 1)$.

The requirement in Eq. (9c) is often referred to as Armijo-type line-search [58], and (9b) is the θ-relative error approximation condition of the exact solution to the linear system:

$$\mathbf{H}(\mathbf{x}^{(k)})\mathbf{p}_k = -\mathbf{g}(\mathbf{x}^{(k)}), \tag{10}$$

which is similar to that arising in classical Newton's method. Note that in (strictly) convex settings, where the sub-sampled Hessian matrix is symmetric positive definite (SPD), conjugate gradient (CG) with early stopping can be used to obtain an approximate solution to Eq. (10) satisfying Eq. (9b). It has also been shown [63, 64] that to inherit the convergence properties of the rather expensive algorithm that employs the exact solution to Eq. (10), the inexactness tolerance, θ, in Eq. (9b) can be chosen in the order of the inverse of the *square root* of the problem condition number. As a result, even for ill-conditioned problems, only a relatively moderate tolerance for CG ensures that we maintain convergence properties of the exact update (see also examples in Sect. 4.1). Putting all of these together, we obtain Algorithm 1, which under specific assumptions has been shown [63, 64] to be globally linearly convergent[1] with problem-independent local convergence rate.[2]

Algorithm 1: Sub-Sampled Newton Method

    Input   : Initial iterate, $\mathbf{x}^{(0)}$
    Parameters: $\epsilon_{\mathbf{g}}$ as in Lemma(2) $\epsilon_{\mathbf{H}}$ as in Lemma(1) and $\sigma \geq 0$
1 **foreach** $k = 0, 1, 2, \ldots$ **do**
2   Form $\mathbf{g}(\mathbf{x}^{(k)})$ as in Eq. (6a)
3   Form $\mathbf{H}(\mathbf{x}^{(k)})$ as in Eq. (6b)
4   **if** $\|\mathbf{g}(\mathbf{x}^{(k)})\| < \sigma\epsilon$ **then**
      | STOP
    end
5   Update $\mathbf{x}^{(k+1)}$ as in Eq. (9)
    end

Theorem 1 (Global Convergence of Algorithm: 1: Inexact Update) *Let Assumptions 2 hold. Also let* $0 < \theta < 1$ *be given. For any* $\mathbf{x}^k \in \mathbb{R}^p$, *using Algorithm 1 with* $\epsilon_{\mathbf{H}} < \frac{1}{2}$, *the "inexact" update direction 6b, and*

$$\sigma \geq \frac{8\tilde{\kappa}}{(1-\theta)(1-\beta)},$$

---

[1] It converges linearly to the optimum, starting from any initial guess $\mathbf{x}^{(0)}$.
[2] If the iterates are close enough to the optimum, it converges with a constant linear rate independent of the problem-related quantities.

we have the following with probability $1 - \delta$:

1. *if "STOP," then*

$$\left\| \nabla F(\mathbf{x}^{(k)}) \right\| < (1 + \sigma)\epsilon_{\mathbf{g}}, \tag{11}$$

2. *otherwise, global convergence results for Hessian sample size hold where*

   (a) *if*

$$\theta \leq \sqrt{\frac{(1 - \epsilon_{\mathbf{H}})}{4\tilde{\kappa}}}, \tag{12}$$

   then $\rho = 4\alpha_k \beta / 9\tilde{\kappa}$,

   (b) *otherwise* $\rho = 4\alpha_k \beta (1 - \theta)(1 - \epsilon_{\mathbf{H}})/9\tilde{\kappa}^2$, *with* $\tilde{\kappa}$ *defined as in (6). Moreover, for both cases, the step-size is at least*

$$\alpha_k \geq \frac{(1 - \theta)(1 - \beta)(1 - \epsilon_{\mathbf{H}})}{\kappa}, \tag{13}$$

   where κ *is defined as in (3).*

Theorem 1 says that, in order to guarantee a faster convergence rate, the linear system needs to be solved to a "high-enough" accuracy, which is in the order of $O(\sqrt{1/\tilde{\kappa}})$.

4.1   Experimental Results

We compare our methods to state-of-the-art methods—SGD with momentum (henceforth referred to as Momentum) [67], Adagrad [24], Adadelta [81], Adam [42], and RMSProp [68] as implemented in TensorFlow [1].

Table 1 presents the datasets used, along with the *Lipschitz* continuity constant of $\nabla F(\mathbf{x})$, denoted by L. Recall that, an over-estimate of the *condition number* of the problem, as defined in [63], can be obtained by $(L + \lambda)/\lambda$. As it is often done in practice, we first normalize the datasets such that each column of the data matrix $\mathbf{A} \in \mathbb{R}^{n \times p}$ has Euclidean norm one. This helps with the conditioning of the problem. The resulting dataset is, then, split into training and testing sets, as shown in Table 1.

Table 1 Description of the datasets

Dataset	Train size (n)	Test size	Features (p)	Classes(C)	Lipschitz Const. (L)
Drive diagnostics	50,000	8509	48	11	3.95
MNIST	38,000	38,000	785	10	28.67
CIFAR-10	50,000	10,000	3072	10	534.92
Newsgroups20	10,142	1127	53,975	20	128.79

We present results for two implementations of second-order methods: (a) *FullNewton*, the classical Newton-CG algorithm [58], which uses the exact gradient and Hessian, and (b) *SubsampledNewton-20*, $|S_g| = 0.2n$, and *SubsampledNewton-100*, $|S_g| = n$, are compared against first-order methods using batch sizes 128 and 20%, respectively. These methods use $|S_H| = 0.05n$. CG-tolerance is set to 10^{-4}. Maximum number of CG iterations is 10 for all datasets except *Drive Diagnostics* and *Gisette*, for which it is 1000. λ is set to 10^{-3} and we perform 100 iterations (epochs) for each dataset.

Tables 2 and 3 present the performance results of the proposed Newton-type methods in comparison with first-order methods for batch sizes 128 and 20%, respectively. In each of these tables we show the plots for *cumulative time vs. test accuracy* in column *1* and *cumulative time vs. objective function (training)* in column *2*. Please note that x-axis in all the plots is in "log-scale."

4.1.1 Drive Diagnostics Dataset

Row 2 of Tables 2 and 3 shows the results for the *Drive Diagnostics* dataset for batch sizes 128 and 20% (of the dataset), respectively. We notice that all Newton-type methods achieve lower objective function in the initial few iterations compared to first-order counterparts. When the batch size is larger, we notice that first-order methods take longer to achieve the same objective function value compared to smaller batch sized counterparts. Note that smaller gradient sample size yields similar results (objective function value and generalization error) throughout the simulations.

4.1.2 MNIST and CIFAR-10 Datasets

Rows 3 and 4 in Tables 2 and 3 present plots for *MNIST* and CIFAR-10 datasets, respectively. Regardless of the batch size, Newton-type methods clearly outperform first-order methods for these two datasets. When larger batch size is used for first-order methods, we notice that these methods take more epochs compared to their smaller batch sized counterparts in reaching same objective function value and generalization error. This behavior is more prominent in *CIFAR-10* dataset, which represents a relatively *ill*-conditioned problem. As a result, in terms of lowering the objective function on *CIFAR-10*, first-order methods are negatively impacted by problem ill-conditioning, whereas all Newton-type methods show excellent robustness. (Note that, for *CIFAR-10*, our proposed methods are $\approx 1000\times$ faster than first-order alternatives irrespective of the mini-batch size.)

4.1.3 Newsgroups20 Dataset

Plots in row 5 of Tables 2 and 3 represent *Newsgroups20* dataset, which is a sparse dataset, and the Hessian is $\approx 1e6 \times 1e6$. We clearly notice *SubsampledNewton-100* yields superior training accuracy compared to all methods (column 1). However, *SubsampledNewton-20* takes more epochs to achieve the same objective function

Table 2 Performance comparison between first-order and second-order methods (batch size = 128)

Time vs. Accuracy	Time vs. Objective Function (training)
First Order Batch Size = 128	
Alg. 1 Gradient Sample Size = 100%	
Alg. 1 Hessian Sample Size = 5%	

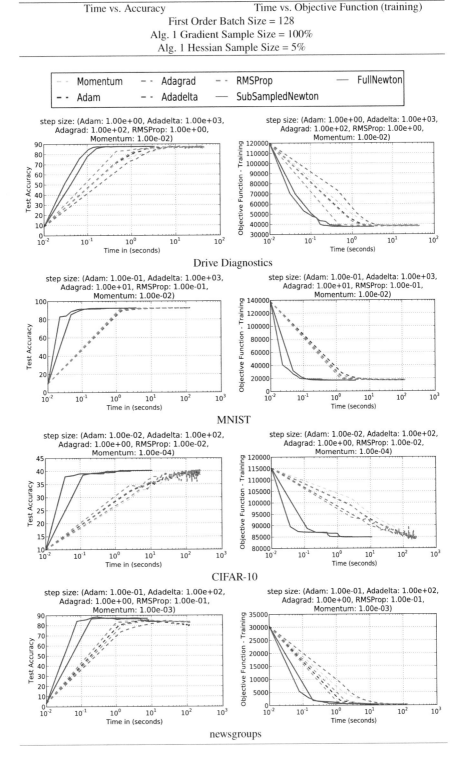

Table 3 Performance comparison between first-order and second-order methods (batch size = 20%)

Time vs. Accuracy	Time vs. Objective Function (training)
First Order Batch Size = 20%	
Alg. 1 Gradient Sample Size = 20%	
Alg. 1 Hessian Sample Size = 5%	

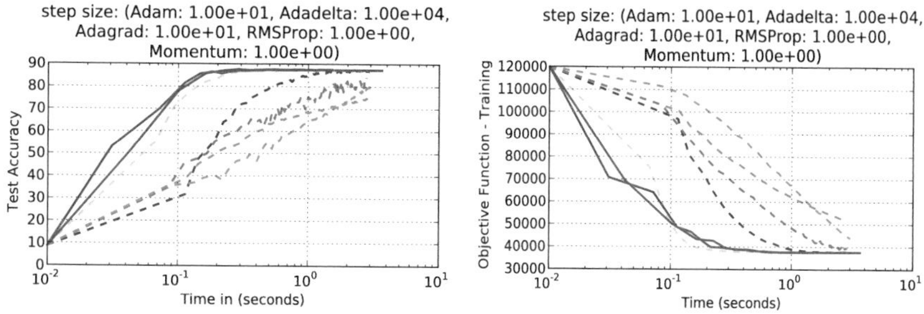

Drive Diagnostics

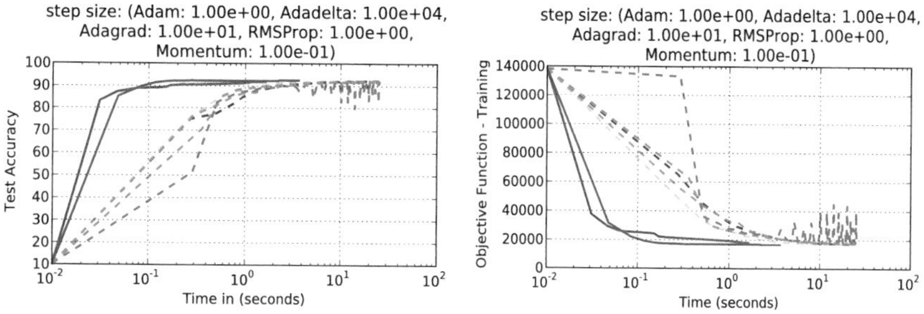

MNIST

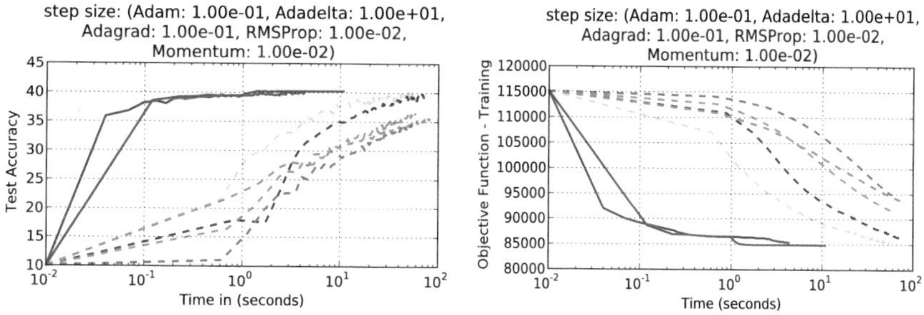

CIFAR-10

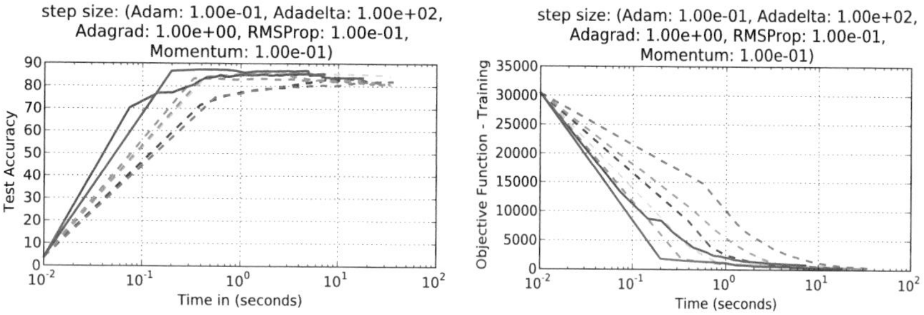

newsgroups

value as its full gradient counterpart, as seen in column 4. This can be attributed to a smaller gradient sample size and the sparse nature of this dataset.

4.2 Sensitivity to Hyper-Parameter Tuning

A major consideration for first-order methods is that of fine-tuning of various underlying hyper-parameters, most notably, the step-size [4, 71]. Indeed, the success of most such methods is strongly determined by many trial-and-error steps to find proper parameter settings. In contrast, second-order optimization methods involve much less parameter tuning and are less sensitive to specific choices of their hyper-parameters [4, 71].

To further highlight these issues, we demonstrate the sensitivity of several first-order methods with respect to their learning rate. Figure 1 shows the results of multiple runs of SGD with Momentum, Adagrad, RMSProp and Adam on

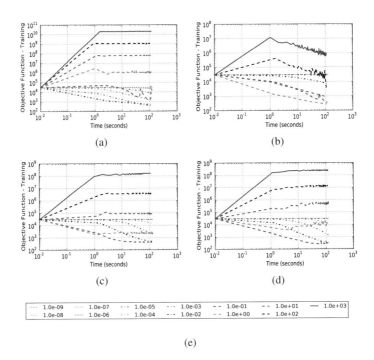

Fig. 1 Sensitivity of various first-order methods with respect to the choice of the step-size, i.e., learning rate. It is clear that too small a step-size can lead to slow convergence, while larger step-sizes cause the method to diverge. The range of step-sizes for which some of these methods perform reasonably can be very narrow. This is in contrast with Newton-type, which comes with a priori "natural" step-size, i.e., $\alpha = 1$, and only occasionally requires the line-search to intervene. (**a**) SGD with Momentum. (**b**) Adagrad. (**c**) RMSProp. (**d**) Adam. (**e**) Step-sizes

Newsgroups20 dataset with several choices of step-size. Each method is run 13 times using step-sizes in the range $10^{-6}/L$ to $10^6/L$, in increments of 10, where L is the Lipschitz constant; see Table 1. It is clear that small step-sizes can result in stagnation, whereas large step-sizes can cause the method to diverge. Only if the step-size is within a particular and often narrow range, which greatly varies across various methods, does one see reasonable performance.

Remark 1 For some first-order methods, e.g., momentum based, line-search type techniques simply cannot be used. For others, the starting step-size for line-search is, almost always, a priori unknown. This is in sharp contrast with randomized Newton-type methods considered here, which come with a priori "natural" step-size, i.e., $\alpha = 1$, and furthermore, only occasionally require the line-search to intervene; see [63, 64] for theoretical guarantees in this regard.

5 Non-convex Optimization

With the goal of avoiding being trapped at saddle points, many first-order alternatives such as Adam and Adagrad, and quasi-Newton methods that use low-rank updates, approximate underlying curvature of the objective function. These methods either require a large number of iterations for convergence (first-order alternatives) or are unstable in practice. Natural gradient based methods were proposed in the early 1960s and have been shown to yield efficient parameter estimates for non-convex applications, but were computationally expensive because of high-dimensionality of deep learning problems. Recently, Martens et al. [30, 50–52] proposed approximation methods to efficiently estimate Fisher matrix (and associated natural gradient direction) and proved that natural gradient based learning methods can yield superior results for non-convex applications. In this section, we describe a stochastic trust-region based method and validate it using real-world datasets for learning convolution neural networks (CNNs).

We describe a technique called **F**isher **I**nformed **T**rust-**RE**gion (FITRE) method, which is inspired by Martens and Grosse [52], Xu et al. [70], and Yao et al. [77] and is formalized in Algorithm 2. At the heart of FITRE lies the stochastic trust-region method using the local quadratic approximation:

$$\min_{\|s\|\leq\Delta_t} m_t(s) = \langle g_t, s \rangle + \frac{1}{2} \langle s, H_t s \rangle . \tag{14}$$

We adopt the approach of [77] and use a stochastic estimate of the gradient g_t and Hessian H_t. [70, 71]. The step-length which is governed by the trust-region radius Δ_t is automatically adjusted based on the quality of the quadratic approximation and the amount of descent in the objective function. In practice, (14) is approximated by restricting the problem to lower dimensional spaces, e.g., Cauchy condition, which amounts to searching in a one-dimensional space spanned by the gradient. Here, we do the same, however by restricting the subproblem to the space spanned by the direction derived from the Kronecker factorization of the Fisher matrix, or its combination with the gradient.

Our choice is motivated by the following observation: when the objective function involves probabilistic models, as is the case in many deep learning applications, natural gradient direction amounts to the steepest descent direction among all possible directions inside a ball measured by KL-divergence between the underlying parametric probability densities. On the contrary, the (standard) gradient represents the direction of steepest descent among all directions constrained in a ball measured by the Euclidean metric [30], which is less informative than the former, though much easier to compute. To alleviate the computational burden of working with the Fisher information matrix and its inverse, Kronecker-product based approximations [51, 52] have shown success in simultaneously preserving desirable properties of the exact Fisher matrix such as invariance to re-parameterization and resilience to large batch sizes. Indeed, many empirical studies have confirmed that the natural gradient provides an effective descent direction for optimization of neural networks [30, 50–52].

5.1 Natural Gradient Computation

We present an overview of the approximations involved in estimating the natural gradient direction. We refer readers to [30, 52] for a detailed discussion on estimation of Fisher information matrix and approximations used in deriving the natural gradient direction.

We define

$$\mathcal{D}\theta := \frac{d\mathcal{L}(y, f(\mathbf{x}, \theta))}{d\theta} = -\frac{d \log p(y|\mathbf{x}, \theta)}{d\theta},$$

where $\mathcal{D}\theta$ is the gradient of the loss function, which is computed using the conventional back-propagation algorithm. Since the network defines a conditional distribution $p(y|\mathbf{x}, \theta)$, its associated Fisher information matrix is given by

$$\mathbf{F}(\theta) = \mathbb{E}\left[\frac{d \log p(y|\mathbf{x}, \theta)}{d\theta} \left(\frac{d \log p(y|\mathbf{x}, \theta)}{d\theta} \right)^{\mathsf{T}} \right] = \mathbb{E}\left[\mathcal{D}\theta \, (\mathcal{D}\theta)^{\mathsf{T}} \right]. \quad (15)$$

Natural gradient is defined as $\mathbf{F}^{-1}(\theta)\nabla h(\theta)$. It defines the direction in parameter space that gives the largest change in the objective function per unit change in the model, as measured by the KL-divergence, which is measured between the model output distribution and the true label distribution. In the context of this discussion, for simplicity, we drop the dependence of $\mathbf{F}$ and h on θ.

5.2 Natural Gradient Using Kronecker Factored Approximate Curvature Matrix:

We define:

$$\mathbb{E}\left[\text{vec}\left(\bar{\mathbf{W}}_l\right)\text{vec}\left(\bar{\mathbf{W}}_l\right)^\mathsf{T}\right] \approx \boldsymbol{\Psi}_{l-1} \otimes \boldsymbol{\Gamma}_l \triangleq \check{\mathbf{F}}_l, \tag{16}$$

where $\boldsymbol{\Psi}_{l-1}$ and $\boldsymbol{\Gamma}_l$ denote the second moment matrices of the activation and pre-activation derivatives, respectively.

To invert $\check{\mathbf{F}}$, we use the fact that: (1) we can invert a block-diagonal matrix by inverting each of the blocks and (2) the Kronecker product satisfies the identity $(\mathbf{A} \otimes \mathbf{B})^{-1} = \mathbf{A}^{-1} \otimes \mathbf{B}^{-1}$:

$$\check{\mathbf{F}}^{-1} = \begin{bmatrix} \boldsymbol{\Psi}_0^{-1} \otimes \boldsymbol{\Gamma}_1^{-1} & & \mathbf{0} \\ & \ddots & \\ \mathbf{0} & & \boldsymbol{\Psi}_{\ell-1}^{-1} \otimes \boldsymbol{\Gamma}_{\ell-1}^{-1} \end{bmatrix}. \tag{17}$$

The approximate natural gradient $\check{\mathbf{F}}^{-1}\nabla h$ can be computed as follows:

$$\check{\mathbf{F}}^{-1}\nabla h = \begin{bmatrix} \text{vec}\left(\boldsymbol{\Gamma}_1^{-1}\left(\nabla_{\bar{\mathbf{W}}_1}h\right)\boldsymbol{\Psi}_0^{-1}\right) \\ \vdots \\ \text{vec}\left(\boldsymbol{\Gamma}_\ell^{-1}\left(\nabla_{\bar{\mathbf{W}}_\ell}h\right)\boldsymbol{\Psi}_{\ell-1}^{-1}\right) \end{bmatrix}. \tag{18}$$

A common multiple of the identity matrix is added to $\mathbf{F}$ for two reasons: First, as a regularization parameter, which corresponds to a penalty of $\frac{1}{2}\lambda\boldsymbol{\theta}^\mathsf{T}\boldsymbol{\theta}$. This translates to $\mathbf{F} + \lambda\mathbf{I}$ to approximate the curvature of the regularized objective function. The second reason is to use it as a damping parameter to account for multiple approximations used to derive $\check{\mathbf{F}}$, which corresponds to adding $\gamma\mathbf{I}$ to the approximate curvature matrix. Therefore, we aim to compute: $\left[\check{\mathbf{F}} + (\lambda + \gamma)\mathbf{I}\right]^{-1}\nabla h$.

Since adding the term $(\lambda + \gamma)\mathbf{I}$ breaks the Kronecker factorization structure, an approximated version is used for computational purposes, which is as follows:

$$\check{\mathbf{F}}_\ell + (\lambda + \gamma)\mathbf{I} \approx \left(\boldsymbol{\Psi}_{\ell-1} + \pi_\ell\sqrt{\lambda + \gamma}\mathbf{I}\right) \otimes \left(\boldsymbol{\Gamma}_\ell + \frac{1}{\pi_\ell}\sqrt{\lambda + \gamma}\mathbf{I}\right) \tag{19}$$

for some π_ℓ.

Algorithm 2: FITRE

Input   :

            - Starting point $\mathbf{x}_0$

            - Initial trust-region radius: $0 < \Delta_0 < \infty$

            - KFAC parameters: damping parameter ($\gamma \geq 0$), moving

average ($0 < \theta < 1$)

Result: $\mathbf{x}_t$ - direction to be used to update model parameters.

foreach $t = 0, 1, \dots$ **do**

       Set the approximate gradient $\mathbf{g}_t$ and Hessian $\mathbf{H}_t$

       /* Compute the approximated Inverse Fisher ×

          gradient, a.k.a *natural-gradient*   */

       Obtain natural-gradient direction $\mathbf{p}_t$, as described in [30, 52]

       Case 1: KFAC

$$\eta_t = \underset{\|\eta \mathbf{p}_t\| \leq \Delta_t}{\arg\min} m(\eta \mathbf{p}_t) = \eta \mathbf{g}_t^{\mathsf{T}} \mathbf{p}_t + \frac{\eta^2}{2} \mathbf{p}_t^{\mathsf{T}} \mathbf{H}_t \mathbf{p}_t$$

$$\mathbf{s}_t = \eta_t \mathbf{p}_t$$

       Case 2: KFAC + Gradient

$$\eta_t = \underset{\|\eta \mathbf{p}_t\| \leq \Delta_t}{\arg\min} m(\eta \mathbf{p}_t) = \eta \mathbf{g}_t^{\mathsf{T}} \mathbf{p}_t + \frac{\eta^2}{2} \mathbf{p}_t^{\mathsf{T}} \mathbf{H}_t \mathbf{p}_t$$

$$\alpha_t = \underset{\|\alpha \mathbf{g}_t\| \leq \Delta_t}{\arg\min} m(\eta \mathbf{g}_t) = \alpha \mathbf{g}_t^{T} \mathbf{g}_t + \frac{\alpha^2}{2} \mathbf{g}_t^{T} \mathbf{H}_t \mathbf{g}_t$$

$$\mathbf{s}_t = \underset{\mathbf{s} \in \{\eta_t \mathbf{p}_t, \alpha_t \mathbf{g}_t\}}{\arg\min} m(\mathbf{s})$$

Set $\rho_t \triangleq \frac{h_t(\boldsymbol{\theta}_t) - h_t(\boldsymbol{\theta}_t + \mathbf{s}_t)}{-m(\mathbf{s}_t)}$, ($h_t(.)$

are evaluated on the same mini-batch as $\mathbf{g}_t$ and $\mathbf{H}_t$).

       if $\rho_t \geq 0.75$ **then**

          | $\mathbf{w}_{t+1} = \mathbf{w}_t + \mathbf{s}_t$ and $\Delta_{t+1} = min\{2\Delta_t, \Delta_{max}\}$

       end

       else if $\rho_t \geq 0.25$ **then**

          | $\mathbf{w}_{t+1} = \mathbf{w}_t + \mathbf{s}_t$ and $\Delta_{t+1} = \Delta_t$

       end

       else

          | $\mathbf{w}_{t+1} = \mathbf{w}_t$ and $\Delta_{t+1} = \Delta_t/2$

       end

end

Algorithm 2 describes a realization of our proposed method in trust-region settings. First, the natural gradient direction, $\mathbf{p}_t$ is computed and used in determining the step-size using the quadratic approximation of the objective function at $\mathbf{p}_t$, whose closed-form solution is $(\Delta / \|\mathbf{H}_t \mathbf{p}_t + \mathbf{g}_t\|) (\mathbf{H}_t \mathbf{p}_t + \mathbf{g}_t)$ (note that gradient, $\mathbf{g}_t$, can also be used to estimate the step-size and may yield a better descent direction in some cases). Once the step-size, η is determined, ρ_t is computed over the same mini-batch to determine the trust-region radius as well as the iterate update. These steps are repeated until desired generalization is achieved. Note that we can compare the efficiency of natural gradient direction with that of the standard gradient and use the appropriate one at each iteration, this is referred to as "KFAC + gradient" in this algorithm.

5.3 Updating KFAC Block Matrices

Block matrices, $\boldsymbol{\Psi}_l$ and $\boldsymbol{\Gamma}_l$, are typically updated using a momentum term to capture the variance in input samples across successive mini-batches. If sample points across the dataset are well correlated, with little variance among the sample points, the inverse block matrices, $\boldsymbol{\Psi}_l^{-1}$ and $\boldsymbol{\Gamma}_l^{-1}$, need not be updated for every mini-batch. "KFAC Update Frequency," the frequency with which the inverse block matrices are updated, is typically decided based on the size of the input dataset as well as the correlatio n among the sample points. For boot strapping the optimizer, we either use a larger sample of the dataset, like 5 × the mini-batch size, or use the very first mini-batch itself for computing the block inverses.

5.4 Experimental Results

Tables 4 and 5 present a comparison of our solver, FITRE with other state-of-the-art methods on the ImageNet dataset using VGG11 convolutional neural networks (CNNs) and Tables 6 and 7 show the results for VGG16 CNN. In these tables, we show the generalization errors plotted against wall-clock time and against number of epochs in Columns 3 and 4, respectively, and negative log-likelihood (NLL) using softmax cross-entropy loss function against wall-clock time and against number of epochs in Columns 1 and 2, respectively. KFAC update frequency is set to 5 (mini-batches) for the first row and for the second row, it is set to 25. Plots in Tables 4 and 6 use *default* initialization, as defined in pyTorch, which is a uniform distribution. Corresponding results using *Kaiming* initialization [31] (this initialization is based on random Gaussian distribution) are shown in Tables 5 and 7.

The following conclusions can be made from the plots for VGG11 (as shown in Tables 4 and 5) and VGG16 (as shown in Tables 6 and 7). (1) FITRE minimizes the likelihood function to a significantly smaller value compared to well-tuned SGD, and at any given wall-clock instant (FITRE yields better NLL value compared to SGD), (2) Kaiming initialization yields superior generalization errors compared to default initialization of the CNNs, (3) contrary to expectations KFAC update frequency of 25 yields better generalization errors relative to more frequent updates, (4) with increasing network complexity, VGG16 compared to VGG11, FITRE yields significantly better generalization errors compared to SGD, showcasing its superior scaling characteristics compared to SGD, and (5) default initialization is relatively immune to ℓ_2 regularization compared to Kaiming initialization.

For VGG16 network with Kaiming initialization and KFAC update frequency of 25 we observe that to attain 50% test accuracy FITRE (with 1e-6 regularization)

Table 4 Comparison of VGG11 using ImageNet dataset and default initialization

Time vs. Negative log-Likelihood	Epoch vs. Negative log-Likelihood	Time vs Test Accuracy	Epoch vs. Test Accuracy

KFAC Update Frequency = 5

- - - SGD (StepSize: 1e-02, BatchNorm: 1, Reg: 1e-04)
- ⋯ FITRE: (DampFactor: 1e-02, MaxTrustRad: 1e+00, KFAC + Gradient: 0, BatchNorm: 1, KFAC Frequency: 5, Reg: 1e-04)
- ⋯ SGD (StepSize: 1e-02, BatchNorm: 1, Reg: 1e-05)
- ⋯ FITRE: (DampFactor: 1e-02, MaxTrustRad: 1e+00, KFAC + Gradient: 0, BatchNorm: 1, KFAC Frequency: 5, Reg: 1e-05)
- - - SGD (StepSize: 1e-02, BatchNorm: 1, Reg: 1e-06)
- —— FITRE: (DampFactor: 1e-02, MaxTrustRad: 1e+00, KFAC + Gradient: 0, BatchNorm: 1, KFAC Frequency: 5, Reg: 1e-06)

KFAC Update Frequency = 25

- - - SGD (StepSize: 1e-02, BatchNorm: 1, Reg: 1e-04)
- ⋯ FITRE: (DampFactor: 1e-02, MaxTrustRad: 1e+00, KFAC + Gradient: 0, BatchNorm: 1, KFAC Frequency: 25, Reg: 1e-04)
- ⋯ SGD (StepSize: 1e-02, BatchNorm: 1, Reg: 1e-05)
- ⋯ FITRE: (DampFactor: 1e-02, MaxTrustRad: 1e+00, KFAC + Gradient: 0, BatchNorm: 1, KFAC Frequency: 25, Reg: 1e-05)
- - - SGD (StepSize: 1e-02, BatchNorm: 1, Reg: 1e-06)
- —— FITRE: (DampFactor: 1e-02, MaxTrustRad: 1e+00, KFAC + Gradient: 0, BatchNorm: 1, KFAC Frequency: 25, Reg: 1e-06)

Table 5 Comparison of VGG11 using ImageNet dataset and Kaiming initialization

KFAC Update Frequency = 5

Legend (KFAC Update Frequency = 5):
- SGD (StepSize: 1e-02, BatchNorm: 1, Reg: 1e-04)
- FITRE: (DampFactor: 0, MaxTrustRad: 1.00, KFAC + Gradient: 0, BatchNorm: 1, KFAC Frequency: 5, Reg: 1e-04)
- SGD (StepSize: 1e-02, BatchNorm: 1, Reg: 1e-05)
- FITRE: (DampFactor: 0, MaxTrustRad: 1.00, KFAC + Gradient: 0, BatchNorm: 1, KFAC Frequency: 5, Reg: 1e-05)
- SGD (StepSize: 1e-02, BatchNorm: 1, Reg: 1e-06)
- FITRE: (DampFactor: 0, MaxTrustRad: 1.00, KFAC + Gradient: 0, BatchNorm: 1, KFAC Frequency: 5, Reg: 1e-06)

KFAC Update Frequency = 25

Legend (KFAC Update Frequency = 25):
- SGD (StepSize: 1e-02, BatchNorm: 1, Reg: 1e-04)
- FITRE: (DampFactor: 1e-02, MaxTrustRad: 1e+00, KFAC + Gradient: 0, BatchNorm: 1, KFAC Frequency: 25, Reg: 1e-04)
- SGD (StepSize: 1e-02, BatchNorm: 1, Reg: 1e-05)
- FITRE: (DampFactor: 1e-02, MaxTrustRad: 1e+00, KFAC + Gradient: 0, BatchNorm: 1, KFAC Frequency: 25, Reg: 1e-05)
- SGD (StepSize: 1e-02, BatchNorm: 1, Reg: 1e-06)
- FITRE: (DampFactor: 1e-02, MaxTrustRad: 1e+01, KFAC + Gradient: 0, BatchNorm: 1, KFAC Frequency: 25, Reg: 1e-06)

Time vs. Negative log-Likelihood   Epoch vs. Negative log-Likelihood   Time vs Test Accuracy   Epoch vs. Test Accuracy

Table 6 Comparison of VGG16 using ImageNet dataset and default initialization

Time vs. Negative log-Likelihood	Epoch vs. Negative log-Likelihood	Time vs Test Accuracy	Epoch vs. Test Accuracy

KFAC Update Frequency = 5

- - SGD (StepSize: 1e-02, BatchNorm: 1, Reg: 1e-04)
- FITRE: (DampFactor: 1e-02, MaxTrustRad: 1e+00, KFAC + Gradient: 0, BatchNorm: 1, KFAC Frequency: 5, Reg: 1e-04)
- - SGD (StepSize: 1e-02, BatchNorm: 1, Reg: 1e-05)
- FITRE: (DampFactor: 1e-02, MaxTrustRad: 1e+00, KFAC + Gradient: 0, BatchNorm: 1, KFAC Frequency: 5, Reg: 1e-05)
- - SGD (StepSize: 1e-02, BatchNorm: 1, Reg: 1e-06)
- FITRE: (DampFactor: 1e-02, MaxTrustRad: 1e+00, KFAC + Gradient: 0, BatchNorm: 1, KFAC Frequency: 5, Reg: 1e-06)

KFAC Update Frequency = 25

- - SGD (StepSize: 1e-02, BatchNorm: 1, Reg: 1e-04)
- FITRE: (DampFactor: 1e-02, MaxTrustRad: 1e+00, KFAC + Gradient: 0, BatchNorm: 1, KFAC Frequency: 25, Reg: 1e-04)
- - SGD (StepSize: 1e-02, BatchNorm: 1, Reg: 1e-05)
- FITRE: (DampFactor: 1e-02, MaxTrustRad: 1e+00, KFAC + Gradient: 0, BatchNorm: 1, KFAC Frequency: 25, Reg: 1e-05)
- - SGD (StepSize: 1e-02, BatchNorm: 1, Reg: 1e-06)
- FITRE: (DampFactor: 1e-02, MaxTrustRad: 1e+00, KFAC + Gradient: 0, BatchNorm: 1, KFAC Frequency: 25, Reg: 1e-06)

Table 7 Comparison of VGG16 using ImageNet dataset and Kaiming initialization

Time vs. Negative log-Likelihood	Epoch vs. Negative log-Likelihood	Time vs. Test Accuracy	Epoch vs. Test Accuracy

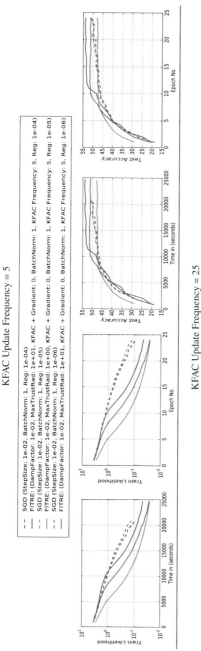

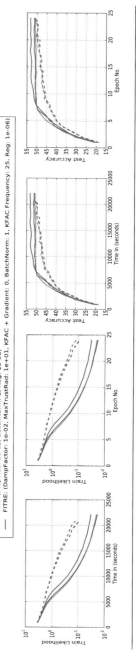

takes $\approx 6500\,\text{s}$ compared to $\approx 20{,}500\,\text{s}$ for SGD (for all regularizations used); a speedup of 3.2 over SGD. Furthermore, when regularization is set to $1e^{-4}$ FITRE achieves 53.5% test accuracy, whereas SGD fails to obtain similar accuracy. Similar arguments can be made for the VGG11 network as well. This shows that even though FITRE is computationally more expensive on a per-iteration basis, it yields significantly better results in shorter time compared to SGD. This can be attributed to better descent direction (SGD's gradient vs. FITRE's natural gradient) and an adaptive second-order approximated learning rate computation within the trust-region framework used by the FITRE.

Contrary to expectation, we notice that for VGG11 CNN and default initialization, FITRE's execution of 50 epochs takes less time compared to SGD for KFAC update frequency 25. FITRE makes two passes over the network (one forward and backward pass for gradient computation and another pass for Hessian-vector product computation used to compute the learning rate in the trust-region framework). One would expect that SGD is at least twice as fast as FITRE on the wall-clock time (on a per-iteration basis). We note that SGD's pyTorch implementation uses *auto-differentiation* to compute the gradient of the given network, whereas our implementation of the FITRE is R-operator based (as proposed by Perlmutter et al. [59]). We note that GPU memory management in pyTorch is not efficient [26, 43]. pyTorch allocates and frees memory very often and tends to persist very little information on the device. Even though FITRE makes two passes over the network and computes inverses of smaller matrices at each layer of the network (for computing the inverse of the KFAC block matrices) our implementation persists relevant information on the GPU memory. Coupled with our efficient implementation of the R-operator based Hessian-vector product, we can significantly reduce the computation cost associated with each mini-batch. In addition, our proposed method is a true *stochastic online* method in which there is no dependence on any part of the dataset other than the current mini-batch during its entire execution, compared to state-of-the-art existing second-order methods [53, 56].

We also note that default initialization is immune to regularization for both networks (VGG11 and VGG16) and for both methods (FITRE and SGD). These two methods show negligible changes in NLL function values (as well as generalization errors) while the FITRE yields superior results compared to SGD for significant part of the execution. At the end of the execution, SGD tends to achieve similar generalization errors compared to FITRE but on minimizing the NLL function FITRE always achieves superior results. However, when using Kaiming initialization, based on random Gaussian distribution, for both the networks, we notice that regularization helps in achieving superior generalization errors for FITRE (with VGG11 network, KFAC update frequency set to 25 and regularization of $1e^{-6}$) compared to SGD. But in all cases, FITRE yields superior results when the underlying model does not use any regularization. Compared to FITRE, SGD is relatively invariant to Kaiming initialization as well, as shown in plots in columns 1 and 2 of Tables 5 and 7. Notice that there is very little change in objective function value throughout the simulations.

KFAC update frequency is a hyper-parameter used to control the frequency with which the block matrix inverses are computed at each layer of the network. These block inverses are used to compute the natural gradient direction eventually for each mini-batch. Since these blocks approximate the *Fisher matrix* of the loss function, they are updated once every few mini-batches. Martens et al. [30, 52] argue that more frequent updates of these block inverses make them too rigid and may lead to overfitting. Using larger values for this update frequency has the effect of a regularizer on the underlying model and helps in avoiding overfitting. As an added advantage, this dependence of the FITRE reduces its computation cost (note also that the computation of block inverses can be delegated to slave processing units, if available, further reducing the computation cost thereby decreasing the time for processing each mini-batch). This is also one of the reasons why our proposed method scales well with increasing network complexity. We note that for VGG16 (with Kaiming initialization), a larger and more complex network compared to VGG11, FITRE yields superior generalization errors as well as minimizing objective function compared to SGD.

6 Distributed Higher-Order Methods

Typically, machine learning problems are associated with massively large datasets during the training phase for learning model parameters. In such scenarios, one must resort to distributed/parallel methods for training models, due to resource constraints on individual nodes. Even in stochastic settings, where only a small part of the dataset is processed at any point in time, time spent in training is a critical parameter contributing to the use of distributed procedures in deep learning. Furthermore, it may be infeasible to accumulate the dataset at a single physical location either due to privacy or resource constraints of the underlying application or system. In such applications communication-efficient optimizers can significantly reduce the training time while optimally using the compute resources. The need for such methods is even more pressing when nodes in distributed systems are connected through high latency networks.

Several distributed optimization techniques have been developed in the recent past to address these concerns. Distributed methods, which are direct adaptations of existing first-order or quasi-Newton methods (i.e., those that parallelize kernel operations such as matrix–vector and dot products), suffer from high communication overhead because of the exchange of model parameters at least once in each iteration among the compute nodes, in addition to inherent communication overhead of the optimizer itself. Ensemble methods in which local optimization procedures compute local solutions (using only locally available data) and a coordinating consensus procedure, which harmonizes local solutions to form a global solution, are more efficient in high latency environments.

In this section, we discuss a communication-efficient method, called Newton-ADMM, based on Alternating Direction Methods of Multipliers (ADMM) framework coupled with sub-sampled Newton-type methods for local optimization as discussed in Sect. 4, along with results using real-world datasets in the context of convex optimization problems of the form (1).

Let $\mathcal{N}$ denote the number of nodes (compute elements) in the distributed environment. Assume that the input dataset $\mathcal{D}$ is split among the $\mathcal{N}$ nodes as $\mathcal{D} = \mathcal{D}_1 \cup \mathcal{D}_2 \ldots \cup \mathcal{D}_\mathcal{N}$. Using this notation, (1) can be written as:

$$\min \sum_{i=1}^{N} \sum_{j \in \mathcal{D}_i} f_j(\mathbf{x}_i) + g(\mathbf{z}) \tag{20}$$

$$\text{s.t.} \quad \mathbf{x}_i - \mathbf{z} = 0, \quad i = 1, \ldots, \mathcal{N},$$

where $\mathbf{z}$ represents a global variable enforcing consensus among $\mathbf{x}_i$'s at all the nodes. In other words, the constraint enforces a consensus among the nodes so that all the local variables, $\mathbf{x}_i$, agree with global variable $\mathbf{z}$. This formulation (20) is often referred to as a *global consensus* problem. ADMM is based on an augmented Lagrangian framework; it solves the global consensus problem by alternating iterations on primal/dual variables. In doing so, it inherits the benefits of decomposability of dual ascent and the superior convergence properties of the method of multipliers. For a detailed discussion on ADMM method, we refer the readers to [9].

ADMM methods introduce a penalty parameter ρ, which is the weight on the measure of *disagreement* between $\mathbf{x}_i$'s and global consensus variable, $\mathbf{z}$. The most common adaptive penalty parameter selection is Residual Balancing [9], which tries to balance the dual norm and residual norm of ADMM. Recent empirical results using Spectral Penalty Selection (SPS) [75], which is based on the estimation of the local curvature of subproblem at each node, yield significant improvement in the efficiency of ADMM. Using the SPS strategy for penalty parameter selection, ADMM iterates can be written as follows:

$$\mathbf{x}_i^{k+1} = \arg \min_{\mathbf{x}_i} f_i(\mathbf{x}_i) + \frac{\rho_i^k}{2} ||\mathbf{z}^k - \mathbf{x}_i + \frac{\mathbf{y}_i^k}{\rho_i^k}||_2^2, \tag{21a}$$

$$\mathbf{z}^{k+1} = \arg \min_{\mathbf{z}} g(\mathbf{z}) + \sum_{i=1}^{N} \frac{\rho_i^k}{2} ||\mathbf{z} - \mathbf{x}_i^{k+1} + \frac{\mathbf{y}_i^k}{\rho_i^k}||_2^2, \tag{21b}$$

$$\mathbf{y}_i^{k+1} = \mathbf{y}_i^k + \rho_i^k (\mathbf{z}^{k+1} - \mathbf{x}_i^{k+1}). \tag{21c}$$

With ℓ_2−regularization, i.e., $g(\mathbf{x}) = \lambda ||\mathbf{x}||^2/2$, (21b) has a closed-form solution given by

$$\mathbf{z}^{k+1} (\lambda + \sum_{i=1}^{N} \rho_i^k) = \sum_{i=1}^{N} [\rho_i^k \mathbf{x}_i^{k+1} - \mathbf{y}_i^k], \tag{22}$$

where λ is the regularization parameter.

Algorithm 3: ADMM method (outer solver)

Input   : $\mathbf{x}^{(0)}$ (initial iterate), $\mathcal{N}$ (no. of nodes)
Parameters: β, λ and $\theta < 1$
1 Initialize $\mathbf{z}^0$ to 0
2 Initialize $\mathbf{y}_i^0$ to 0 on all nodes.
   foreach $k = 0, 1, 2, \ldots$ **do**
3 | (i) Perform Algorithm 1 with, $\mathbf{x}_i^k$, $\mathbf{y}_i^k$, and $\mathbf{z}^k$ on all nodes
4 | (ii) Collect all local $\mathbf{x}_i^{k+1}$
5 | (iii) Evaluate $\mathbf{z}^{k+1}$ and $\mathbf{y}_i^{k+1}$ using (21b) and (21c).
6 | (iv) Distribute $\mathbf{z}^{k+1}$ and $\mathbf{y}_i^{k+1}$ to all nodes.
7 | (v) Locally, on each node, compute spectral step-sizes and penalty
      | parameters as in [75]
   end

Algorithm 3 presents a distributed optimization method incorporating the above formulation of ADMM. Steps 1 and 2 initialize the multipliers, $\mathbf{y}$, and consensus vectors, $\mathbf{z}$, to zeros. In each iteration, Single Node Newton method, Algorithm 1, is executed with local $\mathbf{x}_i$, $\mathbf{y}_i$, and global $\mathbf{z}$ vectors. Upon termination of Algorithm 1 at all nodes, resulting local Newton directions, $\mathbf{x}_i^k$, are gathered at the master node, which generates the next iterates for vectors $\mathbf{y}$ and $\mathbf{z}$ using spectral step-sizes described in [75]. These steps are repeated until convergence.

Remark 2 Note that in each ADMM iteration only *one* round of communication is required (a "gather" and a "scatter" operation), which can be executed in $O(\log(\mathcal{N}))$ time. Further, the application of the GPU-accelerated inexact Newton-CG Algorithm 1 at each node significantly speeds up the local computation per epoch. The combined effect of these algorithmic choices contributes to the high overall efficiency of the proposed Newton-ADMM Algorithm 3 when applied to large datasets.

6.1 ADMM Residuals and Stopping Criteria:

The consensus problem (20) can be solved by iterating ADMM subproblems (21a), (21c), and (21b). To monitor the convergence of ADMM, we can check the norm of primal and dual residuals, $\mathbf{r}^k$ and $\mathbf{d}^k$, which are defined as follows:

$$\mathbf{r}^k = \begin{bmatrix} \mathbf{r}_1^k \\ \vdots \\ \mathbf{r}_{\mathcal{N}}^k \end{bmatrix}, \mathbf{d}^k = \begin{bmatrix} \mathbf{d}_1^k \\ \vdots \\ \mathbf{d}_{\mathcal{N}}^k \end{bmatrix}, \tag{23}$$

where $\forall i \in \{1, 2, \ldots, \mathcal{N}\}$,

$$\mathbf{r}_i^k = \mathbf{z}^k - \mathbf{x}_i^k, \mathbf{d}_i^k = -\rho_i^k(\mathbf{z}^k - \mathbf{z}^{k-1}). \tag{24}$$

As $k \to \infty$, $\mathbf{z}^k \to \mathbf{z}^*$ and $\forall i, \mathbf{x}_i^k \to \mathbf{z}^*$. Therefore, the norm of primal and dual residuals, $||\mathbf{r}^k||$ and $||\mathbf{d}^k||$, converges to zero. In practice, we do not need the solution to high precision, thus ADMM can be terminated as $||\mathbf{r}_i^k|| \leq \epsilon^{pri}$ and $||\mathbf{d}_i^k|| \leq \epsilon^{dual}$. Here, ϵ^{pri} and ϵ^{dual} can be chosen as:

$$\epsilon^{pri} = \sqrt{\mathcal{N}}\epsilon^{abs} + \epsilon^{rel} \max\{\sum_{i=1}^{\mathcal{N}} ||\mathbf{x}_i^k||^2, \mathcal{N}||\mathbf{z}^k||^2\} \tag{25}$$

$$\epsilon^{dual} = \sqrt{d}\epsilon^{abs} + \epsilon^{rel} \max\{\sum_{i=1}^{\mathcal{N}} ||\mathbf{y}_i^k||^2\}. \tag{26}$$

The choice of absolute tolerance ϵ^{abs} depends on the chosen problem and the choice of relative tolerance ϵ^{rel} for the stopping criteria is, in practice, set to 10^{-3} or 10^{-4}.

6.2   Experimental Results

In this section, we evaluate the performance of Newton-ADMM as compared with several state-of-the-art alternatives. In these experiments, pyTorch is used as the software platform and nodes are equipped with NVIDIA P100 GPU accelerators. Table 8 describes the datasets that are used for validation purposes.

6.3   Comparison with Distributed First-Order Methods

While the per-iteration cost of first-order methods (synchronous SGD) is relatively low, they require larger number of iterations, increasing associated communication overhead, and CPU–GPU transactions (because of resource constraints on GPUs, data must be swapped back to the CPU for temporarily releasing global memory

Table 8 Description of the datasets

Classes	Dataset	Train size	Test size	Dims
2	HIGGS	10,000,000	1,000,000	28
10	MNIST	60,000	10,000	784
10	CIFAR-10	50,000	10,000	3072
20	E18	1,300,128	6000	279,998

on the GPUs, partly because of the pyTorch's execution model). In this experiment, we demonstrate that these drawbacks of first-order methods are significant, in the context of MNIST, CIFAR-10, HIGGS, and E18 datasets using four workers for Newton-ADMM and synchronous SGD, both with the GPUs enabled and GPUs disabled. The results are shown in Table 9. We note that GPU-accelerated Newton-ADMM method with minimal communication overhead yields significantly better results—over an order of magnitude faster in most cases, when compared to synchronous SGD.

We present the ratio of CPU time to GPU time for Newton-ADMM and SGD in Table 10. We observe that for both Newton-ADMM and SGD, the CPU–GPU time ratio is proportional to the dimension of datasets. For example, on the dataset with the lowest dimension (HIGGS), the CPU–GPU time ratio is the least for both Newton-ADMM and SGD, whereas on the dataset with the highest dimension (E18), the CPU–GPU time ratios are the highest for both Newton-ADMM and SGD. In all cases, the use of GPUs results in highest speedup for Newton-ADMM. The gain in GPU utilization is compromised by large number of CPU–GPU memory transfers for SGD. As a result, SGD shows meaningful GPU acceleration only for the E18 dataset.

Second, we observe that Newton-ADMM has much lower communication cost, compared to SGD. This can be observed from Table 9. In all cases, SGD takes longer than Newton-ADMM with GPUs enabled. This is mainly because SGD requires a large number of gradient communications across nodes. As a result, we observe that Newton-ADMM is 4.9x, 6.3x, 22.6x, and 17.8x times faster than SGD on MNIST, CIFAR-10, HIGGS, and E18 datasets, respectively.

Finally, we observe that Newton-ADMM has superior convergence properties compared to SGD. This is demonstrated in Table 9 for the HIGGS dataset. We observe that Newton-ADMM converges to low objective function values in just a few iterations. On the other hand, the objective function value, even at 100-th epoch for SGD, is still higher than Newton-ADMM.

6.4 Comparison with Distributed Second-Order Methods

We compare Newton-ADMM against DANE [20], AIDE [61], and GIANT [69], which have been shown in recent results to perform well. In each iteration, DANE [20] requires an exact solution of its corresponding subproblem at each node. This constraint is relaxed in an inexact version of DANE, called InexactDANE [61], which uses SVRG [41] to approximately solve the subproblems. Another version of DANE, called Accelerated Inexact DanE (AIDE), uses techniques for accelerating convergence while still using InexactDANE to solve individual subproblems [61]. However, using SVRG to solve subproblems is computationally inefficient due to its double loop formulation, with the outer loop requiring full gradient recalculation and several stochastic gradient calculations in inner loop. Figure 2 shows the comparison between these methods on the MNIST dataset with $\lambda = 10^{-5}$. Although

Table 9 Training objective function and test accuracy as functions of time for Newton-ADMM and synchronous SGD, both with GPU enabled and GPU disabled, with four workers

Time vs. Test Accuracy	Time vs. Objective Function (training)

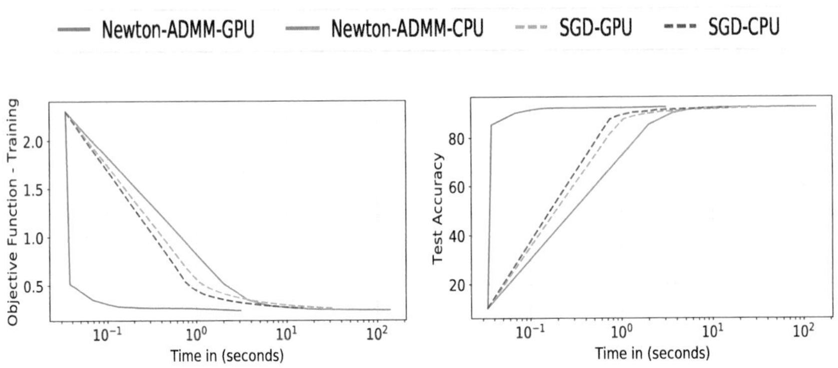

—— Newton-ADMM-GPU   —— Newton-ADMM-CPU   - - - SGD-GPU   - - - SGD-CPU

MNIST

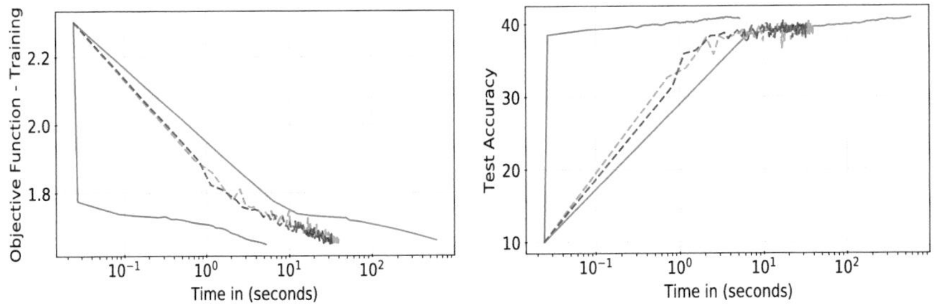

CIFAR-10

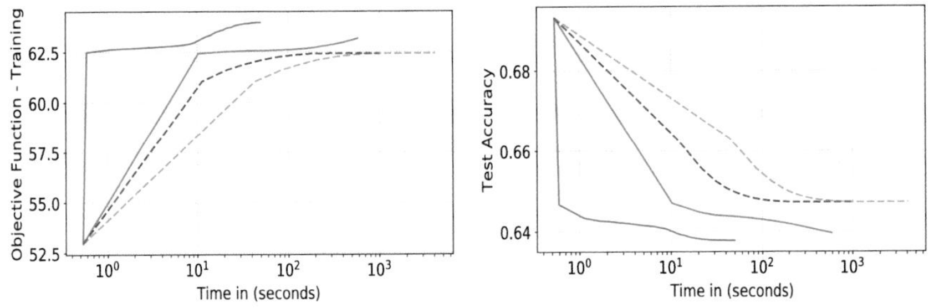

HIGGS

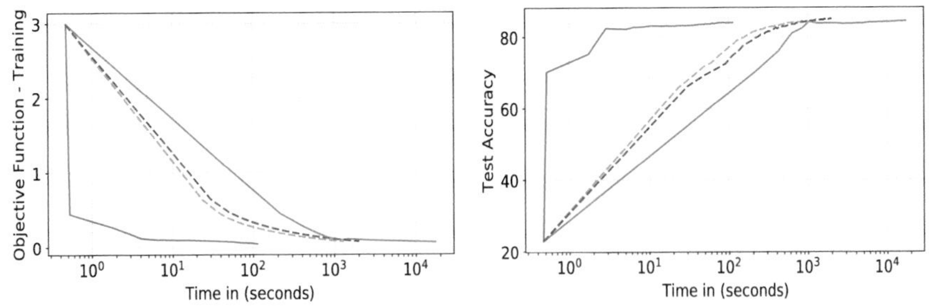

E18

Overall, Newton-ADMM favors GPUs, enjoys minimal communication overhead, and enjoys faster convergence compared to synchronous SGD

Table 10 GPU speedup for Newton-ADMM and SGD	CPU/GPU time ratio	Newton-ADMM	SGD
	MNIST	44.7345904	0.47896507
	CIFAR-10	112.670178	0.8212862
	HIGGS	11.842679	0.26789652
	E18	154.425688	1.54673642

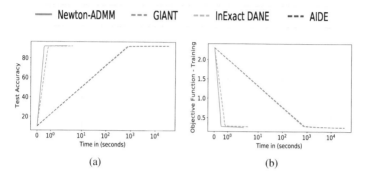

Fig. 2 Training objective function and test accuracy comparison over time for Newton-ADMM, GIANT, InexactDANE, and AIDE on MNIST dataset with $\lambda = 10^{-5}$. We run both Newton-ADMM and GIANT for 100 epochs. Since the computation times per epoch for InexactDANE and AIDE are high, we only run 10 epochs for these methods. (**a**) Time vs. Test Accuracy. (**b**) Time vs. Objective Function (training)

InexactDANE and AIDE start at lower objective function values, the average epoch time compared to Newton-ADMM and GIANT is orders of magnitude higher (*order of 1000x*). For instance, to reach an objective function value less than 0.25 on the MNIST dataset, Newton-ADMM takes only 2.4 s, whereas InexactDANE consumes *an hour and a half.*

7 Concluding Remarks

Optimization techniques for training machine learning models are of significant current interest. Common machine learning models lead to convex or non-convex optimization problems defined on very large datasets. This necessitates the development of efficient (in terms of optimization time), effective (in terms of generalization error), and parallelizable methods.

Most current machine learning applications rely on stochastic gradient descent to solve the underlying optimization problems. In this chapter, we have discussed the use of higher-order methods that rely on curvature, in addition to gradient information for computing the descent direction. We rely on two important concepts: the use of sampling in dealing with the dense Hessian matrix and the use of natural gradient in non-convex optimization. We show that second-order methods are fast (in terms of iteration counts), can be made efficient (in terms of per-iteration computation cost), result in solutions that are generalizable (as determined by test accuracy), are robust to problem ill-conditioning, and do not require extensive hyper-parameter tuning. Finally, we show how these methods can be parallelized using ADMM and formulated to efficiently use GPUs to deliver accurate and scalable solvers.

References

1. M. ABADI, P. BARHAM, J. CHEN, Z. CHEN, A. DAVIS, J. DEAN, M. DEVIN, S. GHEMAWAT, G. IRVING, M. ISARD, ET AL., *TensorFlow: A system for large-scale machine learning.*, in OSDI, vol. 16, 2016, pp. 265–283.
2. Z. ALLEN-ZHU AND Y. LI, *Neon2: Finding local minima via first-order oracles*, in Advances in Neural Information Processing Systems, 2018, pp. 3720–3730.
3. J. BA, R. GROSSE, AND J. MARTENS, *Distributed second-order optimization using Kronecker-factored approximations*, ICLR, (2017).
4. A. S. BERAHAS, R. BOLLAPRAGADA, AND J. NOCEDAL, *An Investigation of Newton-Sketch and Subsampled Newton Methods*, arXiv preprint arXiv:1705.06211, (2017).
5. D. P. BERTSEKAS AND J. N. TSITSIKLIS, *Neuro-dynamic Programming*, Athena Scientific, 1996.
6. L. BOTTOU, *Large-scale machine learning with stochastic gradient descent*, in Proceedings of COMPSTAT'2010, Springer, 2010, pp. 177–186.
7. L. BOTTOU, F. E. CURTIS, AND J. NOCEDAL, *Optimization methods for large-scale machine learning*, arXiv preprint arXiv:1606.04838, (2016).
8. L. BOTTOU AND Y. LECUN, *Large scale online learning*, Advances in neural information processing systems, 16 (2004), p. 217.
9. S. BOYD, N. PARIKH, E. CHU, B. PELEATO, J. ECKSTEIN, ET AL., *Distributed optimization and statistical learning via the alternating direction method of multipliers*, Foundations and Trends® in Machine learning, 3 (2011), pp. 1–122.
10. S. BOYD AND L. VANDENBERGHE, *Convex optimization*, Cambridge university press, 2004.
11. R. H. BYRD, G. M. CHIN, W. NEVEITT, AND J. NOCEDAL, *On the use of stochastic Hessian information in optimization methods for machine learning*, SIAM Journal on Optimization, 21 (2011), pp. 977–995.
12. R. H. BYRD, G. M. CHIN, J. NOCEDAL, AND Y. WU, *Sample size selection in optimization methods for machine learning*, Mathematical programming, 134 (2012), pp. 127–155.
13. A. CAUCHY, *Méthode générale pour la résolution des systemes d'équations simultanées*, Comp. Rend. Sci. Paris, 25 (1847), pp. 536–538.
14. O. CHAPELLE AND D. ERHAN, *Improved preconditioner for Hessian free optimization*, in In NIPS Workshop on Deep Learning and Unsupervised Feature Learning, 2011.
15. J. CHEN, X. PAN, R. MONGA, S. BENGIO, AND R. JOZEFOWICZ, *Revisiting distributed synchronous SGD*, arXiv preprint arXiv:1604.00981, (2016).
16. A. R. CONN, N. I. GOULD, AND P. L. TOINT, *Trust region methods*, vol. 1, SIAM, 2000.
17. A. COTTER, O. SHAMIR, N. SREBRO, AND K. SRIDHARAN, *Better mini-batch algorithms via accelerated gradient methods*, in Advances in neural information processing systems, 2011, pp. 1647–1655.
18. R. CRANE AND F. ROOSTA, *DINGO: Distributed Newton-Type Method for Gradient-Norm Optimization*, in Proceedings of the Advances in Neural Information Processing Systems, 2019. Accepted.
19. B. CRAVEN, *Invex functions and constrained local minima*, Bulletin of the Australian Mathematical society, 24 (1981), pp. 357–366.

20. H. DANESHMAND, A. LUCCHI, AND T. HOFMANN, *DynaNewton-Accelerating Newton's Method for Machine Learning*, arXiv preprint arXiv:1605.06561, (2016).
21. Y. DAUPHIN, H. DE VRIES, AND Y. BENGIO, *Equilibrated adaptive learning rates for non-convex optimization*, in Advances in Neural Information Processing Systems, 2015, pp. 1504–1512.
22. Y. N. DAUPHIN, R. PASCANU, C. GULCEHRE, K. CHO, S. GANGULI, AND Y. BENGIO, *Identifying and attacking the saddle point problem in high-dimensional non-convex optimization*, arXiv:1406.2572v1, (2014).
23. J. DEAN, G. CORRADO, R. MONGA, K. CHEN, M. DEVIN, M. MAO, A. SENIOR, P. TUCKER, K. YANG, Q. V. LE, ET AL., *Large scale distributed deep networks*, in Advances in neural information processing systems, 2012, pp. 1223–1231.
24. J. DUCHI, E. HAZAN, AND Y. SINGER, *Adaptive subgradient methods for online learning and stochastic optimization*, The Journal of Machine Learning Research, 12 (2011), pp. 2121–2159.
25. C. DÜNNER, A. LUCCHI, M. GARGIANI, A. BIAN, T. HOFMANN, AND M. JAGGI, *A distributed second-order algorithm you can trust*, arXiv preprint arXiv:1806.07569, (2018).
26. C.-H. FANG, S. B. KYLASA, F. ROOSTA-KHORASANI, M. W. MAHONEY, AND A. GRAMA, *Distributed second-order convex optimization*, arXiv preprint arXiv:1807.07132, (2018).
27. J. FRIEDMAN, T. HASTIE, AND R. TIBSHIRANI, *The elements of statistical learning*, vol. 1, Springer series in statistics Springer, Berlin, 2001.
28. R. GE, F. HUANG, C. JIN, AND Y. YUAN, *Escaping from saddle points – online stochastic gradient for tensor decomposition*, arXiv preprint: arXiv:1503.02101v1, (2015).
29. P. GOYAL, P. DOLLÁR, R. GIRSHICK, P. NOORDHUIS, L. WESOLOWSKI, A. KYROLA, A. TULLOCH, Y. JIA, AND K. HE, *Accurate, large minibatch SGD: training ImageNet in 1 hour*, arXiv preprint arXiv:1706.02677, (2017).
30. R. GROSSE AND J. MARTENS, *A Kronecker-factored approximate fisher matrix for convolution layers*, arXiv:1602.01407v2, (2016).
31. K. HE, X. ZHANG, S. REN, AND J. SUN, *Delving deep into rectifiers: Surpassing human-level performance on ImageNet classification*, arXiv preprint arXiv:1502.01852, (2015).
32. G. HINTON, *Neural networks for machine learning*, Coursera, video lectures. 307, (2012).
33. S. HOCHREITER AND J. SCHMIDHUBER, *Long short-term memory*, Neural Computation, 9 (1997), pp. 1735–1780.
34. IBM, *Deep blue*. https://www.ibm.com/blogs/think/2017/05/deep-blue/.
35. ———, *IBM project debater*. https://www.research.ibm.com/artificial-intelligence/project-debater/.
36. S. ICHI AMARI, *Natural gradient works efficiently in learning*, Neural Computation, 10 (1988).
37. S. ICHI AMARI, R. KARAKIDA, AND M. OYIZUMI, *Fisher information and natural gradient learning of random deep networks*, arXiv preprint: 1808.07172v1, (2018).
38. G. INC., *Google AlphaGo*. https://ai.google/research/pubs/pub44806.
39. C. JIN, R. GE, P. NETRAPALLI, S. M. KAKADE, AND M. I. JORDAN, *How to escape saddle points efficiently*, arXiv preprint arXiv:1703.00887, (2017).
40. P. H. JIN, Q. YUAN, F. IANDOLA, AND K. KEUTZER, *How to scale distributed deep learning?*, arXiv preprint arXiv:1611.04581, (2016).
41. R. JOHNSON AND T. ZHANG, *Accelerating stochastic gradient descent using predictive variance reduction*, in Advances in Neural Information Processing Systems, 2013, pp. 315–323.
42. D. KINGMA AND J. BA, *Adam: A method for stochastic optimization*, arXiv preprint arXiv:1412.6980, (2014).
43. S. B. KYLASA, F. R. KHORASANI, M. W. MAHONEY, AND A. Y. GRAMA, *GPU accelerated sub-sampled newton's method for convex classification problems*, in Proceedings of the 2019 SIAM International Conference on Data Mining, SIAM, ed., SIAM, 2019, pp. 702–710.
44. K. LEVENBERG, *A method for the solution of certain problems in least squares*, Quarterly of Applied Mathematics, 2 (1944), pp. 164–168.
45. K. Y. LEVY, *The power of normalization: Faster evasion of saddle points*, arXiv preprint arXiv:1611.04831, (2016).

46. M. LI, T. ZHANG, Y. CHEN, AND A. J. SMOLA, *Efficient mini-batch training for stochastic optimization*, in Proceedings of the 20th ACM SIGKDD international conference on Knowledge discovery and data mining, ACM, 2014, pp. 661–670.

47. C.-J. LIN, R. C. WENG, AND S. S. KEERTHI, *Trust region Newton method for logistic regression*, The Journal of Machine Learning Research, 9 (2008), pp. 627–650.

48. D. C. LIU AND J. NOCEDAL, *On the limited memory BFGS method for large scale optimization*, Mathematical programming, 45 (1989), pp. 503–528.

49. D. W. MARQUARDT, *An algorithm for least-squares estimation of nonlinear parameters*, Journal of the Society for Industrial & Applied Mathematics, 11 (1963), pp. 431–441.

50. J. MARTENS, *Deep learning via Hessian-free optimization*, in Proceedings of the 27th International Conference on Machine Learning (ICML-10), 2010, pp. 735–742.

51. J. MARTENS, *New insights and perspectives on the natural gradient method*, arXiv preprint: arXiv:1412.1193v9, (2017).

52. J. MARTENS AND R. GROSSE, *Optimizing neural networks with Kronecker-factored approximate curvature*, arXiv:1503.05671v6, (2016).

53. G. MONTAVON, G. B. ORR, AND K.-R. MULLER, *Neural Networks: Tricks of the Trade*, Springer, 2nd ed., September 2012.

54. W. R. MORROW, *Hessian-free methods for checking the second-order sufficient conditions in equality-constrained optimization and equilibrium problems*, arXiv preprint arXiv:1106.0898, (2011).

55. Y. NESTEROV, *Introductory lectures on convex optimization*, vol. 87, Springer Science & Business Media, 2004.

56. J. NOCEDAL, *Updating quasi-Newton matrices with limited storage*, Mathematics of computation, 35 (1980), pp. 773–782.

57. J. NOCEDAL AND S. WRIGHT, *Numerical Optimization*, New York: Springer, 1999.

58. J. NOCEDAL AND S. WRIGHT, *Numerical optimization*, Springer Science & Business Media, 2006.

59. B. A. PEARLMUTTER, *Fast exact multiplication by the hessian*, Neural Computation, (1993).

60. B. T. POLYAK, *Some methods of speeding up the convergence of iteration methods*, USSR Computational Mathematics and Mathematical Physics, 4 (1964), pp. 1–17.

61. S. J. REDDI, J. KONEČNÝ, P. RICHTÁRIK, B. PÓCZÓS, AND A. SMOLA, *AIDE: Fast and communication efficient distributed optimization*, arXiv preprint arXiv:1608.06879, (2016).

62. H. ROBBINS AND S. MONRO, *A stochastic approximation method*, The annals of mathematical statistics, (1951), pp. 400–407.

63. F. ROOSTA-KHORASANI AND M. W. MAHONEY, *Sub-sampled Newton methods I: globally convergent algorithms*, arXiv preprint arXiv:1601.04737, (2016).

64. ———, *Sub-sampled Newton methods II: Local convergence rates*, arXiv preprint arXiv:1601.04738, (2016).

65. S. SHALEV-SHWARTZ AND S. BEN-DAVID, *Understanding machine learning: From theory to algorithms*, Cambridge university press, 2014.

66. S. SRA, S. NOWOZIN, AND S. J. WRIGHT, *Optimization for machine learning*, MIT Press, 2012.

67. I. SUTSKEVER, J. MARTENS, G. DAHL, AND G. HINTON, *On the importance of initialization and momentum in deep learning*, in International conference on machine learning, 2013, pp. 1139–1147.

68. T. TIELEMAN AND G. HINTON, *Lecture 6.5-rmsprop: Divide the gradient by a running average of its recent magnitude*, COURSERA: Neural Networks for Machine Learning, 4 (2012).

69. S. WANG, F. ROOSTA-KHORASANI, P. XU, AND M. W. MAHONEY, *GIANT: Globally Improved Approximate Newton Method for Distributed Optimization*, arXiv preprint arXiv:1709.03528, (2017).

70. P. XU, F. ROOSTA-KHORASANI, AND M. W. MAHONEY, *Newton-Type Methods for Non-Convex Optimization Under Inexact Hessian Information*, arXiv preprint arXiv:1708.07164, (2017).

71. ———, *Second-Order Optimization for Non-Convex Machine Learning: An Empirical Study*, arXiv preprint arXiv:1708.07827, (2017).

72. P. XU, J. YANG, F. ROOSTA-KHORASANI, C. RÉ, AND M. W. MAHONEY, *Sub-sampled newton methods with non-uniform sampling*, in Advances in Neural Information Processing Systems, 2016, pp. 3000–3008.

73. Y. XU, J. RONG, AND T. YANG, *First-order stochastic algorithms for escaping from saddle points in almost linear time.*, in Advances in Neural Information Processing Systems, 2018.

74. Z. XU, M. A. FIGUEIREDO, X. YUAN, C. STUDER, AND T. GOLDSTEIN, *Adaptive relaxed admm: Convergence theory and practical implementation*, in 2017 IEEE Conference on Computer Vision and Pattern Recognition (CVPR), IEEE, 2017, pp. 7234–7243.

75. Z. XU, G. TAYLOR, H. LI, M. FIGUEIREDO, X. YUAN, AND T. GOLDSTEIN, *Adaptive consensus admm for distributed optimization*, arXiv preprint arXiv:1706.02869, (2017).

76. H. H. YANG AND S. ICHI AMARI, *The efficiency and the robustness of natural gradient descent learning rule*, in Neural Information Processing Systems, 1997, pp. 385–391.

77. Z. YAO, P. XU, F. ROOSTA-KHORASANI, AND M. W. MAHONEY, *Inexact non-convex Newton-type methods*, arXiv preprint arXiv:1802.06925, (2018).

78. J. YU, S. VISHWANATHAN, S. GÜNTER, AND N. N. SCHRAUDOLPH, *A quasi-Newton approach to nonsmooth convex optimization problems in machine learning*, The Journal of Machine Learning Research, 11 (2010), pp. 1145–1200.

79. Y. YU, P. XU, AND Q. GU, *Third-order smoothness helps: Even faster stochastic optimization algorithms for finding local minima*, arXiv:1712.06585v1, (2017).

80. Y. YU, D. ZOU, AND Q. GU, *Saving gradient and negative curvature computations: Finding local minima more efficiently*, arXiv:1712.03950v1, (2017).

81. M. D. ZEILER, *Adadelta: an adaptive learning rate method*, arXiv preprint arXiv:1212.5701, (2012).

82. H. ZHANG, C. XIONG, J. BRADBURY, AND R. SOCHER, *Block-diagonal Hessian-free optimization for training neural networks*, arXiv preprint: arXiv:1712.07296v1, (2017).

83. Y. ZHANG AND X. LIN, *DiSCO: Distributed optimization for self-concordant empirical loss*, in International conference on machine learning, 2015, pp. 362–370.

Printed in the United States
by Baker & Taylor Publisher Services